SCIENTIFIC AMERICAN
Reader

1953

SIMON AND SCHUSTER, NEW YORK

Library of Congress Catalog Card Number: 53–11330
Dewey Decimal Number: 502
MANUFACTURED IN THE UNITED STATES OF AMERICA
BY H. WOLFF BOOK MFG. CO., INC., NEW YORK

TABLE OF CONTENTS

The last generation's work in astronomy has radically revised our picture of the universe. It has been resolved into numberless galaxies, each a congregation of stars vast enough to contain the universe of our fathers. In this mega-universe the classical notion of clockwork order has given way to the finding that processes of growth and decay pervade its farthest reaches just as they do the world near at hand.

Equipped with powerful new concepts and instruments borrowed principally from physics, geology is now reconstructing the earth's history and investigating the forces which have shaped it. The goal is an understanding of the earth as a unitary system, in which the features of the surface are related to the dynamics within.

v

From the nucleus of the atom outward to the galaxies, modern physics gives us an increasingly comprehensible and consistent view of nature. But the nucleus breaks down into too many particles. Physics now asks: What is the nature of these particles? Which of them are fundamental? What holds them together in the nucleus?

The processes of nuclear fusion and fission have been pre-empted by the weaponeer. But they are also processes of nature, which account for the light of the stars and in part for the heat of the earth. They are therefore accessible to the understanding of the citizen, who must grasp their principles in order to judge how they are being applied for either destructive or constructive ends.

By photosynthesis green plants establish the carbon-hydrogen bond that characterizes all the compounds of life; proteins are the most important of these compounds; enzymes are the catalysts of life; the ions of metals like calcium provide the electrical potentials that energize life processes at many points. Thus biology today reduces life to chemistry and physics.

Modern genetics springs from Mendel's laws of heredity and Darwin's theory of evolution. It endeavors, on the one hand, to discover how the genes shape the individual; on the other, to clarify the role which genes play in the perpetuation and diversification of species.

In addition to its obvious pertinence for medicine, the study of viruses has become one avenue toward an understanding of genetics and the origin of life. The control of all the important virus diseases has become a goal for the foreseeable future, but the virus as a subject and a tool for fundamental investigations is only now coming into its own.

Man's trail up from the last branching place on the tree of primate evolution has been made reasonably clear by the investigations reported here. As to his fate in the immediate future—say in the next half million years—modern genetics can now make some shrewd guesses.

Now that we have built and operated some thinking machines, we are approaching a better understanding of the living brain and developing a warmer appreciation of its extraordinary design. Parallel advances on these two fronts—biological and electronic—forecast revolutionary changes in our understanding and application of the rational processes.

All science, indeed all knowledge, presupposes a correlation between the objective world, the senses and the organism's perception of what it senses. The investigations reported here represent a new field of scientific enterprise, which may settle some ancient philosophic controversies.

INTRODUCTION

This Reader *is the product of a unique collaboration of authors, editors and readers. The authors are the scientists in the United States and abroad who have undertaken to explain their work in the pages of* SCIENTIFIC AMERICAN *during the past five years. The editors are the science journalists who joined five years ago in the reconstruction of that century-old publication, with the aim of making it a medium for communication between scientists and their fellow men. The readers are that section of the public whose lively and intelligent interest in science called for the publication of a magazine of science that would fill the gap between the specialized journals of the professional press and incidental coverage of science in the general press*

The three collaborators share the conviction that there ought to be a wider understanding of science in our society. Such understanding, as most of us are uncomfortably aware, is a new obligation of citizenship. Science and technology increasingly affect the public issues of the day; the citizen taxpayer has succeeded the philanthropic foundation as the principal underwriter of the costs of science. However, the needed popularization of science is not primarily a project in mass communication. Illiteracy in science presents its most alarming aspect as it prevails among otherwise educated members of our society. It promotes the antirational, illiberal mood presently ascendant in our culture. It has resulted in the almost complete estrangement of arts and letters from the sciences, which explains why our humanists largely miss the insights that science now offers into so many of their habitual concerns. Among engineers and scientists, all too often ignorant in fields outside their own, it has promoted a sterile insularity that shirks the cultural and social responsibilities of their profession.

The authors, editors and readers of SCIENTIFIC AMERICAN *also urge that to be conversant with science is as much a privilege as it is a responsibility of the times. The world as it is known to science is a far more various and splendid realm than any known to past generations of mankind.*

After five years of experience, the collaborators hold that there are no obstacles to the spreading of a wider understanding of science that, working together, they cannot surmount. Each party to the collaboration has its special interest, and so each makes its particular contribution. The scientists are determined that there shall be no compromise with the integrity of what they have to say, that what they write shall convey their meaning with precision and completeness. The editors are as strongly determined that what they print shall be clear to anyone interested enough to want to understand. As a result, the relationship between author and editor in SCIENTIFIC AMERICAN *is closer and more painstaking than on perhaps any other magazine published today. The readers' critical and insistent demand for both integrity and clarity prevents any slackening of this relationship. The readers, after all, have the last word.*

The 50-odd articles of the Reader *were originally written for publication in the monthly issues of* SCIENTIFIC AMERICAN; *they have the immediacy and tension of the current events they then reported. Brought together here in a book, they gain a new relationship to events and to each other. That relationship becomes clear in the Table of Contents, where the reader will see that the chapters are organized in 12 "Parts." Each Part deals with one of the central topics of investigation now attracting the interest, enthusiasm and energy of the world's scientists. These are the currently most hopeful lines of inquiry, the ones that are yielding, or promise soon to yield, significant new understanding of the world we live in and of ourselves as part of that world. The book thus offers, in addition to information, a broad insight into the objectives, methods and thought processes of modern science. Here, in fairness to the authors not represented in these pages, the editors declare with emphasis that the chapters were not selected as the "best" in the magazine. They include many of the editors' and the readers' favorites, but others as interesting and attractive had to be regretfully omitted because they did not fit the book's plan.*

There is little in the book about those sensations of the day which have identified science so firmly with technology in the public mind. The titles of the 12 Parts indicate, on the contrary, that the interests of science are more closely allied with those of philosophy, history, ethics and esthetics than the engineering bias of much popular reporting would suggest. The structure of the universe and of matter, the origin of life and the reliability of our sensations and perceptions will be recognized as issues that have universal relevance to human life. They are concerned with the ends as well as with the means of our existence.

The titles of the Parts and the chapters reflect another characteristic of scientific inquiry today. The departmental labels of biology, chemistry, physics and the like, which divide science into convenient pedagogic and administrative parcels, are notably absent. After all, it is a single world, which embraces galaxies and fundamental particles, starlight and human thought. As they have delved deeper into their respective fields, scientists have discovered more and more cross connections to the fields of others. The three to seven chapters assembled in each Part of the Reader *report independent work by different scientists, often in entirely different fields. This is typical of the situation today in which the various specialties are moving with increasing concert along common lines of investigation. The unity of nature is now admirably mirrored in a growing unity of science.*

Five years is not a long time in the history of science. These have been years of great progress, however, particularly in the areas covered by the Reader. *Here and there that progress has overtaken the book. There are probably few chapters which would not read a little differently if they had been written afresh for the occasion. The progress of science is, in a phrase of Morris Cohen, "a process of continual self-correction." Nevertheless, the picture of the state of science which this book presents is a current one. It is in the smaller details that the more rapid changes come; corrections in the short run deal for the most part with matters of weights and measures. Footnotes in the* Reader *cite changes that seemed to be of sufficient significance; at a few points, authors have rewritten a paragraph or two. Otherwise, the articles have been modified from the originals only by cutting to eliminate repetition and the discussion of matters not directly relevant to the Parts of the* Reader *where they now appear.*

It is proper here to emphasize again that, with the exception of a few chapters written by professional writers working in close col-

laboration with scientists, the authors of this book are scientists. There are important reasons why the scientists themselves should do our science writing. It is, of course, in the high tradition. Galileo is said to have written his Dialogue not just for astronomers but "principally for gentlemen." As recently as The Origin of Species, a fundamental work of science found wide general readership. To-day, there is special need for this personal contact between the scientist and his fellow citizen. The first-hand accounts presented in the Reader divest the work entirely of that aura of mechanized wizardry which infects the public relations of science and is good for neither science nor the public.

The authors are representative members of their profession. They include some of the foremost scientists of the day, some whose fame waits upon work now in progress and others known only to those who know which particular stone they set. The reader exposed to what these men have to say cannot fail to be stirred by the intensely human quality of their enterprise, by the curiosity, discontent, rest-lessness, energy and aspiration which they bring to their inquiries.

They do not proffer revelations here, nor do they issue magis-terial pronouncements on the truth. On the contrary, they seek to explain how they came to be concerned with the question under investigation, how they framed their theories and designed their experiments, why the present state of evidence is persuasive and what new contacts the results establish with the still unknown. They are concerned, in short, to explain the method as well as the find-ings of science. That method, as they explain and affirm it, consists in the exercise of the power of reason, the faculty which principally distinguishes our species.

The Editors *

* Board of Editors: GERARD PIEL (*Publisher*), DENNIS FLANAGAN (*Editor*), LEON SVIRSKY (*Managing Editor*), ROBERT HATCH, ALBERT G. INGALLS, LAWRENCE P. LESSING, JAMES R. NEWMAN, E. P. ROSENBAUM, JAMES GRUNBAUM (*Art Director*).

SCIENTIFIC AMERICAN *Reader*

I. GALAXIES IN FLIGHT *by George Gamow*

An original and provocative contributor to modern physical theory, George Gamow has delighted and instructed the layman in many articles and books, including *Birth and Death of the Sun, Mr. Tompkins Explores the Atom, One, Two, Three . . . Infinity*. He is professor of theoretical physics at George Washington University.

II. THE MILKY WAY *by Bart J. Bok*

The southern sky, outside the visual field of the great observatories in the northern hemisphere, is the province of Bart Bok. At the Boyden Station of Harvard Observatory in South Africa, his recent work has been devoted to locating the hub of our local universe, the Galaxy of the Milky Way, descrying its spiral arms.

III. THE UNIVERSE FROM PALOMAR
by George W. Gray

A rare exception among the authors in this book, George W. Gray is a journalist, not a scientist. His numerous articles have won him the critical esteem of scientists as well as an admiring audience of laymen. Some of his most important writing has been addressed to a private audience composed of the directorate of the Rockefeller Foundation, for whom he interprets and explains the work supported by the Foundation's grants.

IV. THE DUST CLOUD HYPOTHESIS
by Fred L. Whipple

Professor Whipple has made fundamental contributions to the modern hypothesis of the origin of stars and solar systems about which he writes in this article. Like most Harvard astronomers, he is the author of a book for the layman, *Earth, Moon and Planets*.

V. RADIO STARS *by A. C. B. Lovell*

As professor of radio astronomy at the University of Manchester, the author is concerned with the stars that put out their energy in the radar and radio bands of the electro-magnetic spectrum. During World War II he helped to establish the radar net which played such a decisive role in the victorious Battle of Britain.

EVOLUTION IN SPACE

Introduction

LET US START by asking when and how the universe began. Man in his history has asked millions of questions; he has gained thereby a considerable advantage over the uninquisitive beasts of the field. Many questions have been answered, many have lost their urgency. We know now why the sea is salt and we no longer care how many angels can dance on the head of a pin. But some questions are asked again, from generation to generation, and remain unanswered. This book will consider a number of them, but none more ancient or more insistent than the nightly question: "Where did the stars come from?"

Astronomy seems on the verge of the answer, and that is why astronomy, the oldest and one might think the "purest" of the sciences, is today sparkling with controversy. Of course, other ages have proposed their solutions to the riddle of genesis; it may be presumption that tells us our answer will be conclusive. The articles in this chapter, however, suggest more solid grounds for optimism.

Two main themes run through these five accounts of the present-day search into space. The first is that the universe to its farthest reaches is permeated, like the world near at hand, with processes of change and evolution. This is a relatively new notion in astronomy. The science began as an attempt to codify the immutable laws governing an eternal cosmos. Now the sky watchers tell us of the birth and death of stars; they distinguish bodies that husband their resources and live long from others that squander their substance in brilliant display; they arrange whole systems of stars in sequences of youth and old age. It is a universe of motion: not the familiar cycles that divide our days and nights and define our seasons, but a flight at speeds that approach the limiting velocity of light and in all directions from the observer. We no longer know the universe as a superlatively complex, unfailingly repetitive clock works. It is in order now for an astrophysicist like George Gamow and an astronomer like Fred Whipple to consider, in their essays, how the cosmos began and how, in its later history, it yields a solar system like our own.

That brings us to the second major theme of this chapter. Galileo, Kepler and Copernicus translated the earth and us from the center of the cosmos and placed the sun there in our stead. Later observers, equipped with better instruments, resolved a myriad of suns in the luminous veil of the Milky Way, reducing to a tiny minority the 1,400 that are visible to the unaided eye. Our generation of astronomers, as Bart Bok shows, has found the center of the Milky Way and located the spot occupied by the solar system. It is far out from the hub of this great wheel of stars and inchoate dust and gas. And in the past few years the men on Palomar Mountain, whom George Gray watched at work with the 200-inch mirror, have moved out beyond our home galaxy of the Milky Way to count 10,000 times as many galaxies as Galileo counted stars.

Meanwhile, radio astronomy has opened an entirely new window into space; A. C. B. Lovell tells us of lightless stars, "visible" only to radio antennae.

The subject of modern astronomy is not the solar system or the Milky Way; it is the cosmos itself. But cosmology is now an observational science. Speculation is still in order, as the authors of this chapter show. Their speculations, however, are subject to an ever more securely established body of observed events.

As in other branches of science, the first problem of cosmology is measurement. Its clocks and its measuring sticks are not yet fully calibrated. Therefore many of its calculations are still rough and tentative. The essays in this group were published over a period of five years of active and highly fruitful work. The reader will notice that some of the values change from one essay to another. This should not trouble him, however, for the method remains constant.

All in all, it seems likely that this chapter contains the beginning of the answer to the question: "Where did the stars come from and how is the universe put together?" Of course, when the answer is found, someone will ask what space was like before there was a universe. Our propensity to ask questions may always exceed our ability to observe.

I. GALAXIES IN FLIGHT

by George Gamow

IN THE YEAR 1929 the Mount Wilson astronomer Edwin P. Hubble made a very remarkable discovery. He found that the giant accumulations of stars known as galaxies, which are scattered in great multitude through the vast expanses of the universe as far as the best telescopes can see, seem to be running away from one another at fabulously high speeds. From this observed fact originated the famous theory of the expanding universe. Although the theory is still not finally proved, it seeded a whole generation of fruitful study, not only in astronomy but also in geology, physics and chemistry. It gave us a new start for investigating the age of the universe and the creation of the stuff of which it is made.

The idea of stellar galaxies is a comparatively recent discovery in astronomy. The celestial shapes that we now recognize as galaxies had been observed for a long time as faint nebulosities of various regular forms, but they were generally believed to be simply luminous clouds of gas floating in the spaces between the stars of the Milky Way. Observations with more powerful telescopes, however, resolved these "nebulosities" and showed that they were not clouds but huge collections of extremely faint stars. These giant stellar aggregates were far beyond the outer limits of our own stellar system, the Milky Way; in fact, it soon became clear that they formed systems very similar in shape and structure to the Milky Way galaxy itself.

Tne nearest and most familiar external galaxy is the great nebula in Andromeda, which can be seen with the naked eye as a faint, spindle-shaped speck of light in the upper part (from the Northern Hemisphere) of the constellation of Andromeda. Photographs made with large telescopes show that this galaxy has a rather complicated structure consisting of an elliptical center, or "galactic nucleus," and "spiral arms" flung into the surrounding space from the central body. The photographs also show two nearly spherical nebulosities close by, probably satellites of the central system.

Among the myriads of stars in the arms of the Andromeda Nebula are many pulsating ones, of the type called Cepheid variables. They brighten and fade in a regular rhythm, and their pulsation period provides a method of determining their absolute brightness. By comparing their apparent brightness (which depends on their distance from us) with their calculated absolute brightness, Hubble was able to prove that the Andromeda Nebula is some 680,000 light-years from the Milky Way. To a hypothetical observer in the Andromeda galaxy, the Milky Way would look much the same as the Andromeda system looks to us, except that the spiral arms of the Milky Way are somewhat more open. Our sun, with its family of planets, would be seen through a telescope within the Andromeda Nebula as a rather faint star near the end of one of the spiral arms, some 30,000 light-years from the Milky Way center.

The galaxies generally are shaped like a discus. The Andromeda system looks like an elongated spindle to us because it is tilted to our line of sight, but there are many other galaxies that we see from the top or straight on edge. All large galaxies have the same sort of spiral arms as the Milky Way and Andromeda, but smaller ones are usually armless. The galaxies are scattered more or less uniformly through space as far as our telescopes can probe. The average distance between neighboring nebulae is about two million light-years. The limit of our vision with the 100-inch telescope is about 500 million light-years. Hence in the observable region of space there are some 100 million galaxies.

The 200-inch telescope on Mount Palomar, which doubles the distance we can see into space, will reveal about one billion galaxies. Most galaxies are isolationist, dwelling in remote and solitary splendor, but we find a number that group themselves together to form more or less compact clusters. In the constellation of Corona Borealis, for example, there is a cluster containing some 400 galaxies. Our Milky Way is a member of a small cluster which embraces, among others, the Andromeda Nebula and the two galaxies known as the Magellanic Clouds, which are of a relatively rare type that has no well-defined shape.

The distances of all but the nearest galaxies are so great that even the most powerful telescopes fail to resolve them into individual stars. Astronomers' calculations of their distances depend entirely on their apparent brightness. Hubble, studying a group of about 100 well-known neighboring galaxies, established the fact that on the average they were of about the same size and the same intrinsic luminosity. Using this standard, we can estimate the distances of remote groups of galaxies by comparing their mean apparent brightness with that of nearby galaxies whose distances are known. Such measurements give the value of 7.5 million light-years for the distance of one of the nearest groups of galaxies in Virgo. Similar galactic groups in the constellations of Coma Berenices, Corona Borealis and Boötes are respectively 30 million, 100 million and 180 million light-years away.

Now what was it that gave Hubble the notion that the galaxies are running away from one another and that the universe is expanding? His basic discovery was made with that indispensable tool of the astronomer, the spectrograph, which analyzes the color components of the light coming from stars. Studying the spectra of distant galaxies, he noticed a curious fact: all the lines in their spectra, regardless of the wavelength or color of the line, were displaced toward the red end of the spectrum. Furthermore, the amount of this "red shift" was always directly proportional to the distance of the galaxy from us. The most natural explanation of this shift was that the source of the light was moving away. This is the so-called Doppler effect, of which the classic and most familiar example is the change in pitch of a locomotive whistle as the train approaches us and then speeds away. A light wave, like a sound wave, appears to shift to a longer wavelength when it reaches us from a receding source. And the speed with which the source is moving away is directly proportional to the shift in wavelength. Since the red shift of the galaxies also varied as their distance from us, Hubble concluded that the speed of the receding stars was proportional to their distance; the farther away they moved from one another, the faster they traveled. The red shift of the most distant galaxies that have thus far been observed is 13 per cent, which suggests that they are receding from us at the terrific velocity of 25,000 miles per second.

You must not conclude from this that we stand at the center of the universe and that all the rest of it is running away from us. Picture a slowly inflated rubber balloon with a large number of dots painted on its surface. An observer on one of the spots would be under the impression that the other dots were racing away from him in all directions, and so indeed they would be, but the same thing would be true no matter which dot he was on. In the case of the galaxies, we are dealing with the effect of a uniform expansion throughout all of space.

If you pick an arbitrary point in space, say the Milky Way, and divide the distance of a given galaxy by its recession velocity, you get a figure which represents the length of time that the galaxy has been receding from that point. The strange and wonderful consequence of Hubble's observations is that the figure will be the same no matter what pair of galaxies you pick. Thus it works out that at a fixed, calculable time in the past all the galaxies now so widely scattered were packed tightly together. And the time figure you arrive at is the age of the universe, measured from that instant when the originally highly condensed universal matter was torn apart by the primordial "explosion" that started its headlong expansion.

To get this figure, we must know the exact values for the distances and the recession velocities of distant galaxies. This is less simple than it sounds. The velocities, as we have seen, can be computed from the observed red shift, and the distances, presumably, from the galaxies' apparent brightness. But there is a catch: the apparent brightness of the stars is affected not only by their distance but also by the fact that the light coming from them is redder, and therefore carries less energy, than if the light source were stationary. To illustrate this, suppose for a moment that you are shot at by a gangster operating a submachine gun from the back window of a speeding car. Since the vehicle is receding, the bullets move more slowly toward you than they would from a stationary gun, and they strike your bulletproof jacket with less energy. A receding light source produces exactly the same effect; its emitted light quanta strike the eye with less energy and therefore look redder than they should. An astronomer must make the same correction for the weakening of light intensity as a ballistics expert would make in estimating the muzzle speed of the bullets.

There is a further complication. If the submachine gun shoots, say, one bullet per second, its bullets will strike you at longer and longer intervals as the gun recedes, for each successive bullet will have farther to travel. Similarly, light quanta from receding galaxies enter the observer's eye with less frequency, and this fact calls for another correction of the observed brightness.

Applying both corrections, and taking the most accurate possible observations, Hubble calculated that the universe began to expand less than one billion years ago. This result stands in contradiction to geological evidence, which indicates that the age of the solid earth crust, estimated quite reliably from radioactive decay in the rocks, must be at least two billion years. Since numerous pieces of evidence in various sciences support the two billion-year estimate, Hubble was forced to reconsider the expansion theory and consider the possibility that the red shift was due not to the normal Doppler effect but to some unknown physical factor which caused light to lose part of its energy during its long trip through intergalactic space.

Such a conclusion would ruin many beautiful scientific developments that have flowed from the hypothesis of the expanding universe. It would confront physicists with the difficult task of explaining the red shift in non-Dopplerian terms—which would seem to contradict everything we know at present about light. Fortunately, there is a simple way out of the dilemma which is usually overlooked by the proponents of the "stop-the-expansion" point of view. The point is that Hubble's method of estimating the distances of faraway galaxies assumes that at the moment when they emitted their light they were just as bright as the galaxies we see closer at hand. It must be remembered, however, that the light we see from the distant galaxies was emitted at a fantastically distant time in the past; the light now coming to us from the Coma Berenices cluster, for example, started on its way some 40 million years ago, and the most distant galaxies used by Hubble in his studies are seen as they were almost half a billion years ago!

Do we have the right to assume that the galaxies, which are evolving like everything else in the universe, have kept their luminosity constant over such long periods of time? In view of the known facts about the evolutionary life of individual stars, which maintain their luminosity by the expenditure of nuclear energy, such an assumption would be very strange indeed. Actually, we can remove the entire difficulty in Hubble's time scale by remembering that the nuclear processes that fuel the stars are not endlessly self-perpetuating but are accompanied by a gradual dissipation of the originally available energy. The assumption that an average galaxy loses a mere five per cent of its luminosity in the course of 500 million years would bring the age of the universe to the two billion-year figure demanded by other astronomical, geological and physical evidence. *

---

* Because of new evidence obtained in 1952, all the time and distance figures for galaxies given in this article must be doubled. See the footnote at the end of George Gray's article later in this chapter.

This conclusion finds strong confirmation in recent work by Joel Stebbins and A. E. Whitford at the Mount Wilson Observatory, who have studied the apparent luminosities of distant galaxies on special plates sensitive to red light. To everyone's surprise, they found these galaxies much brighter in the red part of the spectrum than they had previously appeared to be on ordinary photographic plates, which are sensitive mostly to the blue rays. It looked at first as if this phenomenon was due to the same kind of optical scattering which makes the sun look red during dust storms; light from the galaxies, it was thought, was reddened by the clouds of fine inter-galactic dust through which it passed. Calculations showed, how-ever, that to account for the observed reddening would take a fan-tastic quantity of dust—100 times as much as the total amount of matter in the galaxies themselves. Such an assumption would come into serious conflict with many facts and theories about the struc-ture of the universe. It therefore seems more reasonable to suppose that the distant galaxies look redder simply because they actually were redder when they emitted the light which is now reaching our telescopes. This could be explained if we assumed that young galaxies contain more red stars than more mature ones.*

Having made this fiery defense of the right of our universe to expand, let us consider the physical consequence of the expansion theory suggested at the beginning of this article. What physical process was responsible for the present relative quantities of the various chemical elements that make up the universe? Why, for ex-ample, are oxygen, iron and silicon so abundant; and gold, silver and mercury so rare?

We know that, except for the lightest elements (such as hydro-gen, helium, nitrogen and carbon, involved in the sun's nuclear cycle), transformation of one atomic nucleus into another requires tremendous temperatures such as do not exist at the present time even in the hot interiors of the stars. Consequently there can not have been any revolutionary change in the relative abundance of the various elements since the expansion of the universe began. On the other hand, there has been some change, for a number of atoms

---

* More detailed study of the Stebbins-Whitford effect shows that it is confined to the armless galaxies; no excess reddening is found in spirals. This disproves the hypothesis that the reddening might be due to dust and provides direct evidence that the galaxies do evolve in time.

are radioactive and have gradually decayed into more stable elements.

Considering the latter case first, we note, for example, that the lighter isotope of uranium, U-235 (atomic bomb stuff), constitutes only .7 per cent of a given amount of uranium found in nature; the rest is the heavier isotope U-238. The half-life of U-235 is only .7 billion years, while that of U-238 is 4.5 billion years. If we make the reasonable assumption that at the original formation of the universe both isotopes were produced in about equal amounts, the age of the universe figures up to about four billion years. Similar calculations based on the naturally radioactive isotope of potassium (relative abundance—.01 per cent; half-life—.4 billion years) yields the figure of 1.6 billion years. While these figures are only very approximate, they agree roughly in order of magnitude with the age of the universe as estimated from the red shift and other evidence. Thus we have fairly good reason to suppose that the radioactive elements were formed at the beginning of the universe.

Actually, the picture presented by the expanding universe theory, which assumes that in its original state all matter was squeezed together and possessed extremely high density and temperature, gives us exactly the right conditions for building up all the known elements in the periodic system. Recently, Alpher, Bethe and Gamow have attempted to reconstruct in some detail the processes by which the various elements may have been created during the early evolutionary stages of the expanding universe.

Our studies indicate that, under the tremendous temperatures and densities prevailing in the universe during the stage of its maximum contraction, primordial matter must have consisted entirely of free neutrons and protons moving much too fast to stick together and form stable nuclei. As the universe started to expand, this primordial gas began to cool. When its temperature dropped to about one billion degrees, particle condensation began. The growth of heavier nuclei was achieved by adding free neutrons to already existing lighter nuclei. It is known that neutron aggregates are intrinsically unstable unless about half of their particles carry a positive electric charge. Hence they must have emitted electrons until they achieved a state of electrical equilibrium. The electrons fell into orbits around the nuclei and formed electronic envelopes around them; thus atoms were created.

According to our calculations, the formation of elements must have started five minutes after the maximum compression of the universe. It was fully accomplished, in all essentials, about 30 minutes later. By that time the density of matter had dropped below the minimum necessary for nuclear-building processes. All the elements were created in that critical 30 minutes, and their relative abundance in the universe has remained essentially constant throughout the three billion years of subsequent expansion.

by Bart J. Bok

GET AWAY from the glow of city lights and step outdoors in the open country on a clear, dark night. It is like stepping out on a platform in space. Roads, hills and houses hardly make their presence felt; the sleeping earth is hushed, like the audience at a play, by the great show of the heavens. And soon the delighted eye, exploring the vast, twinkling spectacle, is drawn to the most fascinating sight of all—the luminous band of the Milky Way that stretches in quiet majesty all around the sky.

I have lived with the Milky Way for more than a quarter of a century and have never stopped marveling at its beauty. It remains one of the grandest phenomena of nature, and a never-ending challenge to scientific curiosity. What is it made of? Why does it vary so greatly in appearance along the band, showing comparatively dull sections in winter and reaching a height of glory at our latitude in late summer and early fall? These questions, not yet completely answered, are important ones in astronomy. If we can fathom the mystery of the structure of the Milky Way, we shall have learned much about the arrangement of the universe.

Even with a pair of field glasses or a small telescope, one can discern that the Milky Way band is composed of countless stars. They form the body of the galaxy of which our sun is a modest member. As a matter of fact, all the stars in the sky that can be seen with the naked eye, and the majority of those that can be distinguished by the most powerful telescopes, are members of the Milky Way system. The Milky Way, our galaxy (a word derived from the Greek *gala*, meaning milk), has great depth. Its distances are most conveniently measured in terms of traveling times at the speed of light. At this speed, 186,000 miles per second, it would take us only about one seventh of a second to circle the earth, a little more than one second to go from the earth to the moon, about eight minutes to go from the earth to the sun, and about 12 hours to make a comfortable sightseeing tour of the whole solar system, visiting all the planets. But at the same rate we would have to travel more than four years to reach the star nearest the sun—Alpha in the southern-hemisphere constellation of Centaurus; and it would take roughly 100,000 years to pass from one end of the Milky Way to the other. Although there are more than 100 billion stars in the Milky Way, the system is so enormous that only a very minute fraction of the total space it occupies is taken by the stars themselves. We could readily store a million times as many stars in the present volume of the system without the risk of an undue frequency of stellar collisions.

Huge as the Milky Way is, modern telescopes enable us to look beyond its limits and see that distant space is filled with many other galaxies like our own. A single photograph with a large telescope may show easily 1,000 faint galaxies outside the Milky Way.

Why study the arrangement of stars and nebulae in the Milky Way system, when it occupies so insignificant a fraction of the total volume of observable space? One might as well ask: Why study our sun, which is after all one mediocre star among billions? Not only can we examine our home galaxy in much more detail than other galaxies, but the Milky Way is a near and ever-present invitation to investigation.

We would like to know, first of all, the general shape of the galaxy and the point of view from which we are looking at it; in other words, what position our solar system occupies in it. When we survey the visible stars in the sky we note at once that they are concentrated most thickly in or near the band of the Milky Way that arches across the sky from horizon to horizon. If you stand with your hands pointing toward the ends of this arch and look out at the sky at right angles to it, you will find the stars scattered thinly there; they become progressively more concentrated as you move your gaze across the sky toward the Milky Way band. Even more significant is the fact that the fainter (*i.e.*, generally more distant) stars show greater concentration toward the band than do the brighter ones, indicating that the galaxy extends farthest in the direction of the band. This evidence shows that our galaxy has the shape of a flattened disk, or a big wheel, and that we are looking at it from some place in or very near the central plane in the disk. Just as, in looking at a column of marching soldiers, we see many more soldiers and see them stretching farther away in the distance when we look along the length of the column than when we look through a cross section of it, so we see more of our galactic system along the plane of the disk than through a section of it.

Where in this great wheel-shaped galaxy does our solar system lie? Is it at the hub of the wheel, or out toward the rim? The evidence

14

seems conclusive that we are at a great distance from the hub or center of our galaxy. As we look out toward the rim, the Milky Way appears very much brighter (*i.e.*, more extensively populated with stars) in some sections than in others. For example, the section of the Milky Way that includes the constellations of Perseus, Auriga and Orion, which is seen best in winter, is relatively weak, while the section in the direction of the Sagittarius, Aquila and Cygnus constellations, seen best in summer, is so brilliant that parts of it may readily be mistaken for cumulus clouds when observed near the horizon. Photographs to very faint limits show that there are 10 times as many stars per unit area of sky in the Sagittarius cloud as in the richest part of the winter Milky Way. In short, one half of the Milky Way is comparatively thin and dull, the other dense and vivid. Detailed surveys of the weak half show a conspicuous lack of distant objects, star clusters and nebulae; in a careful recent study at the Harvard Observatory not a single object was found in this section that could be placed with certainty at a distance greater than 10,000 light-years from our sun; there may be a few stars beyond this distance, but if so they are certainly spread thinly. On the other hand, the brilliant Sagittarius section abounds in objects that are known to be very far away. All this strongly suggests that we are looking at the galaxy from a position out toward the rim of the wheel, and that the center of our galaxy lies in the direction of the Great Star Cloud of Sagittarius.

On the basis of several wholly independent kinds of evidence astronomers now are certain that our galaxy is a great, wheel-shaped collection of stars rotating in space, with the sun and earth occupying a position about 30,000 light-years from the center. Our sun and the stars near it are whirling in roughly circular orbits around this center at a velocity of about 150 miles a second. So vast is our galaxy that at this fantastic rate of speed the sun takes 200 million years to complete a single swing around the Sagittarius center!

The first serious attempt to study the Milky Way was made by a German-born English astronomer, Sir William Herschel, shortly after the American Revolution. Earlier investigators, notably Thomas Wright, Immanuel Kant and John Mitchell, had made some fair guesses about the shape and structure of the universe, but no pertinent astronomical observations were available to check their guesses. In 1784 Herschel, assisted by his sister Caroline, undertook a systematic survey of the heavens with a telescope 20 feet long. He made accurate counts of the total numbers of stars visible in the field of his telescope, and surveyed 683 such fields. From these observations he derived a diagram of our stellar system which gave it the shape of a flattened grindstone, with the sun located close to the hub. It was a "cloven grindstone," for there is a dark, star-empty rift in a section of the Milky Way between Sagittarius and Cygnus, and Herschel interpreted this rift as a partial void.

During the 19th century surprisingly little further progress was made in studies of the Milky Way. The astronomers of that time made important advances, however, in basic research techniques and precise observations of single stars that were to prepare the way for general theories on galactic structure. The first stellar parallaxes, giving the distances of a few stars, were measured. (The parallax of a star is half the arc measuring its shift in relative position when it is observed from opposite points in the earth's orbit around the sun; from this the star's distance is determined by triangulation.) Stellar photography and the spectroscope were developed as powerful tools in astronomical research; scales of stellar magnitudes were established; the "proper motions" (across the line of sight) and the "radial velocities" (in the line of sight) of many stars were determined.

Toward the end of the 19th century J. C. Kapteyn of Holland began an investigation that resulted in a new theory of the structure of our Milky Way system. He believed that the problem was largely statistical: the system would gradually reveal its structure as astronomers gathered more and more accurate data on the magnitudes, spectral characteristics and motions of the stars. He recommended that they concentrate their various methods of study upon certain sample regions of the sky—the so-called Kapteyn Selected Areas. It took Kapteyn about 30 years to carry out the assignment he had set himself; a little before his death in 1922 he summarized his lifework in a diagram setting forth his picture of the galaxy. Like Herschel's, it showed a flattened system, with the sun close to the center, but Kapteyn introduced a scale of distances that represented the first attempt to indicate the approximate size of the galaxy.

In the early part of this century probably most astronomers agreed with Kapteyn that the next big advance in our knowledge of the Milky Way would come from an analysis of more accurate and more

16

extensive data. Actually the next great illumination of the subject came from a totally unexpected quarter—a study that seemed to have little or nothing to do with the structure of our galaxy. This was the investigation by Harlow Shapley, then a young astronomer at Mount Wilson Observatory, of globular star clusters.

A globular star cluster is a collection of very faint stars, distinguished by its global shape and extreme central density. About 100 such clusters have been observed by astronomers. In 1914 Shapley, following in the footsteps of Solon I. Bailey of Harvard University, began to study faint variable stars in these clusters with the 60-inch telescope at Mount Wilson. Variable stars of this type, known as Cepheid variables, had been studied previously by the astronomer Ejnar Hertzsprung of Denmark and Holland and by Henrietta S. Leavitt of Harvard. The Cepheids, named after their prototype in the constellation of Cepheus, occur in great abundance in the Clouds of Magellan, two star systems that are satellites of our Milky Way system. Each Cepheid variable fluctuates in brightness with a certain definite rhythm or period. Miss Leavitt found that in the Magellanic Clouds all the Cepheids with a given period had the same intrinsic brightness, or what an astronomer calls "absolute magnitude," as distinguished from apparent brightness. It has since been found that this relation is a universal one applying equally to the Cepheids near our sun, the Cepheids in globular clusters and the Cepheids in stellar systems outside our own galaxy, such as the spiral nebulae. With this discovery it became possible to estimate the intrinsic brightness of any Cepheid once its period is known. Comparing the Cepheid's intrinsic brightness with its apparent brightness, measured directly from the photographic plate, an astronomer can readily determine the distance from us of the Cepheid and of the star system of which it is a part.

Shapley, adapting this method to globular clusters, was soon able to determine the approximate distances of about a fourth of the 100 known globular clusters. He also noted a very curious fact that had previously been overlooked: almost without exception, the globular clusters are found in one half of the sky. And even in this half, they are not distributed uniformly; they show a very marked concentration toward the Great Star Cloud of Sagittarius. One third of all the globular clusters fall within an area covering only four per cent of the entire sky!

Thus the center of the globular star-cluster system was conveniently located within a relatively small area. It was logical to identify this center with the center of our galaxy. Shapley estimated that this center, in the direction of the Great Star Cloud in Sagittarius, lay about 50,000 light-years from us.

With this revolutionary work, published in 1918, Shapley did for the Milky Way system what Copernicus had done for the solar system: just as Copernicus had shown that the earth was not the center of the solar system, Shapley showed that our sun was not the center of our galaxy but out toward its outskirts.

Shapley's new ideas did not by any means find immediate general acceptance. Note that Kapteyn, one of the many doubters, placed the sun at the center of the galaxy in his final diagram, published four years after the announcement of Shapley's discovery. At the time there were good reasons to doubt Shapley's conclusions. Kapteyn and others had shown by a straightforward analysis of available star-counts that the number of stars per unit volume of the sky dropped off in all directions away from the sun, which seemed to prove that the sun was at the center. This interpretation could be disputed by assuming that there was an obscuring haze of interstellar material near the central plane of the Milky Way which made the star-counts unreliable, but at the time most astronomers, including Shapley, believed that the evidence was against the existence of any such haze.

Two major discoveries in the late 1920s and early 1930s settled the question. First Bertil Lindblad of Sweden and Jan H. Oort of Holland showed that our galactic system as a whole was in rapid rotation, and that the center of rotation was located at a distance of 25,000 to 30,000 light-years in the direction of Shapley's center for the globular clusters system. Then Robert J. Trumpler of the University of California and Carl Schalen of Sweden found that astronomers had been wrong in supposing that there was no general interstellar haze. They demonstrated that the light of an average star in the Milky Way band at a distance of 5,000 light-years from us was dimmed through interstellar absorption by at least one full magnitude. The revised computations made necessary by this discovery confirmed Shapley's conclusions regarding the direction of the center of the galaxy, though they reduced its estimated distance from 50,000 to 30,000 light-years.

On the basis of the facts then available, Mrs. Bok and I drew an outline of the Milky Way system a little more than 10 years ago. No findings have occurred since then that would make it necessary to change the diagram, nor does it seem likely that any major revision will be required in the years to come. However, while we know the general shape and outline of the system, we have only begun the task of filling in the details.

Our present knowledge of the structural details of our galaxy is summarized in the diagram. The region in which our detailed Milky Way surveys can be said to be more or less complete is indicated by the circle with a radius of 5,000 light-years, centered upon the sun. Inside this circle the evidence seems to point to fairly constant star densities in the directions of Cygnus and Carina, to a steady dropping off of the star density in the direction away from the center of the galaxy, and to an initial dropping off, followed by an increase at much greater distances, in the direction of the center in Sagittarius. The next region, from 5,000 to a little beyond 10,000 light-years from the sun, has been partially explored; here we have studied mostly star clusters and other special objects of high luminosity. In general the structural pattern in this ring seems to be much like that in the inner circle. It looks as though our sun is located in an elongated region of higher-than-average star density. It is tempting to deduce that this region of high density constitutes part of a spiral arm of the galaxy. But we must stress the preliminary nature of this conclusion. The star-counts themselves are subject to further checking, and even at best, the total volume of the Milky Way system that has been explored with any degree of completeness is ridiculously small.

It will, of course, take some time to extend these studies to cover a significant portion of the Milky Way. Our powerful new tools for research are difficult to make and very expensive; they are not yet generally available. And the task of gathering the required accurate basic information with regard to spectra, magnitudes and colors of faint stars is truly gigantic.

It seems absurd, but it is a fact, that we know much less about the detailed structure of our own galaxy than we do about some foreign ones, such as the great spiral nebula in Andromeda and the Magellanic Clouds. In the case of these nearby external systems we can obtain from a single photograph a good over-all view of the arrange-

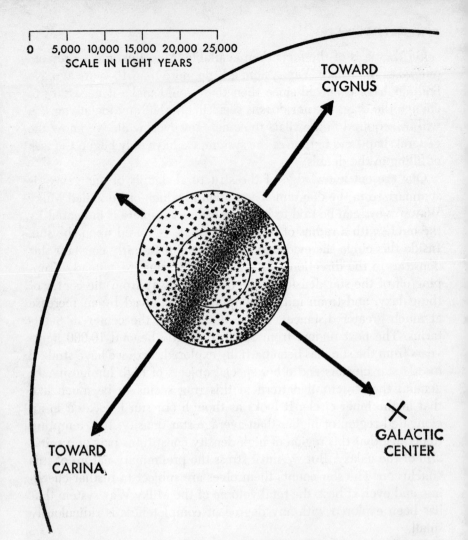

SCALE IN LIGHT YEARS
0 5,000 10,000 15,000 20,000 25,000

TOWARD
CYGNUS

TOWARD
CARINA

GALACTIC
CENTER

Location of the Sun in our Milky Way galaxy has been fixed in a spiral arm about two-thirds of the way out from the galactic center. Exploration of the Milky Way is reasonably detailed to about 5,000 light years out from Sun (inner concentric circle); thoroughgoing investigation has just begun in next region out to 10,000 light years. The observed distribution of stars in these nearby regions supports the deduction that our galaxy is spiral and that the sun is in an arm.

ment of the stars and their velocities, and of differences in relative distribution of, say, the blue and red stars. The Milky Way system, on the other hand, is too close for us to see it whole. It is much simpler to obtain a general impression of the arrangement of a large city from a plane flying overhead at 10,000 feet than from a prison somewhere near the center of the town, or even worse, from one in the suburbs. The astronomer's problem is actually even tougher than this, for he is asked to study the arrangement of the Milky Way from a suburban prison on a day with a pretty heavy fog! The fog, of course, is the great haze of interstellar dust and gas that floats near the central plane of our galaxy.

Yet the astronomer's position is by no means hopeless. There are various stratagems by which he can obtain important clues. One of these is to hunt for the directions of greatest transparency in the system. In the sections of the sky away from the central band of the Milky Way this is not too difficult; at a distance of 10 or 15 degrees from the band there are places where the faint spiral nebulae come into view, which means we are looking right through the haze of our own galactic system into the wide expanse of the universe of galaxies. The chances of finding many transparent regions in the central band of the Milky Way itself are rather slight, but here also we can get some indications of relative transparency, and thereby a measure of the extent of the system in the various directions. For example, since the interstellar haze produces a certain amount of reddening in the light of distant objects, we can locate the regions of greatest transparency by searching for the regions of smallest excess reddening.

Another intriguing problem for the Milky Way astronomer is the study of the central region of our galaxy. The investigation of this region was begun 25 years ago with an examination of variable stars by Shapley and Henrietta H. Swope of Harvard. Inspecting many photographs, they discovered hundreds of variable stars of the Cepheid variety close to or at the center of the Milky Way system. In recent years Walter Baade, using photographic plates with red-sensitive emulsions at Mount Wilson Observatory, has extended this survey to fainter limits. He found one region in the Sagittarius cloud where there were as many as 600 short-period Cepheid variables per square degree of the sky. It will be exceedingly interesting to study the colors of these faint stars for the effect of space-reddening. Already it seems quite likely that Baade's survey has penetrated right to the center of the Milky Way system.

There is also considerable encouragement in the recent discovery that in the direction of the center of the galaxy the obscuring interstellar haze seems to be concentrated mostly in the region relatively near us; beyond our neighborhood the haze appears to thin out. We come to this conclusion not only from studies of the central region itself, but also from inspection of other galaxies. The central regions of practically all spiral nebulae appear to be very much freer from cosmic dust than the outer parts. This gives us hope that once we penetrate the nearby haze, extending perhaps a third of the distance to the galactic center, we shall come to a quite transparent path for the remaining two thirds of the way.

There is another extremely interesting recent discovery that does not help but rather complicates our problem. This is the finding, made clear in studies by Baade, that the various star types are not distributed at random in the sky but seem to be organized in two distinct populations. In the outer regions of a galaxy, such as the region of our sun, the star population is marked by considerable numbers of highly luminous stars, notably the blue-white O and B stars, and by galactic star clusters such as the Pleiades; and there is a great deal of cosmic dust and gas. The central regions of spiral galaxies, on the other hand, apparently have no super-luminous stars, no galactic star clusters and no cosmic dust, but they abound in red giant stars, dwarf stars and short-period Cepheid variables.

The recognition of these two basically different populations in spiral galaxies, one in the outer and the other in the central region, reinforces two earlier conclusions. First, it supports the promise that we may be able to penetrate through the cosmic haze in which our sun is imbedded to study the center of the galaxy. Second, it emphasizes the importance of obtaining accurate data on the spectra and colors of the stars with which we are dealing. Of the many sections of the Milky Way awaiting examination with our new research tools, none seems more ripe for exploration than the center of the galaxy.

III. THE UNIVERSE FROM PALOMAR

by George W. Gray

THE 200-inch Hale Telescope on Mount Palomar, twice as large as its neighbor on Mount Wilson and penetrating twice as far out into space, is perhaps astronomy's first adequate cosmological instrument. George Ellery Hale and his colleagues designed this gigantic light-gathering mirror to study, not the nearby planets or the stars of our galaxy, but the stuff and structure of the Universe. The stuff includes, besides the stars, the clouds of cosmic gas and interstellar dust, from which stars are formed. The structure means the ways in which the stars are organized into clusters and other multiple systems, and how the systems in turn are regimented by the millions to form the gigantic rotating pinwheels known as spiral nebulae. The nebulae appear to be of different luminosities and sizes, and here and there in remote wildernesses of space groups of nebulae are found associated in clusters, with all the members of the group moving in the same general direction and at about the same speed. It would seem, therefore, that these spiral systems are not "island universes," as was once suggested, but are integrated parts of a still larger, all-inclusive structure which we may call the Whole, that is, the Universe.

According to the general theory of relativity, the size of the Universe is determined by the amount of matter it contains. The number of nebulae in a representative sample of space therefore is of critical importance to cosmologists. Mount Wilson's 100-inch telescope is able to photograph out to 500 million light-years. Practically all that mankind has found out so far about the remote nebulae we owe to this superb instrument. And yet its reach proved to be insufficient to stake out a representative sample of the Universe. The giant on Palomar is able to photograph objects at a distance of 1,000 million light-years, thus doubling the space-penetrating power in every direction and enlarging the volume of the observable region eightfold. There are men at Palomar who hope that by studying this enlarged sample they will be able to draw sound conclusions as to the size, structure, composition and nature of the entire Universe.

One of the optimists whom I met at Palomar is Edwin P. Hubble. "It may be wishful thinking on my part and the answer will not be known for a few years, but I believe we shall find that the sample is significantly large and that from it the type of Universe we inhabit can be identified among the long array of possible universes."

The task of determining the physical properties of the Universe and arriving at a picture of the Whole obviously cannot be tackled as a single project; it must be broken down into specific problems. In one way or another, these relate to the two great problems of cosmology: the meaning of the red-shift and the distribution of nebulae.

There is no question of the reality of the red-shift. The experimental proof of its reality is shown by hundreds of photographic plates. But what does the shift mean? Most investigators have assumed that the displacement of lines in the spectra of distant nebulae indicates motion, because motion is known to alter the wavelengths of the light emitted by the Sun and other nearby stars. But there is another possible explanation. It is conceivable that light may simply lose energy in the course of traveling millions of years through space. If it did, all its wavelengths would systematically lengthen and all its spectral lines would shift toward the red or low-energy end of the spectrum, even though the source of the light was standing still or moving at random. It is a question, therefore, whether the red-shift means that the Universe is expanding or merely that light has grown tired. Hubble believes that the 200-inch telescope will help establish the true state of affairs.

The distribution of star systems in the Universe also is a key problem. For if the nebulae are unevenly scattered through space, the task of estimating the size and mass of the Universe is greater than would be the case for a homogeneous Universe It is therefore important to map the positions of nebulae, and for this project the Hale Telescope is the last step in a carefully planned series of observations. At the Lick Observatory on Mount Hamilton a survey is being made of all nebulae down to those of the 18th apparent magnitude—an estimated total of more than a million stellar systems. The fainter nebulae, beyond the range of the Lick telescope, are so numerous, ranging into the billions, that it is impracticable to try to plot the entire sky. The Palomar group will select certain sample areas on the basis of the Lick results and count the nebulae in those areas out to the photographic limit of the 48-inch Schmidt telescope. Then the 200-inch will be used on sample regions uniformly scattered over these selected areas, and again the nebulae will be counted to the telescope's limit. In this way it is planned to determine the distribution of nebulae to different limits of brightness and distance.

The astronomer's chief yardsticks of distance are the Cepheid variable stars, the blue supergiants and the novae, or exploding stars. The intrinsic brightness of these types of stars has been fairly well established by study of their spectra and by other means; they are accessible for study because they occur in our own galaxy, the Milky Way. When one of these beacons is spotted in a spiral nebula, it is possible to determine from the star's faintness the distance to the nebula. With the distance known, it is a simple problem in arithmetic to calculate the whole nebula's real brightness.

Such determinations have been made for several hundred nebulae within about 10 million light-years of the Earth. (Beyond that distance even the 200-inch telescope is unable to resolve individual stars.) Some of these are giant systems, some dwarfs. The calculated average luminosity of each group serves as a distance indicator with which to gauge more remote groups and collections, since it is assumed that the average for a large sample is typical of the average for other large samples of the same kind.

Now such a system is scientifically sound, provided the yardsticks are correct. But recently Joel C. Stebbins and Albert E. Whitford of the Washburn Observatory in Wisconsin, using photoelectric tubes to measure luminosities of very faint stars instead of the old method of measuring photographic images, discovered that many of the old measurements were in error by as much as a quarter-magnitude. And Walter Baade, of the Mount Wilson and Palomar staff, made another discovery that threw all the calculations off. He found that the stars of the Universe, which had all been supposed previously to run true to type, actually had to be divided into two general classes, which he calls Population I and Population II. "Baade's discovery," said Hubble, "is the most important contribution of the last decade to our knowledge of the components of the Universe. It is revolutionizing our thinking about the composition of nebulae, and I am confident it will prove to be a powerful tool in the study of cosmic structure."

Baade was working with the 200-inch telescope when I visited Palomar Mountain. A dozen buildings dot Palomar's peak—powerhouses that provide electricity, water and other utilities; cottages for the mechanics and other workers; the "Monastery," where the astronomers live; the domes of the 48-inch and 18-inch Schmidts. But all these structures and their equipment are auxiliary to the master installation—the 200-inch reflector that rests under the Pantheon-like dome of 135-foot diameter and 137-foot height. Under this great silvery dome they were readying the 200-inch eye for its night's vigil on the Universe.

All day the dome had remained tightly closed to insulate the sensitive giant against solar heat and terrestrial weather. As the Sun sank behind a distant mountain, the roof shutters were rolled back, exposing the vast uptilted framework of the 60-foot tube. At the same time a bank of electric fans began to whir, directing their breezes against the undersurface of the great mirror. Ordinarily the outside temperature on the mountain drops to its night minimum a few minutes after sunset, and the fans quickly cool the 15-ton slab of glass to the same level. By the time the last streak of crimson had faded from the West, the telescope was poised and waiting, ready to peer hundreds of millions of years into the past of the stars.

And the stars were waiting too, in dazzling pageantry. Although Palomar is 2,300 miles north of the Equator, from its elevation the 200-inch mirror can sweep over three quarters of the sky. Only the heavens above the Antarctic zone are beyond its gaze. Palomar's air is so clear and steady that the stars shine with a brilliance never seen down in valleys, prairies or city streets. This starlight is the principal concern of the astronomers, and they take great pains to keep it uncontaminated. Road lamps and headlights are strictly forbidden on the mountain top.

To pass from the star-studded vault without into the darkness within was blinding. There was nothing to do but to hang on to a companion's arm and follow where he led. We finally risked a pocket flashlight, which disclosed that we were groping among a forest of steel piers—the underpinning of the telescope and its observing floor above. Eventually we reached the elevator and a few moments later stepped out on the observing floor.

It was not so dark here. A soft radiance suffused the massive overshadowing structure of the telescope. Looking up, we saw starlight pouring through the open shutter of the dome. Across the floor, under one of the supporting arms of the telescope yoke, a man sat at a table in the semi-darkness watching hands move around a series of illuminated dials. This was Byron Hill, superintendent of grounds, who was serving as instrument-tender for the night. Seventy feet up in the air, seated in a cage at the open end of the telescope tube, was Baade, observer for the night. In the floor of the cage is an aperture that admits the concentrated beam of starlight gathered by the more than 190 square feet of flawless mirror below. At the spot where the curvature of the mirror focuses this beam, Baade had exposed a photographic plate. He was watching through the eyepiece and guiding the steel-glass giant with his fingers on electric controls to keep the star image centered at the cross-hairs. Below, buried in the base of the telescope, a motor of one-twelfth horsepower was slowly turning the great tube toward the West, in step with the rotation of the Earth. The telescope weighs 500 tons, but it swung effortlessly and noiselessly on its oil-padded mounting.

"Thirty seconds to go," sang out Hill, his eye on the clock.

"Ready," answered Baade from his microphone in the cage.

"Twenty-five," droned Hill, "twenty—fifteen—ten—five—four—three—two—and close!"

Baade closed the plate-holder, removed it and inserted another. Then he called down through the speaker: "After we've taken this one I'll come down for a cup of tea—at 12 o'clock."

Baade is 58, a native of Germany who left a private-docentship at Hamburg Observatory in 1931 to accept appointment to the Mount Wilson staff. Over tea, he discussed his work:

"During the war years the city lights of Los Angeles and Pasadena were blacked out, and that was an encouragement to intensified study of the nebulae. Dr. Hubble had found stars in the spiral arms but had been unable to resolve the bright central nucleus of the Andromeda Nebula into stars. Now, with the sky free of artificial light and with improved photographic plates available, I thought I would make a search for them with the 100-inch telescope. I tried blue-sensitive plates, because usually the brightest stars in galaxies had turned out to be blue giants. But these photographs got me nowhere; the central area remained a great unresolved luminous blob.

"Then I had a hunch. It occurred to me that the brightest stars of this region might be red. Red-sensitive photographic plates of high speed had just become available, and I began to use them. The results were immediate. The bright central nucleus broke up into a

27

mass of individual star images, and as I photographed various regions of the nebula, on out to the tips of the outermost spiral arms, I found a change in the pattern. Every time I encountered a spiral arm the blue supergiants showed up, whereas in the center and in the spaces between the spiral arms I could find only red giants."

The Andromeda Nebula has two smaller companion nebulae which do not have the spiral structure; one is nearly circular in shape, the other shows a vague bar-and-disk pattern. No one had been able to resolve either nebula into individual stars. Baade photographed them with his red-sensitive plates and found them resolved into a vast number of red giants. But with the blue-sensitive plates he was not able to get any star images at all.

"From this study," Baade continued, "three results stood out: first, the brightest stars in the hitherto unresolved nebulosities were red; second, the moment one reached the brightest stars in these systems they appeared at once by the hundreds and thousands; third, as far as luminosity was concerned, these brightest red stars were about five magnitudes (that is, 100 times) fainter than the blue supergiants on the spiral arms of the Andromeda Nebula."

Before discussing luminosity comparisons, let us get clearly in mind what is meant by the stellar magnitude scale. The numbers on this scale stand for increasing faintness. Thus Betelgeuse, a bright star, is rated as of the first magnitude, the North Star of the second, and so on down the line. The sixth magnitude is the faintest the naked eye can see; the 21st the faintest the 100-inch telescope can photograph; and the 23rd the limit of the 200-inch telescope. Stars that appear brighter than Betelgeuse are assigned zero and even negative magnitudes: the dog-star Sirius is rated as minus 1½ magnitude and the Sun as minus 26.

These magnitudes stand for the apparent brightness of the stars as we see them, with their light attenuated by distance and the cosmic haze; they do not, of course, measure the stars' real luminosities at the source. As a basis for comparing luminosities, astronomers compute what the brightness of each star would be if it were placed at a standard distance from the Earth. This is called the absolute magnitude. The standard distance is 10 parsecs (about 32½ light-years). If all the naked-eye stars were lined up around the Earth at this standard distance, how our firmament would change! Sirius, which now appears the brightest of the night array, would dim to a magnitude of about 2 and some of the seemingly faint blue stars would blaze into a magnitude of minus 7.

When Baade compared the blue and red supergiants in terms of absolute magnitude, he found that whereas the brightest blue stars, found in the spiral arms, were of the minus-7th magnitude, the brightest red stars, found in the central region and the two companions, were of only the minus-2nd magnitude.

Baade now turned to the famous Hertzsprung-Russell diagram of the classes of stars to see what light it could throw on the sudden emergence of hundreds and thousands of red giants in the nebulae. In this diagram all the stars are plotted according to absolute magnitude and spectral type. Most of the stars of our Milky Way system describe a curve known as the "main sequence." The main sequence begins with blue giants of about the minus-7th magnitude and ranges down through white stars of lesser magnitude, yellow stars, orange stars and finally red dwarfs of about the 19th magnitude. Entirely separate from this curve of the main sequence, the diagram shows a shorter sequence made up almost entirely of red stars; it is known as the "red-giant branch." Examining this group, Baade found that these red giants were a good deal fainter than his new-found red giants in Andromeda; most of those in the Hertzsprung-Russell diagram were around magnitude 0.

"At this point I was stumped," continued Baade, "for it seemed clear that the red giants of the newly resolved systems could not be members of the standard Hertzsprung-Russell diagram. Then I remembered the curious diagram of the globular clusters."

These globular star clusters are satellite systems which surround our Milky Way, apparently hedging it about in all directions. One hundred such clusters have been photographed. Their stars had been plotted by Harlow Shapley according to the Hertzsprung-Russell system, and their red giants had seemed different from the red giants of the main sequence—they were about two magnitudes brighter. In other words, their magnitudes were about minus 2—just like the red giants of the Andromeda Nebula and its two companions!

"Suddenly everything fell into line," related Baade. "There must be, I realized, two populations of stars—one characteristic of the globular clusters and the newly resolved Andromedan systems; the other characteristic of the spiral arms and of our own part of the Milky Way."

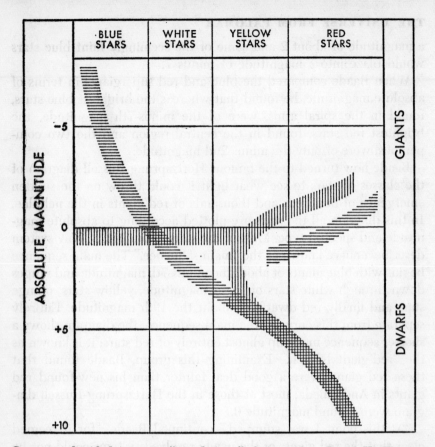

	BLUE STARS	WHITE STARS	YELLOW STARS	RED STARS	
					GIANTS
					DWARFS

ABSOLUTE MAGNITUDE: -5, 0, +5, +10

Stars are classified into two "populations," designated as I and II. The two populations are here compared in a diagram which plots the color of the stars against their brightness or magnitude. The magnitude scale reads upward to the top of the diagram; stars brighter than original standard or first magnitude stars have negative magnitude numbers. The Population I stars, of which our sun is one, are plotted in band of horizontal lines; the plot shows that the brightest stars in this classification are blue. A separate group of Population I stars, at right outside the main sequence, are known as red giants. The brightest stars in Population II (*band of vertical lines*) are not as bright as those in Population I, and they are red in color. This differentiation of stars into the two major groups is essential to working out the structure and evolution of galaxies.

Following the Hertzsprung-Russell scheme, Baade plotted the stars anew, and his diagram, reproduced here, shows how the magnitude factor separates the giants of Population I from those of Population II. Both populations have blue stars, but those of Population I are the brightest giants known, while the blue stars of Population II are not far above the dwarf level. Similarly, both have red stars, but Population II has the top giants of this color. When we reach stars at about the Sun's luminosity, absolute magnitude 4, the two curves merge.

"This is exactly what we should expect if we are dealing with stars of two different age groups," explained Baade. "Inasmuch as we are certain that the conversion of hydrogen into helium is the process that keeps the stars going, we can make rough but reliable estimates of the time period in which a star of given luminosity exhausts its hydrogen supply. The result shows that the fastest spenders are the blue giants of Population I. Even if they were originally composed wholly of hydrogen (with a few impurities to get the transmutation cycle going), they would expend their last hydrogen in about 100 million years. In other words, none of the stars at the top of the diagram of Population I which we observe today can be older than 100 million years. If there were blue stars of this luminosity in our Milky Way 100 million years ago, they must have faded long since and now occupy places of much lower magnitude in the diagram.

"As we move down the scale from the blue giants, the rate at which hydrogen is converted into helium and radiation becomes less and less, until at about absolute magnitude 4 we come upon stars which can not have spent more than a few per cent of their hydrogen. Even if they were formed at the birth of the Universe, which we estimate to have occurred about 3 billion years ago, they still have an enormous fuel reserve in the form of hydrogen. In other words, stars of absolute luminosity 4, such as our Sun, and all fainter stars which were formed at the beginning of the Universe have changed neither their luminosity nor their chemical composition in the 3 billion years that have elapsed. Even if the two populations represent stars of two different age groups—and there are strong indications that Population II is the older—the distinction between them must disappear at about magnitude 4 simply because the Universe is still so young."

If the distinction disappears at about magnitude 4, how can one know that there are different populations in the lower levels?

"That is a very interesting question which we are investigating," answered the astronomer. "If the diagrams of the two populations completely coincide, the distinction indeed becomes meaningless.

But if the diagrams should not coincide precisely, it will make Professor Martin Schwarzschild of the Princeton Observatory very happy. He has obtained very strong spectroscopic evidence that there is a difference in chemical composition between stars of the two populations. Those of Population I seem to have a higher admixture of metals than those of Population II. Since we know that the probable temperatures inside the stars (only a few million degrees) are much too low for the transmutation of lighter elements into metals, we must conclude that the original material from which the stars of Population I were formed contained a higher percentage of metals than the material from which Population II stars were formed.

"Now we know," Baade continued, "that Population I stars are found only in systems that contain dust. In the Andromeda Nebula dust is richly strewn all through the spiral arms, and it is here that we find the blue giants and other Population I stars. Similarly in other nearby galaxies and in our own—the Milky Way—wherever dust clouds are present, there you find Population I stars, and where there is no dust, there is no Population I. Moreover, this dust is richer in metals than is the interstellar gas, because the atoms of the metals easily stick together when a grain of dust is formed of the interstellar gas, whereas the much more abundant hydrogen atoms of the gas won't stick and so evaporate back into space. Professor Schwarzschild therefore guesses—and it is probably a good guess—that the stars of Population I have been and still are being formed from the interstellar dust, whereas those of Population II were formed from the gas at an earlier epoch, before it had time to produce dust particles. If Schwarzschild can prove that even among the stars much fainter than the Sun there are the two types of stars, one with the

Spiral galaxy is shown in top drawing as it appears "face-on," or at right angles to the plane in which its outer stars are concentrated; two lower drawings, which segregate Population I and II stars from one another, show the galaxy "edge-on," or in the plane of the outer disc of stars. Population I stars are indicated by "asterisks"; Population II stars by dots. It can be seen that the brighter and bluer Population I stars, which include our sun, are concentrated in the plane of the galactic disk, and especially in the spiral arms that swing outward from center. The redder and less brilliant Population II stars are concentrated in the spherical nucleus of the galaxy and occur less frequently in the spiral arms. An important feature of the spiral arms, not shown, are the dust clouds in which, it is thought, new stars of the Population I type are still being formed.

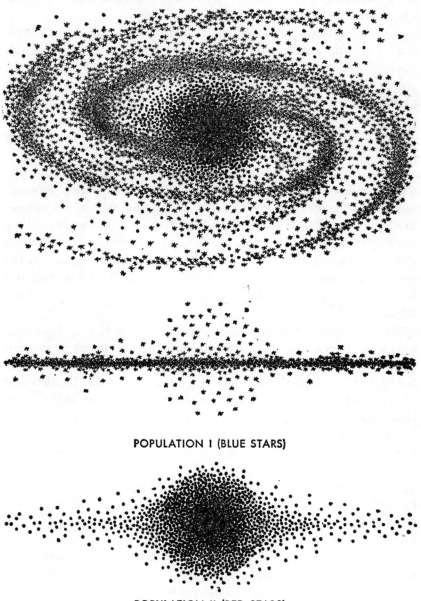

POPULATION I (BLUE STARS)

POPULATION II (RED STARS)

higher abundance of metals (type I) and the other with the lower (type II), he will have won."

It's a dusty Universe. In all the nebulae that have been resolved, dust is abundant in the spiral arms, though absent from the central nuclei. The Sun with its Earth and other planets is believed to be situated far out in one of the spiral arms of the Milky Way, and even a small telescope shows many of the nearby dust clouds, such as the dark patches in Orion and the Coalsac near the Southern Cross. All about us are stars of Population I, born, it is believed, of the cosmic dust.

Among Population II stars explosions are frequent. The Andromeda Nebula continually shows fireworks: in a single year some 15 novae, an average of more than one explosion a month, were found in its photographs. Novae also flare up frequently in Messier 81, the large spiral in the Big Dipper, which like the Andromeda has a large representation of Population II. But they are relatively rare in the two Magellanic Clouds, those satellite galaxies which attend our Milky Way and which are composed almost entirely of Population I stars.

One upsetting result of Baade's discovery is that the Cepheid variable stars, which have been used as yardsticks of distance, turn out to come in two types. Harlow Shapley used the Cepheid variables in the globular clusters that attend our galaxy to measure the distances of the clusters and so estimate the dimensions of the Milky Way. Hubble used Cepheids in the spiral arms of the Andromeda Nebula as yardsticks with which to measure the distance and dimensions of that galaxy. But Baade finds that the Cepheids in the globular clusters are of Population II, whereas those in the Andromeda Nebula are of Population I. Current studies suggest that Population I Cepheids may be about half a magnitude brighter than Population II Cepheids of the same period of variation. If the suggestion is confirmed, it will follow that either the Milky Way is smaller than we have supposed or the Andromeda Nebula larger. In either case our own stellar system will no longer rival the Andromeda in size.

As a consequence of Baade's findings an extensive program of calibration is now under way at Palomar. All the familiar distance indicators are being re-examined, retested and standardized in a grand survey centered on globular clusters and nearby spiral nebulae. The key instrument of this survey is the 200-inch telescope, using photoelectric tubes to measure the magnitudes of the stars it reveals. Two spirals in particular are being sifted for stars of all colors in both populations: the Andromeda Nebula and Messier 81, the large spiral in the Big Dipper. Because the Andromeda is the largest known spiral and is less than a million light-years away, it will yield a wealth of new stellar data of the kind that we are unable to obtain from our own galaxy, for we can never examine the Milky Way as a whole—we cannot see our forest for the trees. Messier 81 is farther away than the Andromeda—its estimated distance is three million light-years—but even so, the 200-inch can break its massed brilliance into individual stars, and the plan is to use Messier 81 to check the conclusions obtained from the Andromedan survey. *

"These studies are crucial," said Hubble. "All of our ultimate problems—the evolution of stars, their association in stellar systems, the nature of the Universe itself—demand that we acquire reliable information concerning distances, luminosities and masses. When these preliminary objectives are won, we shall turn to the cosmological problems with new confidence. We shall then be able to state our observational results positively, within known limits of error or uncertainty, and it will become possible to test and eliminate theories. The long array of possible worlds will be reduced to a few that are compatible with the existing body of knowledge. And possibly, just possibly, we may be able to identify in the shortened array the specific type that must include the Universe we inhabit."

* As the result of further observations by Baade and others, since the first publication of this article, re-calibration of the Cepheid variable distance indicators has substantially confirmed the suspicion that the galaxies are twice as far away from us as had previously been thought. That means that they are twice as large; it also means that they have been traveling for twice as many light years. This correction eliminates the discrepancy between the geological data for the age of the earth and the astronomical calculations for the birth of the Universe. The new tentative figure is four billion years.

IV. THE DUST CLOUD HYPOTHESIS

by Fred L. Whipple

IF THE EARTH should disappear some day in a chain-reacting cloud of dust, a possibility which is mentioned half seriously in these deranged times, philosophers on another planet might find a certain poetic symmetry in its birth and death. Recent astronomical studies have given us reason to surmise that the earth was born in a cloud of dust. This Dust Cloud Hypothesis, as it is called, suggests that planets and stars were originally formed from immense collections of sub-microscopic particles floating in space. Although it is still being developed, the dust cloud hypothesis possesses a plausibility that other theories about the origin of planets and stars have lacked.

The beginnings of our physical universe are necessarily beclouded in the swirling mists of countless ages past. What cosmic process created the stars and planets? Are new ones still being formed? Or were all that now exist made in one fell swoop? Scientists are making progress in their study of these fascinating questions, although they still cannot answer them with certainty.

The dust cloud theory begins with the fact that there are gigantic clouds of dust and gas in the abyss of space that lies between the stars. Observations by the world's astronomers during the past 20 years have proved the existence of these clouds. Interstellar space, formerly supposed to be empty, is now known to contain an astonishing amount of microscopic material. Jan Oort of the Netherlands has calculated that the total mass of this interstellar dust and gas is as great as all the material in the stars themselves, including all possible planet systems. In other words, for every star there is an equal amount of dust and gas dispersed in space. The immensity of this quantity of material is beyond the grasp of human imagination. In the Milky Way alone, it comes to 300 million million earth masses. Yet interstellar space itself is so vast that this dust and gas is scattered more thinly than in the highest vacuum that can be created on earth.

We have a good deal of information about the composition of this nebulous star dust. The gases that we can detect are the ordinary elements with which we are familiar—hydrogen, helium, oxygen, nitrogen, carbon, and so on. Dutch astronomers have recently shown that these gas atoms slowly coalesce into chemical combinations and dust particles. While the structure of the dust particles is uncertain, it appears that most of them are very small—of the order of a fifty-thousandth of an inch in diameter. Evidence of their size and of the fact that they actually are dust particles is afforded by the way in which they scatter the light from distant stars. This scattering produces dark clouds on a photographic plate. The small amount of starlight that filters through these dust clouds is reddened—for the same reason that the sun appears reddened during a dust storm: the long waves of red light are less scattered by small dust particles than are the shorter waves of other colors.

What collects these dust particles into clouds? Lyman Spitzer, formerly of Yale and now at Princeton, suggested that it might be the pressure of light. The pressure of light, which is so exceedingly small that it cannot ordinarily be observed, is strikingly demonstrated in comets' tails, which are formed by the pressure of sunlight forcing fine material away from the head of the comet.

The writer, following the approach suggested by Spitzer, found that under rather unusual, but possible, circumstances the light from stars would tend to force interstellar dust into larger and larger clouds. In the starlight of space, each dust particle casts a shadow. The shadow is minute. Nonetheless, it results in less light shining from the direction of one particle on another nearby, and vice versa. Hence the two particles tend to attract each other, by a force varying inversely as the square of the distance between them. The mathematics of this principle is similar to that of Newton's law of gravitation.

After a few particles are collected into a small cloud, the cloud casts a larger shadow in the starlight on the particles in its neighborhood. These particles are drawn into the cloud, making it larger and larger. If such a cloud is not too much stirred by its motion through other banks of dust and gas, and if too bright a star does not pass through it and scatter the particles by its light pressure, the cloud will continue to draw in dust. Finally it will attain a mass and density sufficient for gravity to become stronger than light pressure. The cloud will then begin to contract. Calculations show that for a dust cloud with the same mass of material as the sun, the two forces would be about equal when the diameter of the cloud was some 6,000 billion miles. This distance is 60,000 times the distance of the

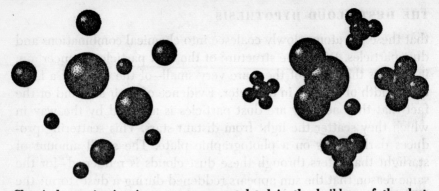

Chemical combination is one process postulated in the build-up of the dust cloud. The process would start with atoms moving free in space, as suggested at left (closeness of the atoms is greatly exaggerated in drawing). Upon collision, atoms of commoner elements would combine to form such simple molecules as cyanogen (CN), methane (CH$_4$) and ammonia (NH$_3$) indicated in cluster at right.

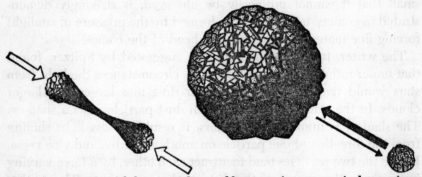

Physical aggregation of dust particles would assume importance in latter phase of the dust cloud growth. Dust particles, made up of many molecules, are pushed together (at left) by gentle pressure of light from thousands of surrounding stars, the relative pressure being lower left in the tiny cone of shadow between them. As aggregations of particles grow in size (at right) gravitational forces would begin to take effect.

earth from the sun. It has been further calculated that such a cloud might develop and collapse into a star in less than a billion years.

These calculations were made before any dust clouds of this nature had actually been observed. During the war, Bart J. Bok of Harvard, while musing one day over some familiar photographs of the Milky Way, noticed some very small, round, dark patches that had not seemed important before. He studied a number of photographs of each region in the Milky Way and found the same dark patches in all the photographs. These were not photographic blemishes but truly

38

dense, dark clouds in space! When Bok estimated their distances and calculated their diameters, he found that the smaller clouds in this group were about the same size as the hypothetical dust cloud for which gravity equals the light pressure on dust. Many were larger but few much smaller.

Bok's discovery suggested the fascinating possibility that the small dark clouds might be stars in the making. That new stars are constantly being formed from cosmic dust seems more than likely. There is no other reasonable explanation for the brightness of certain stars. A star's intensity of radiation, which shows how fast it is burning up its energy in nuclear reactions, indicates its maximum age. Some stars are so brilliant that they could not have radiated for two billion years, the minimum time we can allow since the "beginning." Hence they must have been born later than the solar system. The discovery of these clouds encouraged the writer to study the possibility that not only a star but also a system of planets might condense out of such clouds.

Let us consider our solar system, therefore, as a case study of the formation of a star and its satellites. We have a huge dust cloud, as described above, which has begun to condense under gravity. There will be minor turbulent motions of the material within it—sub-clouds, or streams of dust—that slide by each other or collide. In order to explain the present slow rotation of our sun, we must assume that the motions of the streams in the original dust cloud canceled each other and that the cloud as a whole did not rotate. This point can be illustrated by a well-known parlor game. The victim is persuaded to sit on a piano stool with his arms outstretched and holding some books. The stool is started turning slowly. Its occupant is then instructed to draw in his arms and the books. He now spins so rapidly that he may fall off the stool. This phenomenon demonstrates the law of "conservation of angular momentum," or what might be called "rotational obstinacy." Angular momentum, or rotation around the center, may be variously distributed among the parts of the system, but the total momentum for the system as a whole remains constant. Thus if the great dust cloud from which the sun was formed had had any appreciable rotation to begin with, after its collapse the condensed sun would have rotated with great speed. But actually our sun turns very slowly; it takes nearly a month to make a complete rotation. Consequently the original dust cloud must have been almost stationary.

This, by the way, would not necessarily be true of other dust clouds in our galactic system. The Milky Way itself rotates; hence we might expect many coalescing dust clouds within it to possess a great amount of rotation. In that case, according to our theory, they could not condense into single stars but would form double stars or even clusters. As a matter of fact, most stars *are* double or multiple, so that in this respect the dust cloud hypothesis is consistent with observations. The sun can be considered a somewhat unusual case.

Having accounted for the sun's slow rotation, let us go back to the original dust cloud. Under the force of its own gravity, it has begun to condense. At first it collapses very slowly, because the motions of its internal currents and streams resist its contraction. A group of moving particles is of course harder to collect and compress than one which is standing still. But in the course of millions of years the random motions of the streams within the cloud are damped out by collisions and friction. Meanwhile the cloud contracts more and more powerfully as it becomes smaller, because as its density increases, the force of gravity among the particles increases. The net result, with resistance diminishing and gravity increasing, is that the cloud collapses faster and faster. Its final collapse from a size equal to that of the solar system (*i.e.*, the diameter of the orbit of Pluto, the farthest known planet) would require just a few hundred years. Due to the increased pressure in the contracting cloud, its temperature rises enormously. In the last white-hot phase of its collapse, the sun would begin to radiate as a star. Its central temperature would become great enough to start the cycle of nuclear reactions among carbon, hydrogen and helium which keeps the sun radiating.

Now we must account for the evolution of the planets from the same great dust cloud. We return to the cloud before it has begun to shrink appreciably, and follow the largest stream in the cloud. If the dust in this stream is sufficiently dense, the stream condenses into minor clouds. As these clouds drift along together, roughly in the same direction, they will pick up material less compact than themselves; hence they will grow slowly, feeding on portions of the great cloud. As they grow, the minor clouds, now "proto-planets," begin to spiral slowly in towards the center of the main cloud. They have gained in mass but not in angular momentum, so they move towards the center of gravity, somewhat as a whirling ball on a rubber string, if no force is exerted to keep it whirling, spirals in an ever-narrowing circle as its motion is slowed by friction. Some of the proto-planets move in more rapidly than others, their rate depending on their size and on chance encounters with other streams.

If the great cloud remained spread out forever, all the proto-planets would eventually wind up at its center. But long before some of them have completed their spiral, the main cloud collapses and forms the sun. Its rapid final collapse leaves a number of proto-planets stranded in their orbits, outside the collapsing cloud. Some are trapped too near the center and are pulled in or destroyed in the sun's heat. Others are far enough away to remain intact. They condense and become planets. Some of them may be at enormous distances from the sun. For all we know, there may be planets in our system beyond Pluto, the farthest one that we can see.

When first formed, the planets are hot, perhaps hot enough to be in a molten condition. But since they are relatively small, their heat of contraction is not sufficient to start the nuclear reactions that would make them radiate permanently like a star. Gradually they cool off.

We have described, then, how the dust cloud hypothesis accounts for the origin of the solar system. Now let us see how well our theory accounts for the system's peculiarities.

Any theory about the evolution of our planetary system must explain certain striking characteristics: 1) The planets all move in the same direction and, with one exception, very nearly in the same plane as the earth's orbit, called the plane of the ecliptic. 2) Their orbital paths around the sun are nearly circular. 3) Almost all the planets rotate, or spin, on their axes, in the same direction in which they revolve about the sun. 4) Most of them have moons or satellites—Jupiter has 11—which usually revolve about the planet in the plane of its rotation and in the same direction.

Thus the theory must account for a great deal of regularity in the system. But it must also explain some irregularities. For example, the orbit of Uranus and its system of five satellites, including probably the one recently discovered by G. P. Kuiper at the McDonald Observatory in Texas, is tipped up roughly at right angles to the plane of the ecliptic. Neptune's single satellite revolves backwards, as compared with the rest of the solar system, although Neptune itself turns in the forward direction and is thus properly oriented. Some of the satellites of Jupiter and Saturn also are contrary.

To begin with, the theory explains why the planets generally revolve in the same direction and in nearly the same plane. Their plane and direction are determined by the motion of the original stream from which they were created. The planets' circular paths around the sun are accounted for by the spiraling phase of their evolution. Spiraling reduces the orbit of a revolving body more and more to the circular form.

The spacing of the planets at their present distances from the sun is not explained by the dust cloud hypothesis. This spacing, as every astronomy student knows, follows a regular mathematical relationship known as Bode's Law. It is possible that the planets' distances from the sun were determined by gravitational effects over a long period of time rather than at the very beginning. The great mathematical astronomer Ernest W. Brown, of Yale, doubted that the planet distances could have remained constant for more than 100 million years or so.

The planets' rotation or spinning is adequately explained by the dust cloud theory. As the great cloud condenses, it is denser towards the center than in its outer regions. Thus a proto-planet, when it spirals in, tends to pick up more material on the side that faces towards the center than on the side away from it. This process produces a result something like the rolling of a snowball. The side that picks up more material becomes heavier and is slowed up. The outer side of the planet, being lighter, travels faster and moves forward. Thus the process imparts a forward spin to the whole planet.

On the assumption that the planets actually gained all of their rotation in this fashion, we can estimate how large they must have been when they condensed. The size of a planet when it started to rotate can be calculated from its present speed of rotation and its mass. The results of these calculations are very encouraging to the dust cloud hypothesis, for the calculated diameter of each proto-planet figures out to about the same as or a little more than the diameter of the orbit of that planet's farthest present satellite. This indicates that the satellites were formed while the planets were still distended as clouds. When a planet-cloud collapsed, the outer material collected into satellites or fell to the surface of the planet. Thus the satellites developed in just about the same way as the planets did when the sun-cloud collapsed.

We have no difficulty in accounting for the rapid revolutions of the satellites around their planets. The planets themselves rotate rapidly. We would also expect the satellites to revolve in the same direction as their planets' rotation and in the plane of the equator. In most cases, as we have seen, they do. But what about the exceptions? Neptune's only satellite, the three outer ones of Jupiter and one of Saturn's revolve in a direction opposite from all the others and generally in planes different from those of their planets' equators. The answer is probably very simple: the maverick satellites were not a part of these planets' initial systems but were captured later, when they could no longer be completely controlled. Very likely there were originally many dense minor clouds, or potential planets, which did not develop into full-fledged planets. Some of these were small and outside the main stream. If a small cloud of this sort ran into a planet-cloud before the planet had collapsed, it would be captured and become a satellite. Normally the planet-cloud's rotation would carry the captured cloud along in the same direction, and would reduce the size of the satellite's orbit. But a satellite that was captured by a planet's gravity after the planet had collapsed would be less strongly influenced. If it was revolving backwards when it was captured, it would continue in a retrograde orbit, even though held a prisoner by the planet's force of gravity.

Our hypothesis has another major irregularity to explain. Why are some of the planets so much larger than others, and why are the large planets so much less dense than the earth? The average density of Jupiter, for example, is only a little greater than that of water, while the earth is five and one-half times as dense as water. Saturn, if it could be put into a huge sea, would actually float. The explanation of these differences probably lies, as Henry Norris Russell of Princeton has suggested, in the fact that the giant planets were bigger to begin with. Their size gave them a huge gravitational attraction so that they could hold the light gases, such as hydrogen and helium, which would float away from a less massive planet. Their ability to attract and hold light elements would have a double result: they would grow rapidly, and they would be relatively light in proportion to their volume. On the other hand, a smaller proto-planet such as the earth, which has less ability to hold hydrogen or helium, would soon reach the limit of its growth. Lyman Spitzer has recently shown that hydrogen and helium are escaping from the earth's atmosphere even today.

There is still another odd peculiarity of the solar system that the dust cloud hypothesis seems to explain quite well. This is the fact that the planets which are closest to the sun have relatively few satellites and comparatively thin atmospheres. The explanation is this: When the sun was collapsing to its final form, the energy released by its contraction made it hot. The inner planets—Mercury, Venus, the Earth and Mars—were then in a fairly dense region of the condensing sun. As a result, their atmospheres and surfaces were heated to very high temperatures. Most of their satellites would boil away. Fortunately for the theory, this boiling period was very short —perhaps a few months or years—else the planets would have boiled away also. As it was, the earth and moon probably were not entirely spared by this "bath of fire"; both may have been appreciably reduced in size by the evaporation of their outer rocks. The outer planets, being outside this bath of fire, would not have been much affected, which accounts for the fact that they still have thick gaseous envelopes and a comparative abundance of satellites.

One other puzzling phenomenon, perhaps the most fascinating of all, remains to be explained. This is the great collection of asteroids, or minor planets, found in the region betweens Mars and Jupiter. There are at least 1,600 of these "flying mountains," ranging from a mile or two to three or four hundred miles in diameter. All revolve in the same forward direction and near the plane of the ecliptic, although not as close to it as the planets. Their grouping suggests that the asteroids had a common origin. Are they the building stones of a planet that was stillborn, or the debris of one that was smashed? Present opinion favors the latter possibility. The proof, though not yet complete, is fairly convincing. Harrison Brown at the University of Chicago and Carl Bauer at Harvard have recently shown that meteorites—pieces of interplanetary material that fall on the earth as shooting stars—must be fragments from a broken planet. Other studies by C. C. Wylie in Iowa indicate that these shooting stars or fire-balls swing about the sun in orbits resembling those of some asteroids. Though the chain of evidence at this point is weak, we may reasonably accept the tentative conclusion that meteorites are baby asteroids, and that the asteroids, in turn, are broken pieces of a once completely formed planet. Harrison Brown calculates that the planet was about the size of Mars.

How was the planet smashed? Probably by collision with another planet; there may even have been more than two planets involved. On the dust cloud theory, we can easily assume the formation of an eccentric planet which would cross the path of one in the main stream. If it did, sooner or later we would expect the two planets to collide. The resultant cosmic explosion would produce the scattered asteroids.

We have seen, then, that the dust cloud hypothesis accounts for a great many of the facts about our solar system. The chief difficulty in the theory has to do with the question of how the proto-planets maintained themselves during the early stages. At that period the dust clouds had to be very rare, their average density being more nearly a vacuum than the vacuum in a Thermos bottle. Yet they had to hold together sufficiently to pick up material from the rarer spaces between them, and had to be massive enough to grow and to spiral in towards the sun. Such a situation is difficult to imagine, but there is some theoretical evidence that it is possible.

We must now consider whether the dust cloud theory is more convincing, or has fewer weaknesses, than other theories that have been proposed. The famous Nebular Hypothesis of the French mathematical astronomer, the Marquis Pierre Simon de Laplace, suggested that the sun and planets derived from a great revolving nebula, or cloud. In this respect Laplace's theory sounds somewhat like the present one. But he assumed that the planets were formed from rings of matter left behind when the dish-shaped nebula collapsed. The nebular hypothesis meets with the overwhelming difficulty that it requires that most of the solar system's rotation be carried by the sun. As we have seen, the slowly rotating sun actually contains only a small percentage of the system's total rotation or angular momentum, while Jupiter contains more than half. It is not possible to account for the observed motions by Laplace's theory.

Another famous and extremely important theory, known as the Planetary Hypothesis, was proposed by the geologist T. C. Chamberlin and the astronomer F. R. Moulton, of the University of Chicago, early in this century. They postulated that the planets were formed as the result of a near-collision between the sun and another star. The star came very close to the sun, perhaps even grazing its surface. Huge tides were raised on the sun and great quantities of material were torn from it. This material is then supposed to have condensed into droplets which eventually coalesced into planets. The planetary hypothesis has many attractive features, and has dominated thinking in this field for many years.

One of its principal difficulties, which is also common to other theories that require a stellar collision or near-collision, was pointed out by Henry Norris Russell and proved by Lyman Spitzer. Let us consider the physical state of the hot, gaseous material which, according to this theory, is to be removed from beneath the surface of the sun very rapidly, say within an hour or two. While this gas remains in the sun, it is held by the sun's enormous gravity, some 28 times that of the earth. If it were not extremely hot, the gas would collapse into a very dense mass. It is kept distended by solar temperatures which range from 10,000 degrees F. at the surface to some 40 million degrees at the center. Suppose, then, that we scoop enough material out of the sun to make the planets, allowing for a considerable loss into space. We are drawing out gas at a temperature of perhaps 10 million degrees. At the instant when it is released from the sun's great gravity, the explosive pressure of this superheated gas is fantastically great. Released suddenly, the gas expands in an explosion of almost inconceivable force. Most of the gas is lost forever from the solar system. Furthermore, it is very difficult to conceive of a process whereby the remaining gas would cool and condense into droplets, or collect in masses as large as the planets.

Recently the German physicist C. F. von Weizsäcker has developed a new mathematical theory for the evolution of planets from a gas-and-dust cloud rotating about the sun. His theory can be made to predict the Bode's Law relationship of planet distances from the sun. There is still some question, however, whether the Weizsäcker theory really works. Moreover, it leaves wide open the question as to how the gas cloud came into existence and into motion about the sun.

All of the current theories about the birth of the stars and planets leave much to be desired. The truth is that we are still groping in the haze of a poorly illuminated and ancient past. Perhaps an entirely new advance in science will be required to light our way. On the other hand, it is possible that we have already stumbled on the correct path but cannot see clearly enough to recognize its worth.

V. RADIO STARS

by A. C. B. Lovell

WHEN A COUNTRY that is struggling with a financial crisis and shortages of raw materials decides to spend a million dollars and 2,000 tons of steel on a single instrument for fundamental research, important results must be expected. Great Britain is now making such an investment in a new kind of gigantic telescope. It will be concerned with what may seem a visionary enterprise—the exploration of the universe—but Britain anticipates a rich harvest of discovery from the investment.

The story behind the decision to build this instrument is a thrilling chapter in the history of research. It is the story of radio astronomy. Until 20 years ago our only window into space was the visual region of the electromagnetic spectrum. We knew that our vision was somewhat obscured by dust and vapors clouding the starlight, but it seemed unlikely that outer space had many secrets which our great optical telescopes would not eventually reveal. Then quite by accident a new window was discovered in another part of the electromagnetic spectrum. While studying atmospheric radio disturbances Karl G. Jansky, an electrical engineer at the Bell Telephone Laboratories, picked up radio signals which he decided must be coming from outer space. His now famous discovery was confirmed by the radio engineer Grote Reber, who built in his garden a 30-foot parabolic aerial with which he plotted the first radio map of the sky.

Reber's survey showed that the signals were strongest from the direction of the Milky Way, and that in general the regions of space with the thickest clusters of visible stars emitted the strongest radio waves. But Reber was not able to connect the radio signals with any specific object. He pointed his aerial in the direction of bright stars, extragalactic nebulae and other strong emitters of light, but none of them seemed to be the cause of the radio signals! Reber concluded that the radio waves probably were generated by atomic processes in the hydrogen gas in interstellar space. It was an interesting theory, but apparently not destined to lead to any startling revelations about the universe.

Astronomers at first took little account of the radio experiments. In 1948, however, there came a new development which decidedly quickened their interest. Reber's difficulty had been that his radio telescope had very poor resolution: it could not separate small objects in the sky because it received radiation in a beam several degrees wide. To focus on a small object the reflector or other radiation receiver must be very much larger than the wavelength of the radiation. The wavelength of the light waves collected by an optical telescope is only a few hundred thousandths of an inch. But the radio signals received by Reber's telescope had a wavelength of about six feet, and his 30-foot antenna could receive only a broad beam. In 1948 two experimenters on opposite sides of the world found a way to get better resolution of the sources of the radio signals. They were J. G. Bolton in Sydney, Australia, and Martin Ryle in Cambridge, England. They used a combination of two antennas, placed several hundred yards apart and connected to a single radio receiver. Radio waves coming in obliquely from space reached one aerial slightly before the other and therefore produced an interference effect, either reinforcing or opposing each other. As the earth rotates, this radio "interferometer" sweeps the sky with a fan of fine lobes, thus making it possible to get some idea of the size of the radio-broadcasting region in space; a source smaller than the space between the lobes would produce sharp maxima and minima in the strength of the received signal. To the great astonishment of astronomers, Bolton and Ryle found that at least some of the radio waves were coming from sources small enough to be called "radio stars." Bolton found one in the constellation of Cygnus, and Ryle discovered an even stronger one in Cassiopeia. Subsequently many more radio stars were located.

The strangest feature of these discoveries was that none of the radio stars seemed to coincide with a bright star or any other visible object. The belief soon arose that the radio stars represented a hitherto unknown type of stellar object—dark or only faintly luminous, but with the ability to emit intense radio waves. There seem to be a great number of these radio stars: more than 200 are now known, and it is likely that vastly greater numbers will be found as radio telescopes are improved. In fact, there are grounds for believing that radio stars may be as numerous as the common visible stars.

In 1950 a large radio telescope was built at the Jodrell Bank station of the University of Manchester in England. It is similar in shape to the one originally used by Reber, but 220 feet in diameter, so that it can receive radio signals in a beam only two degrees wide.

48

Its antenna is fixed, however, and it can survey only a small part of the sky. With this telescope R. Hanbury Brown succeeded in recording radio waves emanating from the great spiral nebula in Andromeda and from more distant galaxies. Thus it became evident that radio stars must be common not only in the Milky Way but also throughout the universe.

Determined efforts have been made to unravel this strange mystery of a universe filled with radio-emitting bodies which have no obvious connection with the common stars, and in the last few months a little progress has been made. Among the radio stars detected by Bolton is one in the constellation of Taurus which coincides with an outstanding celestial object known as the Crab nebula. This nebula is believed to be the hot, expanding, gaseous shell of a supernova which exploded in 1054. The position and size of the radio star, the third most intense in the sky, coincide well with the position and size of that gaseous shell. Last summer Brown discovered a radio star in the position of another supernova—the one observed by Tycho Brahe in 1572, the remnants of which are no longer visible in telescopes. Hence it now seems well established that the remains of a great stellar explosion are capable of generating intense radio waves. One more check is needed to place the matter beyond doubt: detection of radio waves from the remains of the third known supernova, observed by Johannes Kepler in 1604. Unfortunately Kepler's object is outside the field of view of the fixed radio telescope at Jodrell Bank, and no other instrument of sufficient size is available to study it.

These three supernovae would account for three radio stars, but what about all the others? To study the situation more closely the astronomers on Palomar Mountain trained their 200-inch telescope on the region of sky containing the two most intense radio stars in Cassiopeia and Cygnus. This search, which began early in 1952, has been very fruitful. Near the Cassiopeia radio star the telescope has revealed a region of diffuse gaseous nebulosity with some strange and still unexplained properties. The result of the investigation of the Cygnus region is even more startling. The Palomar observers Walter Baade and R. Minkowski believe that the Cygnus radio star is caused by the collision of two galaxies!

The general findings so far are indeed remarkable. Of the three strongest radio stars in the sky, one seems to be the remains of a star which suffered a violent death, another appears to represent whole galaxies in collision, and the strongest of all seems to be a very faint region of gas in violent motion.

As the sky has been plotted in greater and greater detail with radio telescopes of improved resolving power, it has become clear that the regions with the greatest concentrations of stars generate the most intense radio waves. Even in our present state of uncertainty regarding the source of the radio waves, this relationship is of the utmost importance to astronomy. Our view of the star-rich central regions of the Milky Way is badly obscured by clouds of minute dust particles in interstellar space. In fact, it has been estimated that this dust must hide over 90 per cent of the stars in the Milky Way from visual detection by even our most powerful telescopes. Naturally this is a severe impediment to the study of the structure of our galaxy. Radio waves, however, can penetrate the dust without absorption and bring to the radio telescopes details of the hidden regions. The radio plotting of the sky is, therefore, a most important task. The work needs high resolution, and we have seen that this requires very large radio telescopes. That is the reason for undertaking the new telescope at Jodrell Bank.

Its design is based on the radio telescope which has been in use there for several years, but it will be bigger. More important still, it will rotate on a track and can thus be trained on any part of the sky. The new instrument will be several times larger than any existing movable radio telescope. Some 12,500 tons of steel and concrete are now being sunk into the ground as the foundation for the instrument. The foundation will support a superstructure of 1,500 tons, mounted on a circular railway and driven by motors which will enable it to track automatically any object in the heavens. Its great antenna will be a steel bowl 250 feet in diameter and 60 feet deep at the center; the 300-ton aerial will pivot on an axis 180 feet above ground level.

The primary assignment of this great telescope will be to survey the heavens, but it will also be equipped for all other types of work in radio astronomy, including radar tracking of meteors.

The telescope will operate over a wide range of radio wavelengths. Until recently most of the work in radio astronomy was done in the range of wavelengths between one and 20 meters. But there has been increasing interest in the use of shorter wavelengths, and in 1951 this field was given a tremendous stimulus by one of those spectacular discoveries that have been so characteristic of work in radio astronomy. It had been suggested that the hydrogen atoms in interstellar space might, as the result of a certain change in energy state, emit radio energy at a wavelength of 21 centimeters. In 1951 radiation of this wavelength was actually detected, first by Harold I. Ewen and E. M. Purcell at Harvard University and then by others. Thus for the first time astronomers had a specific spectral line to work with in the radio spectrum. Slight Doppler-effect shifts in the 21-centimeter radio line will make it possible to determine the relative motion of the earth and the clouds of hydrogen gas in space.

Astronomy has marched forward with the growth in size of its telescopes. The need is always for more light-gathering power and more resolving power. In radio astronomy history will doubtless repeat itself; the building of radio telescopes with more sensitivity and more resolving power should yield striking advances in our knowledge of the universe. High hopes are entertained for the great engineering enterprise now under construction at Jodrell Bank. In combination with visual observations through the giant optical telescopes on the California mountains, it may well open a new era for astronomy.

PART 2

I. THE CRUST OF THE EARTH
by *Walter H. Bucher*

Walter Bucher's particular area of study is the earth's crust and he is considered today the world's leading authority on the subject. He is head of the Geology Department at Columbia University.

II. THE EARTH'S HEAT *by A. E. Benfield*

At present supervisor of the Electrical Laboratories in Harvard University's Division of Applied Science, A. E. Benfield is an important contributor to the current theoretical ferment in geology. He has had his share, however, of the outdoor life which is one of the attractions of this field of science. He was a member of the Harvard-M.I.T. eclipse expedition to Ak Bulak, U.S.S.R., in 1936 and of the Seligman Jungfraujoch expedition to Switzerland in 1938. Dr. Benfield was born in Paris and is probably the only man who owes alumni allegiance to Rugby, M.I.T. and Cambridge.

III. VOLCANOES *by Howel Williams*

Howel Williams, who was born in Liverpool, first became interested in volcanoes as a graduate student, when he mapped the ancient volcanic rocks of Wales. He came to the U. S. in 1926 to study at the University of California. His professional chair at that university now gives him a ringside seat on the "Circle of Fire," the earthquake and volcano zone which surrounds the Pacific Ocean.

STRUCTURE OF
THE EARTH
Introduction

U NLIKE ASTRONOMY, geology is a young science; two hundred years
ago it had not been thought of as a systematic field of inquiry.
Even today, substantial parts of the land surfaces of several continents
are still unexplored and the floors of the oceans, forming about 70 per cent
of the earth's surface, are still largely uncharted. Similarly, there are at
present important gaps in our knowledge of the constitution and processes of our planet's interior.

In the last few decades, however, the pace of discovery in geology
has been quickened by applying the principles and techniques of modern physics to mapping the surface, charting the ocean floors and probing
the opaque crust and the still deeper interior. The now rapidly accumulating facts are leading to a new understanding of the earth and to exciting speculations about its structure and vital processes. In fact, a whole
new approach has been devised to wrest from the globe secrets it has
withheld until now. Geology has become a meeting ground of many sciences.

The articles that follow here are typical of the new attack. The modern
geologist makes fullest use of recent discoveries in radioactivity, crystallography, solid state physics, as well as physical chemistry, in an attempt
to read correctly the evidence of the rocks and to reconstruct the successive
steps in the earth's history. Ultimately he, along with the pioneer geophysicist and geochemist, hopes to define the earth as a dynamic system
in terms that will comprehend the major features of its history and the
fine structure of its present composition.

This ambitious program which geology has set itself is still in a stage of
organization and the new ideas are still tentative. Some of them are surprising. For example, W. H. Bucher points out that granite, long consid-

ered part of the original stuff of the earth, may be nothing of the sort. It may have grown throughout geological history as the end product of the transformation of sandstone, clay and other sediments under the influence of pressure, heat and the infusion of vapors from the depths below. The earth's heat is another subject now under sharp scrutiny. In the traditional view, the globe was originally a molten ball that century by century grows colder. It now seems possible that, on the contrary, the earth's heat is slowly but steadily rising. This idea, upon which A. E. Benfield comments, ties in closely with the dust cloud hypothesis discussed in the previous section. New explanations are being proposed for the creation of the continents, for the long, slow swells of the basement floor beneath our continental landscapes, for the abrupt shelves and deep troughs of the oceans. And volcanoes, which are our most dramatic reminders of the immense forces within, are being studied anew with precision instruments.

The geologist is not unlike the medical diagnostician who deals with the exterior symptoms of a living body whose active processes he cannot directly observe. As the physician uses the stethoscope, microscope and x-ray equipment to extend his insights, so the geologist uses modern physical tools to penetrate into the hidden regions. And as the doctor learns from biochemistry the meaning of chemical processes in the human body, so the geologist turns to geochemistry to interpret for him the lavas and granites and ores he encounters. But medicine is not biochemistry and geology is not geochemistry or geophysics: it is broader and more fascinating than those special studies—an attempt to grasp a living body in all its parts and as a functioning whole.

I. THE CRUST OF THE EARTH

by *Walter H. Bucher*

GEOLOGY is in a period of unprecedented discovery. Systematic explorations of the earth's crust by geophysicists and geologists, working in close cooperation and using new methods of investigation, are yielding wholly new knowledge about the ocean floors and the deeper parts of the continents, and a more precise understanding of regional geology. As a result, traditional views concerning the physical and chemical processes that have produced and are now molding the earth's crust are being challenged. What processes created the contrasting formations of continents and oceans? Are they still going on, forming future continents? In the search for the answers to these fundamental questions there is an atmosphere of fascinated suspense today such as has always marked the high points on the growth curve of a science. This article deals with some of the changes in outlook that are suggested by the work of the past few decades.

The "crust" of the earth is a cool, relatively rigid shell which is probably not much more than 30 miles deep, *i.e.*, less than eight thousandths of the distance from the surface to the earth's center. Beneath this shell, heat and pressure rise to such levels that all rock materials are thought to be plastic. Only a small fraction of the crust's thickness lies exposed. The sections that we can observe have been brought into view by uplift, tilting and erosion on continents and islands. Wherever enough of the crust is revealed, the diversity of materials and the folding and fracturing they have undergone indicate a dynamic activity in the earth that contrasts greatly with the prevailing static concept of the crust as a series of concentric "shells." To explain that dynamic activity is one of the baffling problems of earth physics.

A satisfactory answer will not be possible until we know more about the structure of the deeper parts of the crust. This calls for geophysicists to work hand in hand with geologists—the teamwork of our day. Geology studies only the visible parts of the lands, i.e., about one-quarter of the earth's surface. The three-quarters hidden beneath water and ice, and the deeper parts of the crust everywhere, must be probed indirectly by the geophysical measurement of such quantities as the local value of gravity, the velocity of elastic waves at various levels and the direction and intensity of the local lines of force in the earth's magnetic field.

It is convenient to consider the subcontinent and subocean structures separately. What is the structure of the crust beneath the continents? The first evidence about it is provided by analysis of the travel times of earthquake waves through parts of the crust from known source points. Such measurements show that beneath all the continents the crust consists of an upper part in which the elastic waves travel with relatively low velocity and a lower one through which they pass more rapidly. The magnitude of these differences in velocity proves that they are due to differences in the rock materials at the upper and lower levels. And laboratory studies of the speed with which elastic waves travel through various types of rock give a clue to the type of rock that prevails at each level.

The primary rocks of the earth, the so-called igneous rocks formed by cooling and crystallization from an original molten condition, fall into two general groups. One is relatively rich in silicon (Si) and aluminum (Al), and hence is called "sialic." The other is poor in those elements but rich in magnesium (Mg) and iron (Fe), and this type is called "mafic." The most common sialic rock is granite; the most common mafic rock is basalt. In this article the terms granite and basalt will be used in a broad sense to represent the two groups of igneous rock.

When these rocks are tested in the laboratory, it turns out that elastic waves travel faster through basalt than through granite. Since the speed of earthquake waves is greater in the lower parts of the subcontinent crust than in the upper parts, the deeper parts of the crust must everywhere consist of basalt, and granite must be limited to the upper crust. Large areas of the ocean floors seem to have no granite at all; below the mantle of young sediments basalt forms the surface of the crust.

This distribution suggests that the granite of the earth's crust formed as a kind of "scum" on the fundamental basalt, and for a long time that was the prevailing view. But a peculiarity about the distribution of granite throws doubt on this idea. Not one of the great bodies of granite in the earth is found lying simply on top of basalt. Instead the granite invariably appears in intimate relation with great bodies of older sedimentary rocks, which it seems to have invaded and altered into the so-called metamorphic rocks at high temperatures and pressures.

In this contrast between the structural relations of granite and basalt lies the key to the problem of crustal structure. To see the problem in proper perspective we must take a closer look at the structure of the continents as it is seen at the surface.

All continents exhibit the same major elements of structure. Each contains at least one "Pre-Cambrian shield," a large region of low uplands consisting essentially of ancient rocks made up of metamorphosed sediments and associated intrusive rocks, chiefly granite. This is the "basement complex" of the continents. Formed originally at considerable depths below the surface, it now lies exposed on the continental shields, having been bowed up in the shape of a buckler by regional uplift and then planed down by erosion.

In every continent the basement complex, exposed only in part, passes out of sight beneath sheets of early Paleozoic sediments that were deposited upon it. This "sedimentary platform" overlying the Pre-Cambrian basement constitutes the second structural element. It generally consists of a few thousand feet of limestones, shales and sandstones. Finally in all continents there are belts of folded mountains—great masses of sedimentary rocks, largely of marine origin, compressed into folds and broken by a multitude of fractures.

As early as 1859 the U. S. geologist James Hall observed that when one goes from the plateaus of the sedimentary platform toward a mountain belt, the sediments progressively thicken and the top of the basement complex below them plunges to unknown depths. In the midst of these thick sediments granite comes to light. This granite is not part of the "dead" basement complex but "live" igneous rock that has invaded the sediments in bodies of immense size and has altered them far and wide into the same kinds of metamorphic rocks that are associated with the granites of the continental shields. If the granite-invaded parts of the great new mountain belts were shaved off to sea level, they would be indistinguishable in rock types and structure from corresponding stretches of the continental shields. In fact, the more the complicated structure of the continental shields is studied, the clearer it becomes that they consist of the stumps of former folded mountain belts, formed during the first billion and a half years of the earth's history. If we want to know how the crust beneath the continents was formed, we must study not only these eroded remnants but also the younger folded mountain belts of today.

Before there was an organized science of geology, a discovery of far-reaching importance was made by accident in the Andes of Peru, one of the mightiest of the young folded mountain belts. Around 1740 a French expedition that went to Peru to measure the length of a meridian arc found that the gravitational attraction of the High Andes caused a much smaller deflection of the plumb line, and therefore of the bubble in the surveyor's level, than was to be expected from so large an excess of mass on the earth's surface. Pierre Bouguer, the French mathematician who made the discovery, concluded that the rocks in the mountains and below them to some distance were lighter than their surroundings. He thought this might be due to thermal expansion of the underlying rock.

A hundred years after Bouguer's discovery the English astronomer George B. Airy, having measured the gravitative deficiency of the Himalaya Mountains, suggested the correct explanation: the rocks beneath a mountain belt have a lower density than their surroundings. He postulated that below a mountain belt the light granite rock of the outer crust extends far down into the heavier underlying basalt. From this suggestion came the theory that mountains have "roots." Airy suggested that the young mountains and their "roots" float in their environment like icebergs in water, the lighter mountains projecting higher than the heavier ones.

In recent years efficient new methods of measuring the local value of gravity have shown conclusively that in a general way the value of gravity does decline with altitude faster than the mere change in elevation could explain. See, for example, the accompanying cross-section graph of Europe. The difference between the theoretical value of gravity, assuming a uniform density for all rocks, and the observed values is called the Bouguer gravity anomaly. The discrepancy grows larger as the elevation of the range increases; it reaches a maximum at the crest. This indeed suggests the existence of a mountain "root." Seismic studies provide a clinching proof. At deep levels of the crust beneath the Eastern Alps earthquake waves travel at a noticeably slower rate than at comparably deep levels in other regions, indicating that the light, low-velocity rocks here extend far down to levels normally occupied by denser rocks.

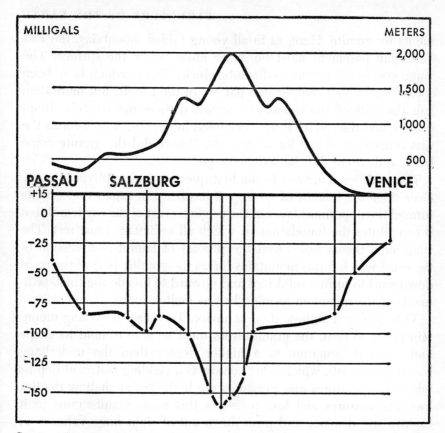

Gravity anomalies found in a survey of the Alps are plotted in this drawing. At each point in the survey, the force of gravity was found to be less than the measurement expected to the relative extent indicated here by the length of each vertical line. The milligal is a unit expressing this departure from the expected measurement. Combined with the knowledge that mountain granite is lighter than the basalt foundation of the crust, this finding of a lesser force of gravity indicates that a root of granite rock must extend to great depths below the mountain.

Do these mountain roots "float" in the heavier basaltic and ultrabasic rock material of the lower levels, or are they more like the roots of teeth anchored in a jaw? That question cannot be answered by any known geophysical method. An answer can come only from an examination of the mountains themselves.

The Sierra Nevada of California is the classic United States example of a young mountain-building zone with a proved granitic root. It also offers one of the best exhibits of the invasion of sediments by granite. Here, as in all young folded mountains, the Pre-Cambrian basement must lie many miles below the surface. The range consists of marine sediments, chiefly shales, which have been compressed into a complicated pattern of tight folds. But more than half the bulk of the sediments deposited there has entirely disappeared, and their place is now occupied by granite, which forms the vast inner core of the Sierra Nevada. Where did the granite come from, and where did the sediments go?

The traditional answer to the first question is deceptively simple. Ever since the infancy of geology it has been an axiom that granite formed the uppermost layer of the original crust of the earth and that it constitutes the foundation on which all sediments must rest. The traditional theory holds that the process of lateral compression of the crust which forms mountains forces the granitic part of the crust downward to form a solid root and upward to invade the thick sediments of the mountain-forming belt as molten rock.

This implies a strange dual behavior. To cause the young mountain range to float, the granitic root must be able to hold its shape and resist deformation to a higher degree than the underlying basaltic materials, which must behave as a yielding matrix at depths where temperatures and pressures are high. Yet at shallow depths, low temperatures and low pressures this same granite must melt rapidly and displace and push down out of sight huge volumes of sediments. No known properties justify the assumption of such a dual behavior. Wherever basaltic and granitic rocks are seen deformed side by side in deeply eroded parts of the earth's surface, it is always found that the granitic rocks have behaved more plastically than the basalt. Moreover, granitic materials have a considerably lower melting point than basaltic rocks.

One must assume, therefore, that when a young mountain system is formed by the compression of the crust of a continental area, the primary granitic part of the crust is squeezed down into and molded with the stiffer and heavier basaltic part of the crust as a plastic mass. As such it would form a root all right, but not one that floats in its environment. The granite would merely make up a larger proportion of the stiff, plastic crust than it does in other places.

We are no longer sure that granite formed an essential part of the original crust of the earth. This venerable axiom of geology is now being challenged in one of the most significant controversies of con-

60

temporary petrography. The argument has to do with the second
question asked above: Where did the sediments go when the granite
took their place?

In the Sierra Nevada granite has replaced the folded sediments up
to an elevation three miles above sea level in a region some 400 miles
long and in some places 70 miles wide. Detailed mapping proves
that the granite was not all emplaced at one time; it invaded the
sediments in a series of successive advances, and the structural relations are not what might be expected if the granite had merely
pushed the sediments aside. It rather appears in their midst as if it
had carved out the space it occupies. Every major granite body is
set in a belt of metamorphosed sedimentary rocks and is surrounded
by a zone in which granite material has infiltrated the surrounding
rocks in an intricate manner. In such border zones granitic material,
ranging in thickness from layers of thousands of feet to paper-thin
seams, is interlaced with layers of metamorphosed sediments and
cuts across them in a network of dikes. Even in the spaces between
these granite seams and dikes, potash and soda feldspars, which are
characteristic minerals of granite, appear scattered through the rock
as well-defined, newly formed crystals or irregular clusters of them.
The crystals occupy space once occupied by sedimentary rock, but
they show no signs of having been introduced forcibly. They must
have been formed by recrystallization of the original sedimentary
material, into which was introduced relatively small amounts of
alkalis and perhaps some silica in the gaseous state or in solution.
Detailed petrographic studies have proved beyond reasonable doubt
that whole bodies of granite have been produced by this process,
known as "granitization."

The evidence for such transformation of sediments and even lavas
into granite is so convincing that no petrographer now denies that
some granite has been produced by granitization. The question is:
How much of the world's granite has been formed by such processes,
and how do they operate?

There are two schools of thought about it. The orthodox hold that
granite is the *agent* of local granitization, the heretics that it is the
general *end-product* of granitization. The former believe that in the
typical case granite bodies are nothing more than parts of the prime-
val granite of the earth's crust which have remelted locally and
reached their present position by mechanically displacing other

rocks; they contend that chemical granitization occurred only incidentally along the borders of the granite bodies. The heretics hold that the original crust of the earth consisted of basalt, and that the granite bodies were created by the transformation of sediments. They argue that the process goes something like this: Whenever disruption of the crust produces belts of folded mountains at the surface and stresses and frictional heat at deeper levels, hot gases and solutions carrying silica, alkalis and other elements rise toward the earth's surface from the lower parts of the crust or the subcrustal layers. These "emanations" change basalt to rocks richer in silica and convert shales and sandstones into schists and gneisses and ultimately into granite. No one objects to granitization on a small scale. But most geologists balk at extending the process to larger bodies of granite. They are awed by the magnitude of the scale on which the necessary processes would have to take place.

Nevertheless we face the stubborn fact that in all mountain belts vast volumes of granite have appeared precisely where vast volumes of sediments have disappeared, in many cases leaving a structure pattern that makes displacement of the sediments by mechanical means highly improbable, if not impossible. Moreover, these large bodies of substituted granite occur only in those parts of the earth's crust (*i.e.*, mountain belts) where intense mechanical deformation has taken place. This deformation must produce heat and create paths in the rock for the rapid diffusion of the "emanations" which are believed to play a part in granitization.

Finally there is the strange fact that at least one half of the earth's crust—beneath the oceans—apparently does not possess a granitic upper layer. We geologists are hard pressed to explain the absence of so much of what we used to think was once a universal part of the crust. Many of us are beginning to think that instead of asking, "Why is granite absent from the oceans?" we should be asking, "Why is granite present in the continents?"

We have been considering changes in the earth's crust in the young folded mountain belts. The sedimentary platforms of the continents reveal, in a more indirect way, that changes and resulting movements of the crust take place even outside these belts. The surface of the platforms is warped into irregular basins separated by low swells. The North American Mid-Continent has several basins (*e.g.*, the Michigan Basin and the Illinois-Kentucky Basin), which measure a few hundred miles in diameter and one to two miles in depth. Of the swells that separate them, the "Cincinnati arch" is the best known.

Little thought has been given as yet to the reason for such warping. The tendency has been to blame the sediments, the idea being that the crust beneath the continents is so weak that it gives even under small local loads, and therefore the weight of sediments laid down on it would cause a downbuckling to form basins. But intensive gravity surveys carried out in recent years have shown that the crust beneath the continents is much stronger than was formerly supposed. There is evidence that even the largest of our Mid-Continent basins cannot owe their existence to the weight of the sediments in them. The sinking of the land surface to form basins must be due to processes that take place at depth. Until we know what these processes are, we cannot hope to solve the largest problem of all—the origin of the ocean basins.

In the smaller basins of the sedimentary platform the old Pre-Cambrian land surface lies bent down to depths of one to four miles below sea level, i.e., depths which we associate with the oceans. Does this mean that parts of the modern ocean floor may be merely downbent portions of old continents? With this question we turn to the structure of the earth's crust below the oceans.

On the coasts of continents bordering the Atlantic, Indian, Antarctic and Arctic oceans, and on some coasts of the Pacific, the continental structure lines run out beneath the sea as if the surface were indeed bent down or had broken off. On a geologic map they end abruptly. The presence of identical or similar marine shallow-water organisms and terrestrial animals and plants on opposite shores of an ocean often suggests that shallow water or land connections once existed where now there is deep sea. Such observations led many geologists in the nineteenth century to the conclusion that in the course of geological history large sections of continents or even whole continents sank to oceanic depths.

In 1846 the U. S. geologist James D. Dana first suggested the opposite view: that the continents have stood high since the earliest part of the geologic record. He thought that the continents were parts of the crust that consolidated early; hence they were thicker and sank less as the earth shrank, while the thinner parts became the oceans. Later, when it became known that basalt is by far the most common rock found on oceanic islands, while the lighter granite is confined to the continents and nearby islands, many geologists came to the same conclusion as Dana, but for a different reason. Since the parts of the earth's crust appeared to be in gravitative equilibrium, they thought that the ocean floors are low because they consist of heavy rock, and the continents are high because they are light. Once a continent, always a continent. But that left unexplained the world of facts that had led others to the opposite conclusion.

About 40 years ago the German geophysicist Alfred Wegener advanced his spectacular continental-drift hypothesis. Contrary to all the available physical and geological evidence, he blandly postulated that the basalt of the ocean floor is so weak that it cannot resist deformation even under the action of almost infinitesimal forces. This would make it possible for granitic continental masses to drift through the basalt of the sea floor like cakes of ice through water. According to Wegener the sea separates great land masses that were once in contact, and the solid basalt of the present sea floor flowed plastically into the spaces it now occupies. His book, a masterpiece of special pleading, drove home to all concerned the need for new, crucial information on the physical properties of rocks and the topography and structure of the ocean floor.

In the last years before World War II Maurice Ewing set out to determine how the granitic crust of the North American Continent connects with the floor of the deep sea along the western border of the North Atlantic Basin (see schematic drawing). Off the Atlantic coast of the United States, the sea bottom at first drops very slowly, forming a continental shelf which reaches a depth of around 360 feet below sea level at its outer edge, some 60 to 80 miles off New Jersey and Maryland. Beyond the edge the bottom drops off relatively fast (about 400 feet per mile) to the deep sea, forming the continental slope.

Ewing faced two questions: 1) What makes the continental shelf? 2) What becomes of the sialic crystalline (granitic) basement as one goes toward oceanic depths? Wegener had suggested that the outer edge of the shelf marked the end of the granitic continental crust —the sharp edge along which it had torn away from the Continent of Europe. If that were true, beneath the shelf the crust should consist of rocks of the basement complex, with no more than a veneer of young sediments, and this structure should end abruptly.

64

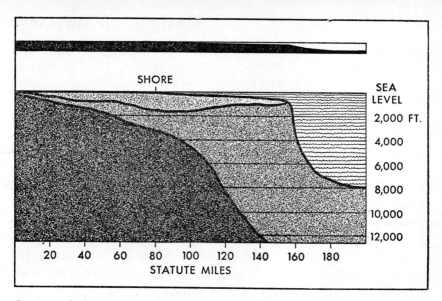

SHORE

SEA LEVEL
2,000 FT.
4,000
6,000
8,000
10,000
12,000

20   40   60   80   100   120   140   160   180
STATUTE MILES

Continental shelf provides important clues to structure of earth's crust. This is a cross-section of the shelf off the coast of southern Virginia. The darkest gray area indicates the pre-Cambrian basement; the middle gray the younger sedimentary rock; and the lightest gray the unconsolidated sediments at the surface. The vertical scale is considerably exaggerated; the actual profile is shown at the top of this diagram.

Ewing showed that the granitic basement does not end at the continental shelf but slopes down continuously to depths of about two miles below sea level. This was a crucial discovery. It demonstrated that here at the edge of a continent the basement of crystalline rocks, which are identical in character with the older rocks of the continental shield, falls away to form an oceanic basin in exactly the same way as the shield bends down to form the Michigan Basin 700 miles inland. Like this basin, the descending North Atlantic basement is covered with sediments, which are the seaward continuation of the formations exposed on the coastal plain. This discovery proves that parts of the deep sea may well consist of continental crust brought to a low level and held there. Presumably this was accomplished by the same unknown processes that produced the basins within the continents.

How much of the Atlantic Ocean floor had such a history? The best answer one can give at present is that probably only a small fraction of the "North American Basin" of the Atlantic Ocean is underlain by crustal material of the continental type. This is contrary to a widely held belief. Old interpretations of the velocity of earthquake waves traveling across ocean floors had suggested that all oceans except the central part of the Pacific are covered by a layer of granitic material not over some six miles thick. But Ewing and his co-workers have developed a new theory about the travel of earthquake surface waves along ocean floors which, while not necessarily disproving the presence of "continental" crustal material on the floor of the North Atlantic Ocean, makes it unnecessary to assume it. And direct seismic refraction measurements recently made by Ewing at over a dozen points in the North Atlantic Ocean have shown only the wave velocities characteristic of basalt.

These considerations make it appear probable that beneath the North Atlantic and under large parts of the other oceans the crust consists wholly of basalt without any sialic material. We recall that Wegener built his hypothesis of the continental drift on the premise that this was the case. But what about the properties of this basalt? According to Wegener's theory, the basalt must be so weak that it yields to very small pressures. This means that the basalt formation would have to be essentially flat; it could not support the weight of hills or mountains. What are the facts?

In the summer of 1947 Ewing started a systematic mapping of the topography of the North Atlantic sea floor with modern echo-sounding devices from the research ship *Atlantis*. This exploration shows in precise detail what had been known before in a general way: that the surface of the sea floor is the exact opposite of what Wegener's hypothesis demands. The floor has a rugged topography. From the plain of the North American Basin at the bottom of the Atlantic more than five miles below sea level there rise large mountains ("sea mounts") which in some cases are more than 6,000 feet tall.

Much the same picture characterizes the floor of the great central region in the Pacific, which apparently consists wholly of basalt and related rocks. Flat-topped sea mounts in large numbers have been found there. The Hawaiian Islands are the top of a great basalt range which rises above sea level from an ocean floor over three miles deep. The topography of the surrounding sea floor suggests that the weight of the range is held up by the elastic strength of the crust, somewhat as a weight is supported by a sheet of ice covering a body of water. This ability to support heavy local loads points to a thick crust of great strength—the exact opposite of the conditions demanded by Wegener.

In short, everything that is now known concerning the configuration and structure of the floors of the oceans proves conclusively that Wegener's hypothesis of continental drift is wholly untenable. It also suggests that Dana was wrong in his idea that the continents were formed by the thicker parts of the crust. Actually the crust beneath the continents appears to be thinner and weaker than that beneath the oceans.

The crust beneath the ocean floors is as much a part of the earth's solid "armor" as that beneath the continents, if not more so. This being the case, the same crustal processes that are at work in the continents should also be in operation in the oceanic areas. Can these processes explain the contrast between the two levels of the earth's surface? The writer believes that they can.

To begin with, the two major types of deformation found on the continents are also seen on the ocean floors. One type of deformation is the succession of basins and swells. The floor of the Atlantic has a basin-and-swell pattern which in principle is the same as that in the sedimentary platform of the North American Mid-Continent, except that in the ocean the pattern is developed on a gigantic scale. The other type of deformation is the one that has given rise to great folded mountain belts. The largest development of this process on earth occurs along the borders of the Pacific Ocean and on the floor of the western half of the ocean itself; there the peaks of submarine mountains form the great island chains of the Pacific. The pattern is characterized by long, narrow, asymmetrical mountain ranges, closely paralleled by deep, narrow trenches along their steeper sides. The trenches are called "furrows" and the mountain ranges "welts." This furrow-and-welt pattern is evident in the island chains of the north and west sides of the Pacific and in the great cordilleras of Central and South America with their accompanying offshore, deep-sea trenches. In the mountain belts around the Pacific have originated over 40 per cent of the earth's near-surface earthquakes, about 90 per cent of those recorded from intermediate depths, and all the deep earthquakes. Here, then, is mountain-making on a vast scale actively going on today.

Perhaps the most significant of the young, active welts is the mighty submarine mountain belt which extends southward from

Japan through the Bonins and Marianas to Palau—a range comparable in length and height to the Himalayas. Among its peaks are the islands of Iwo Jima, Saipan, Guam and Yap.

Suppose there were no ocean and we were standing on the bottom lowlands of the Pacific looking westward toward this towering mountain range. Beyond it westward lies a deep sea plain that extends for more than 600 miles. There the great submarine mountain chains that form the Philippine Islands rise in precisely the same manner from the ocean deep, and behind them, covered by younger sediments, emerges the edge of the Asiatic Continent itself. From this perspective it would seem incomprehensible to us that men should ever ask: "How did the ocean basins come to be?" Instead we might well ask: "How did the continents come to be?"

From this point of view the very expression "ocean basins" becomes meaningless. The continents now can be seen clearly as deformed belts of the earth's surface which have been raised at intervals through geologic time and joined together in various ways. The oceanic areas, on the other hand, must be the undisturbed portions of the earth's surface. They are underlain by the original basaltic crust, covered here and there with a blanket of diverse sediments.

This reasoning deviates considerably from current doctrine. The purpose in presenting it is to indicate possible new directions of thinking and to suggest crucial areas in which these ideas may be tested systematically by geophysicists and geologists.

From all this it is plain that we still know little about the structure of the crust beneath the continents and much less about that below the oceans. All our concepts necessarily rest on extrapolations far beyond the scanty data that can be considered reliable. All are but tentative hypotheses waiting to be tested. Yet we must have hypotheses to test, and we must constantly seek to combine them into a consistent over-all picture that shows the parts in relation to the whole. The essence of a possible picture of the earth's crust that emerges from the observations and thoughts presented in this article may be summed up as follows:

The complexity of the crust beneath the continents is the result of major crustal folding, *i.e.*, the formation of upbowed welts and downbowed furrows that filled with sediments. Compression of these belts drew out the sediment-filled furrows into roots of mountains, and set in motion the physical and chemical processes that transformed part of the sediments into metamorphic rocks and ultimately into granite. The idea that the rocks in the crust lie in essentially horizontal layers is a purely statistical concept which does not reflect the actual complexity of the crustal structure.

The shields and their continuation beneath the sedimentary platforms of the continents are the eroded stumps of earlier folded mountain ranges; their level is therefore related to sea level. Their existence is proof that the position of the sea level with reference to land has not changed radically since Cambrian time. Parts of the old Pre-Cambrian surfaces of the continents have been warped into basins and swells, and the floors of some of these basins have sunk to depths comparable to those of the oceans.

Between the typical continental and oceanic areas lie regions where the basin-forming process has brought sections of the continental granitic crust down to oceanic depths. Crustal deformation, on the other hand, has produced belts of folded mountain ranges from the basaltic crust beneath the ocean floors as well as from continental levels. This process, to which the continents owe their existence, is still going on actively on the borders of the Pacific Ocean and within the ocean's western part.

Such an abstract, overgeneralized picture is but the framework into which the realities of crustal structure must be fitted. It defines the scope of some of the great questions that call for answers, and of the stirring possibilities for work on one of the great frontiers of modern science—the geology of the deeper parts of the earth's crust.

II. THE EARTH'S HEAT
by A. E. Benfield

WHEN A DOCTOR examines a patient, one of the things that is apt to interest him is his patient's temperature. The temperature, together with other tests, helps the doctor to understand what is going on inside the patient's body. A geophysicist similarly hopes to get some clues as to what is going on inside the earth by taking its temperature. The earth of course makes things a good deal more difficult for us, because, unlike the human body, it does not have a self-regulating thermostat to keep its temperature uniform, and we cannot insert a thermometer any deeper than the outer part of its skin. The best we can do is to take the temperature in deep oil wells, mines, railroad tunnels, hydroelectric shafts and so on, sampling at most the top few thousand feet of the earth's crust. The deepest well man has drilled is only about four miles down. It seems safe to say that we will be able to fly through the vast spaces of the solar system and reach some of our neighbor planets long before we find a way to penetrate to the center of our own planet, 4,000 miles away. Nevertheless, our information about the heat of the earth, meager as it is, is helpful. We are adding to it steadily and in time it may help us to determine how mountains are made, what causes volcanic eruptions, how the earth's magnetic field is created, why the great ocean deeps are where they are, and various other intriguing matters that have long concerned geophysicists.

It has been known for many years that as we go deeper and deeper into the earth the temperature steadily rises. This is not true, of course, of the top few tons of feet near the surface—as we notice on descending into the cellar on a warm spring day when the cold of the preceding winter is still in the ground. But the effect of seasonal changes in temperature is seldom appreciable more than 50 feet beneath the surface. Below that the temperature of the earth always increases with depth; at the bottom of some of the deep oil wells in California and elsewhere it exceeds the boiling point of water at atmospheric pressure.

Why does the temperature always increase with increasing depth? There are various answers, depending on various theories as to how the earth was created. The classical view is that the earth originated as a hot body and still retains much of the original heat in its interior. It is easy to understand how this might be so if the earth was formed from a piece of the sun, or from a stellar fragment released by the close approach of two or more stars. There is another theory, variously known as the "dust-cloud hypothesis" and by other names, that the earth was formed by the gradual coalescence of cool dust, gas or small particles in interstellar space. A planet growing in this way might have ended up with a hot surface, due to the melting and vaporization of the fast falling particles as they collided with the planet toward the end of its growth. At the same time the planet's interior would become hot, though possibly not hot enough to melt, due to its compression by the increasing weight of the accreting material accumulating on the surface, and for other reasons.

Yet we cannot be at all sure that the earth actually was very hot when it was formed. In fact, on the basis of the observed abundance of the elements in the earth's crust the University of Chicago chemist Harold C. Urey has recently advanced the theory that the earth may have formed at a relatively low temperature. Subsequently, the earth's temperature may have become considerably warmer, due to the evolution of heat from radioactivity. If we could get deep into the earth and measure its temperature we might find information that could help us to decide what is the correct theory of the earth's origin. We can, of course, study the temperature of the molten lava arising in volcanoes, but we do not know how much the temperature of this has changed on its upward journey, nor from what depth it comes. It used to be thought that lava came from a shallow depth, but recently it has been suggested that it may originate quite deep in the mantle.

We know that in the accessible outermost crust the rate of increase of temperature with depth, called the temperature gradient, differs widely from one place to another. This is true not only for volcanic or hot-spring areas, where we would expect to find wide departures from the "normal," but even in quiet areas far from volcanic activity. Temperature gradients in ordinary quiet areas range anywhere from less than 10 to as much as 50 degrees Celsius (Centigrade) per kilometer. Furthermore, even in a single location the temperature gradient is not always smooth but may change abruptly at some particular depth; for example, in some wells in Cheshire, England, the rate of temperature increase suddenly doubles at a certain level.

What is the reason for these differences in temperature gradients from place to place? One explanation might be that there are differences in the amount of heat flowing from the depths of the earth at the various places. This is certainly partly true. We now know, however, that the observed variations in quiet areas are due largely to differences in the thermal conductivity of the particular rock strata at each place. This would also account for the variations in temperature gradients from one depth to the next; one layer of rock is a better conductor of heat than the other. The heat flowing in a solid depends on the product obtained by multiplying the temperature gradient in the solid by its thermal conductivity.

Measurements on rock samples from wells, mines and tunnels, made during the last twelve years in quiet areas of South Africa, England, Iran and the U. S., have shown that temperature gradients tend to be low where the thermal conductivities of the rocks are high, and *vice versa*, so that the product of these two quantities is fairly constant. It is beginning to be clear, in fact, that apart from special areas like Yellowstone National Park, where some local disturbance causes high temperatures near the surface, the amount of heat flowing out of the earth from below is probably pretty much the same almost everywhere on the continents. However, very little of the earth's surface has yet been investigated, and interesting regional anomalies may well be found in the future.

We do not know how much heat is flowing up from below into the oceans, but Hans Pettersson of Sweden and E. C. Bullard of England have recently begun to make some of the necessary measurements and we should have some information about it before long. Since almost three-quarters of the earth's surface is covered by water, we clearly need such information before we can begin to estimate how much heat is coming from the interior of the earth as a whole.

One thing we do know, of course, is that per unit of area the amount of heat emerging from the earth is very small. On the continents where we have measured it (disregarding such special local phenomena as volcanoes and hot springs), the heat flow amounts to only about one millionth of a calorie per square centimeter of surface area per second. This is several thousand times less than the average heat per square centimeter reaching the earth from the sun; obviously it is the sun, not the heat of the earth, that decides our atmospheric temperature and climate.

It is likely that most of the heat we detect flowing to the surface does not come from the hot core at all but originates in the crust.

We know now that heat is continually being produced in small quantities in all common rocks by the disintegration of radium, uranium, thorium, potassium and other radioactive atoms in them. Radioactivity is particularly pronounced in the granitic type of rocks, of which the continents are largely made. The thickness of the continents' granitic layer is thought to average something of the order of six miles. The amount of heat produced by natural radioactivity in this thickness of granite would account for about half the heat we observe flowing to the earth's surface. (Mountain ranges, where the granitic layer is probably compressed and thickened, should generate more heat than low-lying plains, and the Harvard geophysicist Francis Birch added support to this supposition last year by measurements which showed that the heat flow in the mountains of Colorado is some 60 per cent higher than normal.) Then, to the radioactive heat from granite, we must add that from the basaltic rocks which probably underlie the continents and the oceans. Volume for volume basaltic rocks produce radioactive heat only about half or a third as rapidly as granite, but the basaltic layer is thought to be about twice as thick as the normal continental granitic layer that overlies it. To be sure, we are not absolutely certain about the existence of these postulated layers of granite and basalt, or about the amount of their radioactivity. Nor do we know how radioactive the earth's interior may be, though we have reason to believe from the evidence of meteorites, which are considered by some to be fragments of a broken planet, that it has some radioactivity. (Our ignorance of the amount of radioactivity in the earth's interior is, by the way, another major reason why we cannot estimate its temperature.) At all events, it seems quite possible that the earth is generating radioactive heat faster than it is losing heat to space. Consequently the earth may be gradually warming, though at so slow a rate that we need not be anxious about it.

It has been suggested that radioactivity may be the explanation of volcanic heat, but this is unlikely because lavas usually exhibit rather little radioactivity. However, the possible radioactive content of the earth's liquid core has recently been suggested by Bullard as the possible cause of convection in the core, and perhaps of the mechanism that creates the earth's magnetic field.

Another kind of convection current, of a very slow and intermittent nature, may occur outside the core in the earth's mantle. The mantle of the earth, as a geophysicist uses the term, means not the crust but the 2,000-mile-thick section that lies between the crust and the earth's core. The mantle behaves like a solid for earthquake waves, but it probably resembles a thick viscous liquid more than a real crystalline solid. D. T. Griggs of the University of California and others have suggested that thermal convection currents in the mantle might account for the building of mountain ranges and for certain gravity anomalies associated with some of the ocean deeps.

The theory, illustrated in the adjacent diagrams, is that material at the base of the mantle, near the core, may be expanded by heat so that it becomes light enough to rise. On rising it forces adjacent cooler material to sink and take its place, thus starting a "convection cell" in the mantle.

At the base of the crust, the convection current may tend to drag down a section of the crust, thereby forming a hollow which fills with light sediments. This may explain the curious gravity deficiencies found over some parts of the oceans. Eventually the convection current may bring up enough hot material from below so that the cell stabilizes and the current itself ceases. This would end the downward-pulling force on the crust. Released from the downward pull, the submerged crustal material would rebound upward, like a piece of ice pushed down into a pond and then released. The rising buoyant material, according to the theory, might create a mountain range.

Now there is a way by which this theory of the existence of convection currents in the mantle might be tested. If such currents exist, and if they bring up relatively hot material to the top of the mantle close to the surface, the heat flow to the surface there should be

Convection currents in the mantle of the earth between crust and core have been proposed as the cause of mountain building. The mantle material has great heat-holding capacity and low conductivity. As a result, differential in temperature might build up between the outer, cooler mantle under the crust and the inner depths in contact with the core. Such a differential would provide the energy to set the plastic material of the mantle flowing in convection currents and thus to produce the sequence of events shown here.

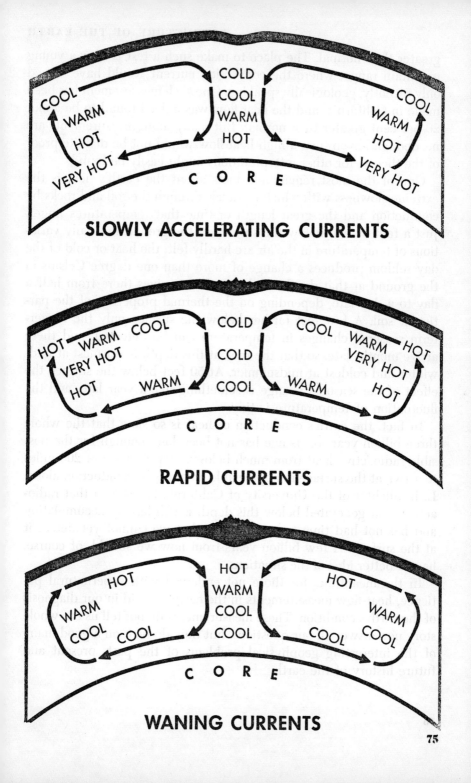

SLOWLY ACCELERATING CURRENTS

RAPID CURRENTS

WANING CURRENTS

greater than normal. The place to make such a test is near a young mountain range, where the convection current would have ceased only recently, geologically speaking. One such measurement has been made in California, and the heat flow was indeed found to be about 20 per cent greater than normal. But many more measurements are needed, and even then a high heat flow would not be definite proof of the theory, for other interpretations might easily be found.

One of the most remarkable facts about the earth's heat is the extreme slowness with which it travels through the soil and rocks by conduction and the great length of time that temperatures linger. Just a foot or two below the surface of the ground the daily variations of temperature in the air are hardly felt; the heat or cold of the day seldom produces a change of more than one degree Celsius in the ground at that depth. And the effect arrives there from half a day to a day late, depending on the thermal properties of the particular soil. A few feet farther below the surface only the longer-term seasonal changes in temperature can be detected, and these arrive months late, so that the rock at this depth is warmest at midwinter and coldest at midsummer. At 50 feet below the surface the effect of the seasonal change is something like a year late, and the fluctuation of temperature is tiny.

In fact, the earth's conduction of heat is so slow that the whole three billion years of its age has not been long enough for the possible radioactive heat from much below a depth of about 200 miles to arrive at the surface in appreciable amounts by conduction alone. L. B. Slichter of the University of California has shown that radioactive heat generated below this depth is still largely accumulating and has not had time to reach us, so that we cannot yet detect it at the surface. A few billion years from now we should, of course, have a better idea of the situation.

In the meantime, for those not endowed with supernatural patience, heat-flow measurements in the earth can aid in our diagnosis of the earth's condition. These measurements do not tell us the whole story nor answer all our questions, but they do bear on a good many of the interesting geophysical problems of the past, present and future history of the earth.

III. VOLCANOES

by *Howel Williams*

URING THE PAST 400 years some 500 volcanoes have erupted from the depths of our planet. They have killed 190,000 people; the most destructive eruption, that of Tamboro in the East Indies in 1815, wiped out 56,000 in one gigantic explosion. Volcanoes have always terrified mankind. Yet it should not be forgotten that they also play a constructive role. It is not merely that volcanic eruptions have provided some of the world's richest soils—and some of our most magnificent scenery. Throughout geologic time volcanoes and their attendant hot springs and gas vents have been supplying the oceans with water and the atmosphere with carbon dioxide. But for these emanations there would be no plant life on earth, and therefore no animal life. In very truth, but for them we would not be here!

What exactly are volcanoes, and how are they formed? Obviously they are symptoms of some kind of internal disorder in the earth. The eruptions we see at the surface are only small manifestations of great events going on below, events about which we can only speculate. We do, however, have some clues to what may be happening—a few tantalizing points of light that make volcanoes a most fascinating field of study.

The first clue lies in the location of the volcanic regions on the world map. We know that active volcanoes are concentrated in parts of the world where earthquakes are most common, particularly where earthquakes have a tendency to originate at a level about 60 miles down in the earth's crust. This suggests that volcanoes are connected with disturbances in the earth at that depth. Secondly, we know that most of the world's volcanoes are in young mountain belts, that is, where the face of the earth has recently been wrinkled and cracked.

Tens of miles below the surface of the earth there is an extremely hot shell of glassy or crystalline material. This solid material becomes liquefied if the pressure on it is reduced or the temperature rises. The pressure may be reduced by the bending or cracking of the rocks lying about it; the temperature may be increased by radioactive heating. In either case, the liquefied material forms a fluid mass, called magma, that is lighter than the overlying rocks, and so tends to rise wherever it finds an opening.

Disturbances of the earth in regions of mountain-building produce conditions favorable to formation of molten magma and its escape to the surface. To be sure, not all young mountains have volcanoes; there is none in the Alps or the Himalayas. These mountains were formed by low-angle thrusting and overfolding of the earth's skin; one layer is piled on another, making a thick cover of rock through which magma does not escape. In mountain belts where volcanoes do occur there is less overlapping of the rock layers; these mountains have steep fractures that go deep into the earth.

A volcano is usually pictured as a cone with a crater at the top which from time to time blasts forth streams and glowing bombs of lava and shattered rock. Actually there are almost as many types of volcanoes as there are landscapes. They range from the explosive kind to the sluggish and gentle, and they come in a great variety of shapes and sizes. The form a volcano takes depends not only on the structure of the earth below it but also on the physical nature of the erupted magma, or lava. One of the most important factors determining the shape and activity of a volcano is the magma's viscosity. This varies greatly; some lavas are so fluid that they flow over the ground at more than 20 miles an hour; others are so viscous that they move at little more than a snail's pace, and even the strong blow of a pick scarcely dents their incandescent surfaces.

Usually the more fluid the magma, the more extensive is the flow of lava, the flatter the resultant edifice and the fewer and weaker the explosive eruptions. A volcano formed mainly by quiet effusions of liquid lava generally has the shape of an inverted saucer. The volcanoes of Hawaii are of this kind, and they illustrate various stages in its growth. During the early stages of formation of such a volcano copious streams of extremely hot and fluid basalt are discharged from two or three intersecting rifts in the rock at the earth's surface. Where the rifts intersect a small summit-crater forms. As the volcano grows to maturity, the summit crater is much enlarged by gradual collapse of its surrounding walls, and lines of pit-craters develop along the rift-zones cutting the flanks of the volcano. The Hawaiian volcanoes Kilauea and Mauna Loa are now in this stage of evolution. Later, in the volcano's old age, new lava flows fill up and obliterate the summit- and pit-craters. Eruptions take place at longer intervals; the lavas become more varied in composition, and, because most of them are more viscous than the earlier flows and therefore stick on the sides of the mountain near the top, the upper part of the volcano becomes increasingly steep. At the same time, because the longer intervals of rest permit development of greater gas-pressure in the viscous magma, explosive activity becomes more frequent and violent. Cones of ash grow in clusters on the higher flanks of the mountain. Mauna Kea and Kohala on Hawaii are now in this stage of old age, and Hualalai has lately entered it.

At the opposite extreme there are volcanoes formed by lava squeezed out of the earth in an exceedingly viscous condition, somewhat like toothpaste from a tube. This produces very steep-sided mountains. Indeed, lava may be so nearly solid when it is thrust up through its "feeding pipe" that it rises as a slender obelisk, like the one pushed to a height of 1,000 feet on top of the dome of the celebrated Mount Pelée in the West Indies in 1902. Lassen Peak in California is another good example of a viscous protrusion.

Other volcanoes, such as Mount Shasta and Mount Rainier in this country, Mount Mayon in the Philippines, Orizaba and Popocatepetl in Mexico and Fujiyama in Japan, are built in part by outpouring of lava and in part by the explosive discharge of fragments of rock. These so-called composite volcanoes have concave slopes that steepen to the summit. Their graceful profiles rise from a wide base to a tall, slender peak. Still other volcanoes are composed wholly of explosion debris. This type of volcano is likely to grow very rapidly, and usually builds a cone with even slopes.

It may take a million years or more to build a giant volcano of the Hawaiian type or one of the composite variety such as Mount Shasta. The viscous kind grows much faster; the steep dome at the top of Mount Pelée, for instance, mushroomed to a height of 1,300 feet within 18 months. But the speed of growth of explosive volcanoes is even more spectacular. The young Mexican volcano of Paricutín was 1,200 feet high on its first anniversary. Monte Nuovo, which grew on the edge of the Bay of Naples in 1538, rose to a height of 440 feet in one day. The record goes to a volcano that sprang suddenly from Blanche Bay on the island of New Britain in 1937. It attained a height of no less than 600 feet within the first twenty-four hours; when it stopped growing several days later, it was 742 feet high.

Volcanoes of the kinds we have been considering so far are all made by discharge of material through a more or less cylindrical conduit in the earth's crust. Such discharge generally produces a cone, a dome or a sharp, slender spine. But there are also volcanoes in which the magma issues from long fissures in the crust. In that case the flood of lava or ash usually produces a plateau, the nature of which depends on the composition of the escaping magma. There are two general kinds of magma. One is represented by basalt—a dark, heavy material, poor in silica and rich in lime, iron and magnesia. The other is a lighter material, rich in silica and alkalies; its most typical variety is rhyolite. A basaltic magma is usually hotter and less viscous than a rhyolitic one.

Between 10 and 20 million years ago colossal eruptions of basaltic lava poured out of a region of fissures in the Pacific Northwest. There was a series of eruptions, sometimes separated by long quiet intervals, so that soils and even forests grew on one flow before being buried by the next. All together some 100,000 cubic miles of fluid lava erupted from the earth and spread over the surface; flow piled on flow until what had been a mountainous terrain was completely buried by a plateau of lava more than 5,000 feet thick and about 200,000 square miles in extent.

The rhyolitic type of fissure eruption, on the other hand, is exemplified by one that took place in 1912 in the Valley of Ten Thousand Smokes in Alaska. In that year swarms of cracks suddenly opened on the valley floor, and a gas-charged, effervescent magma foamed to the surface. It was loaded with droplets and clots of incandescent liquid, which cooled to fragments of cellular glass and lumps of white pumice. So mobile was the mixture that it poured for long distances down the valley in the form of glowing avalanches. Since then many other examples of such deposits have been discovered in this country, notably in Nevada and Utah, on the Yellowstone Plateau, in the Globe district of Arizona, and in the Sierra Nevada and Owens Valley of California. Fissure eruptions of this kind often cause a sinking and downbending of the earth's crust; they account for some of the largest volcanic basins in the world, including those that hold the beautiful lakes of Taupo in New Zealand, of Toba in Sumatra and of Ilopango in El Salvador.

One of the most impressive volcanic structures is the type known as a caldera. Calderas are huge pits, shaped like craters but much larger, usually several miles across. They are also made in a very different way. A crater is the opening through which a volcano discharges its products; it is built during the construction of the cone. A caldera, on the other hand, is a product not of construction but of collapse, for it is created by the cave-in of a crater's sides. In other words, few large volcanoes blow their heads off; usually they are decapitated by engulfment of their tops.

What brings about such a collapse? In composite volcanoes—those built partly of flows and partly of exploded fragments—tremendous explosions of pumice and ash may disembowel the cone and remove support for the volcano's top. The walls of the crater at the summit then founder into the depths. The majestic Crater Lake of Oregon was formed in this way. A 12,000-foot peak which we now call Mount Mazama once stood there. Some 6,500 years ago volcanic eruptions blew 10 cubic miles of pumice out of its subterranean feeding chamber, leaving a caldera six miles wide and 4,000 feet deep. In the gigantic explosion of Krakatoa in 1883, which expelled some four and a half cubic miles of pumice, the tops of the old volcanoes foundered into the ocean. This produced a caldera five miles wide and propelled a catastrophic tidal wave that drowned 36,000 people on the adjacent coasts of Java and Sumatra.

On the present site of Vesuvius there once stood a much higher volcano. It had lain dormant for so long that vineyards extended to the summit. During this long interval of rest gas pressure accumulated in the underlying magma-chamber. In A.D. 79 it suddenly found release in a succession of terrific explosions. First the lighter, gas-rich head of the magma-column was expelled as showers of white pumice. These buried the town of Pompeii. Then came the debris of a heavier and darker magma from lower levels of the feeding chamber. This clinkerlike material, water-soaked from heavy rains, swept down the mountainsides as mud-flows and demolished the town of Herculaneum. During these violent but short-lived eruptions so much magma was emptied from the volcano's reservoir that the top of the mountain collapsed, leaving a huge, semicircular amphitheatre. Today the wrecked volcano is called Monte Somma; Vesuvius is the younger cone that has risen from the floor of its caldera.

In many volcanic eruptions ground water plays an important part, for its sudden contact with rising magma produces steam and violent explosions. This was the cause of a series of strong blasts from the Kilauea volcano in Hawaii in May, 1924. Lava drained from the feed pipes through fissures that opened far down on the sides of the volcano. Many avalanches then tumbled into the pit from the walls and ground water rushed into the empty conduits. The conversion of the water to steam generated enough pressure to blow out the plug of avalanche debris in a series of violent blasts. In 1888 the Japanese volcano of Bandai, which had long been quiescent, erupted with alarming violence. Almost half of the mountain was destroyed and 27 square miles of land were devastated by avalanches resulting from steam blasts that lasted only a few minutes. Presumably ground water had found sudden entry to the hot interior of the dormant volcano.

We have noted that the nature of an eruption depends largely on the viscosity of the magma. The viscosity in turn depends on the magma's composition, its temperature and the amount of gas it holds. The most important factor in producing eruptions probably is the gas. Without gas a magma becomes inert; it can neither flow nor explode. Once the magma, impelled by its relative lightness, has risen from the depths, it reaches a level not far below the surface where the major role in its further advance is played by the effervescence and expansion of bubbles of gas.

What is this gas, this "eruptive element *par excellence*"? In order of importance the gases originally present in the magma seem to be hydrogen, carbon monoxide and nitrogen, with lesser amounts of sulfur, fluorine, chlorine and other vapors. But in the cloud of gas that emerges from a volcano well over 90 per cent is water vapor, with carbon dioxide next in abundance. How much of this water vapor is due to oxidation of hydrogen in the magma, how much is ground water and how much is derived from water-bearing rocks surrounding the magma reservoirs at depth is unknown. Some idea of the prodigious quantities of gas given off from some volcanoes may be gained from the fact that long after the glowing avalanches cov-

History of a volcano is diagrammed, in this case the spectacular cone at the site of Crater Lake, in Oregon. At first the conduits of the volcano were filled with flowing magma. Then a series of explosive eruptions proceeded to disembowel the cone, blowing some 10 cubic miles of pumice from its interior. About 6,500 years ago the volcano collapsed into its feeding chamber. The caldera, as the resulting structure is called, has since filled with water, and a recent eruption has created the cinder cone which tourists know as Wizard Island.

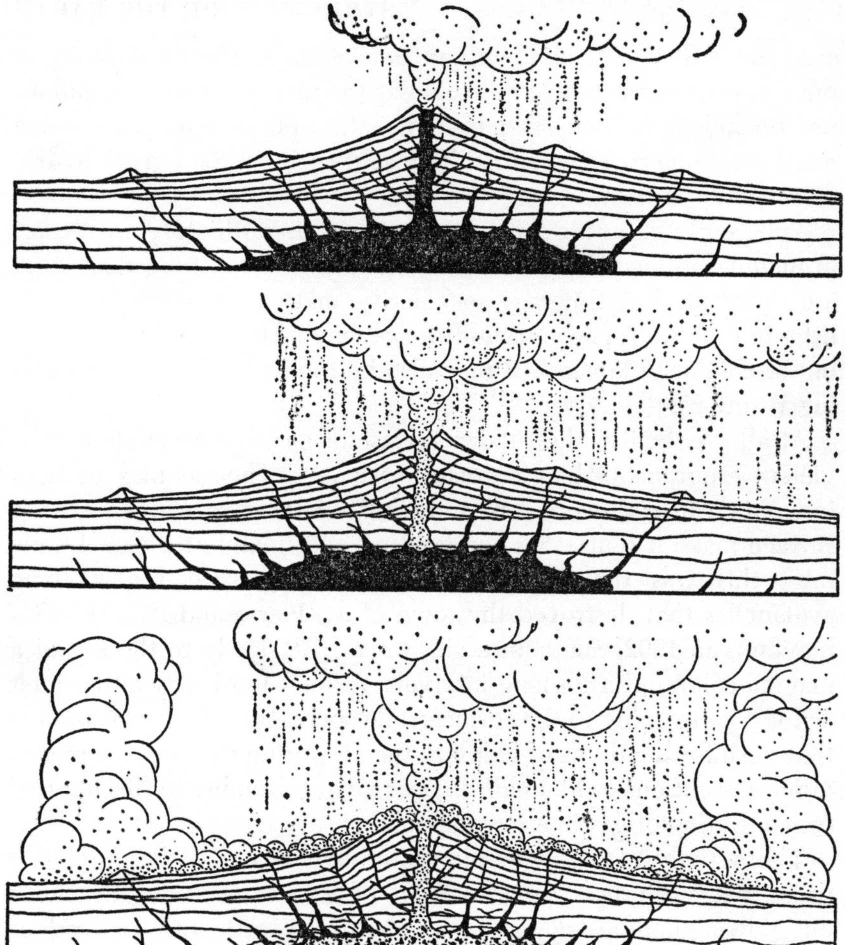

ered the Valley of Ten Thousand Smokes in 1912, the deposits of pumice continued to give off steam at the rate of six million gallons per second and discharged into the atmosphere some one and a quarter million tons of hydrochloric acid and 200,000 tons of hydrofluoric acid per year.

Apparently gases are important in maintaining high temperatures in magma, in keeping volcanoes alive and in awakening those that are dormant. But this is a speculative matter on which we have little information. The volcano-furnace may be kept hot by the burning of combustible gases; it may also be fueled by other heat-yielding reactions.

At all events, it is the sudden release of gas that accounts for the violent eruptions of long-dormant volcanoes. The gas may be held in solution in viscous magma until heat-yielding reactions near the surface make it boil at an accelerating and finally at a cataclysmic rate; this was the way Mount Pelée discharged the glowing avalanches that destroyed the town of St. Pierre and its 30,000 inhabitants in 1902. Sometimes gases may rise slowly to the top of a magma-column during long intervals of quiet until they either melt or blast an opening to the surface. The spectacular fountains of lava that gush for hundreds of feet into the air during the opening phases of most eruptions of Mauna Loa bear vivid testimony to this upward concentration of gas in a magma-column.

The activity of Vesuvius alternates between periods of relative quiet, when it erupts sluggish flows of lava or rhythmically tosses out glowing bombs, and explosions that burst forth with tremendous strength. Vesuvius produced catastrophic eruptions in 1872, 1906, and 1944. During the intervals between these explosions minor eruptions gradually increased the height of the central conelet and therewith the height of the central column of magma. Then the sudden opening of fissures far down the sides of the mountain allowed lava to escape quickly from the lower part of the column. The draining of the column greatly reduced the pressure on the underlying magma, allowed a large amount of dissolved gas to escape suddenly from solution and thereby produced colossal explosions.

One object of any science, perhaps the chief, is to improve our powers of prediction. Is it possible to say when volcanoes may erupt? To some extent, yes. We can get some warning from the seismograph. An increase in the number and intensity of local quakes in a volcanic region is fairly sure to herald an eruption. For 16 years prior to the great eruption of Vesuvius in A.D. 79 the neighboring region was repeatedly shaken. For 20 days before Paricutín was born in Mexico in 1943 the surrounding country trembled from increasingly numerous and vigorous shocks. T. A. Jaggar and R. H. Finch of the Hawaiian Volcano Observatory have foretold when Kilauea and Mauna Loa would erupt, by study of the distribution of quakes caused by fissuring of the ground as magma surged toward the surface. In many volcanic regions such preliminary quakes are accompanied by subterranean rumblings and by avalanches from the walls of craters.

Next to seismic evidence, tilting of the ground around dormant volcanoes is perhaps the most reliable clue to impending activity. The underground movement of magma often causes rapidly changing tilts on the surface. Indeed, active volcanoes almost seem to breathe; they are forever swelling and subsiding as the subterranean magma fluctuates in level. By combining strategically placed tiltmeters and seismographs, it has been found possible to say approximately where, as well as when, an eruption of Mauna Loa would take place. Accurate measurement of the cracks along the rim of the Kilauea crater also serves as a guide, for these cracks are not just superficial openings caused by slippage but mark fundamental planes of weakness that go deep, and when they widen rapidly it usually means that magma is rising beneath the crater floor.

Another hint of imminent eruption may be given by strong local disturbances of the earth's magnetism. These are produced by the rise of hot, non-magnetic magma in the volcanic pipes and by heating of the adjacent wall-rocks. Along with the magnetic changes there are commonly changes in electrical currents in the earth; these are detected, for instance, several hours before every violent explosion of the Japanese volcano Asama.

Some volcanoes behave in a roughly cyclic fashion, so that the likely sequence of events may be foretold. For instance, when the central conduit of Vesuvius has grown to unusual height, the danger of a flank outburst of lava followed by catastrophic explosions from the summit is at a maximum. And once an eruption has begun, it may be possible to predict fairly well what is likely to follow. Thus the late Frank Perret of the Carnegie Institution of Washington, by a careful analysis of the early phases of an eruption of Mount Pelée in 1928, was able to reassure the frightened inhabitants of St. Pierre that there would be no repetition of the awful calamity of 1902.

Given sufficient warning, it is sometimes possible to minimize the damage caused by a volcano's eruption. The first recorded effort of this kind was undertaken during an eruption of Mount Etna in Sicily in 1669. The inhabitants of Catania, a town in the path of the lava pouring down the mountain, made a brave attempt to save their city by digging a channel to divert some of the lava. Unfortunately the new stream moved toward a neighboring town, the angry citizens of which soon put a stop to the efforts of the Catanians. In recent years the U. S. Air Force has tried the experiment of bombing Mauna Loa during eruptions. These tests were directed both at diverting the main lava flow and at breaking down the cinder cone itself in order to dissipate the energy of the eruption into many minor flows.

Naturally a good deal of thought has been given to how the immense energy of volcanoes might be harnessed for man's use. It has been done on a relatively minor scale in several countries, notably Italy and Iceland. In Iceland many buildings are heated by volcanic steam, and by warming fields with steam pipes the country is able to raise crops that normally grow only in more temperate climates. In Italy natural steam has been used to generate electricity since 1904. There is a region in Tuscany where steam from a deeply buried body of magma comes out of the ground through rifts and is also tapped artificially by means of wells. A typical well develops a pressure of about 63 pounds per square inch, and it yields 485,000 pounds of steam per hour at a temperature of about 400 degrees Fahrenheit. In 1941 Tuscany's harnessed volcanic steam generated 100,000 kilowatts of electric power. In addition, a large amount of boric acid, borax, ammonium carbonate, carbon dioxide and other chemicals was recovered from the vapors.

The energy available from the gas vents and hot spring waters of volcanic regions is of fantastic proportions. The hot springs and geysers of Yellowstone National Park, for instance, are calculated to give off 220,000 kilogram-calories of heat—enough to melt three tons of ice—every second. A well drilled to 264 feet in the Norris Basin there developed a steam pressure of more than 300 pounds per square inch. At "The Geysers," 35 miles north of San Francisco, there are wells which, it is estimated, could provide an average of more than 1,300 horsepower each.

Thus far little use has been made of this available energy. There are many technical difficulties, of course, in the way of large-scale utilization of volcanic power, not least among them being the acidity of many of the vapors. But one can expect with confidence that these difficulties will be largely overcome, and that more widespread use will be made of the stores of energy now running to waste.

PART 3

I. THE ULTIMATE PARTICLES *by George W. Gray*

From his discussion in Part I of the infinitely large, George Gray shifts here to a review of modern theory about the infinitely small.

II. THE MULTIPLICITY OF PARTICLES *by Robert E. Marshak*

Robert Marshak, one of the country's leading nuclear physicists, is chairman of the physics department at the University of Rochester. He is the author of the two-meson theory described in his article. In the past, Dr. Marshak has been associated with the Atomic Energy Commission in several key posts; at present he is the director of the Annual Rochester Conference on High Energy Nuclear Physics.

III. THE STRUCTURE OF THE NUCLEUS *by Maria G. Mayer*

Maria Mayer is a physicist with the Argonne National Laboratory (AEC) and the Institute of Nuclear Studies of the University of Chicago.

IV. RADIO WAVES AND MATTER *by Harry M. Davis*

A few months after this article was written, Harry Davis was killed in a swimming accident. His death was a severe loss to science journalism for, as his contributions to this anthology show, he was one of the ablest writers and reporters in the field. Davis had been a member of the science staff of *The New York Times* and science editor of *Newsweek*.

V. THE SOLID STATE *by Gregory H. Wannier*

Gregory Wannier was educated at the University of Basel, in his native Swiss city, and at Cambridge. He came to the U. S. in 1936 on a post doctoral fellowship offered by Princeton University and there studied with Eugene P. Wigner, a pioneer theoretician in the field of solid state physics. Before taking his present position with the Bell Telephone Laboratories, Dr. Wannier taught at several universities both here and in England.

STRUCTURE OF
MATTER

Introduction

THE IDEA that matter is built up of ultimately indivisible particles appeals to common sense. We got the atom from the Greeks, and the atomic theory is as old as natural philosophy. Modern physics is still in quest of the ultimate particle. However, as everyone knows, the atom has long since been smashed into a number of still more fundamental component parts. Just which of these, if any, can be called ultimate and what holds them together in the complex structure of the atom and its nuclear core—these are questions of nuclear physics.

What troubles the physicists is that, for the past 20 years, the high rate of new experimental discovery has put their theoretical system under a heavy strain. Modern atomic theory has satisfactorily extended the concepts of the electromagnetic field, via quantum field theory, to describe the atom in interaction with externally applied forces and with other atoms. The problem is how to relate this reasonably consistent understanding to our new knowledge about the nucleus. In order to accommodate the new findings physicists have had to elaborate the theory in ways that have not always been elegant. Most physicists agree that physics now has too many particles; there were 21 at last count. A number of these particles have never been demonstrated by experiment; they are in the table because their presence is required by theory. There is no assurance that others will not be needed as investigation goes forward. This state of affairs recalls the final stages of Ptolemaic astronomy, when planetary motions were traced in ever more intricate cycles in order to maintain the earth's position at the center of the universe.

No one is particularly happy about our present picture of the atom. The layman finds that he must put aside most of his common-sense notions, including that of the particle itself. Yet, after accepting the proposition that a particle of matter may behave like a wave and that a wave of energy may act like a particle, he still gains no clear notion of what the atom is like. He must take what comfort he can in knowing that the physicists share his difficulty.

In spite of the shakiness of its theoretical foundations, physics continues to make extraordinary experimental progress. The microcosm of the atom grows constantly richer in content and interest. Investigation of the atom, moreover, is central to the advance of knowledge on a broad and varied front. The evolution of stars, the structure of the earth, the biochemistry of cells are fields that have been greatly strengthened by recent progress in physics. The new knowledge and techniques also have been taken up in the avid grasp of technology, and a half-dozen major industries are on the brink of still another revolution.

The ideas discussed in this chapter will be encountered again and again in later chapters of the book. The lay reader can be assured, however, that he does not need a physics degree to grasp these governing ideas as they are presented here. George Gray introduces the present model of the atom, defines the special ways in which such familiar terms as space, matter and energy are used in modern physics and shows how the experimental work, especially that of the atom smashers, is carried on. Robert Marshak then takes the atom and its nucleus apart into its present 21 component particles, explaining what we know about how they fit together and why our understanding is at present incomplete.

Physics faces some of the same difficulties as biology, in that the atom, like the living cell, is a system more complex than the mere sum of its component parts. Just as some proteins behave differently outside the cell, some particles change their identity upon being dislodged from the atomic nucleus. Maria Mayer discusses current theoretical efforts to describe the structure of the nucleus and to account for the particles and forces we have reason to believe are involved. The article by the late Harry Davis tells how physicists have adopted radar technology to look inside the intact nucleus, instead of knocking it apart.

The chapter closes with an article that relates the atom to matter as we know it with our hands and eyes. As Gregory Wannier shows, the application of the essential concepts of atomic physics to the study of crystal structures has recently transformed our understanding of matter in the familiar solid state. We are able now to account for such gross properties as tensile strength or electrical conductivity in the same terms that describe the behavior of atomic particles. The transistor revolution in electronics and certain developments in metallurgy indicate that progress in our understanding of matter may have far more immense consequences than the celebrated arrival of atomic energy.

I. THE ULTIMATE PARTICLES
by George W. Gray

And yet so poor is nature with all her craft, that, from the beginning to the end of the universe, she has but one stuff . . . to serve up all her dream-like variety. Compound it how she will—star, sand, fire, water, tree, man—it is still one stuff and betrays the same properties.

—Ralph Waldo Emerson,
Essay on Nature

SCIENCE TODAY is concerned with a multitude of problems, many of them of a fundamental character, but none is more basic than the search for the ultimate units of matter and energy. Thales of Miletus posed the question 25 centuries ago when he asked: "Of what and how is the world made?" It is still the supreme enigma. The world includes man, and the elucidation of its nature cannot but reveal something of the hidden nature of life and of man.

Our question, therefore, is not alien to our humanity. The ultimate particles which enter into combination to make hydrogen and iron also enter into the construction of bone and muscle, blood and nerve and brain. In studying the constitution of atoms we are studying the fundamental stuff of the universe—of suns and mountains and seas, the black carbon of coal, the green chlorophyll of grass, the red hemoglobin of blood. Indeed, nature knows no such specializations as physics, chemistry, biology and the other categories into which we fit our fragments of knowledge. She knows only the particles and their incessant interactions as expressed in phenomena such as magnetism, radiation, life, death.

A biologist probing the minute architecture of protoplasm must wield his dissecting needle with extreme delicacy, else he may destroy the thing he is trying to explore. But the physicist, wishing to unveil the still more minute architecture of the atom, resorts to artillery. His method is that of banging one particle against another, and the harder the blow, the more revealing is the debris resulting from the violence.

The value of the bombardment technique was shown a half century ago when Wilhelm Konrad Röntgen, experimenting at the University of Würzburg with cathode rays, discovered a mysterious radiation coming from that part of the glass tube against which the stream of cathode rays (electrons) impinged. The discovery of these X-rays in 1895, of radioactivity in 1896, of the electron in 1897 and of radium in 1898–those four golden years which ushered in the heroic age of physics!–brought not only revolutionary knowledge, but also new and powerful research tools.

A few weeks after Röntgen's discovery, J. J. Thomson used X-rays to bombard air and other gases and found that the rays knocked out negatively-charged fragments which eventually were identified as *electrons*. The discovery of radioactivity later led to the finding of alpha particles. It was by using the alpha particles which spontaneously shoot out of radium and other radioactive elements that Ernest Rutherford in 1911 bombarded gold into betrayal of its nucleus. Subsequent batterings showed that all atoms have this central, massive, positively-charged core around which the electrons revolve as negatively-charged satellites. Alpha particles themselves were found to be helium nuclei. In 1919 the alpha-particle barrage turned up another fundamental of structure when Rutherford bombarded nitrogen and found positively-charged *protons* bouncing out of the nitrogen nucleus. Thirteen years later the same artillery helped James Chadwick to blast out a still more elusive nuclear constituent, the *neutron*.

This breakdown of the atom into three component particles did not, at first, unduly complicate the picture of matter held by modern physics. In fact, the combination of electrons, protons and neutrons gave us, momentarily, a highly satisfactory structure. Subsequent research, however, has disclosed more than a dozen additional mass particles, including the positron, the V particles, and the big family of mesons. We must also account for the particles of energy,—the photon, graviton and neutrino. The relationships among these units and the necessity, at least for the present, of recognizing such a variety of them, are discussed in the article by Robert Marshak which follows this survey.

The alpha particles and other discharges from radioactive atoms, and the cosmic rays which continually bombard the earth from interstellar space, have been highly useful as research tools, but both are beyond the control of man. It was the desire to command the conditions of his experiments that led the physicist to devise his own artillery. He wanted to be able to choose the kind of projectile, to put more projectiles into a given barrage, and to regulate the velocity with which the projectiles struck their target.

Twenty years of inventive effort have been devoted to the problem, and several types of apparatus are now available. Each has its distinctive feature, but most of them may be grouped in two broad classes: the direct-voltage accelerators and the resonance accelerators.

In the direct-voltage machines, the projectiles move in straight paths through a long vacuum tube, impelled by the maximum voltage of the discharge. Machines of this type are literally artillery pieces: barrels through which invisible bullets are fired at high velocity.

The resonance accelerators operate on a different principle. In them, projectiles are started at relatively low speeds and by the repeated push of periodic pulses of voltage are brought up to the energy required for the bombardment. Most of the resonance machines accelerate their projectiles in a whirling stream, swinging them around and around in circular or constantly enlarging spiral paths. Examples of these electrical slingshots are the *cyclotron*, the invention of E. O. Lawrence of the University of California; the *betatron*, developed by D. W. Kerst of the University of Illinois; and the *synchrotron*, first suggested in this country by E. M. McMillan of California and independently by V. Veksler of Russia. The most important types of resonance accelerators are explained in the series of diagrams.

There is also a resonance machine, the *linear accelerator*, which makes no use of the whirling principle, but sends its projectiles in straight lines through a tube, starting them slowly and building

Particle accelerators are diagrammed here from top to bottom in ascending order of power (also chronological order). The cyclotron at top pulls and whirls the particles into their spiral trajectory by rapid alternation of the charge on its "dee" plates. The synchro-cyclotron, next in order, varies the frequency of this alternation of charge on its single dee plate to accommodate the increase in the mass of the particles as they approach the speed of light. These two machines accelerate positively charged particles. The betatron whirls electrons up to extremely high energies, for use as primary particles or to generate x-rays. The synchrotron, at bottom, is used to accelerate either electrons or protons. These two machines develop particle energies comparable to those of cosmic rays.

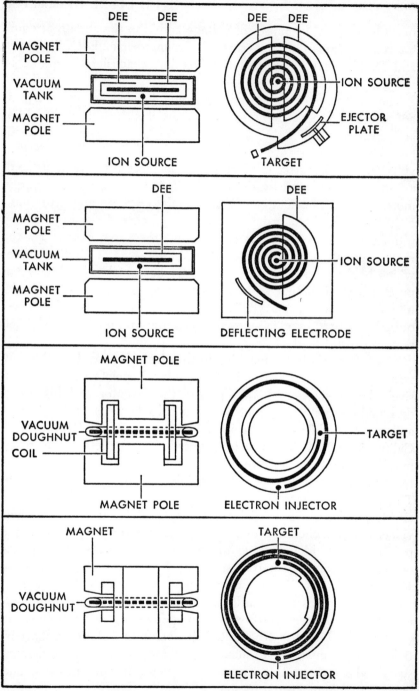

up to high speed by pulses of voltage added to the stream at equal intervals of time as it moves through the long barrel. E. O. Lawrence and D. H. Sloan developed the first linear accelerator about 14 years ago.

The energy of bombardment is rated in electron volts, a unit adopted to express the energy of the particles ejected by the radioactive atoms. The highest-energy projectiles obtained from spontaneous radioactivity are the alpha particles of 10.54 million electron volts (which hereafter shall be noted, in the physicist's shorthand, as "mev") discharged by thorium C'. This is a respectable energy, more than any accelerator was able to deliver until about the mid-1930's; but the bombardiers early realized that the search for the ultimate particles would require many times this power. By the famous equation of Albert Einstein, $E = mc^2$, the binding forces which hold nuclei together could easily be calculated. These computations showed that to break the oxygen nucleus, for example, into its elementary parts 127 million electron volts would be required. It would take 487 mev to smash the iron nucleus, and 1,580 mev to disrupt lead.

The first to project a machine in the class of more than 100 mev was E. O. Lawrence. After building several cyclotrons of progressively larger size and higher power, the inventor and his group designed one with a magnet having a pole diameter of 184 inches, and work on this 4,000-ton apparatus began at the University of California in 1940. The war interrupted its construction, and when building operations were resumed in 1945 the design was modified to apply a newly-recognized principle of frequency modulation. This change was introduced to make a correction for the relativity effect which makes particles increase rapidly in mass as their velocity approaches that of light. By varying the frequency of the pulsations of added voltage to correspond with the lagging rate of the more massive particles, the cyclotron was changed into a *synchro-cyclotron*. Since its completion in the fall of 1946, the synchro-cyclotron has abundantly proved its superiority to the unmodified cyclotron. Designed originally to operate at about 100 mev, the modified machine has accelerated deuterons to 200 and alpha particles to 400 mev. Bombardments with these projectiles have demonstrated the strange transformation of neutrons into protons, and have even manufactured mesons.

This 184-inch synchro-cyclotron at California is the most powerful accelerator now in use, but larger giants are coming. Most of them are proton-synchrotrons, an apparatus which accelerates protons through hundreds of thousands of small impulses. Whereas synchro-cyclotrons appear to have an upper limit somewhere around 750 mev, the synchrotron is able to build up voltages in the billions without disturbing the stability of the paths traveled by the particles or impairing the intensity of the beam. A 1,300-mev proton-synchrotron is under construction at the University of Birmingham, England; in the U. S. one of 3,000 mev is projected by the Brookhaven National Laboratory * on Long Island, and one of 6,000 mev by the University of California. These accelerators will provide projectiles comparable to those produced by cosmic rays. It is believed they may enable experimenters to create protons and neutrons.

The targets of the gigantic artillery pieces of modern physics are unimaginably small. There are several methods of determining the sizes of nuclei, and they are in reasonable agreement in indicating that the atomic core is a globular or oblate structure whose volume varies approximately with its mass. The diameter of the hydrogen nucleus, the lightest, is $\dfrac{15}{100,000,000,000,000}$ of a centimeter, while at the other extreme, the uranium nucleus has a diameter $\dfrac{200}{100,000,000,000,000}$ of a centimeter. Thus the largest nucleus found in nature, although 238 times as massive, has a diameter only 13 to 14 times that of the smallest.

These dimensions are most difficult to visualize. If it were possible to enlarge the uranium atom until its nucleus became just visible, the nearest of its satellite electrons would be revolving in an orbit about six feet from that center; and beyond this orbit would be another, and then another—a total of 92 encircling electrons distributed among seven orbits. Imagine a solar system with 92 planets traveling around its sun, and yet one so small that it would take many billions to form a spot as large as the period at the end of this sentence.

Light has no wavelength small enough to form an image of these

* The Brookhaven synchrotron was completed and put into operation shortly after the first publication of this article.

infinitesimals, and all that the physicist knows of their structure he gathers by inference from circumstantial evidence—the fingerprints, footprints and other clues which the invisible particles leave behind. Many years ago he found that when the projectiles shot out of radium happen to collide with matter under certain conditions, the effects become visible. Since then the physicist has shown great ingenuity in inventing ways to trap matter under the conditions that show up these collision effects. Two arrangements are the Wilson cloud chamber and the Geiger-Müller counter.

The cloud chamber operates by a simple scheme in which the moisture-laden air within it is suddenly cooled by the withdrawal of a piston that expands the volume of the chamber. If a charged particle darts through the chamber at the moment of this expansion, droplets of the supersaturated vapor attach themselves to the air molecules which have been mutilated by collision with the particle, and thus the course of the speeding particle is revealed as a track of cloud. It is as though some invisible demon should plunge through a crowd and the police were to observe its path and estimate its force by the position and number of people knocked to the ground. The physicist is able to get additional information by placing the cloud chamber between the poles of a magnet. Negatively-charged particles then swerve in one direction, positively-charged in the opposite, and the radius of the curvature tells something of the mass of the particle. By installing plates of lead or other dense metals across the chamber, as barriers to slow the particles passing through them, it is possible to measure the particles' energies. The cloud chamber thus is an instrument of unusual versatility.

The Geiger-Müller counter—an invention of two German physicists—reveals the invisible presence by another procedure, although its action also is based on the fact that a charged particle mutilates or ionizes the air molecules with which it collides. If these ionized molecules are positively charged, they will move to a negatively-charged electrode; if negatively charged, to the opposite. The counter is a trap to catch such occurrences. The minute electric current generated by the mutilations is amplified, and can be made to cause a click in a loudspeaker. Or the tiny current may operate a mechanical recorder, or cause a visible pulse on the screen of a cathode-ray tube. There are also other particle-detecting devices—fluoroscopes, electroscopes, specially prepared photographic plates with thick coatings of emulsion. It is by various means that physics has arrived at its present view of the nucleus.

The microcosm of the atom presents a strange spectacle for the imagination. Between the nucleus and the orbit of the nearest encircling electron is a gap vaster in proportion to the sizes of these bodies than the space separating the sun from the orbit of the earth. This abyss within the atom is a perfect vacuum.

Occasionally, if the atom is of uranium, plutonium or some other radioactive species, a readjustment occurs within the nucleus and there is ejected an alpha particle, beta particle or the photon of a gamma ray. When that happens, the vacuum is of course momentarily crossed by the speeding particle, but these transits are rare and are so instantaneous as to be measured in billionths of a second.

Even a stable atom may occasionally suffer the accidental impact of a cosmic ray. If the hit is only a glancing one, it may provoke the nucleus to eject a particle; but if the blow is head on, the energy transfer may be so great as to smash the structure into fragments. Numerous photographs of such explosions have been taken in the course of cosmic-ray research, and more recently in cyclotron research at the University of California. These pictures show a burst of several tracks originating at a common center and radiating in many directions like the points of a star. "Stars" of 20 tracks and more have been photographed, and each marks the path of a speeding nuclear fragment of some atom in the photographic emulsion that was blasted by the impact of a cosmic ray. Each fragment, as it moves away from the place of the explosion, is believed to function as a new nucleus, attracting to itself wandering electrons which become its satellites. Thus an exploded silver atom may become several nuclei of smaller mass and lower charge, each with its surrounding vacuum and attendant electrons.

The vacuum is a theater for the exchange and transformation of energy. Because of its proximity to the central powerhouse of the nucleus, tremendous electromagnetic forces play across this space, and amazing events can transpire there. If a high-energy photon, that of a gamma ray, chances to dart into the vacuum, the interaction of the photon with the electromagnetic field may create two particles, an electron and a positron. Cloud-chamber photographs have been taken of the tracks made by these pairs, both originating at a common point, with the path of the positron curving in one direction, and that of the electron in the opposite. This positron-electron pair is an example of the creation of matter out of energy. The nucleus gives up none of its mass; it serves only as a catalyst to provide a field of electromagnetism for the gamma ray to work on; the ray then surrenders its energy, which instantly reappears in the form of the two material particles.

There is, however, an even deeper underlying reality. Our account of the creation of positron-electron pairs is a rough-hewn portrayal of what is elegantly expressed in the laconic mathematics of P. A. M. Dirac's theory. This theory is concerned with the various states of energy which may be occupied by an electron. The number of such states is infinite. Plotted on a chart they stretch from zero upward, but the chart also points to an equal number of energy states below the zero line. It may strain common sense to speak of the energy of motion as less than nothing, but negative quantities have proved useful in mathematics and the concept must be admitted. The levels of less-than-zero are negative-energy states, but for simplicity (to avoid confusion with the negative electrical charge on the electron) let us call them minus-energy states, and those above zero, plus-energy states. The point is that a negatively-charged electron, any electron, can occupy either a state of minus-energy or one of plus-energy.

But electrons are lazy. Their tendency is to drop into conditions of lower energy, and this would seem to mean a general movement to levels below the zero line. However, there is a law, the Pauli "exclusion principle," which forbids more than one electron to occupy a given energy state; and consequently, after all the minus-energy states were taken, perhaps in some mad cosmic scramble at the beginning of time, there was nothing left for the losers but to accept the lowest available plus-energy states.

It is these electrons in plus-energy states that whirl in the orbits around nuclei, that flow through wires and other conductors to form electric current, that dance in the candle flame to generate light. Indeed, all the electrons that make up the perceptible universe are those which were forced by the exclusion principle into plus-energy states. The others, which presumably are more numerous, are comfortably at rest in their berths of minus energy—withdrawn from the dynamic world, buried in the vacuum.

100

But not completely unobservable, declared Professor Dirac. Theory told him that the chance blow of a gamma ray might knock an electron out of its minus-energy state. In that case, the hole left in the substratum should appear as a particle, a particle having the same mass as the electron but of opposite charge—in other words, a positively-charged electron. This speculation was published in England almost a year before the discovery of the positron. Since then hundreds of tracks of positron-electron pairs have been photographed. The tracks soon end, betraying the early disappearance of both positron and electron as free bodies. This is to be expected. For the electron is usually captured by some nucleus looking for another satellite; and as for the "hole," almost as soon as it appears the nearest electron leaps into it. Then both hole and electron disappear into the vacuum, and their energy darts off as radiation.

Such a picture is difficult to reconcile with ordinary logic. The region surrounding the nucleus was described earlier as a void. Now the same region is portrayed as the repository of innumerable electrons at rest in states of negative energy. Which is right? Can space be both empty and occupied? Apparently the two pictures represent different aspects of the same thing. Nature requires a perfect vacuum to serve as the sea of negative-energy states—and where all the negative-energy states are occupied, there is a perfect vacuum.

Some of the nucleus itself may be vacuum. If the particles which join to form it maintain their individuality as separate units, as seems probable, it is necessary to assume that there is space between them. Whatever its structure, the nuclear material is the quintessence of mass. Most of the world's weight is concentrated in these tiny cores. A cubic centimeter of platinum with its electrons weighs about three-quarters of an ounce; but when the encircling electrons are peeled off and each atom is stripped down to its naked nucleus, we arrive at something that when packed together averages 130 million tons per cubic centimeter. And the same is true of all atoms, for nuclear material whatever its origin is "one stuff and betrays the same properties." The whole of Mount Everest might thus be packed into a cigarette case. W. D. Harkins has calculated that if all the matter making up the earth were collapsed down to its nuclear material, the diameter of our planet would shrink from its present 8,000 miles to about 1,080 feet—with no shrinkage in its weight.

Every particle known to the cloud chamber has been detected coming out of nuclei. Photons come out as gamma rays, electrons in the form of beta particles come out, and so do alpha particles, positrons, neutrons, protons, and mesons. In addition there is the hypothetical neutrino which the physicists assume must come out. Does this mean that the nucleus is a mixture of all these various particles?

The physicists don't think so. In the case of photons, for example, no one has ever seriously suggested that they are constituent particles. Gamma rays are created by the internal forces; when the forces reach such a state of imbalance that the nucleus must change to a lower energy, the shift is made and the surplus energy is ejected as a photon of gamma radiation. In the same way, it is now believed, electrons, positrons and mesons do not exist as components, but under certain stresses are created by the conversion of energy into mass. In this view, which is the prevalent one today, nuclei are composed of protons and neutrons—and nothing else. Protons and neutrons therefore are in a different class from the other particles: they are the nuclear building blocks, and because of this are called *nucleons*. All the other particles are by-products created by the interactions of nucleons.

Interactions are not likely to occur in the case of simple hydrogen, for its nucleus is a solitary proton. The first step toward complexity is the double-weight hydrogen atom whose nucleus, known as the deuteron, consists of one proton joined to one neutron. Helium is the next. Its nucleus is made up of two protons and two neutrons, and therefore represents the union of two deuterons. But the helium core is something radically different from the sum of two deuterons. Measurements show that the binding force which holds this heavier nucleus together exerts the power of 28.20 million electron volts, whereas the binding force of the deuteron is only 2.19 mev.

A reason for the higher binding force of the heavier nucleus has been suggested by Eugene P. Wigner of Princeton. In brief: the more nucleons there are, the greater is the number of bonds holding them together. The accompanying diagrams, depicting the deuteron and the helium nucleus according to the number of particles, illustrate how this works.

The nucleons of the deuteron, since there is only one bond between them, average one-half bond per particle; whereas, with six bonds joining its four nucleons, helium averages one and one-half

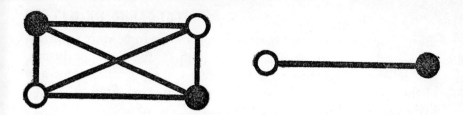

per particle. The effect is to make helium a more tightly knit structure. Its nucleons draw closer to one another, and its mass becomes slightly less than the sum of the masses of two deuterons. This "mass defect," which one finds in all nuclei made of two or more particles, is highly significant, for it represents the amount of matter that has been converted into energy to provide the binding force.

The nature and mechanism of this binding force is one of the great mysteries. Indeed, the quest for the ultimate particles depends, in large measure, on understanding the binding force.

Gravitational attraction operates between particles just as it does between stars and planets. It is possible to calculate the magnitude of this force between two protons by applying Newton's law that gravity varies inversely as the square of the distance. In the close quarters of the nucleus this gravitational attraction of nearby particles for one another is considerable. But it is not nearly great enough to account for the tremendous binding forces of nuclei, ranging from 28.20 mev for helium to 1,780 mev for uranium.

Moreover, there is present in all these nuclei, from helium to uranium and beyond, a disruptive agency more powerful than gravitation. It is the force of repulsion which exists between bodies carrying the same kind of electrical charge. The mutual repulsion of proton for proton is so great that, according to Frederick Soddy's calculation, if a gram of protons could be concentrated at one point on the earth's surface and another gram at the opposite point on the other side of the globe, the repulsion of these two tiny positive charges for one another would be equivalent to a pressure of 28 tons. This is at a distance of 8,000 miles! And yet within the confines of the nucleus protons dwell in such close communion that millions of volts are required to blast them apart.

It was this paradox that led physicists to assume the existence of a force of attraction which is able to overrule the electrical force of repulsion. They call it the *nuclear force*, since it is able to operate only over distances of very small magnitude, such as those within the atomic nucleus. Exact measurements of this force were made by, among others, Merle A. Tuve and a group at the Carnegie Institution of Washington, using the 1.2 million-volt electrostatic generator of the Department of Terrestrial Magnetism. They fired protons at hydrogen nuclei, and gradually increased the voltage until a speed was reached at which the momentum drove the projectile so close to the nucleus that it was attracted rather than repelled. This marked a boundary—the distance at which the nuclear force of attraction began to operate. By applying the principles of wave mechanics, a mathematical analysis of the experimental results was made by Gregory Breit, assisted by E. U. Condon and R. D. Present, and their calculation gave these findings:

1. The electrical force of repulsion ceased to control when the projectile proton got within about 1/12,000,000,000,000 of an inch of the target proton.

2. The nuclear force of attraction which suddenly took control at that distance was approximately 10^{36} times greater than the gravitational force between the two protons.

Experiment has shown that the nuclear force of attraction operates also between neutrons, and between protons and neutrons. The process of reasoning whereby a whole new family of particles—the mesons—was first postulated from the nuclear force and then detected in experiments is also set forth in the Marshak article which follows.

Many physicists are frankly dissatisfied with the complexity of nuclear theory. Some are testing daring new ideas in the search for a simpler and more unified pattern to describe the microcosmic architecture. At Princeton, for example, Dr. Wheeler and his group are considering the possibility that mesons, protons and neutrons may all be built up from electrons, positrons and neutrinos. It is a highly speculative suggestion which physicists view with cautious skepticism, as they do all new attempts at simplification. Dr. Wheeler himself points out that the idea must remain a question until there is fuller examination of electromagnetic and electron-positron pair theories. But he and his associates are definitely exploring the implications of a world built of electrons and positrons as the material units, with neutrinos as the energy units.

The electron-positron pair is the key to the concept. We know that when electron meets positron both are annihilated, with a release of gamma radiation. It has also been proved that the action of a gamma ray on the electromagnetic field can call a pair into existence. Imagine, then, a condition in which pairs of plus and minus electrons are continually being created and destroyed. Imagine these newborn and dying positives and negatives perpetually weaving back and forth within the nucleus, describing a network of paths, and by these movements generating the binding forces.

"It would be a gain if we could get away from the postulate of a special nuclear force," said Dr. Wheeler, "and account for the binding energy in terms of something we already know. Chemists used to imagine a 'chemical force' to account for the affinity of certain atoms and molecules for one another but as knowledge of the electrical structure of atoms advanced, the chemists came to see that affinity was simply a manifestation of electromagnetism. So we think it may turn out to be in the case of the nuclear force."

Of course, there is always the possibility that what Leibnitz called "the pre-established harmony" is beyond the reach of instruments, even beyond the reach of our imaginations. J. B. S. Haldane has given it as his opinion that "the universe is not only queerer than we suppose, but queerer than we *can* suppose." It isn't likely that many nuclear physicists will accept this doctrine of defeatism. They will continue their bombardments, their trapping of cosmic rays, their examination of the debris. The theorists will continue to follow up these experiments with their fascinating analyses, their rationalization of the results, sometimes with predictions of results to come. And eventually the strange assortment of protons, neutrons, mesons, electrons and positrons, with their accompanying photons and neutrinos, will take their places in a completely consistent picture. Such is the hope and the faith of science.

II. THE MULTIPLICITY OF PARTICLES

by Robert E. Marshak

I<small>N THE YEAR</small> 1932, when James Chadwick discovered the neutron, physics had a sunlit moment during which nature seemed to take on a beautiful simplicity. It appeared that the physical universe could be explained in terms of just three elementary particles—the electron, the proton and the neutron. All the multitude of substances of which the universe is composed could be reduced to these three basic building materials, variously combined in 92 kinds of atoms. An atom consisted of a tight nucleus, built of protons and neutrons, and a swarm of electrons revolving around the nucleus like planets around a sun. The number and arrangement of electrons in an atom could explain its chemical properties, and the two nuclear particles, proton and neutron, could account in a very simple manner for other properties of the atom—its weight, its nuclear charge, the isotopes, and so on. From atoms it was possible to arrive at molecules and finally at the macroscopic objects of our workaday world.

Strictly speaking, under the quantum-mechanics theory that energy, like matter, also was made up of tiny discrete units, there were two additional elementary "particles"—the photon, the unit of electromagnetic radiation, and the graviton, the unit of gravitational energy. But the photon and graviton acquire existence only as a result of some rearrangement of electrons, protons and neutrons, and it seemed that the latter three particles could be considered the basis of all physical phenomena. They appeared to be the irreducible units of matter for which scientists had been searching ever since Democritus. The electron was an almost infinitesimal corpuscle with a negative charge, the smallest negative charge ever observed; the proton (the nucleus of the smallest atom, hydrogen) was 1,837 times as heavy as an electron but had only a single positive charge, just balancing the electron's negative one; and the neutron, neutral in charge and with about the same mass as the proton (just two and a half electron masses heavier) appeared to be equally fundamental.

Such was the simple, satisfying picture after Chadwick identified the neutron in 1932. But the period of rejoicing was all too brief. Before the year was out Carl D. Anderson at the California Institute of Technology discovered a fourth elementary particle. In the atomic havoc created in a Wilson cloud chamber by the energetic cosmic rays he found a particle with the same mass as the electron but with a positive charge. It was named, of course, the positron. Soon experimenters were proving the existence of positrons by creating them in the laboratory with high-energy photons, *i.e.*, gamma rays. By interaction with atomic nuclei photons are converted into pairs of electrons and positrons. So physicists now had a fourth particle to confuse the picture. But this was only the beginning.

After the positron was discovered, experimenters began to pay closer attention to the beta particles (electrons) emitted by radioactive nuclei. They were not surprised to find that the beta particles from some of the artificially made radioactive atoms were positrons rather than electrons. But in measuring the energies of beta particles they made a discovery that was very surprising indeed. According to the laws of quantum mechanics, the nucleus can occupy only certain discrete energy states, and when it drops from one energy level to another, emitting a particle in the process, it can impart to the particle only the discrete difference in energy—no more and no less. Consequently all beta particles, whether electrons or positrons, should be discharged from nuclei only with certain well-defined energies, just as a properly operating elevator will let you out at one floor or another but not in-between. When the investigators actually measured the energies of beta particles, however, they found that the energies ranged over the whole continuous spectrum from zero to the maximum possible value.

This was indeed a strange phenomenon. If the quantum theory was correct, it seemed to violate one of the fundamental laws of physics—the principle of the conservation of energy. If a nucleus, in dropping from one energy state to another, could emit a beta particle with some in-between energy value, what happened to the rest of the energy representing the difference between the two states?

Furthermore, the emission of beta particles from nuclei appeared to violate another basic physical principle—the conservation of angular momentum, or spin.

Like the earth and other bodies in space, each atomic body—electron, nucleus and the particles that make up the nucleus—spins like a top around its own axis. The spin, or angular momentum, can be measured. The unit of spin in the atomic domain is Planck's constant divided by 2 *pi;* let us call the unit *h.* The laws of quantum mechanics require that the spin of any particle or system of particles must be quantized: it must be an integral or half-integral multiple of *h.* Thus a system can have an angular momentum of ½ *h* or 1 *h* or 1½ *h* or 2 *h,* but it cannot have a spin of, say, ¾ *h.* When two or more particles are combined in one system, the total spin of the system must lie in the range between the difference and the sum of the spins of the particles composing it.

Now the spins of various particles have been measured. The electron, the positron, the proton and the neutron each has a spin of ½ *h.* The deuteron (nucleus of heavy hydrogen), which consists of one proton and one neutron, has a combined spin of 1 *h,* indicating that the proton and neutron spins are in the same direction. In general all nuclei consisting of an even number of nuclear particles possess integral spin, and all those consisting of an odd number have half-integral spin.

But when we come to the case of the beta particle issuing from a radioactive nucleus, the rule apparently breaks down. The birth of a negative beta particle (electron), for example, evidently is due to the transformation of a neutron into a proton according to this scheme: neutron → proton + electron. Here we have a neutron with a spin of ½ *h* giving rise to two particles, proton and electron, each with a spin of ½ *h.* Where did the extra spin come from, or to put it another way: assuming there was a balancing of opposite spins in the neutron, what happened to the lost spin of half a unit?

Somehow, then, physicists had to save the laws of conservation of energy and conservation of angular momentum. There was only one way to do it: a new particle had to be invented. The man who suggested this solution for the dilemma was Wolfgang Pauli of Zurich. He proposed that when a beta-emitting nucleus erupts, it creates not only a beta particle but another particle which he named the neutrino. The neutrino hypothesis saved all the conservation principles at one fell swoop. The neutrino would carry the energy representing the difference between that of the beta particle and the total energy lost by the nucleus in changing from one discrete energy state to another. The neutrino, assuming it had a spin of ½ *h,* would also restore the lost angular momentum, for the neutron with half-integral spin could transform into three particles the sum of whose spins would also be half-integral.

The neutrino as such has never been detected directly, but by now its properties have been fairly well established by indirect experiments. It is an extremely light particle with a mass less than one two-thousandth that of the electron. A free neutron, which as we have seen is heavier than a proton and electron combined, has been found to decay radioactively with a half-life of about 12 minutes into a proton, an electron and a neutrino. A proton, being lighter than a neutron, can transform into a neutron, a positron and a neutrino only inside the nucleus.

Pauli's hypothesis of the neutrino in 1933 raised the number of elementary particles to five. More were soon to come. Some of the new particles were predicted by theory before they were actually discovered. As George Gray noted in the preceding piece, physicists were seeking an explanation of the extremely powerful forces that hold together the protons and neutrons in an atomic nucleus. In 1935 the Japanese physicist Hideki Yukawa suggested that a new kind of field, consisting of quanta of energy which might take the form of particles of a certain mass, might account for these forces. He pointed out that electrical forces and gravitational forces, the two chief forces previously known, could be explained in terms of the emission and reabsorption of light quanta and gravitational quanta respectively. Since the nuclear forces were of a completely different type—not only much more powerful but acting over much smaller distances than electrical or gravitational forces—it seemed reasonable to Yukawa to introduce a new type of field which would be responsible for the nuclear forces. There is a formula whereby it is possible to calculate the mass of the field quantum exchanged between two particles if the distance over which the force acts is known. On the basis of the extremely short distance (about 10^{-13} centimeters) over which the nuclear force acts, Yukawa estimated that the mass of the field quanta exchanged between two nucleons would be about 200 to 300 times that of the electron. He called these field quanta mesons. The mesons were thought of as the nuclear glue binding together the neutrons and protons in the nucleus. Since there were three types of equally strong bonds in the nucleus (neutron-proton, proton-proton and neutron-neutron) it was assumed that there would be three kinds of mesons, namely, positive, negative and neutral.

Within two years after Yukawa made his proposal investigators in the U. S. actually discovered mesons with masses about 200 times the mass of the electron. The discoverers, Carl Anderson and Seth Neddermeyer at Cal Tech and J. C. Street and E. C. Stevenson at Harvard, found the particles in the atomic debris produced by cosmic rays.

There followed a great number of experiments to determine the properties of mesons. The net result of 10 years of experimentation, however, was the emergence of a very serious discrepancy: namely, mesons were produced with ease and in great numbers by the incoming cosmic radiation in the upper atmosphere, but only rarely were they later absorbed by the nuclei of atoms. The probability of production was 10^{14} times the probability of absorption. This discrepancy contradicted not only Yukawa's meson theory of nuclear forces but also a general principle of physics—the principle of the reversibility of microscopic processes.

In view of the glaring difference between meson production and capture, the author suggested in 1947 that there were really two types of mesons: 1) a heavier variety which possessed the properties Yukawa had postulated and which was responsible for the forces holding atomic nuclei together, and 2) a lighter variety into which the heavier one decayed. It was the latter type, he proposed, that cosmic-ray experimentalists had been observing for 10 years. According to the author's two-meson theory, the heavy mesons are produced by the primary cosmic radiation in the upper atmosphere and decay into the lighter variety in about 10^{-8} seconds—too short a time for them to penetrate very far into the atmosphere. On the other hand, the light mesons into which they decay, and which are the ones chiefly observed at sea level, can be absorbed by atomic nuclei only with great difficulty. Thus one could explain the paradox of the strong production but subsequent weak absorption and interaction of mesons.

Before the two-meson theory was many weeks old, C. F. Powell and his collaborators in England sent to the U. S. actual photographs of the heavier mesons, caught at high altitudes in the Bolivian Andes. Since then many experiments, especially with the high-energy accelerators, have confirmed the idea that the heavier mesons interact strongly with protons and neutrons whereas the lighter ones do not. The heavy particles are called *pi* mesons and the lighter variety are called *mu* mesons.

THE MULTIPLICITY OF PARTICLES

We now know that the *pi* meson exists in three forms—positive, negative and neutral. Only two forms of the *mu* meson—positive and negative—have been found, but a neutral *mu* would be very difficult to detect in any case; it would possess many of the same properties as the elusive neutrino. The charged varieties of the *pi* meson have a mass 276 times that of the electron, and the charged *mu* mesons are 210 times as heavy as the electron. The neutral *pi* meson is a little lighter (about 11 electron masses) than the charged *pi;* this contrasts with the case of the nucleons, where the neutron is somewhat heavier than the proton. The spin of the *pi* mesons is zero, while that of the *mu* mesons is ½ *h.*

A very gratifying aspect of the properties of *pi* mesons is that they obey the right kind of statistics. By the statistics of a particle is meant that property which determines the behavior of an ensemble of particles of a specified kind when they interact or are in close physical proximity. If more than one particle in an ensemble of identical particles can occupy the same energy level, the particle is said to obey Bose-Einstein statistics. On the other hand, if no more than one particle in the group can occupy the same energy level, the particle is said to obey Fermi-Dirac statistics. Particles that obey Bose-Einstein statistics are called bosons; those that obey Fermi-Dirac statistics are called fermions. The electron is an illustration of a particle obeying Fermi-Dirac statistics, whereas the photon is an illustration of a particle obeying Bose-Einstein statistics. Pauli has proved a general relation between spin and statistics: namely, that all particles possessing half-integral spin must obey Fermi-Dirac statistics, whereas all particles possessing integral spin must obey Bose-Einstein statistics. The electron's Fermi-Dirac statistics is responsible for the periodic system of elements and for the separation of solids into the three classes of conductors, insulators and semiconductors. On the other hand, the photon, possessing a spin of 1 *h,* obeys Bose-Einstein statistics, and the Planck radiation law is an immediate consequence of this fact.

Now the *pi* mesons all obey Bose-Einstein statistics. This is gratifying because the field quanta on which electrical and gravitational forces are based also obey Bose-Einstein statistics. Here, then, is a very suggestive kinship between the field quanta (*pi* mesons) connected with forces in the nucleus of the atom and the field quanta of the other two types of forces. As a matter of fact, there even seems to be a certain symmetry in nature: the strongest forces, those in the nucleus, are due to field quanta with finite mass and zero spin; the next strongest, the electrical forces, are due to field quanta (photons) with zero mass and one unit of spin, and the weakest, the gravitational forces, are due to field quanta (gravitons) with zero mass and two units of spin. This would seem to represent progress, even though three new particles (the three *pi* mesons) are added to the list of elementary particles.

But we are thrown into confusion by the entrance of the *mu* mesons. They appear to have no connection whatsoever with nuclear forces. What, then, is their function? Even more confusing is the fact that all the mesons, *pi* and *mu,* undergo spontaneous disintegration. When left alone in free space, a positive or negative *pi* meson decays into a positive or negative *mu* meson and a neutrino within about one 250-millionths of a second. Then the positive or negative *mu* meson decays into a positive or negative electron plus two neutrinos within about two-millionths of a second. The neutral *pi* meson also is unstable and decays into two gamma rays in a very short time indeed—about a hundred-millionth of a millionth of a second.

What is the significance of all these instabilities, and why is the neutrino such a conspicuous partner in many of them? We do not know the answer to these questions. We can assume that the rapid decay of the neutral *pi* meson into two photons probably implies the existence of a negative proton, sometimes called an antiproton. This follows from the fact that the neutral *pi* meson should, by virtue of its strong interaction with nucleons, create a "virtual" proton-antiproton pair, which subsequently annihilate each other as a result of their "virtual" interaction with the electromagnetic field and produce two photons. Consequently, when sufficiently powerful ac-

All of the particles known as of December 1951 are shown in this table. The unit of mass is the mass of the electron; the unit of spin is expressed in terms of Planck's constant *h* (*see text*). Characteristic cloud chamber or photographic emulsion tracks are shown in columns at right. Curvature indicates response to magnetic field in a cloud chamber; secondary tracks at ends of primary tracks indicate breakdown of specified particles into other particles. Several particles leave no tracks and in these cases the space is blank. The white rectangles indicate that tracks for these particles have not yet been observed. Additional particles have been proposed tentatively since this table was compiled.

PARTICLE	CHARGE	MASS	SPIN	STATISTICS	TRACK
NEUTRINO	0	0	½	FERMI-DIRAC	
ELECTRON	—	1	½	FERMI-DIRAC	
POSITRON	+	1	½	FERMI-DIRAC	
POSITIVE MU MESON	+	210	½	FERMI-DIRAC	
NEGATIVE MU MESON	—	210	½	FERMI-DIRAC	
KAPPA MESON	+ or —	1000 ?	½ ?	FERMI-DIRAC?	
PROTON	+	1836	½	FERMI-DIRAC	
ANTIPROTON?	—	1836	½	FERMI-DIRAC	
NEUTRON	0	1838.5	½	FERMI-DIRAC	
ANTINEUTRON?	0	1838.5	½	FERMI-DIRAC	
POSITIVE V-PARTICLE	+	2200 ?	?	FERMI-DIRAC?	
NEGATIVE V-PARTICLE	—	2200 ?	?	FERMI-DIRAC?	
NEUTRAL V-PARTICLE	0	2190	?	FERMI-DIRAC	
PHOTON	0	0	1	BOSE-EINSTEIN	
GRAVITON ?	0	0	2	BOSE-EINSTEIN	
POSITIVE PI MESON	+	275	0	BOSE-EINSTEIN	
NEGATIVE PI MESON	—	275	0	BOSE-EINSTEIN	
NEUTRAL PI MESON	0	268	0	BOSE-EINSTEIN	
TAU MESON	+ or —	975	0 ?	BOSE-EINSTEIN	

celerators are constructed (*e.g.*, the new Bevatron being built at the University of California Radiation Laboratory), it should be possible to produce antiprotons. But if antiprotons exist, antineutrons also must exist, and this adds two more elementary particles to our list.

To add to the confusion, there has been increasing evidence in the last year or two for the existence of other types of elementary particles. Of the new particles described by various investigators some are reported to decay in flight while others are supposed to decay in rest; in some cases the decay products appear to be two particles and in other cases three particles. Some of the new particles appear to be absorbed by atomic nuclei and to lead to nuclear explosions, others to be emitted by atomic nuclei.

The best-established of the new particles seem to be the so-called V-particles, first observed in a cloud chamber by G. D. Rochester and C. C. Butler of England in 1948. They exist in charged and neutral varieties. The neutral V-particle appears to decay in the gas of a cloud chamber into a pair of positive and negative particles, and the charged variety seems to decay into a pair of charged and neutral particles in about one 10-billionth of a second. The V-particles appear to be somewhat more massive than a proton or neutron, because in some instances a proton is a decay product. In other instances one of the decay particles has been identified as a *pi* meson. If it is assumed that the decay products of a V-particle are a neutron or proton and a *pi* meson, the V-particle must possess a mass about 2,200 times that of the electron.

Be that as it may, the situation is obscure at the present time. It is not clear how many new particles will have to be added to our list of elementary particles. The known and probable particles, with their properties, are listed in the table. At least 12 have been definitely established. Seven of these (neutron, proton, electron, positron, neutrino and positive and negative *mu* mesons) are fermions. The remaining five (photon, graviton and positive, negative and neutral *pi* mesons) are bosons. Of the nine probable particles, seven (antiproton and antineutron, three V-particles and two *kappa* mesons) are probably fermions, and two (*tau* mesons) are probably bosons. Already, then, we have a total of 21 elementary particles. And if 21 elementary particles really exist, why not many more?

This is a far cry from the simple picture of the physical universe involving only five elementary particles which seemed possible after the discovery of the neutron in 1932. Apart from the plethora of elementary particles, the fact that so many change from one type to another is extremely disconcerting.

It is difficult to believe that nature is really so complex. To explain the large number of particles of different mass, physicists have looked for some unifying principle, such as, for example, mass quantization. It is also natural to try to regard some of the elementary particles as combinations of others. Attempts in this direction were made directly after the discovery of the neutrino; it was suggested that under suitable conditions two neutrinos might combine to give a quantum of light. More recently Enrico Fermi and C. N. Yang of the University of Chicago attempted to show that *pi* mesons can be regarded as combinations of nucleons and antinucleons. On the other hand, Gregor Wentzel of Chicago has developed a theory in which the *pi* mesons are regarded as compounded out of *mu* mesons. Thus far none of these theories has had any spectacular success, and the field is wide open for a deep-going theory which would elucidate the nature and role of the elementary particles of physics.

III. THE STRUCTURE OF THE NUCLEUS

by Maria G. Mayer

No ONE has ever seen, nor probably ever will see, an atom, but that does not deter the physicist from trying to draw a plan of it, with the aid of such clues to its structure as he has. He needs to construct at least a rough hypothetical model of the atom as a starting point for attempting to understand its behavior. For the atom as a whole modern physicists have developed a useful model based on our planetary system: it consists of a central nucleus, corresponding to the sun, and satellite electrons revolving around it, like planets, in certain orbits. This model, although it leaves many questions still unanswered, has been helpful in accounting for much of the observed behavior of the electrons. The nucleus itself, however, is very poorly understood. Even the question of how the particles of the nucleus are held together has not received a satisfactory answer.

Recently several physicists, including the author, have independently suggested a very simple model for the nucleus. It pictures the nucleus as having a shell structure like that of the atom as a whole, with the nuclear protons and neutrons grouped in certain orbits, or shells, like those in which the satellite electrons are bound to the atom. This model is capable of explaining a surprisingly large number of the known facts about the composition of nuclei and the behavior of their particles.

Let us consider first the shell structure of the electrons, from which our nuclear model derives. The modern exploration of the atom began with the experiments of Ernest Rutherford at Cambridge University early in this century. Rutherford's main experiment consisted in shooting high-speed alpha particles, the nuclei of helium atoms, through metal foils. George Gamow has likened this experiment to the strategy of an overworked South American customs inspector who adopted the expedient of shooting revolver bullets into bales of cotton as a quick method of finding out whether they contained contraband arms. (His bullets must have been shot from a revolver of very high muzzle velocity to penetrate the bales, but this small detail hardly destroys the analogy.) If the bale contained only cotton, the bullets would pass through in approximately a straight line. If, however, the bale contained metal arms, some bullets would ricochet off them and be deflected at a wide angle. Rutherford found in his experiment that most of the alpha particles passed through the metal foil with no appreciable deflection in angle, though of course they were slowed down. But a certain small proportion of his projectiles were markedly deflected. From this it could be concluded that the greater part of the metal foil consisted of light cottonlike material, and that the foil also contained some heavy small targets capable of exerting very strong forces on the alpha particles.

It was essentially from this experiment that the modern concept of the planetary atom emerged. The heavy targets that deflected some of Rutherford's projectiles were the nuclei of atoms. Almost all the mass of an atom is concentrated in this small, positively-charged nucleus. The diameter of the nucleus is only about a hundred thousandth that of the whole atom. The remaining space in the atom is almost empty, and in this space the electrons move in their orbits. Most of Rutherford's alpha particles never came close to a nucleus, but they encountered many of the light electrons as they passed through the outer spaces of the atoms and were thereby slowed down.

With the leads furnished by Rutherford's experiments and by the quantum theory, Niels Bohr and other physicists went on to establish the fundamental laws governing the motion of the electrons around the nucleus. In these studies they worked with the simple hydrogen atom, which has only a single positive charge on its nucleus and a single electron. The hydrogen atom's electron may move in any one of a discrete series of elliptical orbits about the nucleus. Under the quantum rule only these specific orbits are stable, or "permitted," for the electron. The most stable of the allowed orbits is that of lowest energy; in this orbit the electron keeps within about one Angstrom unit (1/100,000,000 of a centimeter) of the nucleus. Here its "angular momentum" is zero. Angular momentum is a measure of the amount of rotation of a particle in an orbit; for a circular orbit it is computed as the mass times the velocity times the radius of the orbit. When measured in appropriate units ($h/2\pi$, with h representing Planck's constant), the angular momentum of a permitted Bohr orbit is always a whole number. At the next higher level of energy an electron is allowed any one of four possible orbits. One of these has zero angular momentum; the other three all have an angular momentum of one unit but differ in the direction of the angular momentum vector. At still higher energy levels the electron may have more than four allowed orbits.

Besides revolving around the nucleus, the electron also spins on its own axis, just as the earth has a daily rotation in addition to its motion around the sun. The electron's spin can take one of two directions. This effectively doubles the number of permitted motions, or kinds of orbit, for an electron, since in any given orbit it may spin in one direction or the other. The angular momentum of the electron's own spin is a half-unit.

In an atom heavier than hydrogen, *i.e.*, with more than one electron, the electrons' motions are of course more complicated. Just as the orbit of the earth around the sun is distorted from a perfect ellipse by the gravitational effects of the other planets, so the motion of each electron around the nucleus of a heavy atom is influenced by the presence of the other electrons. To some extent, however, these effects can be averaged and the motion of a single electron can be considered to be governed by the averaged field of force from the nucleus and the other electrons. A further factor that must be noted is the famous Pauli exclusion principle: namely, that two electrons can never exist in exactly the same orbit of the same magnitude and direction of orbital angular momentum and with their spins in the same direction.

We might liken the electronic structure of an atom to that of an apartment house. The ground floor, or orbit of lowest energy, which is the most desirable from the electron's point of view, has a single apartment. Two electrons of opposite spin can occupy it. The second floor contains four apartments, one of which is at a slightly lower level than the others. This one has zero angular momentum, while the other three have one unit of angular momentum. Each of these four apartments likewise can be occupied by two electrons of opposite spin. Thus the apartment house rises, level by level, with a certain number of apartments at each level, each available for occupancy by a pair of electrons. The atomic physicist has designated the levels (or orbits or shells or whatever one pleases to call them) by a system of numbers and letters: the letters represent the energy levels in terms of angular momentum. The letter s stands for zero angular momentum; p for one unit of angular momentum, d for 2, f for 3, and so on. Thus the ground floor, the lowest orbit that can be occupied by the hydrogen atom's electron, with zero angular momentum, is called 1s; the next orbit, that of the depressed second-floor apartment with zero angular momentum, is 2s; the next, occupied by the three second-floor apartments with one unit of angular momentum, is 2p, and similarly up through the higher orbits.

Now it turns out that atoms are happiest and most stable when all the apartments at a given level are fully occupied by electrons; they do not like vacancies. For example, the single electron of the hydrogen atom is lonely and restless in its 1s apartment, which as we have seen has room for two electrons. Hydrogen readily loses its electron, thereby becoming a positive ion, or picks up a roommate for its lone electron, forming a negative ion. Hydrogen, in other words, is a very reactive element; it easily enters into chemical combinations with other atoms. In contrast, the helium atom, with two electrons, is extremely stable. Its two electrons fill the 1s apartment and form a closed system. There is no room for more electrons in this orbit, and the two electrons are held so tightly by the two positive charges in the atom's nucleus that they are not easily separated from it. As a result helium does not react or combine with other elements. The same is true of the four other "noble" gases in the periodic table of elements—neon, argon, krypton and xenon. Each of these has a closed outer shell, fully occupied by electrons, and does not form ions easily or enter into chemical combinations. Indeed, the stability of the five noble gases is so remarkable that their atomic numbers—2, 10, 18, 36 and 54—may be called the "magic numbers" of the periodic table.

This picture of the atom's electronic structure has been enormously useful to chemistry; by means of it chemists have been able to group the elements in families and predict their chemical behavior. But it tells us little or nothing about the structure of the nucleus. The only piece of information an atom's chemical activity gives us about its nucleus is the number of positive charges, or protons, it contains. The chemical nature of the atom is determined by the number of its electrons. We know that if an atom has a certain number of electrons, it must have the same number of protons in its nucleus to hold them. But how the protons themselves are held together, how they interact with one another and with the uncharged neutrons also present in the nucleus—these are the great mysteries of nuclear physics.

The forces that bind the nucleus together are very great—millions of times greater than those that bind electrons to the nucleus. We do not know how these forces are created. And even if we understood them completely, we would still be confronted with the tremendous difficulty of solving a many-body problem, *i.e.*, calculating the results of these forces upon a large number of protons and neutrons that interact with one another strongly and at extremely short range within the nucleus. The problem has seemed so hopeless that nuclear physicists have preferred to treat the nucleus as a "liquid drop," in which the protons and neutrons essentially lose their identity. (This approach has been very fruitful, particularly in studying high-energy phenomena in the nucleus.)

Yet it is possible to discern some rather remarkable patterns in the properties of particular combinations of protons and neutrons, and it is these patterns that suggest our shell model for the nucleus. One of these remarkable coincidences is the fact that the nuclear particles, like electrons, favor certain "magic numbers."

Every nucleus (except hydrogen, which consists of but one proton) is characterized by two numbers: the number of protons and the number of neutrons. The sum of the two is the atomic weight of the nucleus. The number of protons determines the nature of the atom; thus a nucleus with two protons is always helium, one with three protons is lithium, and so on. A given number of protons may, however, be combined with varying numbers of neutrons, forming several isotopes of the same element. Some isotopes are stable; others decay by radioactivity. Some of the stable isotopes readily add a neutron; others are much less inclined to do so.

Now it is a very interesting fact that protons and neutrons favor even-numbered combinations; in other words, both protons and neutrons, like electrons, show a strong tendency to pair. In the entire list of some 1,000 isotopes of the known elements, there are no more than six stable nuclei made up of an odd number of protons and an odd number of neutrons. The other odd-odd nuclei break down radioactively by emitting a negative or positive electron; this change in charge transforms a neutron into a proton or a proton into a neutron and creates a more stable even-even combination of protons and neutrons.

Moreover, certain even-numbered aggregations of protons or neutrons are particularly stable. One of these magic numbers is 2. The helium nucleus, with 2 protons and 2 neutrons, is one of the most stable nuclei known. The next magic number is 8, representing oxygen, whose common isotope has 8 protons and 8 neutrons, and is remarkably stable. The next magic number of 20, that of calcium. Calcium, with 20 protons, has 6 stable isotopes, ranging in neutron number from 20 to 28. This is an unusually large number of stable isotopes for the lower region of the periodic table.

Among these light elements the relative stability can be determined very accurately in terms of binding energy. The net mass of a nucleus is always smaller than the combined masses of the protons and neutrons of which it is composed. The binding energy is calculated from this "mass defect" by means of Einstein's famous relation $E = mc^2$, with m representing the mass defect and c the velocity of light. Such calculations show conclusively that the nuclei with the magic numbers 2, 8 and 20 have much greater binding energies than their neighbors. But for the heavier elements above calcium the binding energies are not accurately determined, and we must judge their relative stability by indirect evidence. One such piece of evidence is the number of stable (*i.e.*, nonradioactive) nuclei that are found to exist with a given number of protons or neutrons. Another is the relative abundance of a given nucleus in the universe, since it seems reasonable to assume that the most abundant isotopes are the most stable.

By these tests the number 50 joins the list of magic numbers. Tin, with 50 protons, has 10 stable isotopes, more than any other element, and it is much more abundant than neighboring elements in the periodic table. The same is true, to a somewhat lesser degree, of the number 28. Another magic number is 126: an isotope with 126 neutrons holds them much more strongly than one with 127 or 128. Perhaps the most remarkable magic number of all is 82. There are 7 stable nuclei containing 82 neutrons, ranging from isotopes of xenon to samarium. The barium isotope with 82 neutrons accounts for 72 per cent of the abundance of that element, and cerium's 82-neutron isotope represents 88 per cent of all the cerium. Finally, 82 protons means lead, and lead is the stable end-product of the decay of all the heavy radioactive elements that may be found in nature.

There are other indications of the special stability of these magic numbers. For instance, nuclei containing 50, 82 or 126 neutrons do not like to add an extra neutron: their absorption cross-sections for fast neutrons are smaller by several factors of 10 than those of an average nucleus of nearly the same weight.

The list of magic numbers, then, is: 2, 8, 20, 28, 50, 82 and 126. Nuclei with these numbers of protons *or* neutrons have unusual stability. It is tempting to assume that these magic numbers represent closed shells in the nucleus, like the electronic shells in the outer part of the atom. To be sure, the electronic shells form a more distinct dividing line than those in the nucleus: the last electron in a closed shell is held to the atom at least three times as strongly as the last electron in an unfilled shell, whereas in nuclei the energy difference is at most 50 per cent. Yet the situations seem sufficiently similar to justify exploring the possibility that the various particles in the nucleus of the atom may be tied together in much the same kind of shell structure as the electrons.

There is another kind of evidence that supports this shell-structure hypothesis. It has to do with the spin of the nucleus. Many nuclei apparently spin about their axes like a top, just as the earth and electrons do. Since the nucleus carries a charge, its rotation corresponds to an electric current, and it behaves like a tiny magnet. The result is known as the magnetic moment of the spinning body. The spins and the magnetic moments of nuclei and of their building blocks, the neutron and proton, can be measured. It turns out that the spin of nuclei, like that of electrons, is "quantized": that is, when measured in the same units ($h/2\pi$) it is always a whole number or half a whole number. The spin of the proton and of the neutron has been determined to be 1/2-unit. Their magnetic moments, measured in units called nuclear magnetons, also have been accurately determined. The spins of nuclei with an odd number of particles are all half of some whole odd number (*e.g.*, 1/2, 3/2, 5/2 and so on), and the spins and magnetic moments of nuclei with an even number of protons and of neutrons are zero.

Now it is a very surprising fact, not expected from the general theory of the nucleus, that two isotopes with the same odd number of protons but different even numbers of neutrons behave very similarly. They have the same spins, nearly the same magnetic moments and frequently the same kinds of excited states. Take, for example, the 2 isotopes indium 113 and indium 115. Each has 49 protons, but one has 2 neutrons more than the other. Yet both have spins of 9/2, and their magnetic moments are very close in value—5.461 and 5.475

nuclear magnetons, respectively. The extra pair of neutrons in the second isotope does not seem to affect these nuclear properties; in other words, the spins and magnetic moments in this case appear to be due only to the protons. On the other hand, a nucleus with 49 neutrons and an even number of protons (e.g., strontium-87) also has a spin of 9/2, though not the same magnetic moment. As far as nuclear spin goes, 49 neutrons behave just like 49 protons. This seems to be the general rule for the lighter isotopes of mass number less than 120.

On the basis of these observations the German physicist T. Schmidt many years ago made the radical suggestion that it is not the nucleus as a whole that spins. Instead, in nuclei with an odd number of particles the properties of spin and magnetic moment are due entirely to the last odd particle, be it a proton or a neutron. The structure of odd nuclei is thus pictured very simply. The nucleus has a spherically symmetrical core of an even number of neutrons and protons. The core has no spin. Around it revolves the last odd particle. Its motion alone determines the spin and magnetic moment of the nucleus.

It is curious to find that, as the evidence of the magic numbers and of the nuclear spins suggests, protons and neutrons behave with considerable independence of each other in the nucleus. Such independence is unexpected from the standpoint of current nuclear theory.

The angular momentum of a particle's orbit alone is always a whole number. But the particle's total angular momentum is the sum of its orbital momentum and its own spin. Hence the total angular momentum of a nucleus always has a half-integer value, since protons and neutrons, just as electrons, spin about their own axes with an angular momentum of 1/2. This spin can be parallel or anti-parallel to the orbital angular momentum; that is, it can be directed so as to add 1/2 or to subtract 1/2 from the orbital angular momentum. Consequently a measured spin can correspond to either of two different orbits.

That such a picture is not too far from the truth is borne out by the measured values of magnetic moments. From this simple model it is possible to compute the theoretical magnetic moment of a given nucleus. The agreement between the predicted and measured magnetic moments is reasonably good.

Thus the spins and magnetic moments lead to a description of the nucleus in terms of orbits of single particles. One can then picture the building of the structure of the nucleus as the gradual filling up of single-particle orbits by neutrons and protons, in the same way as electrons build the atom. The single-particle orbits for one nuclear particle are determined by the average field of the others. The orbits can be described by the letters used to designate the quantized orbital angular momentum of electrons: s = 0, p = 1, d = 2, f = 3, g = 4, h = 5, i = 6. Since neutrons and protons obey the Pauli principle, the s level has room for just two protons and two neutrons, as in the case of electron orbits; the p level has room for six, and so on. In this scheme the magic numbers should correspond to closed shells; that is, they should indicate the boundaries where one level is filled and the next level is appreciably higher.

The order of levels, or the number of particles in a level, depends on the form of the nuclear potential. The simplest guess we can make about the potential is to compare it to a well: no force is acting on a particle outside the well (i.e., outside the spherical nucleus), and none inside. But at the edge of the well (at the surface of a spherical shell) a strong attraction takes place. The change in attraction may be more or less abrupt. A series of very abrupt transitions would be represented by a well with square edges; a more gradual transition by a well with rounded edges.

The structure of levels that would obtain if the well had square edges has been calculated by the Yale University physicist Henry Margenau. The result is shown in the first of the adjoining tables. The table gives the number of neutrons or protons each level can hold and the cumulative numbers that fill all the levels up to a given point. The same type of calculation for wells with rounded edges divides each level into two or more levels of the same or nearly equal energy; in the table these closely adjacent levels are grouped together, with breaks between the gaps.

This scheme explains perfectly the smaller magic numbers, up to 20. The lowest level can contain no more than two particles. The next level, or p-shell with angular momentum 1, has three different orbits, each of which can be occupied by two nuclear particles with opposite spin. These six particles in the p-shell, plus the two in the ground state, correspond to the stable form of oxygen with eight neutrons and eight protons. The next two levels, 1d and 2s, which lie

ORBITAL ANGULAR MOMENTUM	SHELL	NUMBER OF NUCLEONS IN A LEVEL	NUMBER OF NUCLEONS UP TO A LEVEL
0	1s	2	
			2
1	1p	6	
			8
2 0	1d 2s	10 2	
			20
3 1	1f 2p	14 6	
			40
4 2 0	1g 2d 3s	18 10 2	
			70
5 3 1	1h 2f 3p	22 14 6	
			112
6 4 2 0	1i 2g 3d 4s	26 18 10 2	
			168

Margenau chart of nuclear energy levels supports the shell-structure hypothesis in part. The first three numbers in column at right match the magic numbers. The next four do not; at this point the scheme breaks down.

close together, have room for 12 more nuclear particles, and these bring us to the magic number 20.

From here on the scheme seems to break down. In this table there is no sign of the magic numbers 28, 50, 82 and 126. It is possible, however, to arrive at them by taking into account an effect which is very small and unimportant for electrons but apparently not for nuclear particles.

A nuclear particle can align its own 1/2-unit spin parallel or antiparallel with the angular momentum of its orbit. Thus its total angular momentum can be the orbital angular momentum plus 1/2 or minus 1/2. This splits its orbit into two possible levels. There is a considerable difference of energy between these two levels, brought about by what is called "spin-orbit coupling." The strength of the coupling increases with increasing angular momentum, and one would expect it to decrease with increasing size of the nucleus. Consequently the greatest split in such a pair of levels should occur at the beginning of a group of closely adjacent shells, where rela-

ORBITAL ANGULAR MOMENTUM	SHELL	TOTAL ANGULAR MOMENTUM	NUMBER OF NUCLEONS IN A LEVEL	NUMBER OF NUCLEONS UP TO A LEVEL
0	1s	1/2	2	
				2
1	1p	3/2	4	
		1/2	2	
				8
2	1d	5/2	6	
0	2s	3/2	4	
		1/2	2	
				20
		7/2	8	
		↕		28
3	1f	5/2	6	
1	2p	3/2	4	
		1/2	2	
		9/2	10	
		↕		50
4	1g	7/2	8	
2	2d	5/2	6	
0	3s	3/2	4	
		1/2	2	
		11/2	12	
		↕		82
5	1h	9/2	10	
3	2f	7/2	8	
1	3p	5/2	6	
		3/2	4	
		1/2	2	
		13/2	14	
		↕		126
6	1i	11/2	12	

Mayer chart accounts for the discrepancies by introducing the spins of nuclei (*fractions in the center column*). "Spin-orbit coupling" splits the close-lying levels apart and creates energy gaps where the magic numbers occur.

tively high angular momentum is combined with a relatively small nucleus. Suppose that a wide split of this kind occurs at the 1g level. Then it is no longer correct to treat the 1g level as a single level with room for 18 particles. Instead, it will divide into two levels with room for 10 particles in its lower level. This lower level will be depressed toward the next lower group, consisting of the 1f and 2p shells, as listed in the Margenau table. The 10 particles, added to the 40 up to that point, make the magic number 50. Above this there is an energy gap, created by the spin-orbit coupling.

The second table shows how the various levels are revised when allowance is made for spin-orbit coupling. Now the larger magic

126

numbers come into the scheme: they all occur at the places where spin-orbit coupling has its greatest effect.

This level scheme is in excellent agreement with the observed spins and magnetic moments of odd-numbered nuclei. Among the approximately 90 spins and 73 magnetic moments that have been measured so far there are only four serious disagreements with the theory.

The shell model can explain other features of nuclear behavior, including the phenomenon known as isomerism, which is the existence of long-lived excited states in nuclei. Perhaps the most important application of the model is in the study of beta-decay, *i.e.*, emission of an electron by a nucleus. The lifetime of a nucleus that is capable of emitting an electron depends on the change of spin it must undergo to release the electron. Present theories of beta-decay are not in a very satisfactory state, and it is not easy to check on these theories because only in a few cases are the states of radioactive nuclei known. The shell model can help in this situation, for it is capable of predicting spins in cases in which they have not been measured.

Certainly the simple model described here falls short of giving a complete and exact description of the structure of the nucleus. Nonetheless, the success of the model in describing so many features of nuclei indicates that it is not a bad approximation of the truth.

IV. RADIO WAVES AND MATTER

by Harry M. Davis

I hear beyond the range of sound,
I see beyond the range of sight,
New earths and skies and seas around,
And in my day the sun doth pale his light.
— Henry David Thoreau,
Inspiration

RADAR, the warborn invention that sees in the dark, that pierces clouds and fog and detects tiny targets beyond the range of human sight and hearing, has inspired in many scientists just such a sense of unfolding discovery and of elation as Thoreau expressed in these prophetic lines. Radar is best known as a military weapon and a marvelous aid to navigation, but more fundamentally it has given man a new tool, almost a new sense, with which to explore the universe. With radar he has already reached out to the moon, and with radar's microwaves physicists are now exploring the depths of the atom. Through microwaves physics has entered a new region of the electromagnetic spectrum that illuminates the innermost structure of matter more brilliantly than light itself.

The discovery of the wonderful properties of microwaves reversed the usual order of development in science; it derived from technology, but more and more its fruits are being found in the realm of pure knowledge.

The wartime race among the radar researchers of the opposing powers was in large measure a race toward the shorter and shorter wavelengths—from short waves to microwaves. Each time the frequency frontier was moved to a shorter wave band—first to 10 centimeters, then to 3 centimeters—the definition and accuracy of the image improved. Toward the end of the war, Allied radar researchers reached into a new realm below 3 centimeters, known in the secret listing as the K-band: this should have given better radar than ever. But when the first sets got into the field, mysterious troubles developed. On some days the reception was remarkably good. On other days, the radars seemed to go half blind. Soon it was observed that the K-band radars performed at their worst when the air was damp. A muggy day could play havoc with transmission.

123

The military problem of finding a solution for this difficulty was assigned to physicists at the Radiation Laboratory of Columbia University, which was responsible for developing K-band tubes. The Columbia physicists recalled a prediction by John H. Van Vleck of Harvard that at certain wavelengths microwaves would be absorbed by water vapor. The experimenters found that this was the case: on a humid day in the tropics a radar signal of the predicted wavelength, if beamed at a target only three miles away, would lose 94 per cent of its energy by water absorption alone. This effect was above and beyond the normal absorption due to droplets of rain.

The explanation of this phenomenon lies in the field of quantum theory. According to this theory, each kind of atom or molecule possesses a number of possible states of energy. It occupies only one state at a time, and each state is separated from the one next above or below by a certain difference of energy, called a quantum. To go to the next higher energy state, the atom or molecule must absorb a quantum; to go to the next lower one it must shed a quantum. A molecule will accept only the quantum of energy that can raise it to a higher energy state; by the same token, a wave of energy encountering a molecule must carry the right quantum or it will produce no effect.

Max Planck's and Albert Einstein's famous statement of the quantum theory says that each quantum of energy carried by any given radiation (such as a specific wavelength of yellow light) is directly proportional to the radiation's frequency, *i.e.*, the number of waves per second. Consequently, whether a molecule will absorb a radiation depends upon the radiation's frequency or wavelength. A wave that carries the wrong quanta will sail right by; the molecule is transparent to it. But when the right wavelength comes along, the molecule grabs it and is hoisted to a higher energy level. The molecule is said to be resonant to this wavelength, and the phenomenon of energy-capture is known as resonance absorption. The quantum thus absorbed is subtracted from the energy of the radiation. If enough resonant molecules get in the way of a beam, whether of light or radio waves, its energy may be completely absorbed and the beam is blocked.

In the case of the K-band radar, as it happened, the middle of the wave band accidentally fell squarely on a strong absorption wavelength for water—1.33 centimeters. So the invisible molecules of water vapor in the air effectively absorbed and blocked the radar signals. The remedy was simple: a shift of the radar wavelength away from the water molecules' resonance point.

The Columbia study that developed from this military problem underlined a possibility of which a few physicists had become aware: that the powerful new microwave equipment developed by radar could be used to tune in on the inner secrets of the molecule, the atom and even the ultimately small particles of the atomic nucleus. Almost everything previously known about the structure of atoms had come from study of their visible spectra, that is, from splitting the light emitted by heated atoms into its component lines or wavelengths on the color spectrum. By means of microwaves units of matter can be analyzed in a somewhat similar but frequently more accurate way, and many fundamental new facts have already been learned about the shape, size and arrangement of molecules, the orbits of electrons around atomic nuclei and the spin of protons and neutrons within the nucleus.

At the same time, the discovery of the properties of microwaves has dramatized the unity of the entire electromagnetic spectrum, for these very short radio waves have been found to perform in many ways much like light waves.

Man first observed the spectrum where it was most evident: in the rainbow of visible colors from red to violet. But from the viewpoint of modern physics, visible light is only one octave in the vast keyboard of electromagnetic radiation. It is now known that radiation exists in nature over a great range of frequencies, and that the frequency of oscillation (or, looked at another way, the wavelength) is the only physical difference among gamma rays, X-rays, ultraviolet rays, visible light, infrared or heat rays and radio waves. In each case the frequency multiplied by the wavelength is equal to c, the velocity of light: 30 billion centimeters per second.

The shortest waves of the visible spectrum, those of violet light, are approximately .00004 centimeters in wavelength. The longest visible

Electromagnetic spectrum charts wavelength and frequency relationships of all forms of radiant energy. Long radio waves are at bottom of chart; extremely short gamma rays are at top; entire visible spectrum is encompassed in short segment, less than an octave in width, above center. Radio spectroscopy utilizes the microwave region, where radio waves approach optical frequencies and behave in manner comparable to light waves.

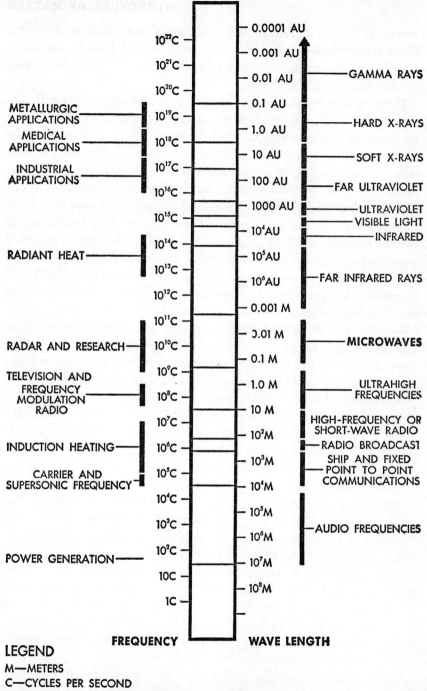

FREQUENCY	WAVE LENGTH	
10^{22}C	0.0001 AU	
10^{21}C	0.001 AU	
10^{20}C	0.01 AU	GAMMA RAYS
10^{19}C	0.1 AU	
	1.0 AU	HARD X-RAYS
10^{18}C	10 AU	SOFT X-RAYS
10^{17}C	100 AU	FAR ULTRAVIOLET
10^{16}C	1000 AU	ULTRAVIOLET
10^{15}C	10^4AU	VISIBLE LIGHT
10^{14}C	10^5AU	INFRARED
10^{13}C	10^6AU	FAR INFRARED RAYS
10^{12}C	0.001 M	
10^{11}C	0.01 M	
10^{10}C	0.1 M	MICROWAVES
10^9C	1.0 M	ULTRAHIGH FREQUENCIES
10^8C	10 M	HIGH-FREQUENCY OR SHORT-WAVE RADIO
10^7C	10^2M	RADIO BROADCAST
10^6C	10^3M	SHIP AND FIXED POINT TO POINT COMMUNICATIONS
10^5C	10^4M	
10^4C	10^5M	AUDIO FREQUENCIES
10^3C	10^6M	
10^2C	10^7M	
10C	10^8M	
1C		

METALLURGIC APPLICATIONS — 10^{19}C

MEDICAL APPLICATIONS — 10^{18}C

INDUSTRIAL APPLICATIONS — 10^{17}C

RADIANT HEAT — 10^{13}C

RADAR AND RESEARCH — 10^{10}C

TELEVISION AND FREQUENCY MODULATION RADIO — 10^8C

INDUCTION HEATING — 10^6C

CARRIER AND SUPERSONIC FREQUENCY — 10^5C

POWER GENERATION — 10^2C

FREQUENCY   WAVE LENGTH

LEGEND
M—METERS
C—CYCLES PER SECOND
AU—ANGSTROM UNITS (1 ANGSTROM = 10^{-8}CM)

waves, at the red end of the spectrum, are about .00008 centimeters. From them we must take a leap by a ratio of a billion to get to the radio wavelengths in the regular broadcast band, which are measured in hundreds of meters.

Even before the invention of radio, James Clerk Maxwell, the gifted 19th-century Scottish mathematical physicist, had shown that its electromagnetic waves were of the same nature as light, but the kinship long seemed more academic than practical. Recent radio and radar developments, however, have whittled down the gap. Most modern radar equipment operates on wavelengths measured in centimeters rather than meters, and some laboratory oscillators have been tuned down into the millimeter region. At these lengths the waves behave as much like light as like ordinary radio. They are transmitted and received not by the conventional wire radio antenna but by a parabolic reflector resembling a searchlight, or by a metallic "lens." Except under very special atmospheric conditions, the ultra-short waves refuse to bend around the curved surface of the earth, as do long radio waves, but proceed in straight lines like a beam of light (a factor which poses a serious engineering problem for short-wave television). Within a microwave set, the waves will not cling to wires but must be channeled inside hollow waveguides resembling square pipes, which physicists call "plumbing."

It is not surprising, therefore, to find that some material substances react with microwaves as other substances react with the visible portion of the spectrum. Just as grass and leaves absorb red light and reject (and thereby reflect) green, one might expect the gases of the air to respond in different ways to the various invisible "colors" of the microwave radio spectrum. Every substance has its favorite energy levels, which are in turn resonant to certain wavelengths. Thus while air is transparent to both the long waves of ordinary radio and the very short waves of visible light, it does not necessarily follow that air will be equally indifferent to waves in the microwave band between light and ordinary radio. As a matter of fact, it is not in general transparent to them. Molecules of oxygen, like those of water vapor, exhibit definite absorption in the microwave region. And many other gases, not found in the air but available in the laboratory, show the same property.

Clerk Maxwell, besides laying the groundwork for the electromagnetic theory of radio and light, also did important work in what was then the new kinetic theory of gases—a theory which correctly supposed that the behavior of a gas in such matters as temperature and pressure was the statistical result of the random movements of enormous numbers of individual molecules. He also introduced the concept that the laws of gases, based on anarchy and statistics, might be evaded if some tiny demon were available that could open and close gates to sort out the faster and slower molecules. "Maxwell's demon" has been a useful fancy in the teaching of thermodynamics and allied branches of physics.

No such demon exists. Yet microwaves offer something akin to it, since they virtually take control of individual molecules and regiment them as no other force can. They can be used to control not only such gross, statistical qualities as pressure and temperature but many subtle properties of gas molecules as well. They may affect the molecules' rotation and orientation, and may even stretch the links between the atoms that constitute the molecule. In at least one case, ammonia, radio waves can literally turn the molecule inside out.

Ammonia, NH_3, is one of the molecules that has been most carefully studied by microwave resonances. Even before radar, in 1934, C. E. Cleeton and N. H. Williams at the University of Michigan employed an early low-power form of centimeter-wavelength magnetron and sent its radiation through a rubber balloon filled with ammonia gas. They found that the balloon greedily absorbed radio waves at 1.25 centimeters, or at a frequency of 24,000 megacycles (24 billion cycles) per second. More recently, with the powerful new radar K-band oscillators and accurate receivers, ammonia's behavior has been observed as minutely as that of a child in the famous Yale clinic of Arnold Gesell.

From the weights of chemicals that combine to make it, chemists have long known that the formula of ammonia is three atoms of hydrogen and one of nitrogen. But what about their arrangement? From various kinds of evidence, it is understood that the three hydrogen atoms form a triangle, with the nitrogen atom in another plane like the apex of a pyramid. If you think of the hydrogen triangle as being in the plane of this page, then the nitrogen atom could be either above or below. The microwave research now shows that it can occupy either position alternately, and the resonant wavelength of 1.25 centimeters is just what is required to keep the N atom

133

oscillating. The physicists describe the effect as "tunneling" and speak of the 1.25 centimeter resonance as due to the "inside out" or "inversion" spectrum of ammonia gas.

There is more to it, of course. Besides turning itself inside out, the ammonia molecule is also vibrating and rotating. In addition, since no molecule can be isolated by itself, the study of ammonia gas involves collisions among the molecules, which transfer energy back and forth. All this can be analyzed by its effects on the inversion resonance.

At atmospheric pressure, when collisions among the molecules are frequent, the resonance band is broad, extending from about 1.1 to 1.5 centimeters, and higher pressure can broaden it still more. Radio men would say that, as a receiver, a container of high-pressure ammonia has poor selectivity. But as the pressure is lowered, and the molecules begin to act more as free individuals, the absorption line sharpens up. It also exhibits a "fine structure"; the absorption line is made up of some 30 distinguishable constituent lines. These have been identified with the various states of rotational energy that the NH_3 molecule can occupy. As it spins faster, its pyramid of four atoms is strained and stretched by centrifugal force, and the amount of radio energy needed to turn it inside out is altered. In a container of gas, the billions upon billions of molecules will include large numbers at each possible rotational speed level, each absorbing energy on its preferred wavelength. Thus when a K-band radar tube has its tuning swept automatically through this region of frequencies, the receiver will display a cluster of adjacent absorption lines.

This kind of test is being applied to all sorts of chemical compounds. Specimens of gases and liquids in tubes and pipes are exposed to super high-frequency microwaves to find their characteristic resonances. By this method more and more is being learned about the qualities of the molecules' structure—the distance between the atoms, the way they are arranged and the electric and magnetic forces among them. Moreover, since each molecule has its own individual pattern of resonant wavelengths, as personal as a fingerprint, microwave tuning offers a new method of identifying chemicals.

The most intensively studied of all atoms has been that of hydrogen. Its optical spectrum has been explored for decades. The reason is that hydrogen is the simplest atom: one electron rotates around a nucleus consisting of a single proton. Science could not hope to make

134

any sense out of the spectra of heavier and more complicated atoms until it could unravel the meaning of the pattern of bright lines of various colors emitted by "excited" atoms of hydrogen. Here were found the celebrated Balmer and Lyman series of spectral lines, in which the sets of frequencies in the various energy states of the atom were found to be related to one another by a surprisingly simple mathematical formula.

Niels Bohr's historic contribution to understanding of the atom was that he succeeded in applying the quantum concept to the spectrum of hydrogen, showing how its succession of visible wavelengths corresponded to parcels of energy emitted as hydrogen's lone electron dropped from one orbit to another. From the springboard of that theory scientists went on to attack the spectra of the larger atoms and their combinations into molecules. Thus a huge edifice of both physics and chemistry has been built upon the understanding of hydrogen's visible color spectrum.

There are limits, however, to the accuracy of measurement of the visible spectrum. Despite great technical improvements in optical spectroscopy, some of the lines are so close together that it is impossible to resolve them—that is, to learn whether one or several pairs of energy levels are involved in some phase of the movements of the hydrogen electron. Willis Lamb of Columbia University decided to tackle this ambiguity by means of the invisible "light" of microwaves. His approach was eminently reasonable. If you want to measure the amount of a liquid accurately, it is logical to use a small unit—a pipette, say, rather than a gallon jug. Similarly, a quantum of red light carries more than 50,000 times as much energy as an X-band microwave. Thus microwaves, representing smaller energy units, should permit finer measurements.

In an ingeniously designed experiment, Lamb and Robert C. Retherford fired hydrogen atoms between the poles of a powerful magnet towards a distant detector. The apparatus was so set up that only the atoms which had their electrons in a certain orbit would register on the detector. A microwave oscillator was then arranged to shoot its parcels of energy across the hydrogen beam. When the microwaves carried the right quantum of energy, they toppled the electron orbit from one level to another and the hydrogen atoms so altered were not recorded on the target. By tuning the microwave oscillator and watching the detector, the experimenters could observe just what quantum of energy was required to change the electron's orbit. The answer, in this experiment, came out at a wavelength of 3.03 centimeters instead of the 2.74 centimeters expected from the celebrated Dirac theory. So refined is the modern theory of the atom that this discrepancy was enough to throw the whole world of physics into its most intensive error-hunt of a decade.

In one sense, all this study of molecules and whole atoms is merely investigating the fringes of the problem. In atomic physics it is the power-packed nucleus that now holds the center of the stage. And perhaps the most exciting explorations made by the microwave apparatus of radar have been those in the nucleus.

As a matter of fact, all the work done on molecules and atoms also involves nuclear effects, even if they are indirect. The nuclear effects can be segregated from those which belong to the outer structure of the atom by comparing the behavior of different isotopes of the same atom. Certain results of this work indicate that some nuclei are spherical, and others are shaped like a cigar or a pancake.

What is observed in these experiments with isotopes is the effect of nuclear mass and shape on the atom or the molecule. Once these effects are fully explored, microwave spectroscopy may serve as a quick means of measuring how much of what kind of isotope is present in a naturally or artificially mixed sample of an element. Hitherto the easily traced radioactive isotopes have been favored by biologists and chemists in their "tagged atom" research. The possibility of identifying isotopes by their microwave spectra should kindle new interest in the stable isotopes, which have advantages from the viewpoint of safety for the experimenter.

Beyond these somewhat indirect dealings with the nucleus, within the past two years an almost incredibly refined procedure has been developed for by-passing the electronic shells of atoms and resonating directly with the nuclei, making them literally dance in rhythm with a radio wave. This method depends on the fact, first surmised in 1924 on the basis of some peculiarities in optical spectra, that most (but not all) nuclei have a characteristic spin—a spin which is separate from that of the molecule or atom as a whole. In a way, this provides a further analogy between the atom and the solar system. Electrons revolve around the nucleus as the earth and other planets revolve around the sun. But the sun itself also spins on its axis. Just so, the nucleus spins on its axis. Again, the spinning earth and sun

136

are magnets, with north and south poles. So, too, is the spinning nucleus.

In the world of the atom, these spins cause an interaction between the magnetic fields of the satellite electrons and the field of the central nucleus. These interactions show up in the hyperfine structure of the lines in the visible spectrum. But the distance between these hyperfine visual lines is so small that it is difficult to measure with any accuracy. And so here is another opportunity to employ the more delicate measurements of radio waves.

If the magnetized nucleus reacts to the magnetic field of moving electrons, why should it not react also to the forces of a powerful external magnet? Two groups of experimenters—at Harvard and at Stanford—saw this as the basis for a new kind of resonance experiment. The forces applied to the substance under study were two: 1) a steady magnetic field, which could be varied in strength; 2) an oscillating magnetic field or radio wave, which could be varied in frequency. The idea was that the steady magnet would cause the tiny nuclear magnets to line up with the applied magnetic forces, like minuscule compass needles. Then the oscillating force of the radio wave would give some of the nuclei the impetus to flip over in the opposite direction, like compass needles pointing the wrong way. The frequencies which accomplish this are in the realm of short-wave radio.

As the experiment was set up at Harvard by E. M. Purcell, H. C. Torrey, and R. C. Pound, the innate magnetism of the proton was studied by using such hydrogen-rich substances as a drop of water or a blob of paraffin. The sample, in a small glass tube, was surrounded by a coil of wire carrying the radio-frequency current, which in turn was placed between the poles of a powerful magnet. As the magnet strength or the radio wavelength was changed, a sharply defined point was found at which the coil of wire showed an increase of electrical resistance. The conclusion was that at this particular combination of magnetic field and wavelength the test substance was able to absorb energy. That this was not a molecular phenomenon was made clear by the fact that resonance took place at the same wavelength regardless of whether the material tested was liquid water, ice, paraffin, or some other substance—so long as it contained hydrogen.

An interesting variation was independently worked out at Stanford by Felix Bloch, William W. Hansen and Martin Packard. Bloch, who has conducted most of this research, had served during the war at both the Los Alamos atomic-bomb laboratory and at the Radio Research Laboratory at Harvard, home of counter-measures against enemy radar.

He worked on the theory that a spinning proton, when its essential magnetism is subjected to both an outside magnetic field and a radio wave, will precess (wobble) like a top, so that its axis at any moment is not exactly in line with the magnetic field created by the outside magnet but is at an angle to it. This, he reasoned, should cause the spinning particle to give out radio waves at right angles to those it received from outside. Exactly that happened. When a radio wave was sent into a proton-rich material from one direction, an answering signal came out in another direction and was picked up on a radio receiver tuned, in one instance, to 42.5 megacycles. As the receiver, not unlike those used to intercept enemy radar signals, was swept through the frequency band, it showed a sharply tuned resonance peak on a radarlike oscilloscope.

By a different although somewhat related method, a research group at Stanford has measured a magnetic moment for the fundamental uncharged particle, the neutron. The unexpected magnetic property of the neutron is one of the important recent clues to the mysterious inner nature of all matter. It is easy to see why a spinning electrified particle should have magnetism, for the principle is similar to that of any electromagnet or electric motor. But the existence of magnetism in the neutron has posed a new problem. It indicates that this particle, which was named neutron because it was supposed to be electrically neutral, actually belies its name and, like a miniature atom in itself, must have some complex distribution of positive and negative charge.

Microwaves may have another important scientific use quite aside from those in the laboratory. In all the experiments described in this article the quantity measured is frequency—the number of electrical oscillations per second. Stated in another way, the quantity measured is time—the duration of one oscillation. The measurement is extremely accurate, already approaching one part in a million. And it can readily be compared with conventional timepieces; ultimately the oscillations of microwaves can be tied in with an electric clock running on 60-cycle alternating current.

These clocks at present are regulated by astronomical observation of the earth's daily rotation and its movement around the sun. A clock regulated by microwaves would have as its ultimate standard the rotation of molecules or atomic nuclei. The question arises: which is a more reliable timekeeper, the spin of the earth or the spin of an atomic particle? The evidence leans toward the fundamental and universal constants of the atom. The spin of the earth changes; the moon, by gravity and by the friction of the tides, is gradually slowing it down.

Already a number of laboratories are working on methods of measuring time by radio. At the moment, a tube of rarefied ammonia gas is favored as an experimental standard because of the excellent sharpness of its radio spectral lines. Just as the green line in the visual spectrum of mercury 198 may become the ultimate standard of length, a line in the radio spectrum may become the ultimate standard of time.

Thus, by a combination of optical spectroscopy with the new radio spectroscopy, science is on its way to obtaining not only a new insight into the structure of all matter but an immutable reference for its measurements of space, time and the dimensions of the universe.

V. THE SOLID STATE

by Gregory H. Wannier

EVERYONE DEVELOPS in the course of his life a general notion of what a solid body is: that which supports when sat on, which hurts when kicked, which kills when shot. We have also known for a long time certain laws of the behavior of solid bodies: the laws of free fall, of refraction, of elasticity and so on. Yet none of those laws really describes the nature of the solid state. Long after the behavior of gases had been formulated and explained in terms of molecular action, the essential character of the solid state remained a secret. No suggestion of an answer was forthcoming to the question how the same molecules could behave so differently in a solid and in a gas. In fact, the question was not even clearly asked for a long time, and an unasked question is especially difficult to answer.

The physics of the solid state is a new science, developed only within the present century. It is often the unusual and startling aspects of a subject that wake it from its slumber. Two such aspects can be singled out as having opened up the physics of the solid state at the turn of the century: the structure of crystals and the conduction of electricity by metals.

Crystals have never suffered from the lack of glamour which so long delayed progress in the physics of the solid state. The travelogues of the ancients are liberally sprinkled with mentions of wondrous gems. The classification of crystals was one of the chief occupations of Arab scientists. To these early observers, crystals seemed to be exceptional forms of solids. Nineteenth century research disclosed, however, that this view was not correct. It is only the *well developed* crystals that are exceptional. A very large number of solids which do not appear to be crystalline at first sight are seen to be composed of tiny crystals when examined under the microscope. The list includes all rocks and all metals; only a small group of recalcitrant solids such as glass fail to show any trace of crystallinity.

Crystals are homogeneous solids bounded by plane faces. Many of them are strikingly symmetrical in shape; for example, the little cubes formed by rock salt and the hexagonal prisms of quartz, shown in the stereoscopic drawings. Even in some of those that are not fully symmetrical, the asymmetry is often due simply to the fact that cer-

140

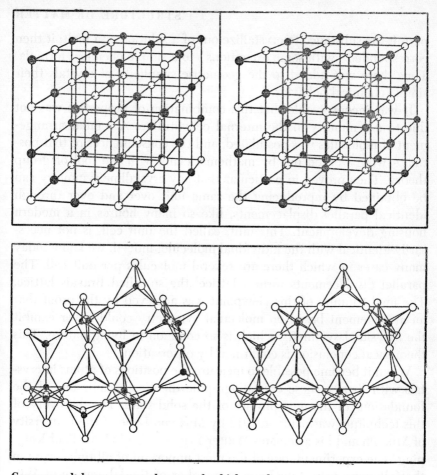

Common salt has a rectangular crystal, which can be seen even in tiny crystals of table salt. This reflects cubical arrangement of its atomic lattice, in which bonds (black lines) hold constituent sodium (black) and chlorine (white) atoms at right angles to one another. Quartz, in contrast, has a characteristic hexagonal crystal, determined by the angles of the bonds between its constituent oxygen (white) and silicon (black) atoms. Images in these stereoscopic drawings may be fused, after some practice, by viewing at reading distance with eyes adjusted as if to fuse their images of a more distant object, or with card held edge-on between eyes to present only one image to each eye.

tain planes have been shifted parallel to themselves. If the angles between the planes are taken as the basic indication of structure, the same structure always goes with the same chemical species. For this reason crystals acquired great value to the chemist as a means of identifying and preparing his chemical compounds. He had only to permit the substance to crystallize out of a solution or a melt; it then corresponded (with some exceptions) to a simple chemical formula. Thus there was added to the geometric simplicity of crystals their simple chemical behavior.

It was inconceivable that this combination of properties was not due to a symmetry in the internal constitution. A regular arrangement of molecules was postulated, and the consequences of that postulate were worked out by mathematicians with the tool of group theory. They proved in particular that any crystalline symmetry can be obtained by reproducing the same unit over and over through identical parallel displacements, like so many houses in a modern housing development. This unit, called the unit cell, is not necessarily identical with the individual molecule; indeed, we know today many cases in which there are several molecules per unit cell. The parallel displacements form a lattice, the so-called Bravais lattice. The final triumph of this viewpoint was achieved by the proof that an arrangement based on molecular symmetry could never exhibit the five-sided symmetry which is so common in living nature. This theoretical conclusion is confirmed by observation.

When it became possible to measure the positions of atoms in crystals with X-rays, these crystallographic theories became one of the foundations of the new physics of the solid state. The discovery of this technique was made in 1912 by Max von Laue of the University of Munich and his associates Walter Friedrich and C. M. Paul Knipping. The experiment, one of the most important of all time, arose in the following way. Von Laue, a member of the laboratory of Wilhelm Röntgen, the discoverer of X-rays, was inclined to consider these rays as electromagnetic waves; that is, of the same nature as radio waves or visible light, but of much shorter wavelength. To prove that they were waves he had to construct or find gratings much finer than any grating ever made by man. At this point he happened to become acquainted with studies then being made of crystals, and he perceived that here was a grating made by nature which should serve his purpose. At his suggestion Friedrich and Knipping directed a beam of X-rays at a crystal, and looked for reflections from the regular crystal planes which existed in the interior. After some unsuccessful trials these reflections were found.

Within a few years the technique of diffracting X-rays by means of crystals was taken up by others, notably W. H. and W. L. Bragg, father and son, in England. Today it has become a standard tool for analyzing crystal structure, while its original purpose—determination of the wavelength of X-rays—has receded into the background. The simplest example, and the first to be analyzed with this tool, is the sodium chloride crystal, NaCl. The partners Na and Cl occupy the corners of a cubic lattice; they alternate in this position, like the black and white squares of a checkerboard. Each Na is surrounded impartially by six Cl's at equal distance, and *vice versa.*

Thus crystallography provides a geometric analysis of the solid state which is unusual in its beauty and perfection. But it is not yet physics. Johann Kepler's laws of planetary motion, which had a similar beauty, were not physics but astronomy; Newton transformed them into physics by finding the law of force to which the planets were subject. In the same way physicists asked what forces made the atoms in crystals arrange themselves as they did, and what dynamic phenomena took place in crystals. They learned that the forces responsible for the formation of atoms, molecules and crystals are electrical, which placed solids and molecules on a similar footing.

It is therefore appropriate to examine the chemical binding forces in molecules as a preliminary to the more complicated case of solids. The basic unit of these forces is the negatively charged electron. Even before the discovery of the electron by the Cambridge physicist J. J. Thomson in 1895 the existence of units of electrical charge had been detected in solutions. When a salt such as sodium chloride is dissolved in water, the molecule dissociates into sodium atoms, charged positive, and chlorine atoms, charged negative. These charged carriers, called ions, are identified by the passage of an electric current, in which a transport of charge is linked in a fixed way with deposition of matter—atoms of sodium on one electrode and atoms of chlorine on the other. Thomson's discovery that the charge of the electron was identical with the known unit of charge of ions suggested immediately that electrons were constituents of the atom and that ions were characterized by an excess or a defect of such particles.

At first the electrons in the atom were thought of as negative raisins in a positively-charged cake; then it was shown that the positive charge, far from being a cake, is concentrated in an extremely small space within the atom where it forms the atomic nucleus, about which the electrons cruise very much like the planets around the sun. Next it was seen that the charge of this nucleus determines the chemical element. Thus the atom of copper consists of a nucleus bearing 29 positive charges surrounded by 29 negative electrons; from it an ion with two extra units of positive charge (Cu^{++}) is derived by the surrender of two electrons. The manner in which these electrons are bound to the nucleus was elucidated only after considerable difficulty. It was learned that the electrons were restricted to certain definite orbits, representing energy states, around the nucleus. A modification of mechanics, called quantum or wave mechanics, had to be evolved to pick out the possible orbits; between these possible states the electrons can make transitions, emitting or absorbing the energy difference in the form of light.

The picture was reasonably well completed by the discovery in 1927 that the electron spins on its own axis, much as the earth does, and that only two states of rotation are possible. With this discovery, it became possible to formulate the so-called Pauli exclusion principle, which states that two electrons cannot be in the same quantum state including spin. The Pauli principle gives the electrons around the atom a shell structure, and so explains valency. Thus the lightest atom, hydrogen, which consists of a nucleus bearing one positive charge and an electron, has the valency 1, corresponding to the fact that it can surrender its electron to an atom desiring one, as it does in hydrochloric acid (HCl). The next atom in order of weight, helium, has two charges and two electrons. This arrangement is so stable that the electrons do not usually leave the atom and helium does not form chemical compounds. The next atom, lithium, has three electrons, only two of which can enter the stable helium shell, while the third, called the valence electron, is easily detached. Consequently lithium has valence 1 and forms salts such as lithium chloride, in which lithium bears a positive charge; a solution of this salt in water will conduct electric current, with lithium coming out at the negative pole. Beryllium, the next element, has two valencies, and so on up the scale of elements. The chemically inert gas argon, with 10 electrons, fills the second shell. The element immediately preceding argon, fluorine, lacks one electron in the second shell and is an extremely reactive substance with one valence. The valence is of the electronegative type; that is, fluorine tends to grab an electron to complete its second shell. In consequence a salt such

144

as lithium fluoride, in which the lithium atom passes one electron to the fluorine atom, is very stable, because both atoms are now surrounded by complete electron shells. After argon comes sodium, with 11 electrons, one of which is detached easily; this means that it will act very much like lithium. Thus the elements fall into groups of similar behavior according to the structure of the outermost shell. The group of atoms having just one electron outside a closed shell is called the alkali metals. This group occupies a key position in the theory of the solid state, as will be seen.

Now let us return to the structure of crystals. The simple lattice of the sodium chloride crystal is composed of ions. Ionic crystals had revealed something of their electrical character as early as the 18th century. Jewelers, testing the durability of the colorful gem tourmaline by putting it in the fire, noticed that foreign particles collected on it, and they named it "the stone which attracts ashes." Careful study showed that the gem became electrified when it was heated, always in the same crystallographic direction, positive charges appearing on one crystal face and negative charges on its opposite. More than a century later French scientists followed up this discovery of a "pyroelectric" effect with demonstration of the "piezo-electric" effect—the electric polarization of certain crystals under pressure. These polarizations result from shifts in the equilibrium positions of the ions; the ions return to their original positions when the heat or the squeeze is removed. The shift of positive ions is different from that of negative ones. Today piezoelectricity is sufficiently well understood to have made possible the design of artificial crystals which can replace natural quartz in some applications.

The ionic crystals, and sodium chloride in particular, were not only the first solids whose structure was analyzed, but also the first chemical compounds whose chemical binding energy was accounted for by physical principles. This feat was accomplished by the German scientists Fritz Haber and Max Born. Haber suggested that the chemical binding of sodium and chlorine in the rock salt crystal was accomplished simply by the electric attraction of Na^+ and Cl^-. If this hypothesis was true, the chemical binding energy could be computed in electrical terms, and Born was able to do so. His calculation was the first computation of a chemical quantity from physical premises. The importance of this step cannot be exaggerated. One can

145

infer from it that very probably all chemical binding is electrical in the last analysis, and that it should be possible to implement this view by direct computation.

To explain how solids hang together, however, one missing link remains to be forged. The ionic concept does not account for the binding of atoms in all chemical compounds. In the molecules of hydrogen (H_2), oxygen (O_2) or chlorine (Cl_2), for example, neither of the two atoms which compose the molecule carries an effective charge. Similarly in many metallic crystals all the atoms are known to be in the same condition, and in consequence, we cannot possibly assume half of them to have positive charge and half negative. The chemical bond in such cases is called covalent. It is one of the successes of quantum mechanics that it can explain the covalent bond; because of its importance for the theory of the solid state as well as for chemistry we must examine that bond in some detail.

Quantum mechanics, sometimes called wave mechanics, represents the final formulation of a schizophrenic viewpoint which has evolved in 20th century physics. Earlier scientists heatedly debated whether light was made up of particles or waves. This conflict finally was resolved by the assertion that it is both. There is a discrete unit of light called a quantum, but if we try to locate this particle, we must apply wave theory. The height of the wave at any point gives the probability of finding the light quantum at that spot. Wave mechanics is concerned with the second stage of this reasoning. If light waves are also particles, then particles are also waves. The wave pattern associated with a particle gives the probability of finding it at a given spot: on the wave crests the probability is high, in the troughs it is low. The wavelength of this "probability wave" is obtained from the formula devised by the French physicist Louis de Broglie; the wavelength equals Planck's constant divided by the mass times the speed of the particle. It follows from this formula that a particle cannot be located with certainty within a space much smaller than one de Broglie wavelength. Therefore you cannot confine the particle without doing work, because in order to do so you have to reduce its wavelength. This reduction can be accomplished in de Broglie's formula in only one way: by increasing the speed of the particle. An increase in speed means an increase in energy, which has to be supplied from somewhere.

For electrons this feature of quantum mechanics is particularly important, because the mass of an electron is far less than the mass of an atom and hence the wavelength tends to be larger. The compression of an electron into a space of atomic dimensions therefore requires significant amounts of energy. If an electron is given a chance to spread itself over several atoms it will do so, and, between two possible states of matter, the one in which the electron has that chance will be the more stable, other things being equal. The question remains: What limits this process? Why do not all electrons spread out their waves indefinitely? The answer is Pauli's exclusion principle. Each possible wave pattern is a quantum state in the sense described for atoms, and according to the exclusion principle only two electrons (for the two directions of spin) can spread themselves out in the lowest energy state, or wave pattern. The third and fourth electron must go into the next higher state, and so on. The tendency to form covalent bonds can be described as the tendency of the electrons to extend their wave patterns, and thus to reduce their energy of motion, as far as this is compatible with the Pauli exclusion principle.

Take as an example the simplest molecule, H_2, formed by two atoms of hydrogen. The system contains only four constituents: two positively charged nuclei and two electrons. When the two atoms are far removed from each other, the electric forces confine each electron to its own nucleus. If the two atoms are brought closer, the wave corresponding to each electron can be spread over the two nuclei, lowering their energy. The Pauli principle can be satisfied by giving the electrons opposite spin. In consequence the energy of the system as a whole is lowered, and the two atoms cannot be separated again without furnishing a certain amount of energy; this is the chemical binding energy of the molecule H_2.

The covalent two-electron bond, the bond achieved by two electrons of opposite spin sharing their wave function, is the most frequent bond of chemistry. It is much more usual than the ionic bond in which an electron is transferred from one atom to another. Its widest application is in organic chemistry. On the other hand, very few crystalline solids are built exclusively on this principle. The classical examples are diamond, germanium and gray tin, a nonmetallic substance. In these crystals the number of nearest neighbors of any given atom just equals its valence. This situation is most perfectly exemplified in the structure of diamond, made up of four-valent carbon. But in general covalent bonding enters into the theory of crystals in a secondary way, in conjunction with other forms of binding.

Implicit in all this is the fact that covalent as well as molecular and ionic crystals are essentially nonconductors of electricity. To be sure, the charged ions in ionic crystals can conduct electricity, but generally their conductivity is small. Electric conduction demands mobile electrons. In these three types of crystals the electrons are all locked into certain quantum states. Metals represent a fourth type of crystal, in which the number of quantum states must be greater than the number of electrons available to fill them, giving the electrons freedom to switch and to move around. For example, in metallic lithium each atom is surrounded symmetrically by eight other atoms. The number of valence electrons per atom is one. If there were no electron spin, one quantum state on each atom would be filled. However, because of spin the number of available states is double the number of electrons, thus giving them a possibility of motion. Because of the electronic charge this motion can be observed as electric current.

In 1900, five years after the discovery of electrons, the German physicist Paul Karl Ludwig Drude suggested that they were the agents which conduct electricity in metals, and he constructed a theory in which he assumed the electrons in the metal to be free, like the molecules in a gas. His theory assumed that the flow of current in a metal depended on two factors: the applied electric force and the resistance offered by the numerous collisions between the electrons and the atoms in the metal. Drude's formula was not immediately verifiable because it contained an unknown—the mean free path of the electrons between collisions. However, his theory could be used to calculate the conductivity of heat as well as that of electricity. The ratio between the two conductivities could be computed theoretically and compared with experiment. The result was that theory and experiment checked each other very well.

Yet Drude's theory raised more questions than it answered. One troubling question had to do with the matter of specific heat. If the electrons form a gas, they must obey the laws that apply to gases. The specific heat of a gas is easily computed from theory, because it tells exactly what energy is required to impart to the gas molecules a given speed. These same considerations, applied to a gas of free electrons, predicted that the specific heat of a metal should be larger by a substantial amount than the specific heat of an insulator. But no such difference was found. This contradiction, together with the fact that no criterion was forthcoming to tell when electrons are free and when they are not, stopped further progress. The Drude theory remained a theoretical fragment of doubtful validity which sometimes gave right answers and sometimes did not.

The development of quantum mechanics and the enunciation of the exclusion principle resolved the apparent contradiction within the Drude theory. The solution was found in 1928 by Arnold Sommerfeld of the University of Munich. Applying quantum mechanics to the hypothetical electron gas of Drude, he showed that electrons, even at room temperature, are in a condition where all low-lying quantum states are tightly occupied. A gas in such a condition is called a degenerate gas. If we try to heat our degenerate electron gas, we find that most electrons are incapable of accepting energy because all neighboring states are occupied. Only a small number of electrons of highest energy can accept heat in the usual way. In many ways the degenerate electron gas inside a metal can be compared to a container filled with water in which the energy is simulated by the height, and Pauli's exclusion principle by the impossibility of two drops being in the same place. What corresponds to the surface level is the so-called Fermi level for electrons. Below the Fermi level all states are filled; above they are all empty. If the container is agitated slightly, only the surface drops can jump up in the air; the other drops are hindered by the ones above them.

A final clarification of the matter was worked out by Felix Bloch, Eugene P. Wigner and Frederick Seitz, now respectively at Stanford and Princeton Universities and the University of Illinois. They studied chiefly sodium. Think of the metal as arising from a gradual pushing together of independent atoms. As the atoms approach one another, they arrive at a distance at which the valence electrons can jump from atom to atom. De Broglie's relation then enters into play: each electron tries to reduce its energy by spreading its wave function uniformly over the entire metal. But according to the Pauli exclusion principal only two electrons in the entire metal can do so. The other electrons have to accept states with shorter wavelengths and a correspondingly smaller binding energy. Thus the electrons are piled on top of each other just as in the degenerate electron gas. The totality of states available to these electrons occupies a band of energy, and the filling occurs to a certain level. In the case of sodium there are exactly twice as many possible states as electrons, because there is one electron per atom outside a closed shell and two possible directions of spin. The metallic binding is a consequence of this sharing of electrons, and the conductivity a consequence of the incomplete filling of the available energy levels.

Returning to the water-in-the-container picture, we would say that the electrons in sodium are analogous to a bottle half filled with water, whereas the electrons in a valence crystal like diamond correspond to a full bottle. Finally the picture of free electrons is analogous to water in an open container. So long as the container, whether open or closed, is only partly filled, a slight tilting of the bottle—which simulates the effect of an applied static electric field—brings about a rearrangement of the electrons, emptying certain states and filling others. The filled bottle, on the other hand, shows no response. Thus a close analogy exists between free electrons and electrons in a partially filled band. It is clear that we are dealing with an analogy only; there is no guarantee that it is a perfect one. Indeed, we know it is not quite perfect, for to make it work one sometimes has to ascribe to electrons in a metal an artificial mass which is not their true value. But apart from this the analogy works rather better than one would have a right to expect.

The band into which the quantum state of the valence electron in sodium spreads out is only an instance of a more general feature applicable to all quantum states of an atom when they are brought together. In a solid each atomic state becomes a band, with the lower ones generally narrower than the upper ones. Somewhere within this system lies the Fermi level. Below this level all energy states are filled; above it all are empty. If this level falls within a band, the body is a conductor; if it falls in a gap between bands, it is an insulator. We may say that the solid is similar to a bottle consisting of separated sections connected by narrow tubes. The water level in this bottle may lie within a section, in which case a slight tilting of the bottle can rearrange the regions occupied by the water near the surface (conductor), or it may lie within one of the narrow tubes, in which case tilting will have no effect unless it is very strong (insulator).

The division of solids into conductors and insulators has become complicated in recent years by the observation that all the so-called

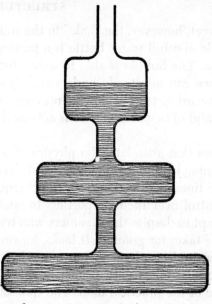

Energy levels in a crystal are explained in the text as analogous to the chambers in a bottle of this peculiar design, with sections of varying capacity connected by thin tubes.

insulators conduct electricity to some extent. Some of them, called semiconductors, have enough conductivity to be technically interesting. Their conductivity is electronic, but in detail it is quite different from the metallic kind. Metallic conductivity increases as the body is cooled, because of the absence of impeding thermal agitation. Semiconductors, on the other hand, lose their conductivity at low temperature and gain upon heating. The explanation, given by the English physicist Alan H. Wilson, falls into the water-bottle picture developed here. If the tilting or the agitation in the bottle is violent enough, water may be brought up into the empty part, in compensation for which a bubble will appear in the full bulge. In other words, an electron appears in the empty band and a "hole" on the top of the full band. Such pairs, of electrons and holes, are the carriers of electricity in a semiconductor. Since their number will increase rapidly with temperature, so, therefore, will the electrical conductivity of the material.

Assessing this conductivity one has a very easy time with the "excited" electrons. The Drude theory will apply to them in its original form, because they form a gas which this time is not degenerate; that is, the quantum states adjoining a given one are usually empty.

One must not forget, however, the "hole" in the full band. The analogy with a bubble in a full water bottle is a particularly happy one in this connection. The bubble is able to move just as freely as the water in a way we can intuitively understand; it acts as if gravity were directed upward instead of down. In exactly the same sense a "hole" in a full band of negative electrons acts as if it had a positive charge.

Like all sciences that arise late, the physics of solids has the disadvantage of running behind technology. People have used and must use solids all the time, and in doing so, must acquire some sort of technological control over their properties. In such a situation the practical man is apt to despise the dreamers who try to "understand" features which he takes for granted. It looks, however, as if the time for this kind of attitude is about to run out. Theoretical analysis is beginning to produce solids with properties beyond the "common sense" experience of the practical man. The transistor is only one of the spectacular results that are destined to flow from this painstaking study of what has long seemed a dense and commonplace subject.

I. ATOMIC POWER *by Leon Svirsky*

Leon Svirsky is Managing Editor of SCIENTIFIC AMERICAN.

II. REACTORS *by Lawrence R. Hafstad*

Lawrence Hafstad is director of the Reactor Development Division of the Atomic Energy Commission. Before taking this position in 1949, he was director of research at the Applied Physics Laboratory of The Johns Hopkins University.

III. THE ENERGY OF STARS *by Robert E. Marshak*

Physics has gained many of its principal insights into the nature of matter and energy by study of their manifestations in stars and interstellar space. Robert Marshak's own work in the field of astrophysics provided important background for his article on the structure of matter in Part III as well as for this article on atomic energy.

IV. THE HYDROGEN BOMB I *by Louis N. Ridenour*

During World War II, Dr. Ridenour was assistant director of the Radiation Laboratory at M.I.T., the radar headquarters. He has since served as dean of sciences at the University of Illinois and as chief scientist of the U. S. Air Force; he is now vice president and director of research of International Telemeter Corporation, engaged in developing a system for metered subscription television.

V. THE HYDROGEN BOMB II *by Hans A. Bethe*

Hans Bethe attained world eminence in physics in the early 1930's as a principal elucidator of atomic fusion. He did not then anticipate that this cosmic process would be demonstrated on a terrestrial scale in the H-bomb. The prospect of an H-bomb rests essentially upon the successful engineering of the opposite process of atomic fission, in the A-bomb. Dr. Bethe also played a major role in this development, as chief of the theoretical physics section of the Los Alamos Laboratory. Dr. Bethe left his native Germany in 1933, taught for two years in England and then came to the U. S. He has been a professor at Cornell since 1937.

VI. TRACERS *by Martin D. Kamen*

At the age of 39, Martin Kamen is young enough to have secured an education that qualified him as a founder of the new field of biophysics, that fruitful confluence of physics, chemistry and biology. He and his co-worker, the late Samuel Ruben, struck pay-dirt when, in 1940, they isolated carbon 14, the most valuable tagged atom. Dr. Kamen was inevitably employed by the Manhattan District during the war. Since then he has continued his tracer studies of photosynthesis, nitrogen fixation and other biological processes at Washington University, St. Louis.

ATOMIC ENERGY

Introduction

IN OTHER chapters of this book the general reader will doubtless encounter ideas and fields of investigation that are entirely new to him. But the book is not apt to fall into the hands of anyone who does not already know that spectacular amounts of energy can be released by "smashing" the atom.

Atomic energy has aroused an unparalleled public interest and has been more widely discussed than any other single development of the past 10 years. Unfortunately, most of the interest and much of the discussion have been compounded of fear, superstition, exaggeration and wishful thinking. It is said, for example, that a sufficiently large hydrogen bomb could ignite the atmosphere and turn the earth into a small star; it is suggested that ocean liners will eventually circle the globe on a supply of nuclear fuel no larger than an aspirin tablet; it is supposed that the scientific theory and the technical means for releasing and applying atomic energy are so profoundly difficult that they can be kept secret from anyone not privy to the findings of the Atomic Energy Commission. All this is nonsense, but the aura of enigmatic omniscience that cloaks our atomic projects encourages the citizenry to believe impossibilities at an unprecedented rate.

This chapter has been prepared, therefore, as an objective review of the available facts. It contains nothing that even approaches classified material and nothing that will be new to any competent physicist. But it does present an authoritative and comprehensible review of substantially all the information that the layman needs to form intelligent opinion in an area where he has so often had reason to doubt the reliability of his information and has been frustrated by obscurities.

The atom releases energy under two conditions: when the nucleus of a heavy element (notably uranium) splits, and when two or more nuclei of a light element (notably hydrogen) combine. The first reaction is called fission; the second is fusion. The discussion here opens with a review of the principles of nuclear fission reactors, followed by a description of several types of reactors now in operation. The A-bomb is itself a reactor; it yields its budget of energy, however, in a fraction of a second. In the reactors with which this part of the book is concerned, the critical mass of the heavy metal is made to yield its energy at a controlled rate and at a predetermined temperature. Various ways of effecting this controlled reaction have been worked out in detail; the next step—the converting of the heat engendered to electricity—is a matter of technology, albeit somewhat bizarre technology. There is at present no scientific impediment to the production of useful atomic power; the question of whether or not the nuclear reactor is to emerge as an important energy generator depends on the solution of engineering, economic and political problems that are only now beginning to be explored.

The light and heat of the stars are produced by the fusion of hydrogen nuclei (protons) into the nuclei of heavier elements. The third essay in this section explains the two reactions, the carbon cycle and the proton-proton cycle, by which this is accomplished. The two pieces which then follow discuss the prospect that man may successfully engineer the fusion principle into an H-bomb. Between stars and engineering, the authors are able to give the reader some small comfort. The earth, it seems, cannot be ignited, however prodigious the H-Bomb set off for the purpose. It also follows that today the fusion reaction cannot be maintained under control as a source of power. Though speculation already suggests methods that may give fusion far greater economic significance than fission, it must for the present be considered exclusively as a province of the weaponeer. The essays on the H-bomb reaction, therefore, are preoccupied with social, moral and military considerations.

This section closes on the more optimistic subject of tracers. Tagged atoms are already standard implements of research, medicine and industry, solving problems that were inaccessible to earlier methods. They are the rare and often unstable isotopes of common elements—carbon, nitrogen, potassium, etc.—and, for the most part, they are produced by exposing samples of such elements to the emanations of nuclear reactors. As a weapon, the nuclear reactor is a world problem; as a power generator, it is still a question. But as the source of tagged atoms it has already proved itself an invaluable aid to modern science and technology.

156

I. ATOMIC POWER

by Leon Svirsky

ISTORY MAY DATE the atomic age from the explosion of the first atomic bomb at Alamorgordo on July 16, 1945, but mankind will be far happier to date it from the day when the first plant is built to harness nuclear fission for constructive power. We are moving, haltingly, and indirectly, to be sure, toward that day. As a sideline to its main business of manufacturing atomic weapons, the Atomic Energy Commission has taken experimental steps toward developing power reactors and already has answered some important questions. The physics of atomic power and the problems that must be solved are neither so recondite, nor so secret, but that any intelligent citizen can understand them.

The entire atomic energy enterprise rests ultimately on one basic reaction—the splitting of uranium 235, one of the lighter isotopes of the heavy element uranium, as the result of its capture of a neutron. Actually it is not U-235 itself that splits but an extremely short-lived daughter, U-236, formed when the neutron is added to the parent. U-236 is so unstable that it cracks almost instantly, within millionths of a second, into two nearly equal parts, recognizable as lighter elements in the middle of the periodic table. These product elements are not always the same; some 40 to 45 different species of atoms have been identified as fission products. The combined weight of the two atoms into which the uranium atom splits is less than that of the parent; the lost mass of the annihilated matter is converted into energy, mainly in the form of gamma rays and the kinetic energy of the flying fragments. And the fission products themselves are highly unstable, i.e., radioactive, giving off particles and energy until they decay into stable forms. The energy released by the fission of a single uranium atom is 200 million electron volts. For the purposes of a chain reaction, however, the most important product of the fission process is the release of free neutrons for the production of further fissions.

Uranium fission is not a man-made phenomenon; it can occur in nature. U-235, though a rare species, is present in all natural uranium in the proportion of one part to 139 parts of the common isotope U-238. Occasionally a U-235 atom in uranium-bearing rocks may capture a stray neutron. It fissions, releasing energy and new neutrons. But the probability that these neutrons will be captured by other U-235 atoms and produce further fissions is so small as to be practically nonexistent. And the reasons for this are at the heart of all the problems and difficulties in developing a practical power reactor.

The first reason is that neutrons are exceedingly eligible for capture by nearly all kinds of matter. Most elements, especially the heavier ones, have a strong affinity for neutrons and greedily absorb them. In the rocks any neutrons produced by an accidental fission have an almost infinitely greater chance of being absorbed by the abundant other materials present than by another rare U-235 atom. Consequently the first step in building a chain-reacting pile, or reactor, is to refine natural uranium, removing all the impurities that would absorb neutrons profitlessly. A chain reaction can be maintained in a pile of uranium oxide, because oxygen is a poor absorber of neutrons, but in that case also the compound must be exceedingly pure.

The second problem is that, in uranium itself, fissionable U-235 is at a double disadvantage in its competition for neutrons with U-238. Not only are the U-238 atoms much more abundant but they intercept neutrons at a more likely speed. When neutrons are released by a fission, they are traveling at very high velocity. At this high speed no type of atom has a high probability of capturing them. U-238 absorbs neutrons of intermediate speeds; there is a certain resonance velocity at which it gobbles them up. U-235, on the other hand, is partial to slow neutrons; it captures them most readily when they move at so-called thermal velocities, that is, the normal rate of vibration of the atoms in a solid. Obviously the neutrons are likely to be absorbed by U-238 at intermediate speeds before they can slow down to the thermal speed favored by the rare fissionable atoms. One way to get around the difficulty, of course, is to get rid of the U-238 and use almost pure U-235, as in the bomb. U-235 *can* capture fast neutrons (though with a lower probability), and if there is little U-238 present to take them out of circulation, the chain reaction can proceed. But the separation of U-235 is costly; practical economics demands that a reactor operate, if possible, on natural uranium. A chain reaction can be established in natural uranium by slowing down the fast neutrons quickly so that as many as possible will arrive at the thermal speed favored by U-235 before they can be absorbed by U-238. This is the function of the moderator in a reactor. Its job is to brake the neutrons from millions of electron volts (the physicist's measure of neutron speed) to less than three hundredths of an electron volt within a foot or two.

The material used as moderator (1) must not absorb neutrons and (2) must be light in mass. The neutrons are slowed by a series of collisions with the nuclei of the atoms in the moderator. The reason why these atoms must be light can be illustrated by comparing the particles to billiard balls. A billiard ball that hits a much more massive object than itself (e.g., a large iron ball) rebounds with little loss of speed; it is slowed most when it collides with a body of about its own size, such as another billiard ball. In a reactor the nuclei of light atoms are effective in slowing neutrons because they are comparatively close to the weight of a neutron, which has about the mass of a hydrogen atom.

The materials that come closest to fulfilling the specifications for an ideal moderator are pure carbon (graphite), the light metal beryllium, and heavy water—ordinary water will not do because common hydrogen is an avid neutron-absorber, whereas the heavy hydrogen isotope (H^2) in heavy water, which already has a neutron, is not.

In the actual construction of a reactor, thin rods of pure natural uranium perhaps an inch in diameter, encased in aluminum (a weak absorber of neutrons) to prevent oxidation, are inserted into blocks of graphite. Most of the fast neutrons produced by fissions in such a uranium rod escape from the rod into the graphite, and by the time they have traversed this buffer to their next encounter with uranium they are at the thermal speeds that favor their selective capture by U-235. This is the pile type of reactor, so-called because it is a block built up as a lattice of uranium and graphite. While the name "pile" has commonly been used for all types of chain-reacting systems except bombs, "reactor" is now preferred as a more inclusive term, covering the newer types. A reactor using heavy water as moderator, for example, is not strictly a pile but an assembly in which rods of uranium are immersed in a tank of heavy water.

The third basic factor that bars a chain reaction in nature and controls the design of a reactor is the escape of neutrons from the system itself. To retain enough neutrons to keep the reaction going, the reactor must be built to a certain minimum size, so that the volume in which the neutrons are held is large in proportion to the surface from which they may escape. Obviously the necessary size depends on the shape of the system. A sphere, having the smallest surface area for its volume, is the most efficient shape and permits the smallest critical size. A cube is nearly as good. The minimum size also is governed by several other considerations—the degree of enrichment of the uranium with U-235, the kind and amount of moderator, the amount of other materials inserted into the reactor, and so on. The critical size of a reactor consisting of almost pure U-235, such as the bomb, may be very small, perhaps the size of a softball; on the other hand some of the low-energy reactors are bulky—of the order of 40 feet in diameter including their seven-foot-thick concrete shielding. Thus size itself is no measure of the power of a reactor.

In an atomic bomb the escape of neutrons presents a special problem, for the neutrons and exploding atoms must be held together long enough for the chain reaction to get well under way before the whole system blows apart. This is accomplished by enclosing the bomb in a casing of very dense material that retards the bomb burst and reflects some neutrons back into the system. This suggests one of the obvious avenues for research in the future development of power reactors: the finding of more effective reflectors of neutrons. The invention of a suitable reflecting material that would hold most of the neutrons within the working part of a reactor would not only reduce the critical size of the reactor but would greatly decrease the shielding now required to protect personnel against escaping neutrons. Thus it would enhance the possibility of building compact nuclear engines for ships, planes and vehicles.

The simple principles thus far described are the basis for the design and operation of all present reactors. In operation the pile type of reactor is controlled simply by inserting bars of a material, usually cadmium, that readily absorbs neutrons. By taking neutrons out of circulation, they can bring the chain reaction into equilibrium or stop it entirely. The bars slide into pockets in the interior of the reactor. To start the reactor, all the bars are withdrawn. The chain reaction develops rapidly; its control would be extremely difficult but for the fortunate fact that some of the neutrons released by fission—about one per cent—are delayed in emission from a few seconds to a minute. This gives the system enough inertia so that the operator, working with dials that operate the motors, has ample time to bring the controls into play as the reactor heats up. One bar suffices to regulate the level of the chain reaction. When the reaction reaches the desired level, the operator stabilizes it by inserting the bar to a length that soaks up enough neutrons so that from each fission one and only one neutron is captured by another U-235 atom to continue the reaction. The power level of the reactor is determined by the number of atoms fissioning at any moment at that equilibrium stage. Reactor technicians designate this rate of operation as the "neutron flux," meaning the number of neutrons passing through a given section of the pile per second. In most reactors the ionization chambers, as a safety measure, control extra rods which would drop automatically into the pile if the human operators failed and the reactor rose to a dangerous level. A pile could not possibly approach the energy of a bomb; at worst, if all controls failed and the reactor blew up, it would simulate a rather bad steam-boiler explosion.

If U-235 were the only fissionable material, there would be no hope for uranium power plants except as an experimental curiosity; the isotope is too rare to be seriously considered as a fuel for economical, large-scale use. The hopes for nuclear power lie in the fact that it is possible to manufacture two other fissionable materials: 1) plutonium, derived from U-238, and 2) U-233, a synthetic uranium isotope derived from the heavy natural element thorium.

Plutonium is produced by this series of reactions: A U-238 atom, upon absorbing a neutron in a reactor, becomes U-239. This short-lived isotope (half-life: 23 minutes) promptly emits a beta particle (electron) from its nucleus, gains an electron in its outer shell, and is transformed to the artificial element neptunium 239. Neptunium also is unstable (half-life: 2.3 days) and it in turn expels a beta particle, plus gamma rays, and becomes plutonium 239. Plutonium is fissionable in the same way as U-235; when it captures a neutron it splits in two with a vast yield of energy, possibly greater than that from the fission of U-235.

The thorium chain is this: Thorium 232 absorbs a neutron and becomes the short-lived isotope thorium 233 (half-life: 23 minutes), which emits a beta particle and is transmuted to protoactinium 233 (half-life: 27 days), which in turn loses another beta particle and becomes uranium 233. U-233, like U-235 and plutonium, fissions by capture of a neutron. To start this series of reactions, thorium would have to be mixed in a pile with U-235 as the source of neutrons. Thus the manufacture of both plutonium and U-233 depends basically upon U-235 as the spark plug. But the great promise of these re-

161

actions is that they add comparatively abundant U-238 and thorium to U-235 as potential fuels, and the fissionable materials made from them, once formed, become additional sources of neutrons.

As we have seen, a reactor using natural uranium is deliberately designed to make most of the neutrons by-pass absorption by U-238 in order to maintain the chain reaction. But some neutrons inevitably are absorbed by some U-238 atoms, and there is always at least a small surplus of neutrons that makes this possible without stopping the chain. Thus even in a reactor running at a very low power level a little plutonium is created in the uranium rods. To increase the production of plutonium, the power level must be raised: obviously the greater the number of fissioning atoms, the more plutonium will be made. This is the basis of operation of the plutonium-producing piles at Hanford. They are built of pure natural uranium and graphite, like other thermal reactors, but run at considerably higher power levels. Periodically the uranium rods are removed and the plutonium in them is extracted by chemical methods.

By now the dimensions of the power problem begin to become apparent. It is easy enough to calculate that one pound of fissionable fuel contains 10 million kilowatt-hours of energy and is the equivalent of 2.6 million pounds of coal. It is even no great feat to design the outline of a fission power plant: One simply constructs a hot reactor enclosed by a reflector and shield, pipes a coolant (cooling fluid or gas) through it to extract the heat, and transfers the heat to a boiler and steam turbine or a gas turbine. But there are certain difficulties about details.

For instance, the reactor must be raised to temperatures far above any yet attempted under control. Even the "hot" piles at Hanford operate well below the boiling point of water. By attaching the necessary heat-exchanging equipment to one of these reactors and allowing it to go to a high level, power might be generated to light an electric-light bulb or perform a slightly more burdensome task. But it would be the labor of a mountain to produce a mouse. And it would ruin the pile.

The paradox of the vast energy in uranium is that its very concentration constitutes one of the prime problems in extracting power efficiently from it. Its power potential is fantastic; in a bomb the temperature produced is measured in millions of degrees. But pound for pound uranium is the most expensive of all possible fuels, and its use is prohibitive unless a way is found to apply an appreciable part of the power it is capable of producing in a small space. This raises a number of serious difficulties. One is the heat question. Obviously, to approach any reasonable efficiency in the extraction of nuclear energy means going to very high temperatures. But the vulnerability of materials to heat imposes a relatively low limit on practical operation. The most resistant known materials will tolerate no more than 2,000 degrees Fahrenheit. Reactor designers must consider the effects of heat on all the reactor ingredients—pipes, coolant, moderator, controls, reflectors, the fuel itself. Even more troublesome than the heat problem is the fact that the high neutron flux that develops in a hot reactor also is destructive to materials, especially metals. And materials that may tolerate high temperatures do not withstand particle radiations. All these lessons were expensively learned in the experience with the Hanford piles, which approached a breakdown after the war.

Still another materials problem is that introduced by contamination of the reactor with the pipes and coolant necessary to extract its heat. They divert neutrons from the chain reaction. Even if materials that absorb few neutrons are found for these purposes, at best they will impair the reactor's delicately balanced neutron economy.

Thus the materials question injects a whole galaxy of new problems into power-reactor construction. The answers will require not only engineering studies but new fundamental knowledge in physics and chemistry. Kenneth S. Pitzer, director of the AEC's Division of Research, observed: "We need entirely unprecedented materials, not merely improvements of those already known."

Yet materials constitute but one of many hurdles that must be cleared to make a power reactor economically feasible. Just as important is the conservation of the costly uranium fuel. A reactor run at a high level quickly becomes poisoned by its own fission products. They absorb neutrons. Moreover, in a natural uranium pile operated for power the proportion of U-235 would soon drop to a level where it could not maintain a profitable chain reaction in the contaminated fuel. The poisoned and diluted uranium probably would have to be removed after but a small part, perhaps one per cent, of its fissionable atoms had been used. Should the uranium be reprocessed by chemical removal of the fission products and returned to the pile? That would be a costly process, because the chemical treatment of the highly radioactive material must be carried on by remote control. Should the reactor be enriched with booster portions of fissionable material? Should it be operated on slow neutrons, intermediate neutrons (less moderator) or fast neutrons (no moderator)? These are the principal questions that the Commission is investigating.

The answers to most of them may be provided by tests of an idea on which AEC is staking most of its hopes—a "breeder" reactor that will produce new fissionable atoms as fast as it uses the old ones up. The new installation at Arco, Idaho, is of this type and the Commission has described it as a "milestone in the history of atomic power." The theory rests on the fact that when a U-235 atom fissions, at least two neutrons are produced. The precise number is a secret, but the AEC has stated that the average is between two and three; the possibility is not excluded that it is more than three. Suppose that a reactor could be built so efficiently that from every fission two neutrons were available for useful capture. One would be taken by a U-235 atom and produce a fission and energy. The other would be absorbed by a U-238 atom and produce an atom of fissionable plutonium. Thus the reactor, while producing energy for power, would constantly replenish the supply of fuel and source of neutrons. If more than two usable neutrons could be realized from each fission, the reactor would actually produce more fissionable material than it consumed. The AEC aptly named the idea "Operation Bootstrap."

In such a reactor a blanket of natural uranium around the fissioning material would absorb the excess neutrons and be converted into fuel. By recovering the fissionable plutonium so manufactured and adding natural uranium to the reactor from time to time, it would be possible to keep the reactor going indefinitely. And thorium would serve as well as uranium, for the original capital investment of fissionable U-235 or plutonium could be used to convert thorium into U-233, which in turn could continue the transformation of thorium into fuel.

Gordon Dean, Chairman of the AEC, has warned that the Arco reactor does not mean that "economic power from atomic fuel is here." It is still true, as the AEC's General Advisory Committee of scientists reported earlier, that the "engineering difficulties associated with breeding are enormous." To achieve the goal of two usable neutrons per fission would require reducing neutron losses in and from the reactor to an extremely low level. On the other hand, the requirements for a power reactor—high operating level, coolant system, and so on—tend to increase the loss of neutrons. Moreover, the chemical recovery of the new fuel bred in the uranium or thorium blanket would be no small job. The radioactive metals must be treated behind heavy shielding by remote control; recovery of the fissionable material must be virtually complete at each stage to maintain the level of original capital in the pile; and the treatment must be repeated many times until the fuel is completely consumed. There is also the problem of the removal of fission products from the power-producing part of the pile.

All these manifold problems help to explain why the AEC hesitated in its first years to undertake specific power projects. A reactor is expensive to build: about $25 million in round figures. In the absence of an emergency such as justified the bomb, or of public awareness or concern about the problem, the Commission has been slow to decide on costly, uncertain experiments. Early in 1949, however, the Commission determined to make a beginning. It allotted $120 million to the reactor program and initiated several experimental plants for the production of power to serve both military and industrial needs. The Arco reactor is to date the most promising result of this investment.

Is there enough available uranium in the world to make the struggle to develop its power potentialities worth while? Investigations indicate that the answer is clearly yes. If breeding works, all of the fairly abundant U-238 and thorium that can be extracted from the earth can be converted into fissionable fuel. The known deposits of uranium ores of commercial grade (at least one per cent uranium) contain an estimated 100 million pounds of uranium—in terms of contained energy that would be enough to supply the entire power needs of the world for at least 50 years at the present rate of power consumption from all sources. In addition, the earth contains an equal amount of available thorium. Uranium and thorium together represent a potential wealth of power greater than the world resources of petroleum and perhaps even of the earth's coal.

In view of the multitude of unsolved problems that stand in the path to nuclear power, attempts to forecast whether uranium will compete as a fuel with coal obviously can be little more than guessing games. Early post-bomb estimates by engineers calculated the cost of nuclear power at from 4 to 10 mills per kilowatt-hour, which would place uranium almost on a par with coal and give it an advantage in areas remote from coal sources. But these computations admittedly were based on extremely meager data. There are persons close to the atomic energy program who believe that at best the power project will be an interesting and costly experiment, possibly demonstrating that power reactors may be usable for certain special purposes, but unlikely to prove that uranium is feasible as a common source of energy.

Yet to dismiss the possibility of the constructive use of atomic power at this stage would be as pointless as it would have been a century ago to dismiss Michael Faraday's discovery of the induction of electric currents. The one certainty in science is that it is unpredictable, and the solution of the atomic power problem may well lie along roads still unseen. The discovery, for example, of large-scale uses for the troublesome radioactive fission products would go a long way toward reducing the power cost; properly safeguarded, they might perhaps be used for energy, for killing bacteria, as industrial tracers, and so on. Some investigators envision a homogeneous reactor—one in which the fuel is mixed in solution with the moderator —that would produce power, manufacture fissionable material, reprocess the fuel and remove wastes—all in one continuous circulating system. Some even dream of short-cutting most of the problems by finding a way somehow to convert the nuclear radiations in a reactor directly into electricity instead of heat.

At all events, no one in the AEC doubts that the U.S. should push the further exploration of atomic energy with all possible speed, if for no other reason than that other nations are doing so. The British, who are already heating one of their atomic research centers with nuclear energy, are believed by some to be ahead of the U.S. in the study of the power problem.

The reactor program is the heart of the whole atomic energy enterprise—if one thinks in terms of peace rather than war. From that point of view the "atomic age" is still in the future. As of 1949, the AEC hoped to produce a functioning power reactor in five to 10 years and an economically practical one in 20 years. Whether it will achieve that aim may depend as much on politics as on research. For it is idle to pretend that atomic bomb-making does not seriously impede the development of atomic energy for peaceful purposes. "Guns-or-butter" issues arise at many points—in the budget, the use of uranium, the recruitment of personnel, security restrictions, the control of research.

II. REACTORS

by Lawrence R. Hafstad

NUCLEAR REACTORS are the machines for converting the energy of nuclear fission into forms that can be turned to useful purposes. They are large, complicated, expensive and controversial. They provide man with the most concentrated energy source thus far devised, and in the imagination, at least, they have unlimited possibilities. The uses proposed for them have ranged from performance of the world's drudgery to the powering of rockets for space travel. Some of the proposed uses are strictly figments of the imagination. But reactors themselves, as a new type of machine with new potentialities, are a functioning reality, and it is now possible to discuss some of them publicly in considerable detail. Much of the information on the construction and performance of eight early reactors, five of which are in the U. S., has been declassified by agreement between the U. S., Britain and Canada.

In the preceding article, Leon Svirsky discussed the principles of nuclear power and indicated some of the problems that face the atomic engineer. To show how the principles are being applied, how the problems are being met, this report will briefly describe three of the declassified U. S. reactors.

Not the least striking feature of reactors is the great variety of possible designs. This is one of the things that makes them so different from the power plants that we have been used to. Conventional coal burning furnaces, for example, are all generally recognizable as belonging to a common genus; they may differ in size and details, but essentially they are built and operated on the same fundamental plan. Imagine, however, a furnace variation which for fuel used carbon dioxide, say, instead of coal, which burned its fuel in a medium of water instead of air and which operated in a kettle instead of the usual firebox. That would certainly be a radical variation, but it is no more radical than the variations among the nuclear furnaces called reactors. The three reactors to be described here, all low-power machines used mainly for research, represent three very different types which indicate the wide range of possibilities.

The world's oldest nuclear reactor is the uranium-graphite pile at the University of Chicago in which a chain reaction was first achieved in 1942. This pile, later taken apart and rebuilt at the Argonne National Laboratory near Chicago, is a cube-shaped mass of blocks of graphite containing lumps of natural uranium or uranium oxide in a lattice arrangement; the oxide was used because only a small amount of pure uranium metal was available when the pile was built. The graphite blocks, which act as the moderator and the structural bricks of the pile, are 4⅛ inches square in cross section and of various lengths, usually 16½ inches. Some of the blocks have holes bored in them at intervals 8¼ inches apart. In these are inserted the uranium lumps, each a cylinder 2¼ inches in diameter and about six pounds in weight. The uranium-loaded blocks are called "live." The solid graphite blocks without uranium are known as "dead" graphite; they are used for spacing and neutron-reflecting purposes. The pile was built up, layer by layer, with alternating layers of live and dead blocks. It became "critical"; *i.e.*, started a self-sustaining chain reaction, when the 50th layer was laid on. Four layers of dead blocks were then laid on top as a reflector, and over this went a cover of six inches of lead and about four feet of wood. The sides of the pile were surrounded with a similar reflecting barrier of at least 12 inches of dead graphite and with protective walls of concrete five feet thick.

When completed, the pile measured 30 feet wide, 32 feet long and 21.5 feet high and had a total weight of more than 1,400 tons. It contained approximately 52 tons of uranium, in 3,200 lumps of the pure metal and 14,500 lumps of uranium oxide.

Openings were provided in the pile for five control rods 17 feet long. They are bronze strips covered with cadmium, an excellent absorber of neutrons. Three of the rods are automatic safety controls; they are equipped with 100-pound weights that quickly pull them into the pile in the event of electrical power failure or other emergency. The operating level, or power, of the pile is controlled by a single rod which is moved in or out to regulate the neutron flux. Because it lacks a cooling system, the maximum safe operating power of this pile is only 200 watts.

The research reactors at Oak Ridge and Brookhaven, and the great production piles at Hanford, are roughly similar to this first reactor. Their mass is largely graphite, and they are fueled by natural uranium. But they have systems for cooling, either by air or water, and therefore can run at higher power levels. They also have devices

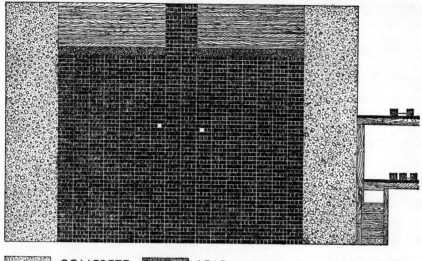

CONCRETE   LEAD
WOOD   GRAPHITE

Natural uranium-graphite reactor in which the first self-sustaining nuclear chain reaction was achieved is shown in this simplified cross section. The reactor was originally built at the University of Chicago and later rebuilt at Argonne National Laboratory. It is basically a large cube of graphite blocks, some of which contain small cylinders of uranium or uranium oxide. An outer layer of graphite blocks containing no uranium serves as a neutron reflector. The shaft of blocks sticking up from the top of the main cube in the cross section is the "thermal column," a means by which slow neutrons may be permitted to diffuse out of the reactor for experimental purposes. The wooden structure at the right supports control rods and stringers that may be inserted into the cube; the holes in the front of the cube permit the insertion of safety rods. The uranium cylinders are not shown.

for loading and unloading fuel without tearing down the reactor. In addition they incorporate a number of other improvements and refinements required to provide dependable control for their much greater power and to serve their different purposes.

The second type of reactor is the heavy-water unit at Argonne. The first version of this machine, which used natural uranium, was also the first of its type in the world. It was built by the Manhattan District during the war because of the remote chance that graphite piles might not be able to produce fissionable material in the amounts needed for bombs. In 1950 it was modified to use enriched uranium. In this reactor heavy water, instead of graphite, is the moderator that slows fast neutrons emerging from the fission of U-235 atoms to

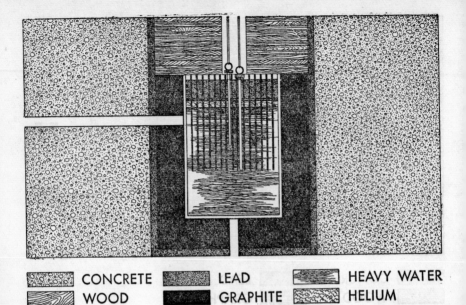

| CONCRETE | LEAD | HEAVY WATER |
| WOOD | GRAPHITE | HELIUM |

Heavy-water reactor at Argonne National Laboratory is shown in a simplified cross section based on written descriptions released by the Atomic Energy Commission. In this reactor the neutrons produced by fission are moderated by heavy water instead of graphite. The heavy water is contained in an aluminum tank approximately six feet across and nine feet high. Suspended in the tank are 120 uranium rods. Around the tank is a layer of graphite to reflect escaping neutrons. Around the graphite is a layer of lead-cadmium alloy and a heavy shield of concrete. At the top of the tank is a layer of lead; beneath it is a thin layer of cadmium, and between the cadmium and the heavy water is a layer of helium. Toward the right side of the tank is a control rod. In the middle of the tank is the "experimental thimble," a device for inserting experimental materials into the reactor. Above the tank is a layer of wood and steel. The heat generated by the reactor must be dissipated; this is accomplished by circulating the heavy water through a heat exchanger (*pipes at left and bottom*). Among the details not shown by this cross section are two horizontal columns of graphite that pass through the concrete shielding. One of these is to insert experimental materials into the reactor; the other, to permit a beam of slow neutrons to pass out of it. The reactor can handle 32 neutron irradiations simultaneously.

the thermal speeds necessary for their capture by U-235 and the continuation of a chain reaction. The following details apply to the reactor as it was run on natural uranium; the principles are unchanged for the present version. In an aluminum tank six feet in diameter and about nine feet high, filled with about 6½ tons of heavy water, are suspended 120 uranium metal rods 1.1 inches in diameter and six feet long. The rods are sheathed in aluminum for protection against

170

corrosion. They are arranged to form a square lattice with 5⅝ inches between their centers. This reactor is smaller than the graphite pile because heavy water is more effective than graphite in slowing neutrons and does not absorb them as readily. Its total weight of uranium is less than three tons, in contrast to the 52 tons in the graphite pile.

The heavy-water reactor tank is surrounded on its bottom and sides by a two-foot neutron reflector of graphite blocks. Around the reflector, in turn, are a four-inch casing of lead-cadmium alloy and then a shield of concrete eight feet thick. The reactor's top shielding consists of a layer of helium gas, a thin layer of cadmium metal, a one-foot layer of lead bricks and finally a four-foot layer of blocks of wood and steel.

Eleven small openings pierce the shield and reflector. These portholes, equipped with removable shielding plugs, are used to measure neutron intensity within the reactor, to insert materials for exposure to neutrons and to let beams of radiation emerge for experiments outside the reactor. There are also three large openings. One is a pocket roughly 30 inches square containing 20 hollow graphite receptacles in which materials can be placed for irradiation. Another is a graphite-plugged aperture five feet square which permits the passage from the reactor of a beam of slow neutrons. The third is a four-inch aluminum pipe inserted in the tank through the top. Into this pipe, known as the central experimental thimble, samples of material may be lowered for exposure to the most intense radiations of both fast and slow neutrons. When all the reactor's openings are used, 32 irradiations can be performed simultaneously.

To cool the reactor, the heavy water circulates between the tank and an external heat exchanger at the rate of 200 gallons per minute. The cooling system is able to handle about 300 kilowatts of heat. Normally the heavy water is at a temperature of 104 degrees Fahrenheit when it emerges from the tank; after it has passed through the heat exchanger, it is about 14 degrees cooler.

The operator measures and controls all the major aspects of the reactor's operation at panels in a nearby room. To start the machine he pushes a button: this lifts the 32-pound safety rods out of the heavy water. Other buttons control two motor-driven rods that regulate the reactor power level. To shut off the reactor the operator presses another button that causes the safety rods to fall back into the tank. If controls fail to function, if there is an interruption in the electricity supply or if for any reason the power level rises too high, the safety rods automatically fall into the "in" position by gravity. As further provision for an emergency shutdown there is a large quick-opening valve through which the heavy water can flow out into a storage tank. Without the moderator, the chain reaction promptly stops.

The third declassified reactor is the famous "water boiler" at Los Alamos. Two small research reactors have been modeled on it. One of these is at North American Aviation, Inc., Downey, California; the other is at North Carolina State College, Raleigh. These are homogeneous reactors, meaning that the fuel is distributed uniformly throughout the moderator instead of being in lumps or rods. The fuel, a uranyl salt, is dissolved in heavy water, and this solution, called "soup" by the reactor scientists, is both the fuel and the moderator. As a chain reaction develops, the solution heats up; hence the name water boiler.

The first version of this reactor was a low-power model known as LOPO. It had no shielding and reached a power of only a twentieth of a watt. In December, 1944, LOPO was replaced by HYPO, a higher-powered model of six kilowatts which went critical with 1.8 pounds of U-235. Its fuel is "enriched"; that is, the uranium in the uranyl salt contains about one part of fissionable U-235 to six parts of U-238, instead of the 1-to-140 ratio in natural uranium.

The heart of this reactor is a one-foot stainless-steel sphere filled with the "soup," a solution of uranyl nitrate. Then comes a reflecting shell, or tamper, consisting of an inner layer of beryllium oxide and a thicker outer layer of graphite. Around this assembly is the shield: four inches of lead, 1/32-inch of cadmium and five feet of poured concrete.

At the working face of the reactor a square tunnel pierces the shield. This tunnel is plugged with graphite to form a column for the passage of slow neutrons. The graphite column and the reflector have a number of ports for experimental irradiations. In addition there is a tube one inch in diameter that penetrates into the heart of the reactor itself. Through this tube, which Los Alamos workers call the "Glory Hole," materials can be thrust into the sphere for exposure to a very high flux of neutrons.

The operation of HYPO presents some special problems. Cooling is accomplished easily enough by circulating water, at the rate of

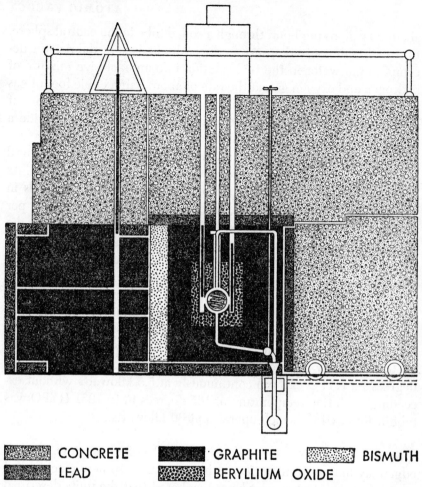

CONCRETE	GRAPHITE	BISMUTH
LEAD	BERYLLIUM OXIDE	

Homogeneous reactor, or "water boiler," at Los Alamos Scientific Laboratory combines fissionable material and moderator in one solution: a uranium salt dissolved in heavy water. The solution is contained in a one-foot steel sphere near the center of this simplified cross section. Around the sphere is a neutron reflector made up of a layer of beryllium oxide and a layer of graphite. Inserted in the beryllium oxide to the left and right of the sphere are the reactor's control rods. Within the sphere is a coil of pipe through which water is circulated to cool the reactor. At the lower left in the cross section is a tunnel of graphite for experiments requiring exposure of materials to slow neutrons. For experiments requiring exposures to a high flux of fast neutrons materials can be inserted through the "Glory Hole" into the heart of the reactor. Hanging through the concrete shielding at the upper left is a cadmium curtain that can be dropped across the graphite tunnel. This reactor was first built as a low-power apparatus without shielding; it was called LOPO. Later its power was stepped up and shielding was installed; it is now called HYPO.

about 50 gallons per hour, through a coiled tube in the central sphere. But the highly ionized atoms produced by fissions in the reactor decompose the water in the fuel solution into an explosive mixture of hydrogen and oxygen gas, at the rate of about half a cubic foot of gas per hour. Moreover, the radiations also produce a small volume of highly radioactive gases. Consequently the design has to provide a means for constantly diluting and flushing out these gases.

Under normal operating conditions the water boiler has a natural self-regulating action that prevents it from getting out of control. As the reaction rate and the temperature rise, the solution expands in volume, the U-235 atoms move farther apart and the reactivity per unit volume therefore declines. This automatically slows down the reaction and holds it within safe limits. HYPO has a safety control rod, nonetheless, and two regulating rods for accurate control of its flux. Since extremely accurate regulation requires almost continuous adjustment of the control rods and this can become tedious over long periods, the reactor has been provided with an automatic pilot that can hold the power at any desired level with an accuracy within a few tenths of one per cent. An experienced operator can bring HYPO up to full power in a few seconds. If cooled by water at about 45 degrees F., HYPO can run continuously at 5.5 kilowatts without exceeding a solution temperature of 185 degrees F. In 1951 HYPO was rebuilt into SUPO, which operates at 30 kilowatts.

So much for the description of the declassified reactors. Knowledge may help citizens to feel at home with these new machines of our civilization. But it must be remembered that the units described here were among the first built. By now they represent mere "taking-off points" for the better reactors that have been built since and for further improved ones that are under construction or envisioned for the future. The program for developing better reactors still faces many unsolved technical problems. We need materials that will stand very high temperatures and intense radiations and that will not absorb too many neutrons; we need more efficient shielding and control gear with very rapid response times; we need ways of attaining heat-transfer rates higher by an order of magnitude than those conventionally used in power plants; we need better ways of handling fission products in the reactor and of disposing of radioactive wastes. The national laboratories of the Atomic Energy Commission have been occupied with these problems since 1945. Much progress has been made toward solutions, but just how much is difficult to assess without some pilot and full-scale trials. We are now entering this new period of reactor development.

The AEC has five major reactor projects under way. First, there is an experimental reactor designed primarily to explore the possibilities of the breeding reaction discussed earlier by Leon Svirsky. Operation of this reactor has demonstrated on a very small scale that a reactor can produce at least as much fissionable material as it consumes. This reactor installation also contains a small, experimental turbo-generator power plant. Though uneconomic, it produced on December 20, 1951, the first electric power from nuclear energy. The second project is the materials-testing reactor, designed for the highest neutron flux of any research reactor. It was completed early in 1952 and put to use making much-needed tests of materials and components considered for use in new and better reactors. Third, there is the slow-neutron reactor for propelling submarines. A full scale land-based prototype was completed in the spring of 1953 at the National Reactor Testing Station. This is the first practical application of nuclear power. Fourth is a second submarine reactor designed to operate in the intermediate neutron energy range, in which we have had less experience. Construction of a prototype is in progress. And fifth, there is a homogeneous reactor of small pilot size, about as large as the experimental breeder, which has already been operated to demonstrate the feasibility of this promising type of design. It, too, supplies heat for an experimental power plant.

III. THE ENERGY OF STARS

by Robert E. Marshak

THE STARS of the universe hurl energy into space on a scale so gigantic that the numbers in which it is measured are almost meaningless to human understanding. The sun, for example, which is by no means a giant among the stars, radiates energy at the rate of half a million billion billion horsepower —of which only a small fraction is intercepted by the earth. Ever since man began to acquire some conception of the magnitude of the stars' radiations, he has been fascinated by the problem of how the stars have been able to maintain these tremendous outpourings of energy for so many millions of years. Thus in *Gulliver's Travels* the inhabitants of Laputa were worried lest "the sun daily spending its rays without any nutriment to supply them, will at last be wholly consumed and annihilated—which must be attended with the destruction of this earth and of all the planets that receive their light from it."

When astronomers began to make a scientific attempt to explain the source of stellar energy more than a hundred years ago, their first idea was that a star's energy derives from gravitational contraction. The mass of an average star is so huge that a large amount of energy could be released by the gradual shrinking of the star. Investigators eventually found, however, that this process could not account for the rate at which a star generates energy; in the case of the sun the gravitational source of energy would have been exhausted in a time much shorter than the age of the earth. Then, after the discovery of radioactivity at the turn of the present century, speculation turned to that process as a possible source of stellar energy. But it soon became clear that there was a conclusive objection to this theory: the rate of radioactive disintegration of atoms is not influenced by temperature, whereas it was shown by various kinds of evidence that the rate of generation of energy by a star depends directly on its internal temperature.

It was the study of the probable physical conditions in the interiors of stars that finally provided the clue to the origin of stellar energy. By 1930 the researches of the great British astrophysicist Sir Arthur Eddington and others had shown that the stars are very hot masses of gas with interior temperatures of 15 to 30 million degrees centigrade. At such high temperatures all molecules are decomposed into atoms, the atoms themselves are stripped of their electrons, and only the atomic nuclei can remain intact. Moreover, the nuclei travel at such high speeds that they overcome the nuclear forces of repulsion and occasionally crash into one another.

The next question was: How much energy could such collisions release? By 1938 it became known that the forces holding together the protons and neutrons inside an atomic nucleus were about a million times as strong as the electrical force holding the electrons in an atom. This means that if the atomic nuclei in the stars interact with one another in such a way as to tap their inner sources of energy, the reaction will release a million times as much energy as when the electrons of atoms interact in energy-producing chemical reactions.

What might these nuclear reactions be? Not all nuclear reactions yield energy, just as not all chemical reactions produce energy. For example, a medium-weight nucleus like that of iron cannot be made to release energy by any process we know, either by splitting the nucleus or building it up. There are two general types of reaction that can release nuclear energy. One is the splitting of a heavy nucleus such as that of uranium 235 into two nuclei of almost equal weight. The other is the combination of two or more light nuclei such as those of hydrogen into a heavier nucleus. To account for the vast production of energy in a star, we must assume a process that involves a sizable fraction of the atomic nuclei present in the star. This rules out the possibility that the stars are fueled by the fission of uranium or other heavy elements, since the stars contain relatively few heavy nuclei. So we can safely assume that the stars produce energy by the combination of light elements through the collisions of their swiftly moving nuclei.

The building-up mechanism requires very high temperatures. It is relatively easy to make small numbers of atomic nuclei undergo nuclear reactions in the laboratory by bombarding them with particles accelerated to high energies by a cyclotron or similar machine. But the amount of energy created is insignificant when compared with the input of energy needed to accelerate the bombarding particles. This is due to the low percentage of hits by the relatively few projectiles on their tiny nuclear targets. At the tremendous temperatures in the interior of a star, however, a large fraction of the nuclei possess the high velocities necessary to produce nuclear reactions. Moreover, this condition continues indefinitely, for the great mass of hot material surrounding the interior of the star prevents the interior from losing too much of its heat by radiation. The energy released from the nuclear reaction produces the high temperature in the star, and the high temperature in turn enables the atomic nuclei to come together and release their energy. Therefore thermonuclear reactions can go on indefinitely, and a self-sustaining reaction is produced.

When astrophysicists had got this far, the way was prepared for formulating a theory. The problem was to reconstruct a system of chain reactions among light nuclei that would account for the rate of energy output by the stars. Its solution was achieved principally by the physicist Hans A. Bethe of Cornell University. The reasoning was such a superb application of scientific method that it deserves a detailed account.

It was evident, as a result of an analysis first performed by R. d'E. Atkinson of the U. S. and F. G. Houtermanns of Germany, and later extended by George Gamow and Edward Teller of the U. S., that the probability of a nuclear reaction decreased sharply with increasing charge; that is, it would be very much greater among nuclei of small electrical charge than among those of larger charge. Thus protons, the nuclei of hydrogen atoms, with but a single positive charge, would be expected to play the most important role in the production of energy. Bethe and, independently, the German physicist C. F. von Weizsäcker, carefully examined all the known nuclear reactions involving protons. The relevant nuclear reactions could be divided into two classes: 1) those in which only protons are consumed; 2) those in which other nuclei also are consumed. Bethe and von Weizsäcker were able at once to eliminate the second type of reaction as a source of energy for most stars. Among the light elements, only lithium, beryllium and boron would react fast enough with protons to give any appreciable energy production. (Helium cannot react with hydrogen, for the resulting nucleus is unstable.) Reactions consuming lithium, boron and beryllium might be responsible for the energy production in some young stars, such as the red giants, but these elements are too rare to supply energy for more than a very limited period for the older stars such as the sun. Thus a lasting source of energy could be found only among the reactions that consume protons alone.

There are essentially two possible sets of such reactions. One starts with the direct combination of two protons, and is known as the proton-proton set of reactions. The other, using carbon nuclei as catalysts, is called the carbon cycle. Both sets of reactions lead to the formation of helium nuclei out of protons; both are outlined here in step-by-step diagrams.

Let us consider the carbon cycle first. It starts when an energetic proton deep in the interior of the star collides with a nucleus of carbon 12, the common isotope of carbon, with such force that it overcomes the electrical repulsion of the carbon nucleus. According to Bethe's calculations, under the temperature-density conditions in the interior of the sun each proton undergoes such capture on the average about once in a million years. When the carbon nucleus captures the proton, it is transformed into a nucleus of radioactive nitrogen 13, and simultaneously emits a gamma ray, which eventually finds its way out of the star with a degraded energy. About 10 minutes after the radioactive nitrogen 13 nucleus is formed, it spontaneously emits a positron (a particle with the same mass as that of the electron but with opposite charge) and becomes stable carbon 13. After moving around for some time in the hot gas in the interior of the star, the carbon 13 nucleus in turn captures a proton in about 200,000 years to form a stable nitrogen 14 nucleus plus a gamma ray. About 30 million years later, on the average, the wandering nitrogen 14 nucleus captures a proton and is changed into a radioactive oxygen 15 nucleus plus a gamma ray. In about two minutes the unstable oxygen 15 gives off a positron and becomes nitrogen 15. About 10,000 years later comes the climactic reaction in the cycle: the nitrogen 15 nucleus captures another proton. But this time the result is not a simple capture accompanied by gamma-ray emission. Instead, the product splits into two large parts—a nucleus of carbon 12 and a nucleus of helium 4.

Since the creation of a positron is equivalent to the destruction of an electron, the carbon cycle in effect converts four protons and two electrons into a stable helium nucleus. But the helium nucleus possesses only 99 per cent of the mass of the four protons from which it was formed. The one per cent difference in mass has been released in the form of energy, and the energy release is enormous. It is measured in terms of the Einstein formula for the equivalence of mass and energy, in which the energy produced in ergs is equal to the loss of mass in grams times the square of the velocity of light in centimeters per second ($E = Mc^2$).

One of the key facts about the carbon cycle is that the carbon nuclei are not consumed; for every carbon nucleus that enters the cycle one is reproduced at the end. This is very important for the economy of stellar-energy production, since hydrogen is much more abundant than carbon in stars, thereby providing nuclear fuel for a longer period of time. In most stars hydrogen is by far the most abundant element, with the possible exception of helium.

The proton-proton set of reactions synthesizes a helium nucleus by another complex series of transformations. According to calcula-

Carbon cycle has been postulated as a primary source of stellar energy. The steps shown here are more fully described in the text on page 2, 179 ff. This caption will explain the symbols used in the opposite diagram and the others which illustrate the discussion of atomic energy in this chapter. The atomic nuclei only are shown in these diagrams because, at the temperatures at which these reactions occur, atoms are stripped of their planetary electrons. Black circles signify protons, the positively charged nuclear particles which give atoms their chemical identity or atomic number; e.g., carbon, with six protons, as distinguished from nitrogen, with seven. White circles signify neutrons, the electrically neutral nuclear particles which, added to the protons in the nucleus, give an atom its atomic weight or isotope number; e.g., common carbon 12, with six protons and six neutrons, as distinguished from carbon 13, with six protons and seven neutrons. Thus, in the top panel, carbon 12 is transmuted into nitrogen 13 by collision with a proton, or hydrogen nucleus. This reaction yields one quantum of energy in the form of a gamma ray, shown as a wavy arrow. The notation "10^6 years" is shorthand for the statement that, under the conditions prevailing in the interior of the star, each carbon 12 atom will enter this reaction once in that period of time and is thus a way of stating how much of the carbon in the star is involved in this reaction at any given moment. In the next panel, upper right, unstable nitrogen 13 emits a beta particle, or positron, a positively charged particle of tiny mass which represents the positive charge of one of the protons. With the conversion of this proton into a neutron, the atom transmutes into carbon 13. In the carbon cycle, summarized briefly, four protons, or hydrogen nuclei, are fused to yield an alpha particle, or helium 4 nucleus (panel at upper left) plus energy. It is called the carbon cycle because carbon 12, acting as a kind of nuclear catalyst, provides a matrix for the fusion of the protons and comes out of the reaction intact. The energy yield is considerable; the carbon cycle accounts for a good deal of starlight.

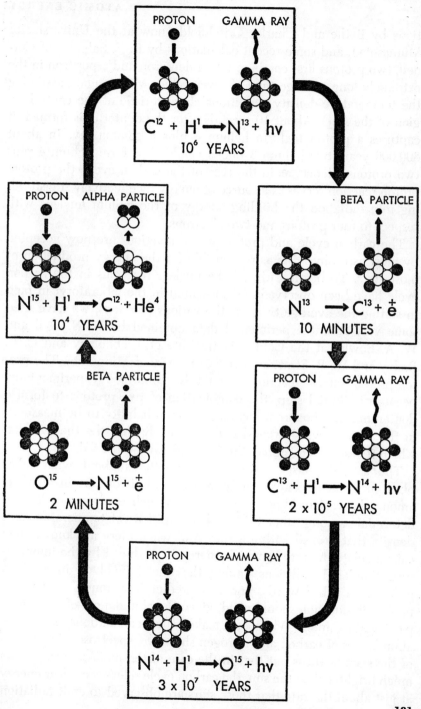

tions by Bethe and Charles Critchfield (now at the University of Minnesota), and some recent calculations by E. E. Salpeter of Cornell, two protons first combine into a deuteron and a positron in the extremely long average time of seven billion years (again assuming the temperature-density conditions characteristic of the central region of the sun). About 10 seconds after the deuteron is formed, it captures a proton to form helium 3 plus a gamma ray. In about 300,000 years two helium 3 nuclei combine to form helium 4 plus two protons. Hence, as in the case of the carbon cycle, the proton-proton reaction serves as a source of energy by effectively transforming into radiation the binding energy of the helium nucleus with respect to four protons and two electrons.

The carbon cycle and proton-proton reactions are now accepted by most astrophysicists as responsible for the energy production of most stars. To begin with, all the nuclear reactions in the carbon cycle have been observed and measured in the laboratory; in fact, in stating the average times of the various reactions, we have used some very recent experimental data obtained by R. N. Hall and W. A. Fowler at the California Institute of Technology and C. L. Bailey and W. R. Stratton at the University of Minnesota. The proton-proton set of reactions cannot be confirmed experimentally, because its initial step, the combination of two protons to form a deuteron and a positron, has too low a probability to be measured in the laboratory. However, it is very likely that the theory upon which its estimated probability is based can be trusted.

The astrophysicist's test of any hypothesis about what takes place in a star depends mainly on the analysis of certain relations among the various types of stars. From such comparisons he can determine, among other things, the conditions of temperature and density that prevail within a star. And when these are known, one can calculate the rate at which energy is produced by the temperature-stimulated collisions among the nuclei involved in a given process. In other words, one can predict how much energy the process in question should yield, given the abundances of the nuclei participating in it. When one makes such calculations, using the abundances of carbon and nitrogen shown by spectroscopic analysis of the stars, it turns out that in the case of the main-sequence stars much brighter than the sun, the carbon cycle releases nuclear energy at just about the rate that those stars are observed to emit radiation

H^1 + H^1 → H^2 + $\overset{+}{e}$	7 x 10⁹ YEARS
H^2 + H^1 → He^3 + hv	10 SECONDS
He^3 + He^3 → He^4 + $2\,H^1$	3 x 10⁵ YEARS

Proton-proton reaction is another primary source of cosmic energy, now thought to be even more important than the carbon cycle, illustrated on page 181. Graphic symbols used in this diagram are explained in the caption to that illustration, and the steps in the reaction are fully described in the text on pages 180-182. Like the carbon cycle, this reaction may be described as the fusion of four hydrogen nuclei, or protons, to make a helium 4 nucleus, or alpha particle, with the incidental return of two protons back to the cycle and the yield of a considerable budget of energy. In the first panel above, two protons fuse and, with the emission of a positron carrying the positive charge one of the protons, yields a heavy hydrogen nucleus, or deuteron. Two seconds later, as shown in the second panel, the deuteron fuses with another proton to yield a helium 3 nucleus, and a gamma ray quantum. In the last step, after 300,000 years two helium 3 nuclei fuse to yield a helium 4 nucleus and two protons. Via the carbon cycle and the proton-proton reaction, 564 million tons of solar hydrogen are converted each second into 560 millions tons of helium. The lost four million tons are converted into radiant energy.

energy. Thus the theoretical predictions agree with the observations, providing good evidence that the carbon cycle is actually the source of energy in these stars. In the case of main-sequence stars fainter than the sun, the same calculations indicate that the source of energy is not the carbon cycle but the proton-proton set of reactions. In the sun itself, the proton-proton reaction makes the dominant contribution to the observed rate of energy production.

Now the significant fact is that on the basis of such calculations no other nuclear reaction or set of reactions comes close to giving the observed rate of energy release for a star in the temperature range of 15 to 30 million degrees. For example, at a temperature of the order of 20 million degrees, nuclear reactions involving boron, the element next lighter than carbon, would give an energy production at least 10,000 times too large, while the next heavier element, oxygen, would yield an energy output about 100,000 times too small.

Thus the carbon cycle and the proton-proton set of reactions fit the observed facts in a unique way: they consume only the most abundant element in the stars—hydrogen—and they give the right energy evolution at temperatures consistent with the equations of stellar equilibrium for the various main-sequence stars.

IV. THE HYDROGEN BOMB I

by Louis N. Ridenour

I N THE PRESIDENT'S first announcement of the Hiroshima atomic bomb, it was stated that the bomb drew its energy from the same source that fuels the sun and the stars. This statement is true only in the loosest sense. To be sure, a uranium-fission bomb, like the sun, derives energy from the transformation of one atomic species into others, but the types of reaction involved in the two cases are quite different.

As Robert E. Marshak explained in the previous article, there is excellent reason to believe that the energy source of most stars, including the sun, is a rather complicated chain of nuclear transformations whose end result is to form one atom of helium out of four atoms of hydrogen. In the sun and the stars, then, the source of energy is a chain of so-called thermonuclear reactions that ends in fusing four light hydrogen nuclei into the heavier and more complicated helium nucleus. The old-fashioned atomic bomb, on the other hand, uses as its explosive the nuclei of some of the heaviest elements known to man: uranium and plutonium. Energy is released when one of these heavy nuclei splits, or fissions, into two lighter, simpler nuclei. Thus the lightest atoms liberate energy if they are combined into heavier atoms; the heaviest atoms liberate energy when they are split into lighter ones. Only near the middle of the periodic table of the chemical elements do we find atomic species that are fully stable, in the sense that energy cannot be liberated either by combining them into heavier atoms or by splitting them into lighter ones.

To paraphrase a remark of George Gamow, we live in the midst of an atomic powder magazine, where immense amounts of energy are locked in every bit of matter. Why, then, are we safe? Why does common matter, possessed as it is of tremendous stores of energy, seem so inert, so permanent? The answer is that in order to liberate the energy of a fusion or a fission reaction we must ourselves invest some energy, just as we must expend energy to strike a match.

In the case of nuclear fission, the energy investment is small. Fission occurs when an atom of uranium 235, plutonium 239, or uranium 233 captures a neutron, even one of very slow speed. But these neutrons must be somehow produced, since they do not occur free in nature, and are in fact unstable. The nuclear "chain reaction" that is exploited in atomic bombs and in nuclear reactors like those at Hanford is possible only because the fission reaction itself releases neutrons, the very particles that are needed to keep it going.

If the lump of uranium or plutonium in which these neutrons are liberated is large enough, the neutrons released by one fission process will cause other fissions before they escape from the lump, and the process will go on faster and faster until an atomic explosion has been produced. The lump of uranium is then said to have exceeded the "critical size."

This limitation on critical size dictates the design of fission bombs. Detonation of such a bomb requires the rapid assembly of an over-critical mass; as soon as this is assembled, it blows up in about half a millionth of a second. The greatest ingenuity is needed to achieve an instantaneous condition exceeding the critical by as much as a few per cent; no amount of ingenuity has yet allowed the design of an efficient fission bomb so much as two or three times critical size. Thus there are inherently narrow limits to the size of a fission bomb: as it begins to exceed the critical, it explodes at once; if it is smaller, it cannot be exploded at all.

At the opposite end of the scale, among the light elements, the explosive conversion of mass into energy is not so easily achieved in terrestrial laboratories. To cause two light atoms to fuse into a heavier one, we must overcome powerful forces of electrical repulsion, since each nucleus is positively charged. Up to now this has been accomplished by scientists only laboriously, with poor efficiency and on an extremely small scale, by striking target atoms with a fast-moving beam of atoms accelerated in an electronuclear machine, or "atom-smasher."

In the centers of the stars fusion reactions go on all the time, because the temperature there is some 20 million degrees centigrade. The average energy of an atom at this temperature is only some 1,700 electron volts, but some atoms have several times this energy, and collisions between atomic nuclei are so frequent that fusion reactions are produced in substantial numbers. We cannot maintain stellar temperatures on the earth, but we can produce them for very small fractions of a second. In the explosion of a uranium or plutonium bomb, the central temperature of the exploding mass has been estimated as high as 50 million degrees C. At such a temperature fusion reactions in a dense mass of light atoms occur often enough to liberate significant amounts of energy.

Obviously the fusion reactions that are likely to be most effective for producing energy are those that will go best at relatively low collision energies, since even the highest temperatures reached in the explosion of an atomic bomb correspond to rather modest bombarding energies from the laboratory standpoint. For this reason, the stellar-energy reaction cycle is out of the question; this cycle involves the fusion of hydrogen with heavy atoms such as carbon and nitrogen, and therefore proceeds relatively slowly even at temperatures of millions of degrees. It has been known for some time, however, that fusion reactions between the rarer, heavier isotopes of hydrogen can take place much more rapidly at substantially lower temperatures. Formulas for three of these are given in the charts.

A few years ago the reaction that would have been chosen as the most promising for a hydrogen bomb was the fusion of two atoms of deuterium, or hydrogen of mass 2, containing one proton and one neutron. This results in the formation of helium 3, with the emission of a neutron and the release of about four million electron volts of energy. Gamow has calculated the thermonuclear energy release from this reaction; the results were given in his 1946 book *Atomic Energy*. He remarked that if the reaction took place at a temperature of something over one million degrees C., "a small charge of deuterium could be used as an explosive with tremendous destructive power."

Nowadays we know that a more effective reaction can be obtained with hydrogen 3, known as tritium, a radioactive but long-lived isotope of hydrogen that has one proton and two neutrons in its nucleus. Tritium not only reacts faster than deuterium at low temperatures, but also liberates more energy when it does so. A fusion reaction of tritium with deuterium produces helium 4, with the emission of a neutron; its energy yield is 17.6 million electron volts. Tritium can also fuse with tritium, yielding helium 4, two neutrons and 11.4 million electron volts of energy; the cross section for this reaction is not available in the published literature, but it is probably large.

Ponderable amounts of tritium have been and are being made by the Atomic Energy Commission in its huge facilities. The designer of a fusion bomb clearly would start with a fission bomb of uranium

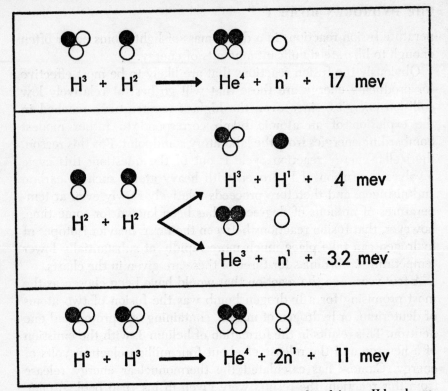

$$H^3 + H^2 \longrightarrow He^4 + n^1 + 17 \text{ mev}$$

$$H^2 + H^2 \longrightarrow \begin{cases} H^3 + H^1 + 4 \text{ mev} \\ He^3 + n^1 + 3.2 \text{ mev} \end{cases}$$

$$H^3 + H^3 \longrightarrow He^4 + 2n^1 + 11 \text{ mev}$$

Heavy hydrogen fusion reactions which might be employed in an H-bomb are compared here (*see* caption to illustration on page 180 for explanation of graphic symbols). At top, hydrogen 3, familiar to newspaper readers as tritium, reacts with hydrogen 2, or deuterium, to yield a nucleus of helium 4 plus a free neutron plus 17 million electron volts (hence "mev") of energy. The reaction of two deuterium nuclei, shown in the middle panel, may move in one of two directions. It may produce a tritium nucleus, plus a hydrogen 1 nucleus (or proton); or it may yield helium 3 nucleus plus a neutron. The energy yield in either case is about the same. The reaction of two tritium nuclei, at bottom, yields a helium 4 nucleus (or alpha particle) plus two neutrons plus 11 mev.

or plutonium, the explosion of which would produce the high temperatures required for the thermonuclear fusion reaction. To the fission bomb he would add a certain mixture of deuterium and tritium to fuel the fusion process. The final energy release of the bomb —its total deadliness—would be determined by the amount of deuterium added. To say that the fusion bomb would be 2, 7, 10, 100 or 1,000 times as devastating as the conventional fission bomb is to speak from ignorance; the effective size of a fusion bomb will depend upon the intentions and the skill of its designers.

Note, however, that there is here no concept like that of critical size. The size of the bomb depends, and depends exactly, on the amount of the reacting elements built into it. The fission detonator itself must be made overcritical in order to explode, but the happenings thereafter will depend on the amount of fuel provided for the fusion reaction.

Here, then, are the technical conclusions that one must draw about the fusion bomb:

First, it can be made.

Second, there is no limit, in principle, to the size of a fusion bomb. It cannot be smaller than a fission bomb, since it must use a fission bomb as detonator, but it can be many times, perhaps thousands of times, bigger.

Third, while fission can be controlled in an orderly way to produce useful power in a reactor, the fusion reaction offers no prospect at the present time of any use except in terms of an explosion. We cannot find in the development of the fusion bomb any such peacetime values as are inherent in the development of nuclear fission. Except where the uses of peace demand the detonation of an explosive equivalent to, say, a million tons of TNT, there is no use for a fusion reaction. Thus when we discuss the "hydrogen bomb" we are clearly speaking of a weapon, and a weapon only.

Fourth, we are speaking of a very special type of weapon—one that is appropriate only to the destruction of large targets. A weapon of this sort is clearly of much greater significance to other nations, such as the U.S.S.R., than it is to us. We have several large targets; the U.S.S.R. has only one or two.

In view of all this, it seems a little curious that the fusion bomb should have been proposed—as it apparently was—in terms of a reply to the Soviet achievement of the fission bomb. To the detached observer, it would appear to be potentially a Pyrrhic reply, involving as it does the production of a weapon uniquely suited to the destruction of the great cities around which our own economy and our own civilization are built.

V. THE HYDROGEN BOMB II

by Hans A. Bethe

EVERYBODY who talks about atomic energy knows Albert Einstein's equation $E = Mc^2$: *viz.*, the energy release in a nuclear reaction can be calculated from the decrease in mass. In the fission of the uranium nucleus, one tenth of one per cent of the mass is converted into energy; in the fusion of four hydrogen nuclei to form helium, seven-tenths of one per cent is so converted. When these statements are made in newspaper reports, it is usually implied that there ought to be some way in which all the mass of a nucleus could be converted into energy, and that we are merely waiting for technical developments to make this practical. Needless to say, this is wrong. Physics is sufficiently far developed to state that there will never be a way to make a proton or a neutron or any other nucleus simply disappear and convert its entire mass into energy. It is true that there are processes by which various smaller particles—positive and negative electrons and mesons—are annihilated, but all these phenomena involve at least one particle which does not normally occur in nature and therefore must first be created, and this creation process consumes as much energy as is afterwards liberated.

All the nuclear processes from which energy can be obtained involve the rearrangement of protons and neutrons in nuclei, the protons and neutrons themselves remaining intact. Hundreds of experimental investigations through the last 30 years have taught us how much energy can be liberated in each transformation, whether by the fission of heavy nuclei or the fusion of light ones. In the case of fusion, only the combination of the very lightest nuclei can release very large amounts of energy. When four hydrogen nuclei fuse to form helium, .7 per cent of the mass is transformed into energy. But if four helium nuclei were fused into oxygen, the mass would decrease by only .1 per cent; and the fusion of two silicon atoms, if it ever could occur, would release less than .02 per cent of the mass. Thus there is no prospect of using elements of medium atomic weight for the release of nuclear energy, even in theory.

The main problem in the release of nuclear energy in those cases that we can consider seriously is not the amount of energy released —this is always large enough—but whether there is a mechanism by which the release can take place at a sufficient rate. This consideration is almost invariably ignored by science reporters, who seem to be incurably fascinated by $E = Mc^2$. In fusion the rate of reaction is governed by entirely different factors from those in fission. Fission takes place when a nucleus of uranium or plutonium captures a neutron. Because the neutron has no electric charge and is not repelled by the nucleus, temperature has no important influence on the fission reaction; no matter how slow the neutron, it can enter a uranium nucleus and cause fission. In fusion reactions, on the other hand, two nuclei, both with positive electric charges, must come into contact. To overcome their strong mutual electrical repulsion, the nuclei must move at each other with great speed. Ridenour explained how this is achieved in the laboratory by giving very high velocities to a few nuclei. This method is inefficient because it is highly unlikely that one of the fast projectiles will hit a target nucleus before it is slowed down by the many collisions with the electrons also present in the atoms of the target.

The only known way that energy can be extracted from light nuclei by fusion is by thermonuclear reactions, *i.e.*, those which proceed at exceedingly high temperatures. The prime example of such reactions occurs in the interior of stars, where temperatures are of the order of 20 million degrees centigrade. At this temperature the average energy of an atom is still only 1,700 electron volts—much less than the energies given to nuclear particles in "atom smashers." But all the particles present—nuclei and electrons—have high kinetic energy, so they are not slowed down by colliding with one another. They will keep their high speeds. Nevertheless, in spite of the high temperature, the nuclear reactions in stars proceed at an extremely slow rate; only one per cent of the hydrogen in the sun is transformed into helium in a billion years. Indeed, it would be catastrophic for the star if the reaction went much faster.

The temperature at the center of a star is kept high and very nearly constant by an interplay of a number of physical forces. The radiation produced by nuclear reactions in the interior can escape from the star only with great difficulty. It proceeds to the surface not in a straight line but by a complicated, zigzag route, since it is constantly absorbed by atoms and re-emitted in new directions. It is this slow escape of radiation that maintains the high interior temperature, which in turn maintains the thermonuclear reactions. Only a star large enough to hold its radiations for a long time can produce significant amounts of energy. The sun's radiation, for example, takes about 10,000 years to escape. A star weighing one-tenth as much as the sun would produce so little energy that it would not be visible, and the largest planet, Jupiter, is already so small that it could not maintain nuclear reactions at all. This rules out the possibility that the earth's atmosphere, or the ocean, or the earth's crust, could be set "on fire" by a hydrogen superbomb and the earth thus be converted into a star. Because of the small mass of the bomb, it would heat only a small volume of the earth or its atmosphere, and even if nuclear reactions were started, radiation would carry away the nuclear energy much faster than it developed, and the temperature would drop rapidly so that the nuclear reaction would soon stop.

If thermonuclear reactions are to be initiated on earth, one must take into consideration that any nuclear energy released will be carried away rapidly by radiation, so that it will not be possible to keep the temperature high for a long time. Therefore, if the reaction is to proceed at all, it must proceed very quickly. Reaction times of billions of years, like those in the sun, would never lead to an appreciable energy release; we must think rather in terms of millionths of a second. On the other hand, on earth we have a choice of materials: whereas the stellar reactions can use only the elements that happen to be abundant in stars, notably ordinary hydrogen, we can choose any elements we like for our thermonuclear reactions. We shall obviously choose those with the highest reaction rates; the range of our choice is indicated in the accompanying table.

The reaction rate depends first of all, and extremely sensitively,

Possible hydrogen fusion reactions are charted here to illustrate an important consideration in design of a hydrogen bomb. The various reactions as shown at left may be understood by reference to the caption on page 180, which explains the graphic symbols, and the caption on page 183, which explains the most significant of these reactions. The middle column gives the energy yield of each reaction, expressed in millions of electron volts, or mev. The column at right gives the time required for each reaction at exceedingly high temperatures such as are generated by a fission bomb. It can be seen that the heavy hydrogen reactions proceed at a much faster rate. Since it is difficult to maintain the necessary temperatures for more than a brief instant on earth, the heavy hydrogen reactions are likely to be preferred. Among these, it can be seen that the tritium-deuterium reaction not only yields the greatest energy, but requires the briefest instant of time.

Reaction	Energy	Time
$H^1 + H^1 \longrightarrow H^2 + \overset{+}{e}$	1.4 mev	10^{11} YEARS
$H^2 + H^1 \longrightarrow He^3 + h\nu$	5 mev	.5 SECOND
$H^3 + H^1 \longrightarrow He^4 + h\nu$	20 mev	.05 SECOND
$H^2 + H^2 \longrightarrow He^3 + n^1$	3.2 mev	.00003 SECOND
$H^2 + H^2 \longrightarrow H^3 + H^1$	4 mev	.00003 SECOND
$H^3 + H^2 \longrightarrow He^4 + n^1$	17 mev	.0000012 SECOND
$H^3 + H^3 \longrightarrow He^4 + n^1 + n^1$	11 mev	?

on the product of the charges of the reacting nuclei; the smaller this product, the higher the reaction rate. The highest rates will therefore be obtainable from a reaction between two hydrogen nuclei, because hydrogen has the smallest possible charge—one unit. (The principal reactions in stars are between carbon, of charge six, and hydrogen.) We can choose any of the three hydrogen isotopes, of atomic weight one (proton), two (deuteron) or three (triton). These isotopes undergo different types of nuclear reactions, and the reactions occur at different rates.

The fusion of two protons is called the proton-proton reaction. It has long been known that this reaction is exceedingly slow. As Robert E. Marshak states in his article the proton-proton reaction takes 100 billion years to occur at the center of the sun. Ridenour points out that the situation is quite different for the reactions using only the heavy isotopes of hydrogen: the deuteron and triton. A number of reported measurements by nuclear physicists have shown that the reaction rates for this type of fusion are high.

What would be the effects of a hydrogen bomb? Ridenour notes that its power would be limited only by the amount of heavy hydrogen that could be carried in the bomb. A bomb carried by a submarine, for instance, could be much more powerful than one carried by a plane. Let us assume an H-bomb releasing 1,000 times as much energy as the Hiroshima bomb. The radius of destruction by blast from a bomb increases as the cube root of the increase in the bomb's power. At Hiroshima the radius of severe destruction was one mile. So an H-bomb would cause almost complete destruction of buildings up to a radius of 10 miles. By the blast effect alone a single bomb could obliterate almost all of Greater New York or Moscow or London or any of the largest cities of the world. But this is not all; we must also consider the heat effects. About 30 per cent of the casualties in Hiroshima were caused by flash burns due to the intense burst of heat radiation from the bomb. Fatal burns were frequent up to distances of 4,000 to 5,000 feet. The radius of heat radiation increases with power at a higher rate than that of blast, namely by the square root of the power instead of the cube root. Thus the H-bomb would widen the range of fatal heat by a factor of 30; it would burn people to death over a radius of up to 20 miles or more. It is too easy to put down or read numbers without understanding them; one must visualize what it would mean if, for instance, Chi-

194

cago with all its suburbs and most of their inhabitants were wiped out in a single flash.

In addition to blast and heat radiation there are nuclear radiations. Some of these are instantaneous; they are emitted by the exploding bomb itself and may be absorbed by the bodies of persons in the bombed area. Others are delayed; these come from the radioactive nuclei formed as a consequence of the nuclear explosion, and they may be confined to the explosion area or widely dispersed. The bombs, both A and H, emit gamma rays and neutrons while they explode. Either of these radiations can enter the body and cause death or radiation sickness. It is likely, however, that most of the people who would get a lethal dose of radiation from the H-bomb would be killed in any case by flash burn or by collapsing or burning buildings.

There would also be persistent radioactivity. This is of two kinds: the fission products formed in the bomb itself, and the radioactive atoms formed in the environment by the neutrons emitted from the bomb. Since the H-bomb must be triggered by an A-bomb, it will produce at least as many fission products as an A-bomb alone. The neutrons produced by the fusion reactions may greatly increase the radioactive effect. They would be absorbed by the bomb case, by rocks and other material on the ground, and by the air. The bomb case could be so designed that it would become highly radioactive when disintegrated by the explosion. These radioactive atoms would then be carried by the wind over a large area of the bombed country. The radioactive nuclei formed on the ground would contaminate the center of the bombed area for some time, but probably not for very long because the constituents of soil and buildings do not form many long-lived radioactive nuclei by neutron capture.

Neutrons released in the air are finally captured by nitrogen nuclei, which are thereby transformed into radioactive carbon 14. This isotope, however, has a long half-life—5,000 years—and therefore its radioactivity is relatively weak. Consequently even if many bombs were exploded, it is not likely that the carbon 14 would become dangerous.

The decision to proceed with the development of hydrogen bombs has been made. I believe that this decision settles only one question and raises a hundred in its place. What will the bomb do to our strategic position? Will it restore to us the superiority in armament that we possessed before the Russians obtained the A-bomb? Will it improve our chances of winning the next war if one should come? Will it diminish the likelihood that we should see our cities destroyed in that war? Will it serve to avert or postpone war itself? How will the world look after a war fought with hydrogen bombs?

I believe the most important question is the moral one: Can we who have always insisted on morality and human decency between nations as well as inside our own country, introduce this weapon of total annihilation into the world? Whoever wishes to use the hydrogen bomb in our conflict with the U.S.S.R., either as a threat or in actual warfare, is adhering to the old fallacy that the ends justify the means. The fallacy is the more obvious because our conflict with the U.S.S.R. is mainly about means. It is the means that the U.S.S.R. is using, both in dealing with her own citizens and with other nations, that we abhor; we have little quarrel with the professed aim of providing a decent standard of living for all. We would invalidate our cause if we were to use in our fight means that can only be termed mass slaughter.

We believe in personal liberty and human dignity, the value and importance of the individual, sincerity and openness in the dealings between men and between nations. All this is in great contrast to the methods which the Soviet Government uses in pursuing its aims and which it believes necessary in the "beginning phase" of Communism—which by now has lasted 33 years. We believe in peace based on mutual trust. Shall we achieve it by using hydrogen bombs? Shall we convince the Russians of the value of the individual by killing millions of them? If we fight a war and win it with H-bombs, what history will remember is not the ideals we were fighting for but the methods we used to accomplish them. These methods will be compared to the warfare of Genghis Khan, who ruthlessly killed every last inhabitant of Persia.

What would an all-out war fought with hydrogen bombs mean? It would mean the obliteration of all large cities and probably of many smaller ones, and the killing of most of their inhabitants. After such a war, nothing that resembled present civilization would remain. The fight for mere survival would dominate everything. The destruction of the cities might set technology back a hundred years or more. In a generation even the knowledge of technology and science might disappear, because there would be no opportunity to

196

THE HYDROGEN BOMB II

practice them. Indeed it is likely that technology and science, having brought such utter misery upon man, would be suspected as works of the devil, and that a new Dark Age would begin on earth.

It is ironical that the U. S. of all countries should lead in developing such methods of warfare. The military methods adopted by this nation at the outset of the Second World War had the aim of conserving lives as much as possible. Determined not to repeat the slaughter of the First World War, during which hundreds of thousands of soldiers were sacrificed in fruitless frontal attacks, the U. S. high command substituted war by machines for war by unprotected men. But the hydrogen bomb carries mechanical warfare to ultimate absurdity in defeating its own aim. Instead of saving lives, it takes many more lives; in place of one soldier who would die in battle, it kills a hundred noncombatant civilians. Surely it is time for us to reconsider what our real intentions are.

One may well ask: Why advance such arguments with reference to the H-bomb and not atomic bombs in general? Is an atomic bomb moral and a hydrogen bomb immoral, and if so, where is the dividing line? I believe there was a deep feeling in this country right after the war that the use of atomic bombs in Japan had been a mistake, and that these bombs should be eliminated from national armaments. This feeling, indeed, was one of the prime reasons for President Truman's offer of international control in 1945. We know that the negotiations for control have not led to success as yet. But our inability to eliminate atomic bombs is no reason to introduce a bomb which is a thousand times worse.

When atomic bombs were first introduced, there was a general feeling that they represented something new, that the thousandfold increase of destructive power from blockbuster to atom bomb required and made possible a new approach. The step from atomic to hydrogen bombs is just as great again, so we have again an equally strong reason to seek a new approach. We have to think how we can save humanity from this ultimate disaster. And we must break the habit, which seems to have taken hold of this nation, of considering every weapon as just another piece of machinery and a fair means to win our struggle with the U.S.S.R.

I have reviewed the moral issues that should deter us from using hydrogen bombs even if we were sure that we alone would have them, and that they would contribute to our victory. The true situation is rather the reverse; we can hardly hope to keep a monopoly on hydrogen bombs. At the moment we are not allowed to know whether or not the U. S. has built a successful H-bomb and the Russians will certainly not tell us how they are progressing on this project. It is clear, however, from the fact that they have built the fission bomb that they have the knowledge, technology and resources to build the fusion bomb. When the time comes that both the U. S. and the U.S.S.R. have this weapon, we shall be more vulnerable than the Russians. We have many more large cities that would be inviting targets, and many of these lie near the coast so that they could be reached by submarine and perhaps a relatively short-range rocket. I think it is therefore correct to say that the existence of the hydrogen bomb will give us military weakness rather than strength.

But, say the advocates of the bomb, what if the Russians obtain the H-bomb first? If the Russians have the bomb, Harold Urey argued in a speech just before President Truman decided to go ahead with our Hydrogen project, they may confront us with an ultimatum to surrender. I do not believe we would accept such an ultimatum even if we did not have the H-bomb, or that we would need to. I doubt that the hydrogen bomb, dreadful as it would be, could win a war in one stroke. Though it might devastate our cities and cripple our ability to conduct a long war with all modern weapons, it would not seriously affect our power for immediate retaliation. Our atomic bombs, whether "old style" or hydrogen, and our planes would presumably be so distributed that they could not all be wiped out at the same time; they would still be ready to take off and reduce the country of the aggressor to at least the same state as our own. Thus the large bomb would bring untold destruction but no decision. I believe that "old-fashioned" A-bombs would be sufficient to even the score in case of an initial Soviet attack with H-bombs on this country. In fact, because of the greater number available, A-bombs may well be more effective in destroying legitimate military targets, including production centers. H-bombs, after all, would be useful only against the largest targets, of which there are very few in the U.S.S.R.

So we come finally to one reason, and only one, that can justify our building the H-bomb: namely, to deter the Russians from using it against us, if only for fear of our retaliation. Our possession of the bomb might possibly put us in a better position if the U.S.S.R. should present us with an ultimatum based on their possession of it. In other words, the one purpose of our development of the bomb would be to prevent its use, not to use it.

If this is our reason, we can contribute much to the peace of the world by stating this reason openly. This could be done in a declaration, either by Congress or by the President, that the U. S. will never be the first to use the hydrogen bomb, that we would employ the weapon only if it were used against us or one of our allies. A pledge of this kind was proposed in a press statement by 12 physicists, including myself, in 1950. It still appears to me as a practical step toward relief of the international tension, and toward freedom from fear for the world. The pledge would indicate our desire to avoid needless destruction; it would reduce the likelihood of the use of the hydrogen bomb in the case of war, and it would largely eliminate the danger that fear of the H-bomb itself would precipitate a war.

We have proposed unilateral action rather than an international treaty on this pledge. We have done this because negotiations with the U.S.S.R. are known to be long and frustrating. A unilateral pledge involving only this country could be make quickly, and it could not again lead to the disappointment of a breakdown of negotiations. On the other hand, we certainly would not want to exclude a pact with the U.S.S.R. on this subject. This might be the first point on which the two countries could agree, and this in itself would be important.

Obviously the pledge can only be a first step. What we really want is a workable agreement on atomic energy, as part of our efforts toward a lasting peace. President Truman voiced the fears of many of us when he stated that there is no security in agreements with the Russians because they break them at will. He referred to the agreements of Yalta and Potsdam in 1945. Since then we have learned much about Soviet methods, and the Russians have found that we do not retreat as easily as they apparently imagined in 1945. This more realistic mutual appraisal makes it much more likely that we could now come to arrangements which neither side would regret afterward.

The situation in atomic energy has changed, both because of the Soviet development of the A-bomb and because of our decision on the H-bomb. To leave atomic weapons uncontrolled would be against the best interests of both countries. If we can negotiate seriously with the U.S.S.R., the scope of the negotiations should probably be as broad as possible. But the situation would be greatly eased even if we could agree only to eliminate the greatest menace to civilization, the hydrogen bomb.

VI. TRACERS
by Martin D. Kamen

THE RECENT APPLICATION of isotopes as tracers for investigating life processes is now widely recognized as one of the most significant developments in the long history of the biological sciences. The researches made possible by this technique already have yielded substantial contributions to fundamental biology. Beyond these, the visions opened by the technique have both surprised and stirred biologists. It is as if, looking into some quiet forest pool, one were to find its microscopic animal life suddenly endowed with visibility, revealing a vast activity of movement, interchange and transformation hardly indicated by the seeming calm and stability of the surface.

The power of the tracer technique is readily illustrated by a simple experiment. Suppose that a beaker is filled with a solution of sodium chloride—common table salt—in water, and the beaker is then divided into two compartments by means of a permeable membrane. Obviously material is diffusing back and forth through the membrane, yet no ordinary chemical or physical means can demonstrate this movement, for the system is in a state of equilibrium. The solution on both sides of the membrane is identical; each half of the beaker has the same number of positively-charged sodium ions and negatively-charged chlorine ions.

Now suppose some of the chloride is removed from compartment A and replaced with an equal volume of specially prepared chloride containing only the lighter isotope of chlorine—Cl^{35}, of atomic weight 35. Normal chlorine, a mixture of two stable isotopes, has an atomic weight of 35.5. Thus the new material, though chemically indistinguishable from the chloride it replaced, is lighter. In consequence the diffusion of the new chlorine ions into compartment B, as well as the rate of this diffusion, can be detected merely by observing the rate at which the weight of the solution in B decreases. Precisely the same result is obtained by labeling solution A with a radioactive, or unstable, isotope of chlorine instead of a stable one. The only difference is in the method of detection: in one case the diffusion is detected by weighing samples of the B solution, in the other by using an instrument to record radiation from the B solution.

This experiment demonstrates, at the simplest level, the principle

201

of all tracer work. The two interacting systems, A and B, may be two pieces of solid salt—in which case the membrane is the surface of contact between them—or two pieces of metal, or a cell and its surroundings; the diffusing material may be a metal, a gas, a protein, an organic product of metabolism; the membrane may be a plant surface, an animal's skin, a cell interface. In all cases the process under study is traced by labeling one of the constituent atoms of the diffusing material with an isotope and determining its fate. When, for example, the diffusing material is a labeled hormone and the membrane a cell interface, the investigator is conducting a research into the absorption and retention of hormones by a living system.

Just as in the case of the two salt solutions, isotopic tracers can be used to distinguish material entering a cell from that already present, even though no net change in the chemical composition of the cell occurs. Thus biologists for the first time are afforded a method for investigating directly the important problem of how a cell governs the uptake of material from its environment. The great sensitivity of tracer methods also makes it possible to study the exchange of material within the cell, and the constant intercourse among the agents and products of metabolism which circulate in the organism.

Any account of the genesis of this fundamental advance, which thus far has been the most beneficial and useful fruit of nuclear science, must begin with the early 19th-century English physician and chemist William Prout. In 1816 he suggested that all the elements might be built of the lightest atom, hydrogen. On the Prout theory, it was expected that the atomic weight of every element would be a whole number multiple of the atomic weight of hydrogen. It turned out, however, as the weights of the elements became accurately determined, that most of them did not obey any such law. Thus chlorine, which should have had an atomic weight of 35 or 36, actually had a weight 35.5 times hydrogen. Consequently the Prout hypothesis was abandoned. It might not have been if physicists had realized that the "irreducible" elements they measured so confidently were *mixtures* of atoms, and that the atoms themselves did follow the rule of integral atomic weights when compared with hydrogen.

The twin discoveries of X-rays and radioactivity in 1895 and 1896 changed the whole perspective. By means of new instruments capable of determining the properties of single atoms, it was soon found that most elements were actually families of atoms, the members of which, although chemically identical, differed from one another in certain physical properties, notably weight and stability. Chlorine, for example, was found to be a mixture of two types of atoms with weights of 35 and 37, present in the proportion of three to one respectively. Among the heavy elements were discovered a considerable number of radioactive varieties; thus a radioactive product of thorium, called Thorium C, was shown to be a member of the bismuth family, with a weight of 212 instead of 209—the weight of stable bismuth atoms. It was the British physicist-chemist Frederick Soddy, a collaborator of the great Ernest Rutherford, who in 1912 gave these variant atoms of the same element the name of isotopes, from the Greek *isos* (same) and *topos* (place), meaning that they occupied the same place in the periodic table.

Once the existence of isotopes had been discovered, chemists proceeded to attempt to separate them. The experiments they devised for this purpose naturally were based on the physical differences among isotopes that resulted from their differences in mass—such as different volatility, mobility in gases and liquids, and so on. No one was more active in this type of research than a young Hungarian chemist, George Hevesy, who had come in 1913 to Rutherford's fertile laboratory at Manchester, England. In a series of researches notable for their ingenuity and precision, Hevesy and his collaborators showed that the isolation of isotopes in noticeable quantities required the most arduous kind of laboratory procedures, involving many thousands of repeated distillations or diffusions. These observations, supported by those of other researchers, indicated that in ordinary chemical processes no appreciable differences in the isotopic composition of samples of an element would be noticed until the chemical methods for determination of atomic weight approached a precision of the order of one part in 10,000 to 100,000. Thus it was established that for all practical purposes ordinary chemical manipulations would produce no change in an element's isotopic composition.

Hevesy reasoned that if one could somehow change the isotopic composition of any sample of an element, the sample could then be distinguished from the normal element. In other words, an element could be labeled by altering its isotopic content, and the labeled material could be followed through any chemical reaction. In 1923 Hevesy made his first famous tracer experiment. Because radioactive isotopes were available only in the heavy elements, he chose to begin by studying the intake of lead by plants, and he used a radioactive isotope of lead, Thorium B, to label the material. He bathed the rootlets of young plants with solutions of lead nitrate mixed with Thorium B nitrate as the tracer. After intervals ranging from one hour to two days, he burned various parts of the plants and determined the amount of Thorium B present in each part by measuring the radioactivity of its ashes.

This experiment in plant nutrition was the beginning of the isotopic-tracer method in biology. The method was severely limited at first by the fact that all work had to be done with the heavy elements, such as lead, bismuth and mercury. No feasible method existed for the separation of stable isotopes in quantity. But the discovery of the neutron, of heavy hydrogen and of artificial radioactivity in the 1930s, and the subsequent development of large cyclotrons and atomic piles, solved these problems. By the middle 1940s there was available a practically unlimited supply of radioactive isotopes of nearly all the elements. Meanwhile Harold C. Urey and his collaborators at Columbia University, in a series of remarkable researches, developed methods for bulk separation of the stable isotopes of the important biological elements.

The assaying instruments are of two general types. In experiments using radioactive isotopes, the measuring instrument is the electroscope, the electrometer or the now familiar Geiger-Müller counter. The latter consists essentially of an ionization chamber and an electronic apparatus which perform the functions of detecting and amplifying each radioactive disintegration of an atom; the number and rate of disintegrations is a measure of the amount of labeled material present. In experiments using stable isotopes, the measuring instrument most commonly used is the mass spectrometer, which ionizes the atoms in a sample, swings them through a magnetic field and separates the isotopes by means of their differing masses. Thus the atoms of each isotope are deposited at a separate point on a collecting plate and the concentration of the labeling isotope in the sample determined.

The biochemist's use of tracers is focused primarily on this basic problem: How does a given molecule play its part in the metabolism of a living organism? More specifically, what is the mechanism by which the molecule is mobilized either as a source of energy or as a contributor to the structure of living cells? In the investigation of this problem, some of the outstanding contributors have been the late Rudolph Schoenheimer and David Rittenberg of Columbia University, Harland G. Wood of Western Reserve University, and Vincent du Vigneaud of Cornell University. Their principal tools have been the stable isotopes hydrogen 2, carbon 13 and nitrogen 15.

The result of these researches may be summed up in the important general finding that a living system is a finely balanced complex of chemical reactions which, like the salt solution in the beaker, is in a continual state of flux. Metabolism is not a simple, one-way process. The substances of cells are constantly being built up and broken down from a "metabolic pool" of chemically active fragments of molecules that circulate in the organism. Schoenheimer aptly likened the adult organism to a military regiment:

"[It has] a size which fluctuates only within . . . limits, and a well-defined, highly organized structure. The individuals of which it is composed are continually changing. Men join up, are broken, and ultimately leave after varying lengths of service. The incoming and outgoing streams of men are numerically equal, but they differ in composition. . . . Recruits may be likened to the diet; their retirement and death correspond to excretion."

This analogy is admittedly incomplete; it fails, for example, to depict the chemical interaction of body constituents in the living system. Yet it remains an admirable illustration of the meaning of the "dynamic state" in biology.

The tracing of a metabolic process, as already indicated, involves two basic steps: 1) labeling a food or another substance fed to the organism, and 2) analyzing the intermediate and ultimate products that may be formed from this substance to determine the amount, if any, of the tracer isotope present. In biological research it is sometimes preferable to use a stable isotope rather than a radioactive one, to avoid the danger of damaging the organism by radiation effects.

The labeling process itself is merely a matter of chemical preparation. The compound to be fed to the organism is synthesized chemically in the usual way; the only difference is that for one of the components in the molecule a single isotope, or an unusual proportion of that isotope, is used instead of the natural element. Thus if the atom to be labeled is nitrogen, the compound is prepared with nitrogen 15. Supplies of separated isotopes are now available commercially or from an institutional laboratory.

The tracing of the labeled material, however, is somewhat less simple than a game of hare and hounds. The labeled compound must be prepared in such a way that the label is not lost, either because of excessive dilution in the pool of the same material that is already present in the organism, or because of interference by biochemical processes not connected with the one being studied. Moreover, the success of the experiment often depends on an accurate estimate of how thoroughly the labeled compound mixes with the same material in the organism; if the mixing is incomplete, it becomes difficult to judge the significance of the concentration of labeled material finally found in the cells where it is used. A further difficulty enters with regard to the purity of the sample; in tracing the intermediate steps in metabolism it is usually no simple problem to isolate the molecules carrying the label in a state that is sufficiently pure to permit an unambiguous analysis of the meaning of their labeled content.

Let us consider now a classic example of tracer research. This study, of the type known as "precursor-product" research, in which it is shown that a certain compound, B, is derived from a precursor compound A, was conducted by John C. Sonne, John M. Buchanan and Adelaide M. Delluva at the University of Pennsylvania Medical School. It was designed to determine the sources of the various carbon atoms found in uric acid, a product of the breakdown of protein in birds, snakes, lizards and invertebrates. The structure of uric acid is shown at the top of the diagram. Each atom is numbered for identification according to its position in the molecule. The objective of the experiment was to learn the respective origins of the carbons numbered 2, 4, 5, 6 and 8.

The investigators used pigeons as the test animal. They fed or injected into a group of birds several simple carbon compounds which might be the original sources employed by the pigeons to synthesize uric acid. These compounds were labeled with carbon 13 in various ways. Thus one of the compounds, lactic acid, which has the formula $H_3CCHOHCOOH$, had three carbons available for labeling; the carbon 13 atom could be placed in the carboxyl group

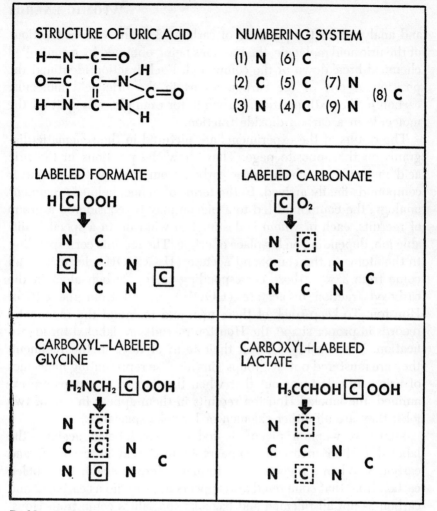

STRUCTURE OF URIC ACID

NUMBERING SYSTEM

(1) N (6) C
(2) C (5) C (7) N
(3) N (4) C (9) N
(8) C

LABELED FORMATE

LABELED CARBONATE

CARBOXYL–LABELED GLYCINE

CARBOXYL–LABELED LACTATE

Problem in this example of "precursor-product" research is to trace carbon atoms in uric acid back to the compounds that contribute them. Complete structure of uric acid is given at top left; in later diagrams its hydrogen and oxygen atoms are omitted for clarity. Numbering system is one used in text. When compounds with various carbon atoms labeled (squares) were fed to birds, they appeared at various points in uric acid structure. Dotted squares indicate only traces of labeled carbon were recovered in uric acid.

(COOH) or in the first two positions, the so-called alpha and beta carbons.

After the birds had been fed these compounds for a day or two, uric acid was recovered from their excretions. It was then purified and analyzed for the presence of carbon 13 in the various fractions of the uric acid molecule, the fractions being obtained by a controlled chemical breakdown of the compound. Each fraction identified the position of one of the carbon atoms in the uric acid molecule; carbon in the sixth position, carbon-6, for example, was split off the molecule in a carbon dioxide fraction.

The results of the experiments are pictured in the schematic diagrams on the opposite page. They show the positions in the uric acid molecule at which tracer carbon from each of the precursor compounds finally arrived. In the terms of Schoenheimer's regiment analogy, the compound fed to a pigeon may be considered a group of recruits, each of whom is destined to wind up in a specific suitable job, depending on his place of origin. The recruits corresponding to the atoms in the compound acetate (H_3CCOOH), let us say, all come from Texas; those corresponding to the carbon atom in the carboxyl fraction of acetate (COOH) are all from the city of Houston. To keep track of these men and maintain the regimental records in proper shape, the Houston recruits are labeled for identification. When soldiers finish their term of duty in the regiment, they are mustered out by groups (in this case representing molecules of uric acid). It turns out that when the departing groups are examined, the labeled Houston recruits in them appear in one of two jobs; they are all either riflemen or bazooka specialists.

Similarly, when a pigeon is fed carboxyl-labeled acetate, the labeled carbon invariably appears in uric acid as carbon-2 and carbon-8. When a pigeon is fed formate (formic acid), the labeled carbon from that compound also appears in uric acid as carbon-2 and carbon-8—not all riflemen and bazooka specialists come from Houston. On the other hand, carboxyl carbon does not always appear in the same positions; the carboxyl carbon derived from the amino acid glycine, for example, becomes carbon-4 in uric acid. It is as if the fate of each recruit depends not only on the type of city but also on the state from which he comes. Thus the most important basic finding developed by this research is that the compounds fed to a living organism are all used in different ways.

Measurements of the tracer carbon in uric acid established that carbons-2 and -8 could be derived from formate or from the carboxyl group of acetate, that carbons-4 and -5 could be contributed by the amino acid glycine, that carbon-6 came mainly from carbon dioxide but was also derived in small amounts from other compounds.*

Yet so intricate is the mechanism of metabolism that the appearance of a tracer isotope in the final product may be deceptive. For example, all compounds labeled with tracer carbon in the carboxyl group are subject to breakdown in the body, yielding labeled carbon dioxide. As a result, when the labeled carbon is found in a later product it is difficult to tell whether the product's carbon is normally derived from the original compound or from carbon dioxide. Often the solution of the problem requires close attention to the time factor in the diffusion of the isotope.

An outstanding example of this type of research was a series of experiments recently conducted by David Shemin and David Rittenberg at Columbia University. They were studying the manufacture of hemin by the human body. Hemin is the iron-containing blood pigment that combines with globin, a protein, to form hemoglobin, the substance that transports oxygen in the blood. The researchers had established that in the rat the nitrogen of hemin is derived mainly from the amino acid glycine (NH_2CH_2COOH). To trace the process in a human subject, they labeled glycine with nitrogen 15 and fed small amounts of the compound (a total of one and one-half ounces) to the subject for three days. Then at regular intervals they obtained from the subject samples of a hemin derivative called protoporphyrin and examined it for evidence of the tracer nitrogen.

Now the result one would normally have expected, in view of what was known about the dynamics of metabolism in general, is somewhat as follows: The labeled glycine would mix with the glycine already present in the circulation and within a relatively short time would become available for incorporation into the new hemin constantly being produced in the body. As new red cells were formed, labeled hemin would appear in the circulating blood. Meanwhile, unlabeled hemin would be removed from the blood as older red cells were destroyed. Thus the proportion of labeled hemin in the blood would increase steadily as long as labeled glycine was fed. Shortly after the subject returned to a normal diet containing un-

* A few months after the first publication of this article, Buchanan and his co-workers found that no carbons from acetate were incorporated into uric acid. The mistake arose because labeled formate was present, by accident, in the carboxyl-labeled acetate prepared for the experiment.

labeled glycine, the situation would be reversed. Now unlabeled nitrogen would be coming into the hemin and labeled nitrogen would be going out. If the red cells were constantly exchanging material with their surroundings—a condition characteristic of practically all metabolic processes—the proportion of labeled hemin would soon start to drop. The curve recording the abundance of labeled hemin in the blood during the course of the experiment would follow a certain familiar pattern: it should rise sharply at first, level off for a brief period, and then decline.

The actual result was quite different. After the feeding of labeled glycine was stopped on the third day, the concentration of labeled hemin did not level off but continued to increase for nearly 25 days! Then it became stabilized for a long period. Not until the 100th day did the concentration begin to drop.

The deduction to be drawn from these facts was clear. It appeared that hemin, unlike other active tissue, was comparatively stable. Instead of being continually broken down and rebuilt, red cells, like human beings, evidently have a definite average lifetime during which they remain intact. Normally they are destroyed or die only after reaching a certain age. The observations indicated that the average human red cell has a life span of about 130 days. The new red cells that were formed while labeled glycine was available retained the label until they died, and the labeled nitrogen then released was not used again in the manufacture of hemin.

Obviously this finding has considerable significance to medical science. Tracer nitrogen can be used to study the life span and destruction of hemin and red cells in various types of human blood disorders. Such researches have already begun and they have a bright future.

It must not be supposed that precursor-product researches are the most important type of investigation accessible to tracer techniques. They are only one aspect of the central problem in biochemistry: the elucidation of the mechanisms by which the body regulates and integrates its constant breakdown and synthesis of materials. In any given biochemical process, an important role is played by linked chains of reacting agents which act as intermediates in accepting and passing along certain necessary atomic fragments supplied by material taken into the body. Each step in such a process is controlled in general by an enzyme. All enzymes appear to be proteins. The entire process is self-contained and self-regenerating: it proceeds in a cyclic fashion. The tracer approach is particularly useful in ferreting out possible intermediates and participating molecules that are not obviously involved or are not observable by conventional chemical methods alone.

The great subtlety of the tracer approach is well illustrated by still another technique. Sometimes the biochemical product being studied does not exist in the organism in sufficient quantity to be isolated or analyzed. In such cases some of the product is added to the labeled precursor when the latter is fed to an organism. This unlabeled "carrier" material adds to the body's supply of the product; it is fed in a quantity sufficient to permit analysis of the product, but not too great to dilute the label beyond detection.

The mixing of labeled with unlabeled material is extremely useful in analytical biochemistry. Suppose, for example, one desires to ascertain whether a given compound is present in a cell extract; let us say the problem is to determine the percentage of glycine present in a mixture obtained by hydrolysis (breakdown by water) of a cell protein. This is a formidable problem by conventional chemical methods, for complete recovery of the glycine in a pure state is required and amino acids are not easy to separate from one another.

The tracer method solves it easily. We add to the extract a carefully measured quantity of labeled glycine. The labeled sample mixes with the glycine in the extract. Now we can measure the amount of glycine originally present by measuring the dilution of the label, which depends only on the relative amounts of the original glycine and the labeled addition. Suppose, for example, we add 10 milligrams of glycine containing 10 units of labeled carbon to one gram of the original protein containing an unknown amount of g ycine. We now isolate and purify a small sample, say one milligram, of the mixture of labeled and unlabeled glycine, and measure the concentration of labeled carbon in it. We find that the concentration is one half of one unit. In the labeled glycine that was added the concentration was one unit per milligram. Thus it is clear that the added glycine doubled the amount originally present, which means that the extract contained 10 milligrams of glycine. Here, then is a method of analyzing any compound for one of its components without separating out all of the component in a pure state: we need only measure the labeled content of a known amount of added material and the labeled content of a purified sample of the combined mixture; a simple formula then gives the amount of the component being measured. This method is being developed with many variations as one of the most useful analytical tests in biochemistry.

These examples suggest, but by no means completely define, the immense new frontiers opened by the isotopes in biochemistry. And biochemistry is only one of the fields of application for tracers. Their usefulness is equally impressive in physiology. An inquiry in which tracers have been especially helpful is the investigation of the biological role of elements that living cells use in vanishingly small amounts, such as boron, molybdenum, manganese, copper, and so on. The functions of these elements have been little understood, for biologists have lacked reliable techniques for studying them. By the use of labeled samples which make it possible to detect microscopic amounts of these materials, research workers have now begun to develop considerable data on their absorption, retention and excretion by the organism.

In medicine, the application of the tracer method is still only in its infancy, but much has already been done. The use of radioactive isotopes in treating certain blood disorders and some cancers is now routine. Radioactive phosphate has been shown to have definite advantages over X-rays in the treatment of polycythemia vera, a disease of the red blood cells, and of some types of leukemia, the cancerlike blood disorder. Radio-iodine is becoming a popular prescription for the control of hyperthyroidism.

In medical research, tracer studies have already made several invaluable contributions. Investigations with tracer iron of the conditions for survival of human red cells have improved the method of storing blood for transfusions. Tracers have made possible accurate measurements of the blood volume in the body under a variety of conditions, ranging from the normal condition to that of patients in extreme shock. They have permitted studies of disturbances in iron physiology that attend pregnancy. They have shed light on the dynamics of inflammation; tracers may even make it possible to locate internal sites of inflammation in the circulatory system without recourse to surgery. Labeled sodium has been used to diagnose disturbances of circulation in the small peripheral blood vessels, and to test the value of various drugs used to dilate or open the vessels.

Valuable as these direct medical harvests are, it seems clear that the most profound results of tracer research in the coming years will be achieved at the fundamental level of biochemistry and physiology. Even the most cautious observers agree that this research promises incalculable benefits to mankind.

PART 5

I. PHOTOSYNTHESIS *by Eugene I. Rabinowitch*

The name of Rabinowitch is attached to two important enterprises in science. The first is the subject of this essay, on which he has also written the indispensable review volume bearing the same title. His other preoccupation is the *Bulletin of the Atomic Scientists,* which he and a number of other Manhattan District colleagues founded in 1945. That highly influential forum "for science and public affairs" continues publication largely through the fortitude of Eugene Rabinowitch, who manages not only to get it to press 10 or 12 times a year but also to find underwriters for its annual deficit. Born in St. Petersburg (Leningrad) in 1901, Dr. Rabinowitch is qualified for his work on the central problem of photosynthesis by a diverse background in biochemistry and physics acquired at Berlin, Goettingen, Copenhagen, London and M.I.T. He is now research professor in the department of botany at the University of Illinois.

II. PROTEINS *by Joseph S. Fruton*

Before going to Yale in 1945, where he is now chairman of the department of biochemistry, Joseph Fruton was for 11 years a member of the staff of the Rockefeller Institute for Medical Research. Dr. Fruton was born in Poland; he came to the U. S. as a youth and took his B.A. and Ph.D. degrees at Columbia University.

III. ENZYMES *by John E. Pfeiffer*

Like George Gray and Harry Davis, John Pfeiffer is a science journalist, rather than a scientist. He has been science editor of *Newsweek,* science director of the Columbia Broadcasting Corporation and a member of the staff of SCIENTIFIC AMERICAN. Mr. Pfeiffer is at present writing a book on the brain; his project after that will be a volume on modern astronomy, to be written on a Guggenheim Fellowship.

IV. CALCIUM AND LIFE *by L. V. Heilbrunn*

L. V. Heilbrunn is known to the rising generation of physiologists as one of its great teachers. *An Outline of General Physiology,* which he wrote in 1937, is rated as indispensable by both teachers and students of the subject. He is now at work as co-editor (with Friedl Weber) on an encyclopaedic treatise on protoplasm, which is to be published by Springer-Verlag of Vienna under the title of *Protoplasmatologia.* Dr. Heilbrunn is professor of zoology at the University of Pennsylvania and a trustee of the Marine Biological Laboratory at Woods Hole, Massachusetts.

ORIGIN OF LIFE

Introduction

BIOLOGY is compartmented into a host of special sciences. There are departments for the study of physiology, genetics, and all the variety of life processes; departments, with appropriate Latin names, for the study of life in ponds, on mountain tops, in caves and deserts and all the other nooks and crannies it has managed to invade; and departments to match each phylum, order and genus into which the plant and animal kingdoms are divided. It is in biochemistry, presumably, that biology at last gets down to the study of life itself. The biochemists are engaged in the task—notoriously subtle—of taking the clock to pieces to find the tick. In our time biochemistry has traced the tick of life back to the point where it can be described in terms of molecular recombination and the exchange of energy.

The investigations reported in this section cover a broad range of questions about the physics and chemistry of life. How does photosynthesis carry raw materials across the chlorophyll bridge between the inorganic reservoir and the organic machine? How are proteins synthesized out of the simple elements and what is the structure of their fantastically complex molecules? Where do the life processes obtain the energy to force their reactions against the "chemical gradient"? How do enzymes quicken the breakdown of living matter and its recombination into new forms? What roles do the inorganic minerals—notably calcium—play in organic systems? Each of these questions presents a different approach to the main question: What is the origin of life and how is it maintained in the hierarchies of complexity from the virus to man?

None of the authors of this chapter would suggest he has the answer to the main question. They show us the way to another kind of understanding. We begin to yield our secure grip on the distinction between animate and inanimate and to comprehend living things as a part of the continuum of matter and energy, as natural as galaxies. We find a new universality in the processes of life when we discover that the same enzyme mediates the metabolism of both yeast and brain cells. And we

215

learn that life is just as various in the realm of biochemistry as it is in the other departments of biology—too rich in variety and invention to yield one over-all definition in answer to our question.

Meanwhile, biochemistry has begun to play another vital role in our lives. There is perhaps no realm of pure science in which the successes of research are so quickly translated into practical benefits. From the beginning of his history, man has attempted to manipulate the processes of life for his own ends, but the interference has been largely on a hit or miss basis. The knowledge of how life is engineered in terms of physical and chemical relationships has given him a much more direct control of his environment.

I. PHOTOSYNTHESIS

by Eugene I. Rabinowitch

ALL THE ANIMALS on land and in the sea, including man, are but a small brood of parasites living off the great body of the plant kingdom. So far as we know, green plants alone are able to produce the stuff of life—proteins, sugars, fats— from stable inorganic materials with no help but the abundantly flowing light of the sun. This is the process called photosynthesis. Each year the plants of the earth combine about 150 billion tons of carbon with 25 billion tons of hydrogen, and set free 400 billion tons of oxygen. In endlessly repeated cycles the atoms of carbon, oxygen and hydrogen come from the atmosphere and the hydrosphere (the world sea) into the biosphere (the thin layer of living things on the earth surface and in the upper part of the ocean). After a tour of duty that may last seconds or millions of years in the unstable organic world, they return to the stable equilibrium of inorganic nature. Few are aware, incidentally, that perhaps as much as 90 per cent of this giant chemical industry is carried on under the surface of the sea by microscopic algae.

The organizations of atoms in the biosphere are distinguished from those of the inorganic world by two characteristics: chemical complexity and high energy content. In the inorganic state they are simple molecules of carbon dioxide (CO_2), water (H_2O), carbonic acid (H_2CO_3), carbonate and bicarbonate ions (CO_3^{--} and HCO_3^{-}). In striking contrast is the complexity of even the simplest organic compounds, such as glucose ($C_6H_{12}O_6$)—not to speak of the enormous and intricate structures which are the molecules of proteins. It is this complexity that permits the almost infinite variability of organic matter. One thing, however, all the multifarious organic molecules have in common: they are all combustible, *i.e.*, they have an affinity for oxygen. When oxidized, they release an average of about 100 kilocalories of heat for each 10 grams of carbon they contain. Thus all organic matter contains a considerable amount of "free" energy, available for conversion into mechanical motion, heat, electricity or light by gradual or sudden combination with oxygen. Such oxidations are the mainspring of life; without them, no heart could beat, no plant could grow upward defying gravity, no amoeba could swim, no sensation could speed along a nerve, no thought could flash in the human brain. Certain lower organisms can exist using sources of chemical energy not involving free oxygen, such as fermentation, but these are "exceptions that prove the rule."

Photosynthesis by plants is the process by which matter is brought up from the simplicity and inertness of the inorganic world to the complexity and reactivity that are the essence of life. The process is not only a marvel of synthetic chemical skill, but also a *tour de force* of power engineering. When plant physiologists and organic chemists study photosynthesis, they are struck most of all by the feat of manufacturing sugar from carbon dioxide and water. When physicists or photochemists contemplate the same phenomenon, they are awed and intrigued by the conversion of stable, chemically inert matter into unstable, energy-rich forms by means of visible light.

Not only are scientists unable to duplicate photosynthesis outside the living plant cell; they do not know of *any* halfway efficient method of converting light energy into chemical energy. If we knew the chemical secret of photosynthesis, we could perhaps by-pass plants as food producers and make sugar directly from carbonates and water. If we knew its physical secret, we could perhaps by-pass the "storage-battery" function of plants and produce chemical or electrical energy directly from sunlight. We might decompose water, for example, into an explosive mixture of hydrogen and oxygen that could be used as a source of heat or power.

The story of the little we know about photosynthesis begins with Joseph Priestley, who announced in 1772:

"*I have been so happy as by accident to hit upon a method of restoring air which has been injured by the burning of candles and to have discovered at least one of the restoratives which Nature employs for this purpose. It is vegetation. One might have imagined that since common air is necessary to vegetable as well as to animal life, both plants and animals had affected it in the same manner; and I own that I had that expectation when I first put a sprig of mint into a glass jar standing inverted in a vessel of water; but when it had continued growing there for some months, I found that the air would neither extinguish a candle, nor was it at all inconvenient to a mouse which I put into it.*"

In these words Priestley, religious reformer, philosopher and spare-time naturalist of the Age of Enlightenment, described one of the most momentous observations in the history of experimental biology:

the discovery of the capacity of plants to produce oxygen. Seven years later Jan Ingen-Housz, a Dutch physician of the Empress Maria Theresa, added a second important fact that *"this wonderful operation is by no means owing to the vegetation of the plant* (as Priestley thought) *but to the influence of the light of the sun upon the plant."* In 1782 a Swiss pastor, Jean Senebier, found that the presence of "fixed air" (carbon dioxide) was needed for photosynthesis; in 1804 another citizen of the learned city of Geneva, Nicholas Theodore de Saussure, found that water also contributes its elements to the synthesis.

Julius Robert Mayer, the German doctor who discovered the law of the conservation of energy, wrote in 1845:

"Nature set herself the task to catch in flight the light streaming towards the earth, and to store this, the most evasive of all forces, by converting it into an immobile form. To achieve this, she has covered the earth's crust with organisms, which while living take up the sunlight and use its force to add continuously to a sum of chemical difference. These organisms are the plants."

The overall physico-chemical equation of photosynthesis was thus established more than 100 years ago:

$$\text{Carbon dioxide} + \text{water} + \text{light} \rightarrow \text{organic matter (sugar)} + \text{oxygen} + \text{chemical energy.}$$

Ever since it has been clear that photosynthesis, being the only ultimate source of food, is the source of all animal energy; indirectly, through wood, coal, peat and oil, it is also the source of practically all our technical energy. However, despite innumerable studies, we still do not understand photosynthesis.

The biochemist feels that he "understands" a chemical process in the living cell if he knows its successive stages, the intermediate compounds that are formed and the enzymes (biological catalysts) that make the individual stages possible. This knowledge he achieves by taking the biochemical apparatus apart and putting it together again. His ultimate aim is to imitate a biochemical process, such as respiration or conversion of carbohydrates to fats, in the laboratory and to describe each step in detail by chemical equations. Considerable success has been achieved in this direction in many metabolic reactions, but in the case of photosynthesis, we know very little about the individual reaction steps and even less about the catalysts which make them possible. We can prepare extracts from plant cells containing chlorophyll or other pigments that are present wherever photosynthesis goes on. But not only are these extracts incapable of photosynthesis; we cannot find in them any catalytic or photochemical properties clearly related to the probable steps in photosynthesis. We may then decide that chemical methods of fractionating the plant-cell contents are too drastic, and attempt to take the cell apart mechanically. We take a giant green cell, such as that of some algae, and prick it with a needle in an attempt to reach its interior. Immediately photosynthesis ceases. The cell still respires, it is alive, but oxygen liberation and carbon dioxide absorption have stopped. Thus we find ourselves in the position of being asked to find out how an automobile motor operates without being permitted to lift the hood.

Until quite recently, there was no known way of dismantling the biochemical apparatus and studying its parts separately. It was an "all or nothing" situation: at one moment we had a living cell engaged in complete photosynthesis; at the next it was an agglomeration of broken parts or isolated chemical components, with no indications of what role, if any, they had played in photosynthesis while the cell was whole and alive.

Within the last few years the situation has changed. The problem looks less forbidding. Some progress has come from an improved general understanding of the mechanism of chemical, in particular photochemical, reactions. Some has come from the improvement of old experimental methods and the development of new ones: exact analysis by electrochemical and pressure-measuring devices, the use of radioactive tracers, quantitative spectrophotometry. None of these methods, not even the glamorous radioactive tracers, provides an immediate solution to the secrets of photosynthesis, but all of them together promise progress toward the understanding of photosynthesis *in vivo* and its imitation *in vitro*.

From measurements of the rate of photosynthesis under different conditions, the English plant physiologist F. F. Blackman concluded as early as 1905 that photosynthesis is not a single photochemical reaction, but must include at least one "dark" reaction (one which is not affected by light). As the intensity of illumination is increased, the rate of photosynthesis (as measured, for example, by the volume of oxygen produced each minute) does not increase indefinitely but approaches a saturation state in which a further increase of light intensity has no effect. This suggests a two-stage process in which only one stage can be accelerated by light. The reasoning may be illustrated by the following analogy. If a million men are to be transported overseas in two stages—first by train to the harbor and then by ship to their destination—the provision of more and faster trains will accelerate the transportation only up to the point where all available ships are used to capacity. Thereafter the further improvement of rail transportation merely jams up the harbor. Conversely, it will serve no useful purpose to provide more ships than can be filled by the arriving trainloads. In photosynthesis there is a stage or stages accelerated by light (corresponding to the railroad journey), and another stage or stages independent of light (the ship voyage). The rate of the latter may depend on how many enzyme molecules—equivalent to ships—are available in the plant cell. It is useless to accelerate the light reaction beyond the capacity of dark reactions to transform the products of the light reaction.

The division of photosynthesis into a photochemical stage and a dark one is brought out clearly by experiments with flashing light. After a plant is exposed to a brief light flash lasting, say, for .0001 second, the liberation of oxygen continues in the dark for about .02 second; more exactly, a dark interval of about .02 second is necessary to obtain the maximum oxygen production per flash. The experiment measures directly the time required for the completion of the slowest dark reaction in photosynthesis. It is equivalent to the time our ships need to complete the ocean crossing and return to the harbor. It has also been found that there is a limit to the amount of photosynthesis that can be brought about by a single light flash: the maximum yield is about one molecule of oxygen for 2,000 molecules of chlorophyll present in the cell. This is surprising. One would expect that during a short flash each chlorophyll molecule would have a chance to perform its function once, producing one molecule of an intermediate product. Consequently the maximum production would be one molecule of oxygen for each chlorophyll molecule, or for a small number of them. James Franck of the University of Chicago suggested an explanation of the paradox: the maximum yield per flash depends not on the number of chlorophyll molecules but on the number of molecules of the enzyme involved in the second stage (*i.e.*, on the number of ship berths, rather than train berths). In other words, the flash can produce as many intermediate molecules as there are chlorophyll molecules, but comparatively few of them will succeed in completing the subsequent dark stage to produce oxygen.

But why cannot the intermediates wait at the harbor while ships (the catalytic enzymes) ferry some to the other shore and return for a second, third or fourth load? Franck's explanation is that the intermediate photoproducts are unstable. Unless they are immediately processed by a "finishing" catalyst, they disappear by "back reactions" before the catalyst is ready for a second load.

Thus we have the following outline of photosynthesis. It consists of a light stage and a dark stage. The light stage produces unstable intermediates; the dark stage stabilizes them by conversion into the final products, oxygen and carbohydrate. The rate of photosynthesis is limited by the bottleneck of a dark reaction which can process only one molecule, or at most a small number of molecules, of intermediates per 2,000 molecules of chlorophyll each .02 second.

From our general knowledge of the nature of chemical reactions, particularly those involved in metabolic processes, we can make a guess as to the probable nature of the light stage in photosynthesis. Plant respiration, the reverse of photosynthesis, involves two types of reactions: those which break the carbon chains in the large organic molecules, and those which remove hydrogen atoms from association with carbon and, with the catalytic help of enzymes, transfer them to oxygen, thus forming water. In photosynthesis the same two types of processes must be involved, but running in the opposite direction—the transfer of hydrogen from water to carbon dioxide, and the building of carbon chains. Of these two types of reactions, the transfer of hydrogen is the one that *liberates* energy in respiration, hence this must be the one that *stores* energy in photosynthesis. The energy that is stored comes from light. Consequently the light reaction in photosynthesis in all probability is a hydrogen transfer from oxygen to carbon "against the gradient of chemical potential," meaning from a more stable to a less stable form. To use a mechanical picture, in respiration the hydrogen atoms run downhill; in photosynthesis, the impact of light quanta (discrete "atoms" of light), absorbed by chlorophyll, sends them uphill.

Let us illustrate this reversible process by mixing a solution of the dyestuff thionine with a solution of ferrous sulfate. In intense light, the color of the dye disappears in a second or less; in the dark, the color immediately returns. This is an example of how an oxidation-reduction reaction can run in one direction in the dark and in the opposite direction in light. In light, ferrous iron reduces the dye to a colorless form and is itself oxidized to ferric iron; in the dark, ferric iron oxidizes the dye back to the colored form and is reduced to ferrous iron.

The energy content of the final products of photosynthesis—sugar and oxygen—is well known; it is represented by the amount of heat produced when sugar is burned to carbon dioxide and water. The energy is 112 kilocalories per gram atom (one gram multiplied by an element's atomic weight) of carbon. This, then, is the minimum energy that has to be supplied by light in photosynthesis. To reduce (hydrogenate) a molecule of carbon dioxide to the "reduction level" of sugar, four hydrogen atoms must be transferred to the molecule:

$$CO_2 + 4H \longrightarrow > C(H_2O)_2$$

To move these hydrogen atoms "uphill" from water to carbon dioxide, each of the four atoms must receive a push equivalent to at least one fourth of 112 kilocalories, or 28 kilocalories per gram atom of hydrogen. These pushes must be supplied by light.

Niels Bohr and Albert Einstein showed in 1913 that light is absorbed by atoms or molecules in the form of quanta of definite energy content, which is proportional to the wavelength of the light. Red light, which is strongly absorbed by chlorophyll, has quanta with an energy content such that it provides about 40 kilocalories per gram atom of the absorbing atoms. Obviously one such quantum is not enough to transfer four hydrogen atoms (requiring 112 kilocalories). A minimum of 2.8 quanta is needed in theory, not to violate the law of conservation of energy; considerably more is likely to be required in practice, because of inevitable "friction losses" in all natural processes.

In 1923 Otto Warburg, the German cell biologist, first attempted to measure the "quantum yield" of photosynthesis—the number of quanta required to reduce one molecule of carbon dioxide. This implied measuring exactly the light energy absorbed and the volume of oxygen produced. In order to obtain the maximum possible yield, it was advisable to work in very weak light to avoid saturation effects. The measurements therefore were very delicate. The results were striking: Warburg found an absorption of four quanta per molecule of oxygen! This indicated that one quantum is used to move each hydrogen atom; the ratio is 28:40 = an efficiency of 70 per cent.

Warburg's result, however, did not remain unchallenged. Other groups of researchers were unable to confirm Warburg's observations. Instead they found yields of eight or 10 quanta per oxygen molecule; some values were as low as eight, but none was lower. Warburg and co-workers, on the other hand, have recently observed quantum requirements as low as 2.8, corresponding to 100 per cent conversion of light into chemical energy! The question is still unsettled; the weight of evidence favors the higher value: eight or more quanta per molecule of oxygen. Even at this value, however, the 35 per cent yield in energy conversion by plants is very respectable—considering that we do not know of any reaction produced by visible light *outside* the plant cell which would convert as much as 10 per cent of absorbed light into chemical energy. If some economical means could be found to capture and convert even 10 per cent of light energy, the discovery could produce a greater revolution in our power economy than can be expected at present from atomic energy.

One plausible picture of how chlorophyll may use eight light quanta to move four hydrogen atoms from water to carbon dioxide is this: A chlorophyll molecule absorbs a quantum and is raised to an "excited," energy-rich state. It is then able to pull a hydrogen atom away from water (or from a product derived from water by a dark, enzymatic reaction). In this reduced form, chlorophyll takes up another light quantum and uses its energy to force the same hydrogen atom on a reluctant "acceptor," such as carbon dioxide or a compound derived from carbon dioxide by a dark reaction. It is as if a workman, suspended halfway on the face of a building, fortified himself with a drink, hauled a construction piece up from the ground, and then, fortified with a second drink, threw this piece up to the roof.

It has long been assumed that chlorophyll is the only agent that can perform this trick. It has been well known that all green plants also contain yellow or orange pigments (carotenoids, identical or similar to the pigments of carrots and egg yolk), and that many algae contain red or purple pigments. But all plants capable of photosynthesis were found to contain chlorophyll, and chlorophyll alone

224

among the plant pigments absorbs red light. Since photosynthesis proceeds satisfactorily in pure red light, light absorption by chlorophyll must be *sufficient* to bring about photosynthesis, and from that experimental fact there is only a short step to the assumption that it is the *necessary* prerequisite.

Recently, however, the position of chlorophyll has been challenged. Indications have been found that the light energy absorbed by the yellow pigments also is utilized in photosynthesis and that among some red algae the light absorbed by red pigments is even more effective than the light absorbed by chlorophyll. On the other hand, experiments using fluorescence to trace the course of light energy indicate that whenever yellow or red pigments contribute to photosynthesis, they do so by transferring the light energy they have absorbed to chlorophyll. This seems to be true even when the light absorbed by the red pigments is more effective than that absorbed by chlorophyll itself. Chlorophyll thus retains its key position as the one "photocatalyst" in photosynthesis, the other pigments being merely "photon-catchers."

So we are beginning to get a somewhat clearer idea of the events in the light stages of photosynthesis, and recently we have also gained a little information about the dark stages. The total process, as we have noted, proceeds in two separate sequences: 1) the oxidation of water, which releases free oxygen, while hydrogen becomes attached to some intermediate "acceptor"; 2) the hydrogenation of carbon dioxide to produce carbohydrates. Each sequence of reactions apparently has a separate catalytic system. The two sequences and their relation to each other are pictured in an accompanying diagram, which shows the separate sequences as two legs, with chlorophyll as the bridge between them. One sequence (the left leg) begins with molcules of gaseous carbon dioxide. These are first bound or fixed in a form suitable for reduction, perhaps by enzymatic formation of an organic acid. The bound carbon dioxide is then reduced by hydrogen atoms supplied in light by chlorophyll, which has recovered the hydrogen from water in the other sequence. The reduction, in turn, is followed by other enzymatic transformations which lead to a carbohydrate molecule.

In the right leg we first have a similar binding of the water molecule, followed by its dehydrogenation in light, and then the enzy-

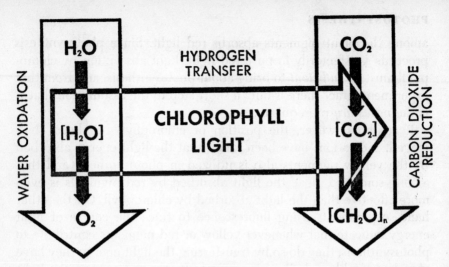

Photosynthesis can be visualized as an H-shaped diagram. On vertical leg at left, water gives up hydrogen, releasing oxygen. Hydrogen is then transferred to carbon dioxide by agency of light and chlorophyll. Reduction (hydrogenation) of carbon dioxide produces carbohydrate (*bottom of right leg*). Brackets indicate intermediate steps and compounds that are unresolved.

matic conversion of the residue into free oxygen, perhaps through the intermediate formation of a peroxide, similar to but apparently not identical with hydrogen peroxide.

Some of these reactions are now being studied with the help of isotopic tracers. We are concerned with the fate of three kinds of atoms—hydrogen, carbon, oxygen. The heavy non-radioactive hydrogen, deuterium, (H^2), has been available since before the war; the weakly radioactive tritium (H^3) is not yet generally available. Three isotopes of carbon are usable: the short-lived C^{11}; the long-lived C^{14}; and the stable, non-radioactive C^{13}. C^{14}, which the atomic pile at Oak Ridge has made widely available, is by far the most useful. To our great sorrow no radioactive isotope of oxygen is known; the stable isotope O^{18} offers the only means of studying the fate of this important element. Tracer carbon is an appropriate tool to study the reduction of carbon dioxide to carbohydrate. Tracer oxygen could be equally useful for the study of the oxidation of water to oxygen. Tracer hydrogen could help to trace the processes in the bridge between these sequences, including the primary photochemical process.

The application of carbon tracers to photosynthesis, first by Samuel Ruben (who died in a research accident during the war); later by

Melvin Calvin and co-workers at the radiation laboratory of the University of California and by Hans Gaffron and E. W. Fager at the University of Chicago, has produced many interesting but often confusing results. A large variety of compounds appear "tagged" with radiocarbon after brief exposure to radioactive carbon dioxide both in light and, to a lesser extent, in darkness. This situation was somewhat simplified when we learned that many metabolic intermediates, in animal as well as plant tissues, can exchange ordinary carbon for radiocarbon when exposed to radioactive carbon dioxide. (More precisely, they decompose, losing ordinary carbon dioxide, and are reformed by taking up tagged carbon dioxide from the air.) This exchange takes place rather slowly and, once it was recognized, could be reduced to insignificance by performing experiments very rapidly. For example, a suspension of algae is brought in contact with radioactive carbon dioxide, exposed immediately to light for a second or so and then instantly killed in boiling alcohol.

Paper chromatography, an important experimental innovation based on relative adsorption rates (chromatography is discussed in detail in the essay by Joseph Fruton), is used to identify the minute amounts of compounds that have taken up radiocarbon in these fast experiments. Once their nature is known, it is possible to isolate them also by the more orthodox methods of analysis. The most important of these compounds has proved to be a phosphoric acid ester of glyceric acid (PGA). Glyceric acid is a well-known compound with the composition $C_3H_6O_4$, more than half-way between carbon dioxide (CO_2) and sugar ($C_6H_{12}O_6$) in respect to the length of the carbon chain (C_3) and the level of reduction (1½ hydrogen atoms per oxygen atom). After one second of exposure to light and radioactive carbon, as much as 95% of the radiocarbon taken up can be located in this compound. Another C_3 compound, pyruvic acid ($C_3H_4O_3$) appears immediately afterwards as the reaction continues, perhaps indicating the first step in the transformation of phosphoglycerate. Other compounds—with four, five, six and seven-membered carbon chains—appear in rapid succession. The trail (or trails) along which the conversion of CO_2 to sugars proceeds, is then soon lost in their jungle and remains at present a matter of more or less plausible speculations. One question is particularly puzzling: if glyceric acid is the first product of the uptake of carbon dioxide in photosynthesis, what is the compound from which it is formed? On paper, it must be a very simple compound, with the composition $C_2H_6O_2$; but so far, it has eluded all attempts to find it.

The isotope O^{18} was employed in a study of photosynthesis by Samuel Ruben and Martin Kamen before the war, and a very significant result was obtained. Using CO_2 and H_2O containing heavy oxygen, they showed that all the oxygen liberated in photosynthesis originated in water; none came from carbon dioxide. (This is a fine example of information that only isotopic tracers can provide!) Their finding was consistent with the hypothesis that photosynthesis is fundamentally a transfer of hydrogen atoms from water to carbon dioxide, with the oxygen left behind.

Of all our recent glimpses into the mysterious mechanism of photosynthesis, none appears more promising than the one which was made possible by a discovery made in 1937 by R. Hill of Cambridge University. It had been known for a long time that dried and powdered leaves, when suspended in water and illuminated, sometimes release a small amount of oxygen, although of course they produce no carbohydrates. Hill found that the oxygen production could be increased and sustained for an hour or more if the suspension was provided with a supply of ferric oxalate or some other ferric salt. Later studies by others showed that ferric salts could be replaced by quinone or by certain dyes. These compounds have one thing in common: they are all rather strong oxidants. They accept hydrogen atoms much more readily than carbon dioxide does. The most plausible interpretation of the results is that when leaves are disintegrated, a product is obtained which still contains the chlorophyll bridge and the enzymatic system required to produce free oxygen (the right leg in our schematic diagram), but which has lost the left leg's enzymatic system. The suspension therefore can oxidize water and liberate oxygen in light, but it cannot reduce carbon dioxide and produce carbohydrate. Without the aid of enzymes, the carbon dioxide is unable to perform its job of "accepting" hydrogen, but the reaction is kept going by substituting a more willing acceptor (*e.g.*, ferric iron) for carbon dioxide.

Thus we have, in effect, photosynthesis without carbon dioxide! Microscopic studies yield further pertinent evidence. The photosynthesizing cells of almost all plants contain chlorophyll (and accompanying pigments) in microscopic bodies called chloroplasts. Closer observations have revealed that the pigments are further concentrated within the chloroplasts in tiny "grana," almost too small to be seen under ordinary microscopes but beautifully revealed under the electron microscope. Analysis of the Hill suspension shows that its particles are whole or broken chloroplasts or isolated grana. The grana, then, are the "bricks" in the catalytic structure of photosynthesis which permit the liberation of oxygen from water in light but do not contain the enzymes needed to take up and reduce carbon dioxide. The essential independence of the two enzymatic systems thus receives striking confirmation.

The "Hill reaction" is perhaps the widest crack that has yet appeared in the earlier picture of photosynthesis as a unique and indivisible process. We have lifted the hood and taken out the motor and it still runs, even though we do not know how to connect it again to the drive shaft.

It has not yet been possible to perform the converse of this feat: *i.e.*, to eliminate the right leg of the photosynthetic apparatus and keep the left leg functioning. However, something closely related to this has been found to occur in nature: organisms capable of reducing carbon dioxide in light, but unable to use water as a reductant. As substitutes, these organisms use hydrogen sulfide, thiosulfate, or even free molecular hydrogen.

Certain species of bacteria, purple or green in color, contain a pigment called bacteriochlorophyll which is closely related to the chlorophyll of green plants. They thrive in sulfur waters or other media containing reducing agents. Cornelius B. van Niel, the Dutch microbiologist now at Hopkins Marine Laboratory in California, has shown that they can build their organic matter from inorganic materials in light. He suggests the following general chemical equation for their photosynthesis:

$$2RH_2 + CO_2 \xrightarrow[\text{light}]{\text{bacteriochlorophyll}} > [C(H_2O)]_2 + 2R.$$

This equation is similar to the one usually given for photosynthesis of green plants, but it is more general, since R can stand for many different radicals, consisting of a single atom or a chemically unsaturated group of atoms. If R is taken as representing an oxygen atom, we have plant photosynthesis; if it is taken to represent an atom of sulfur, we have the photosynthesis of "sulfur bacteria," and so on.

With one stroke van Niel's interpretation of the chemical activity of purple bacteria has removed photosynthesis by green plants from its entirely unique position in biological chemistry and placed it alongside other types of "photosynthetic" processes. Does this discovery indicate that the purple and green bacteria are predecessors of green plants, relics of a time when life was restricted to those places on earth where inorganic reductants were present? A time, perhaps, when the earth's crust was less well stabilized chemically than it is now, and hydrogen sulfide, sulfur, or perhaps even free hydrogen were available in much greater abundance?

Further exciting vistas are opened by the similarity of the photosynthetic purple bacteria to some colorless bacteria which are capable of reducing carbon dioxide by means of the same or similar reductants but without the help of light. They use instead the chemical energy liberated by enzymatic oxidation of these reductants by the oxygen of the air. This phenomenon is called bacterial chemosynthesis; it, too, may be a relic of the more primitive forms of life. Hans Gaffron found in 1939 that if certain unicellular green algae are deprived of oxygen, they cease to be capable of ordinary photosynthesis but become capable of reducing carbon dioxide in light if hydrogen is provided as a substitute reductant to replace water! It looks as if lack of air causes these algae to simulate purple bacteria.

In photosynthesis, we are like travelers in an unknown country around whom the early morning fog slowly begins to rise, vaguely revealing the outlines of the landscape. It will be thrilling to see it in bright daylight!

II. PROTEINS
by Joseph S. Fruton

There is present in plants and in animals a substance which . . . is without doubt the most important of all the known substances in living matter, and, without it, life would be impossible on our planet. This material has been named Protein.

S O WROTE Gerard Johannes Mulder, a Dutch agricultural chemist, in 1838. It was in his scientific papers that the word "protein," from the Greek *proteios*, meaning of the first rank, made its first public appearance. The word had been suggested to him by the great Swedish chemist Jöns Jacob Berzelius (who also introduced to chemistry "catalysis," "polymer" and other important terms). Mulder and his great German contemporary Justus von Liebig thought that protein was a single substance—a basic structural unit existing in the same form in materials as diverse as egg white and blood fibrin. This was soon shown to be an error; the number and variety of proteins was found to be legion. But Mulder has certainly been proved correct in his emphasis on the importance of proteins to life.

The proteins are one of the three principal organic constituents of living matter (the fats and carbohydrates are the others), but in the importance and diversity of their biological functions they stand alone. They represent nearly one half of the body's dry matter. (About 70 per cent of the body is water.) Of the total body protein, more than a third is found in the muscles: the protein myosin forms the fibers that are the fundamental contractile elements in muscular movement. The bones and cartilage account for another 20 per cent; here the protein collagen contributes to the structural stability of the skeleton. And the skin has about 10 per cent of the body protein, the skin protein keratin serving to protect the interior tissues against attack from the external environment.

Perhaps the most important of the proteins are the enzymes. These substances are present in only minute amounts in comparison with myosin, collagen or keratin, but they are indispensable for the promotion and direction of the body's myriad chemical reactions. They are discussed in detail later in this chapter.

Some of the hormones also are proteins. These remarkable products of the secretory activity of the endocrine organs are carried by the blood in infinitesimal amounts to the tissues, where they play a decisive role in the regulation of the pace and direction of metabolism. Still other proteins are the antibodies of the blood, which defend the organism against viruses (themselves proteins) and the harmful substances produced by disease-causing bacteria. Finally the genes, the basic units of heredity, are believed to contain a particular type of protein called nucleoproteins.

Where there is such diversity of function, there must be a corresponding diversity of chemical structure. The number of identified proteins is extremely large, and growing rapidly. To learn what proteins are present in living systems, to examine their chemical structure, to explain their biological functions in terms of their structure —these are among the most fundamental problems of modern biochemistry. When the answers to them are found, we shall have a much more precise definition of what has been termed "the physical basis of life."

To study the chemical structure of a particular protein it is necessary to destroy the cellular organization characteristic of life and to extract the protein with a suitable solvent, such as a dilute salt solution. This procedure inevitably brings into solution many of the other proteins present in the cell, and the task of separating the desired protein from the unwanted materials becomes a test of the experimenter's skill and, very frequently, of his good fortune. Proteins are extremely fragile chemical structures. This imposes serious restrictions upon the kind of laboratory procedures the chemist may use in separating them from one another. By careful control of factors such as salt concentration, alcohol concentration, acidity and temperature, fairly selective precipitation of a given protein may be achieved; today it is often possible to isolate a single protein from the dozens or even hundreds present in a tissue extract. Many individual proteins have been obtained in the form of crystals which may be recrystallized at will, thus leading to further purification. Although crystallinity *per se* is not a satisfactory criterion of a protein's purity, the availability of crystalline proteins has for the first time given to the biochemist reproducible material for the study of the chemical nature of these substances.

All proteins are made principally of carbon, hydrogen, oxygen, and nitrogen. It is the nitrogen, representing from 12 to 19 per cent of the molecule, that is the special mark of a protein. Most proteins also contain small amounts of sulfur, and many have some phosphorus. Over a century ago Mulder, noting these very small proportions of sulfur and phosphorus in his crude protein preparations, concluded that the protein molecule must be huge, since each molecule had to contain at least one atom of these elements. Proteins, in other words, are "macromolecules." Not until modern methods of measuring their molecular weights were developed, however, was it possible to determine just how large they are.

The most reliable and convenient method is to whirl them in an ultra high-speed centrifuge, a technique devised by the Swedish physical chemist, The Svedberg. The proteins are spun in a centrifuge at speeds up to 70,000 revolutions per minute, which develops a centrifugal force as much as 400,000 times that of gravity. In such a field the large protein molecules move outward from the center of rotation with selective speeds: the larger they are, the faster their motion. An ingenious optical apparatus measures the rate of this molecular sedimentation, and the molecular weight can then be calculated.

Now these measurements show that the smallest known protein is about 13,000 times as heavy as a hydrogen atom, *i.e.*, its molecular weight is about 13,000. The largest known proteins have molecular weights of the order of 10 million. To determine the structure of molecules of such sizes is obviously quite a formidable problem. One can get some idea of how formidable it is by comparing a protein with a nonprotein organic molecule. A particularly complex example of the latter is one of the penicillins, which has a molecular weight of 334 and the formula $C_{16}H_{18}O_4N_2S$. This molecule is simplicity itself in comparison with the typical milk protein lactoglobulin, whose molecular weight is about 42,000 and whose approximate formula is $C_{1864}H_{3012}O_{576}N_{468}S_{21}$.

The structure of the penicillin molecule was worked out only after years of joint labor by the great chemists of the U. S. and England. The usual method of attacking such a task in organic chemistry is 1) to establish the proportions of the various elements in it, 2) to develop a working hypothesis about the arrangement of these atoms by a process of trial and error, and finally 3) to test the hypothesis by trying to synthesize the molecule from known substances by known chemical reactions. By this classical procedure organic chemists within the past 100 years have found the formulas of about 500,000 organic compounds, including many that are made by living organisms. But in a protein the number of atoms is so large that it has not been possible to establish its molecular structure by this method.

What the protein chemist can do at present is to cleave the protein molecule into the smaller molecules of which it is composed—the amino acids. The protein is cleaved by treatment with acids or alkalis; because water enters into the reaction, the process is called hydrolysis. When the protein has been broken down into its amino acids, the chemist can then obtain some clues to its composition, because the atomic structures of the amino acids themselves have all been determined by the classical methods of organic chemistry.

The amino acids formed by hydrolysis of a protein have certain structural features in common: each has an acidic carboxyl group (COOH) and a basic amino group (NH$_2$) or imino group (NH). Both the acidic and basic groups are attached to the same carbon atom, the so-called alpha-carbon. Since a carbon atom has four chemical bonds, this same alpha-carbon has two other units linked to it. One of these is invariably a hydrogen atom. What distinguishes the amino acids from one another is the fourth group attached to the alpha-carbon. This group, the so-called side chain, differs in each amino acid.

The simplest amino acid, glycine, was isolated in 1820 by the French chemist Henri Braconnot. He obtained it by acid hydrolysis of gelatine. The list of known amino acids from proteins has now grown to 22. It is not likely that many new ones will be added to it. Every protein amino acid except glycine can exist in two geometrical forms, one the mirror image of the other; by convention these are designated the "L" and "D" forms. Only the "L" type of the amino acids is obtained by the hydrolysis of proteins.

During the past 80 years an intensive effort has been devoted to the development of experimental methods for the accurate quantitative determination of the relative amounts of the various amino acids formed by hydrolysis of a protein.

Because the various amino acids are structurally similar in all respects except the nature of the side-chain group, the problem has been to find chemical processes that will select and isolate them on the basis of this rather subtle mark of identification. In the past few years this goal has been achieved, and it is now possible to say that the problem of protein analysis has been solved, at least in principle.

The most valuable contribution to the solution was the development of new chromatographic techniques for the separation of amino acids. Chromatography itself was invented by the Russian botanist Michael Tswett in 1906. It got its name from the fact that it was first used to separate pigments. Tswett was interested in isolating the chlorophyll pigments of green leaves. He conceived the idea that they might be separated quickly by taking advantage of their differing rates of adsorption by an adsorbing material. As he himself described it, "if a petroleum ether solution of chlorophyll is filtered through a column of an adsorbent (I use mainly calcium carbonate which is stamped firmly into a narrow glass tube), then the pigments . . . are resolved from top to bottom into various colored zones, since the more strongly absorbed pigments displace the more weakly absorbed ones and force them farther downwards. This separation becomes practically complete if, after the pigment solution has flowed through, one passes a stream of pure solvent through the adsorbent column. Like light rays in the spectrum, so the different components of a pigment mixture are resolved on the calcium carbonate column . . . and can be estimated on it qualitatively and quantitatively. Such a preparation I term a chromatogram and the corresponding method, the chromatographic method."

Tswett realized that "the adsorption phenomena described are not restricted to the chlorophyll pigments, and one must assume that all kinds of colored and colorless chemical compounds are subject to the same laws." It was many years before this brilliant intuition of Tswett was appreciated. Since 1930 chromatographic techniques have been developed to separate colorless as well as colored chemical compounds. It was the English chemists A. J. P. Martin and R. L. M. Synge who found a way to apply the technique to the separation of amino acids. They introduced the use of a starch column as the adsorbent. From this idea William H. Stein and Stanford Moore of the Rockefeller Institute for Medical Research later worked out a beautiful method for the precise quantitative analysis of all the amino acids formed when a protein is hydrolyzed.

So far only a few proteins have been studied by this method, but the results attained are sufficient to indicate its great importance in protein chemistry. Nevertheless, it has not by any means solved the problem of protein structure. What this advance has accomplished is to bring the proteins to the historical stage reached by the simpler organic molecules a century ago, when it became possible to calculate the relative proportions of the atoms constituting an organic compound. From this, organic chemists were able to go on to discover the arrangement of the atoms in an organic molecule. In the same way protein chemists are now in a position to proceed with greater confidence to consider the spatial arrangement of amino acids in a protein molecule.

The next question concerns the nature of the linkages between the individual amino acids. The most widely accepted hypothesis is one proposed independently by Emil Fischer and the German biochemist Franz Hofmeister in 1902. They suggested that the amino group attached to the alpha-carbon of one amino acid is joined to the carboxyl group attached to the alpha-carbon of another. This union is accompanied by the elimination of the elements of water from the molecules that unite. It is this bond that is broken when the elements of water are introduced in acid hydrolysis. The bond is called a "peptide linkage," and the Fischer-Hofmeister hypothesis is known as the peptide theory.

The theory has been supported by so much experimental evidence that its essential truth seems highly probable. Support for the theory came from work on artificially synthesized peptides, i.e., groups of amino acids linked together by peptide bonds. In this Fischer was the pioneer; he pointed out that "if one wishes to attain clear results in this difficult field, one must first discover a method which will permit the experimenter to join the various amino acids to one another in a stepwise manner and with well-defined intermediary products." Much research has been done during the past half-century, and is still continuing, to develop methods for the laboratory synthesis of peptides. One of the greatest achievements came in 1932 with the invention of the "carbobenzoxy" method by a distinguished pupil of Fischer, Max Bergmann, who was then director of the Kaiser Wilhelm Institute for Leather Research in Dresden and later came to the Rockefeller Institute for Medical Research.

In living systems proteins are hydrolyzed by enzymes such as pepsin, trypsin and chymotrypsin. These catalysts act to speed up the hydrolytic reactions, thus making it possible for them to take place at the ordinary temperatures and under the normal acidity condi-

236

tions of the organism. According to the peptide theory, these enzymes cause the hydrolysis of peptide bonds. If this theory is correct, then the same enzymes should hydrolyze simple peptides synthesized in the laboratory. For a long time protein chemists made intensive but vain efforts to create synthetic compounds that could be hydrolyzed by the enzymes, and their failure was interpreted by some as evidence against the peptide theory. In 1937, however, the author, working in Bergmann's laboratory at the Rockefeller Institute and using the carbobenzoxy method of peptide synthesis, succeeded in forming synthetic compounds which were specifically hydrolyzed at their peptide bonds by these enzymes. This finding strongly supported the Fischer-Hofmeister theory.

An additional support for the peptide theory is the finding that when the hydrolysis of protein is interrupted before the protein is entirely converted to amino acids, peptides can occasionally be isolated. The isolation of peptides obviously is not easy, for we have here the same difficulty of separating the components of a complex mixture that we encounter in the case of amino acids. The problem is, if anything, even more complicated, because the number of different peptides into which a protein may be split is considerably larger than the number of possible amino acids, and the amount of each peptide is very small. The new methods of chromatography appear well suited to the fractionation of peptides, and many investigators are now using them. Another valuable new approach to the problem has recently been provided by Lyman C. Craig of the Rockefeller Institute. He has developed a separation method based on the same general principles as are the familiar laboratory procedures for the extraction of a chemical substance from one solvent, such as water, by another solvent, such as ether. It is to be expected that this promising technique and the chromatographic method will form the main experimental lines of attack in the investigation of peptides obtained from proteins.

The brilliant work now being done by the British chemist Frederick Sanger at Cambridge University provides further grounds for optimism. Sanger is studying the structure of the important protein hormone insulin. He has subjected insulin to the action of the reagent dinitrofluorobenzene, a substance that combines readily with the alpha-amino groups at the ends of insulin's peptide chains. The result of this combination is a compound called dinitrophenylinsulin

(DNP-insulin). All the end alpha-amino groups in the compound are occupied by dinitrophenyl (DNP) groups. When the protein is subjected to hydrolysis by strong acid, all the peptide bonds are cleaved, but the linkages between the DNP group and the alpha-amino groups of the end amino acids are essentially unaffected. In other words, each end amino acid remains linked to a DNP group. Since the DNP group confers upon any compound in which it is present a distinctive yellow color, the DNP-amino acids can readily be separated by means of chromatography, and their structure can be determined.

By this method of analysis Sanger has shown that the basic structural unit of insulin is composed of four peptide chains, of which two have glycine and two have phenylalanine as the end amino acids. He has also offered strong evidence for the idea that the four peptide chains are held together by bridges consisting of two sulfur atoms; when these disulfide bridges are broken by a relatively mild chemical treatment, the peptide chains are separated from one another. The next step, which is now occupying Sanger's attention, is to determine the sequence of the amino acids in each of these four chains. For this purpose the four DNP-peptides are subjected to partial, rather than complete, acid hydrolysis of the peptide bonds. The result is a series of yellow DNP-peptides that can be separated by chromatography. If the elucidation of the structure of these peptides, now under way, is successful, the next step will be to attempt to construct a picture of the insulin molecule from the nature of the fragments.

The studies discussed thus far suggest that the protein molecule is a threadlike structure of several hundred amino acids, linked to one another through peptide bonds and strung out to form a chain (or several chains joined by disulfide bridges) of considerable length. There is good evidence that this description actually applies to insoluble proteins such as keratin or silk fibroin and to a few soluble ones, notably the myosin of muscle and the fibrinogen of blood.

But most of the known proteins are not threadlike or fibrous. The enzymes, the protein hormones and all the blood proteins except fibrinogen are globular. They are soluble in water or salt solutions, but this characteristic solubility may readily be lost or decreased by subjecting the proteins to relatively small increases in temperature (up to 140 degrees Fahrenheit) or to mild acidity. This alteration

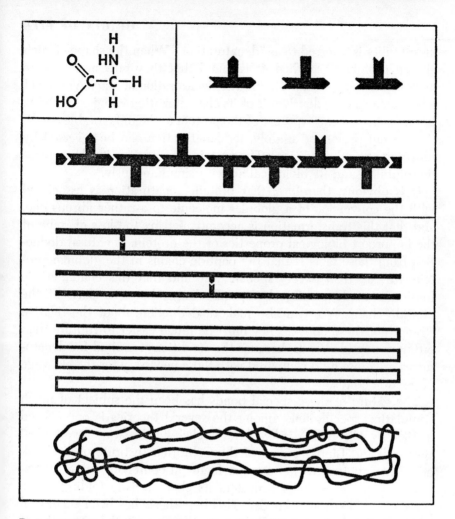

Protein structure is shown in this series of simplified sketches. *At upper left* is
atomic diagram of the amino acid, glycine. The simple hydrogen at
right is the distinguishing characteristic of this compound; the atomic
structure to the left of its bond is common to all aminos. *To right* of
glycine are the three symbols used to indicate amino acids in peptides;
ends are shown acidic (pointed), neutral (flat) and basic (notched).
Immediately below appears a series of amino acids arranged in a peptide
chain; unions are acidic end to basic. Line below peptide is a symbol
for longer chain of amino acids. *Middle panel* diagrams a fibrous protein
constructed of polypeptide chains. Some of the chains are joined by linkages between amino acid side chains. *Next panel* shows how native globular protein is made up of organized polypeptide chains. This structure
must be visualized as occupying three dimensions. *At bottom*, a globular
protein has been denatured. The skein of polypeptide chains is disorgan-
ized and the structure of the native protein is thus destroyed.

in solubility is referred to as "denaturation." When the shape of such altered proteins is studied, it is found that they now approximate more closely the firous proteins. The denaturation of an enzyme or of a hormone usually deprives it of its characteristic biological activity. In some cases, if the exposure of the protein to the unfavorable conditions is not prolonged unduly, its denaturation can be reversed by restoring normal conditions. The protein then regains its characteristic solubility and its biological activity simultaneously.

It is obvious, therefore, that protein denaturation is associated with the conversion of a globular molecule to a rather fibrous one, and that this transformation is accompanied by the loss of some of the important biological properties of the protein. A natural deduction from these facts is that the peptide chains in the globular protein are coiled in a very specific way and that this characteristic folding is made possible by specific bonds between parts of the peptide chains. We can also make a deduction about the relative strength of these bonds. They must be much easier to rupture than ordinary peptide bonds, because the conditions required for denaturation are quite mild compared with those necessary for the cleavage of peptide bonds.

The nature of these special bonds has been the subject of much stimulating speculation. Among the several theories is one offered in 1936 by Alfred E. Mirsky of the Rockefeller Institute for Medical Research and Linus Pauling of the California Institute of Technology. They suggested that a major factor in conferring upon the extended peptide chain of a protein its characteristic folding is the presence of "hydrogen bonds." This hypothesis has been successful in accounting for many of the known differences in the properties of "native" and denatured proteins. According to the theory, there are a multitude of bonds formed by the "sharing" of a hydrogen atom of an amino group with an oxygen atom of a carboxyl group. Taken individually these hydrogen bonds are weak, but in a protein molecule with several hundred amino-nitrogen atoms and a correspondingly large number of carboxyl-oxygen atoms, these weak bonds reinforce one another so that a stable structure results.

To the concept of the protein molecule as a long polypeptide chain or chains composed of many amino acids must therefore be added the idea that in each kind of protein the parts of the peptide chain have a characteristic internal arrangement which is responsible for that molecule's particular chemical and biological properties. Consequently the problem of protein structure involves not only the already formidable task of establishing the arrangement of the amino acids in the peptide chain, but even more difficult questions as to the nature and position of the bonds that are broken during denaturation.

Although the artificial creation of a protein molecule still lies beyond the powers of the chemist, it is no problem at all for the living organism. The living cell, whether of an amoeba or of a mammalian liver, performs the task of protein synthesis with rapidity and precision. Many organisms can use proteins foreign to their make-up, break them down to the component amino acids or peptides, and use the fragments to create their own characteristic proteins. Moreover, the proteins of a living organism are not laid down and kept intact throughout its life; rather there is a ceaseless breakdown and resynthesis of body proteins. In a sense, therefore, the problem of life is the problem of how living systems make proteins and how they constantly counteract the tendency toward protein degradation. Thus the study of the mechanisms by which cells synthesize proteins is perhaps the most challenging task of biochemistry.

A logical starting point for this investigation is the comparatively simple question of how living systems put two amino acids together to form a peptide bond. Many laboratories in this country are actively engaged in the exploration of this question. Although no clear-cut answer can yet be given, there are several hints as to its possible solution.

Among the views that have been entertained is the theory that in living cells the formation of peptide bonds is effected by the same enzymes that cause the breakdown of the peptide bonds after death. The principal support for this hypothesis has come from the demonstration that protein-splitting enzymes can indeed link two amino acids together to form a peptide bond. But this process will occur only under certain specific conditions. The most important of these is the necessity of counteracting the natural tendency of the protein-splitting enzymes to effect the hydrolysis of peptide bonds. A simple experimental procedure for achieving this reversal of hydrolysis is to choose a reaction which will result in formation of an insoluble peptide that comes out of solution as fast as it is formed. By taking advantage of this fact, it is possible to show without question that protein-splitting enzymes can catalyze the synthesis of peptide bonds.

The attractive feature of the theory is the fact that these enzymes exhibit a striking specificity of action on peptide linkages. In the case of the protein-splitting enzymes, the specificity of enzyme action depends largely on the nature of the amino acids that participate in the formation of the peptide bond. What is more, these enzymes act only at peptide linkages that involve amino acids of the L-type, which we noted earlier to be characteristic of the protein constituents. For example, one enzyme may catalyze peptide synthesis only when the amino acid that contributes the alphacarboxyl group for formation of a peptide is L-tyrosine or L-phenylalanine. Replacement of either of these two amino acids by any other amino acid, as far as tests made so far show, prevents action by the enzyme. Indeed, biochemists know of no other group of biocatalysts that compares with the protein-splitting enzymes in their selective action on peptide bonds. By virtue of this sharp specificity, therefore, these enzymes are well fitted to direct, precisely and reproducibly, the complex sequence of successive peptide syntheses required for the formation of a protein.

To observe their synthetic action, however, it is necessary to remove the product from the reaction. In other words, work must be put into the system to counteract the natural tendency of the enzyme to hydrolyze the product after it is synthesized. It follows that if enzymes do actually perform peptide synthesis in cells, this process must be coupled to another reaction that provides the necessary chemical energy. There is excellent reason for believing that the chemical energy comes from the breakdown of foodstuffs such as glucose by oxidation or fermentation, but it has not been possible as yet to demonstrate how the breakdown of glucose is linked directly with peptide-bond synthesis in cells.

Much attention has been paid in recent years to the suggestion of Fritz Lipmann of the Massachusetts General Hospital that peptide bond synthesis involves the intermediate formation of amino-acid derivatives of phosphoric acid. The source of these phosphoric-acid intermediates, according to Lipmann's theory, is a "high energy" phosphate carrier such as adenosine triphosphate (ATP). The latter substance has been shown to play a decisive role in the exchanges of chemical energy that occur during the metabolic breakdown of sugars. Work in several laboratories during recent years has provided experimental evidence that phosphate-containing intermediates may indeed be involved in the biological formation of certain amides, such as hippuric acid or glutamine. These amides are closely related structurally to the peptides; they differ from the latter only in that the CO-NH bond links an amino acid to a non-amino acid group. The intervention of ATP in the synthesis of the amide bond of glutamine, first demonstrated by the late John F. Speck of the University of Chicago, is of especial interest because this glutamic acid derivative is widely distributed in the tissues and fluids of animals and plants.

Although the experimental data offered in support of the Lipmann theory are impressive, they do not yet present a picture that would account satisfactorily for the specificity of peptide-bond formation. Each of the two theories discussed above thus contributes to a different, but equally essential, facet of the problem; it may well be that the two theories are complementary, rather than mutually exclusive. Such a view is supported by work begun in Bergmann's laboratory in 1937 and continued by the author at Yale University. These experiments have shown that protein-splitting enzymes will catalyze the hydrolysis and synthesis of peptide bonds. They have also shown that the same enzymes will cause reactions in which one of the two components contributing to a peptide bond may be replaced by another, without the need for the introduction of appreciable chemical energy but with the same specificity exhibited in synthesis and hydrolysis.

If these results should prove to have general significance, it would mean that an amide containing an amino acid derivative linked to ammonia (*e.g.*, glutamine) could exchange the ammonia for an amino acid or even a peptide. The energy for this process would come from the synthesis of glutamine, which, as Speck has shown, may involve a phosphate-containing intermediate. The specificity of the enzyme that catalyzes the exchange of the amide-nitrogen for the alpha-amino nitrogen of an amino acid or peptide would then determine the nature and sequence of the amino acids linked by peptide bonds in the final product. As a further consequence of this hypothesis, it would follow that a simple peptide composed of two or three amino acids would be transformed, in the presence of a suitable enzyme, into a longer chain by the replacement of one of the amino acids by a peptide. Energy would be required for the formation of the simple initial peptide, perhaps via a phosphate-containing intermediate, but the further course of peptide synthesis would be under the directive control of the highly specific enzymes that act as peptide bonds.

Another avenue of approach to the problem of how peptide bonds are formed is to seek out biological systems that exhibit unusual requirements for certain peptides, as compared with their demand for the individual amino acids of which these peptides are composed. If a bacterial cell, for example, uses a peptide for growth more efficiently than it does the amino acids, that would suggest that the rate of synthesis of the peptide controls the rate of utilization of the amino acids for protein formation. This approach is being explored in studies of the bacterial metabolism of peptides at Yale University by Sofia Simmonds in collaboration with the author. They may be expected to provide valuable biological material for the unequivocal testing of the various hypotheses relating to the mechanism of peptide-bond formation.

From all this it must be abundantly evident that the decisive discoveries in the study of the biological synthesis of proteins still lie in the future. Whatever the answer concerning the enzymatic mechanism of peptide-bond formation turns out to be, clearly it will provide only a part of the picture of the total process. What, for example, is the nature of the forces that confer upon the biologically interesting proteins, such as the enzymes and hormones, their characteristic physical, chemical and physiological properties? A denatured insulin molecule, though rendered inactive as a hormone, presumably still contains the same amino acids as the active molecule, and the peptide linkages that join these amino acids apparently have not been broken measurably. How, then, is the peptide chain molded in the living cell so as to form an active hormone with its specific attributes? Are we dealing here with an intricate mechanism whereby a model of the finished product is available as a matrix upon which the fragments are assembled?

These questions cannot be answered as yet, but it is well to remember that biochemistry is a relative newcomer among the scientific disciplines. Its growth has been meteoric, and it is exerting a decisive influence on the future development of all aspects

244

of biology and their applications to medicine and agriculture. In the last analysis all the problems of biology meet in the unsolved problems concerning the structure and the mode of action of proteins. In groping for new experimental avenues into this great unknown, the protein chemist is thus probing into the basic questions of life. Whether he succeeds or not, he cannot help being filled with a sense of awe and humility in the face of what has justly been called the noblest piece of architecture produced by Nature—the protein molecule.

III. ENZYMES

by John E. Pfeiffer

S OME THREE MILLION of your red blood cells die every second. Or, to look at it another way, three million red cells are born every second, because the body continuously calls up reserves to keep the total count the same. The entire red-cell population is replaced in about three months, and cycles of birth and death turn even faster among the molecules in the plasma of the blood. This rapid molecular turnover goes on in relatively solid tissues as well as in the circulating blood. Deposits of fat, which were once believed to serve as warehouses for the storage of food reserves, are more like department stores during the Christmas rush. They seethe with biochemical activity, decomposition and synthesis neatly balancing each other so that within a few months entirely new fat deposits are created. The same goes for connective tissue, tendons and ligaments, blood-vessel walls, muscles. Swift changes occur even in the bones as the links of molecular chains are split and welded again in the ceaseless round of metabolism.

Yet the cycles of breakdown and synthesis proceed in the face of an apparent paradox. The great majority of biochemical reactions do not take place spontaneously. When they are tried in laboratory glassware, most chemical constituents of life combine or decompose at a rate far too slow for the pace of metabolism. The average protein must be boiled for 24 hours in a solution of 20 per cent hydrochloric acid to be thoroughly broken down. The body does the same thing in four hours or less, and without high temperatures and strong acids.

The phenomenon that makes life possible is catalysis—the action of certain substances that speed up chemical reactions thousands of times without themselves being changed. Industrial chemistry uses catalysts in the cracking of petroleum, in the synthesis of ammonia, and in many other processes; organisms use them to help build tissues and to degrade foodstuffs to simpler materials, as in the case of the four-hour breakdown of proteins. The catalysts of life are called biocatalysts, or enzymes, and the rise of biology has come with an increased understanding of what they are, what they accomplish, and how they work.

Enzymes are unaffected by the reactions they produce; they are destroyed only by wear and tear or poisoning. They operate in

246

amazingly small concentrations. A single cell has been estimated to contain about 100,000 enzyme molecules to accelerate its 1,000 to 2,000 chemical reactions—an average of only about 50 to 100 molecules for each process. A single molecule of the enzyme that splits hydrogen peroxide into water and oxygen (and creates the white foam when the antiseptic is placed on a wound), can transform more than 5 million peroxide molecules a minute. Other enzymes transform from 1,000 to more than 500,000 molecules in the same time.

Investigation has shown that these biochemical middlemen play a significant role in every vital process. They are key substances in the photosynthetic reactions that build plant tissues from water, carbon dioxide and sunlight. Enzymes turn leaves red and yellow in the fall, make the freshly cut surface of an apple or potato brown, convert grape juice into wine, and grain mash into whisky. But the chemical processes in which they participate are so obscure that it has taken centuries to elucidate a few fundamental principles, and every month new facts are published which promise to alter many currently accepted ideas.

The first enzyme was obtained in pure form and identified chemically by James B. Sumner of Cornell University in 1926. For nine years Sumner had been working to isolate an enzyme from the jack bean. The enzyme was urease, which decomposes the metabolic waste product, urea. Urease, like many other enzymes, is named after the substance on which it acts. The problem was to find a solvent that would dissolve urease and not other chemicals, and a substance that would then precipitate the enzyme.

The final process was, in Sumner's words, "absurdly simple." One day in April of 1926 he mixed jack-bean meal with acetone, a solvent that had been suggested by his former biochemistry professor at Harvard, and allowed the solution to filter overnight. Next morning he examined a drop of the filtrate under the microscope and saw something he had not seen before—tiny octahedral crystals. Then he centrifuged the crystals out of solution, concentrated them and found that the new solution possessed very strong urease activity. That afternoon Sumner telephoned his wife the news that was to win him a Nobel prize 21 years later: "I have crystallized the first enzyme." Urease turned out to be a protein with a molecular weight of 483,000. (One unit of molecular weight equals the weight of one hydrogen atom.)

Once the correct procedure had been found, it was easy to isolate urease. But there is no single process for the crystallization of all enzymes. To purify pepsin John H. Northrop of the Rockefeller Institute for Medical Research in Princeton, who with Wendell M. Stanley of the same institution shared the 1947 Nobel prize with Sumner, needed a far more complex process. He announced the method in 1930, nearly a century after pepsin had first been identified in the digestive juices. One of Northrop's chief problems was to precipitate pepsin without destroying it. It is a protein (as are all of the 40-odd enzymes isolated to date) and the molecular structure of protein is both exceedingly complex and maintained by a balance of electrical forces so delicate that a slight change in the chemical environment may distort it into a tangled mass which cannot be restored to the original pattern. That is what happens, for instance, when an egg is boiled and the white coagulates. To overcome this "denaturing," Northrop developed a precipitation technique that has since been used to crystallize several other enzymes.

The fermentation of sugar, yielding alcohol, is an admirable illustration of the detailed chemical processes engineered by enzymes. In the early days of enzyme study it was thought that this was accomplished by the single catalyst, zymase, and the glucose-to-alcohol reaction was represented by the following uncomplicated formula:

$$C_6H_{12}O_6 \xrightarrow{\text{zymase}} 2\,CO_2 + 2\,CH_3CH_2OH$$

This is a chemical statement to the effect that one molecule of glucose, catalyzed by zymase, yields two molecules of carbon dioxide and two of ethyl alcohol. But if fermentation were such a single-step process, it has been calculated, most of the resulting energy would appear as useless heat. Actually fermentation directly involves at least 12 enzymes—and it took hundreds of research workers from more than a dozen countries to unravel nature's scheme for altering glucose.

The molecule of glucose is built around a chain of six carbon atoms, the splitting of which is a crucial step in the fermentation process. Before this step can be taken, however, glucose must be suitably prepared for its destruction. Three enzymes transform glucose for the splitting, and nine more are involved in the remaining steps which lead to ethyl alcohol. This outline of the process is a poor reflection of the detailed chemical processes that propel it. These are presented in greater detail by the accompanying drawing.

This splitting of the carbon chain of glucose involves one of the most significant of all biochemical cycles. It requires a large amount of chemical energy, and the source is adenosine triphosphate, or ATP. The energy is obtained from one of ATP's potent phosphate groups, indicated by the two right-hand P's in the following simplified formula:

$$\text{Adenosine}-(\text{P})\frown(\text{P})\frown(\text{P})$$

The two right-hand phosphate groups are attached by chemical bonds which yield 12,000 calories of energy when they break. (The left-hand group yields only 2,000 calories.) The bonds may be considered a sort of "cement" of electrons holding the phosphate groups. Oscillating back and forth at high speed, the electrons endow the phosphate groups with an extra reactivity. The specific usefulness of ATP is that enzymes can transfer its phosphate groups to other substances—along with the energy of their oscillating bonds.

The energy that drives the cycles of fermentation is first obtained by splitting off the right-hand phosphate group and attaching it to glucose; this step, catalyzed by the enzyme hexokinase, leaves adenosine diphosphate (ADP) and a glucose phosphate which is catalyzed to a fructose monophosphate. Transforming this latter substance to fructose diphosphate means splitting another section from another ATP molecule.

At this stage the process leaves two ADP molecules wandering about with missing groups. Unless they are rebuilt as ATP, the entire cycle will grind to a stop. The deficit is made up during the next steps of fermentation when the six-carbon chain is finally split and two inorganic (non-ATP) phosphates are taken up into an intermediate compound. These inorganic phosphates, however, are low in energy content, and ATP will only accept the high-energy variety. The cell therefore uses an enzyme called triosephosphate dehydrogenase to remove two hydrogen atoms, which transform low-energy phosphates into high-energy phosphates.

The potent phosphate groups are then split off by another enzyme and attached to the two dismembered ADP molecules, forming ATP. The energy supply is thus sustained—but the cell does better than that. During the next stages two extra high-energy phosphate groups are created and passed back to other ADP molecules. This bonus may be used to accelerate fermentation or to provide the energy needed for the growth and reproduction of yeast cells.

There is still one biochemical loose end. During the manufacture of high-energy phosphate bonds, two hydrogen atoms have been lost. These are picked up by a special hydrogen carrier called coenzyme I. This substance now cannot participate in later reactions and, again, the entire cycle would break down unless it included a mechanism for freeing coenzyme I of its hydrogen. The opportunity comes at the very last step of fermentation, after pyruvic acid has been converted to acetaldehyde and carbon dioxide. Carbon dioxide goes off as a gas. Acetaldehyde, which remains, is just two hydrogen atoms short of being ethyl alcohol, the final product of the fermentation. The missing atoms are naturally presented by the hydrogen-bearing coenzyme I, and the latter is restored to perform its function.

This completes the fermentation process and some of its interrelated systems. Ethyl alcohol can then be taken internally and used to interfere with human enzyme systems. Incidentally, for every 99 parts of ethyl alcohol, yeast produces one part of fusel oil, a mixture of various higher alcohols which is not only responsible for most of the flavor of liquor but also for hangovers.

While some biologists traced the intricate cycles of fermentation, others studied the mechanism of muscle. Gradually, first from fragments of evidence and finally from an imposing structure of knowledge, both groups began to realize that the workings of yeast and muscle cells were very much alike. In fact, the processes that change malt and hops to beer, and those that provide the energy for an Olympic sprinter have 14 steps—and 11 of the 14 are exactly the same for the two types of process. The workings of the great ATP cycle and the wheel-within-a-wheel coenzyme I cycle are the same in both cases. One important difference is that in muscle contraction pyruvic acid is broken down to lactic acid instead of ethyl alcohol. The lactic acid is then carried by the bloodstream to the liver, where the reverse of the 14-step process builds it into animal starch, or glycogen. (Muscles cannot utilize glucose.) Another important difference is that the breakdown of glycogen yields three instead of two "bonus" ATP molecules. Muscles attain an efficiency of 60 per cent or better, as compared with the 50 per cent efficiency reached in modern steam turbines.

250

ATP not only supplies energy in muscle contraction, but also plays an important role in the workings of the nervous system. Nerve cells build up one of their essential chemicals, acetylcholine, with the aid of the enzyme choline acetylase, and the synthesis requires energy from ATP. That it may also be associated with the enzyme systems necessary for the movement of single-celled organisms is indicated in at least one case, the wriggling of sperm towards the unfertilized egg.

Water is the medium for the majority of biochemical processes. In water the molecules of life are in ceaseless thermal motion, occasionally reacting when they collide with one another. Essentially the function of an enzyme is to increase the rate of reaction. In a solution without enzymes the chance that a molecular collision will result in a reaction may be a trillion to one. If the appropriate enzyme is present, the probability will be much increased. To use the gambling term, enzymes lower the odds. The question is how they perform this mathematical feat by chemical means.

Any explanation of the phenomenon must account for certain experimental facts. One of the most obvious is that a given enzyme does not speed reactions among all the molecules of protoplasm. If this were the case, the result would be biochemical chaos. Actually enzymes are highly specific, producing reactive collisions only among the molecules of selected compounds. These compounds are generally known as substrates.

A spectacular example of enzyme specificity involves molecules that are made up of exactly the same atoms in different structural arrangements. Such close chemical relatives are known as isomers. In 1860 Pasteur discovered that tartaric acid, a by-product of wine fermentation, exists in two forms. When a beam of polarized light was transmitted through crystals of tartaric acid, some crystals turned the plane of polarization to the right, while others turned it to the left by exactly the same amount. Since both types of tartaric acid are identical in chemical composition, the difference must be in the arrangement of their atoms.

It has been shown that such pairs of crystals—called dextrorotary and levorotary—are found among many compounds, and are related to one another as an object to its mirror image or as a right-hand to a left-hand glove. Enzymes can make the subtle distinction between isomers. The muscle enzyme lactic dehydrogenase, for example, acts on levorotary lactic acid but has absolutely no effect on its mirror image, dextrorotary lactic acid.

Some enzymes are even more selective. The so-called hydrolytic enzymes, as an example, are involved in the following general type of reaction:

$$A - B + H_2O \longrightarrow AOH + BH$$

Here A — B represents a molecule consisting of two parts connected by a chemical linkage. Some enzymes will break down any molecule with a particular linkage regardless of the nature of the linked structures; others demand not only the right linkage but also the right part, say the B structure. Still other enzymes, the most specific of all, operate only on molecules that satisfy the three-way requirement that both the A and B parts and their linkage must be of a particular kind.

A more detailed explanation of what specificity means is furnished by the phenomenon of competitive inhibition. The enzyme succinic dehydrogenase catalyzes the breakdown of succinic acid and nothing else. Its effectiveness is considerably reduced, however, if malonic acid, the structure of which closely resembles that of succinic acid, is added to the solution. Experiments show that malonic acid, while not being changed itself, apparently attaches itself to the enzyme and takes it out of circulation by occupying a position on its molecule that would normally be filled by succinic acid. In other words, malonic acid seems to compete with succinic for an active region of the enzyme molecule.

These and other experiments suggest an attractive analogy to explain specificity. It is a theory which some protein chemists label "philosophy," although they concede that philosophy can be useful. The enzyme molecule can be visualized as a "lock" with notches and indentations of a particular pattern; the substrate molecule, in this case succinic acid, is the "key," and its configurations mesh into the enzyme pattern. Malonic acid, a very similar key, will fit the lock, but not perfectly. The fit is good enough to keep out succinic acid, but not good enough to unlock the door—hence the door stays closed. Perhaps the most brilliant experimental evidence for such a concept, adduced by E. S. G. Barron and his associates at the University of Chicago, involves three closely similar substances: acetic acid, monofluoroacetate and monochloroacetate. The only difference among

252

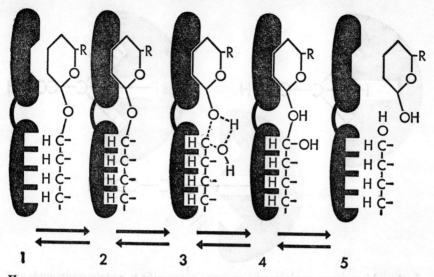

How enzymes may work is illustrated by this theoretical example. Here the enzyme, drawn in black, has two active parts which fit the molecule on which they act (1). When they combine with the molecule (2), they deform it in such a way (3, 4) that it is broken down (5) into two molecules, in this case sugar and alcohol.

them is that one hydrogen of acetic acid is replaced by a fluorine atom in monofluoroacetate, and by a chlorine atom in monochloroacetate. The links between each of the three atoms and the rest of the molecule to which they are attached are of different length. For hydrogen the link is 1.09 Angstrom units (one Angstrom unit equals one hundred millionth centimeter); for fluorine it is 1.41 A.; for chlorine it is 1.76 A.

An enzyme catalyzes the oxidation of acetic acid (a process involved in fat metabolism) and presumably its molecule contains "notches" into which the acid fits. The addition of monofluoroacetate to the solution completely inhibits the oxidation, meaning that this compound also fits the enzyme molecule. Monochloroacetate, however, appears not to fit, for it has no effect on the enzyme's ability to oxidize acetic acid. In other words, the tiny difference in length between the link of the fluorine-containing inhibitor and that of the ineffective chlorine compound (.35 A., or about 1/762,000,000 inch) is enough to prevent a sufficiently close lock-key fit.

The lock-key theory implies that there is some sort of fleeting union between enzyme and substrate, an implication which has been backed by many experiments. As a matter of fact, the spectroscope

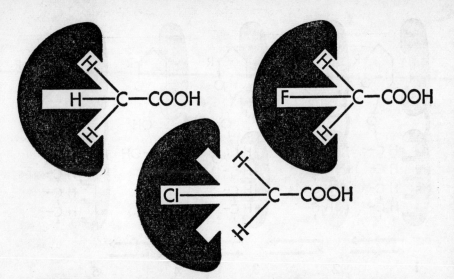

Lock and key theory explains why enzymes are specific to certain compounds and how their action can be inhibited by similar compounds. Here at left acetic acid fits an enzyme (black) and the reaction works. When a fluorine atom is substituted for a hydrogen atom, forming monofluoroacetate, the fit is close enough to occupy the enzyme and to block its reaction with acetic acid. However, when a chlorine atom replaces the fluorine atom (monochloroacetate), the key fits so poorly that the enzyme does not accept it. This compound therefore does not rob acetic acid of its enzyme.

has permitted biologists to "see" the union taking place. This was attempted for the first time 12 years ago by Kurt G. Stern, then at Yale University. Using the enzyme catalase and a hydrogen peroxide derivative as substrate, he observed first the spectral light pattern characteristic of catalase and then a new pattern, presumably that of the enzyme-substrate union. A short while afterward, however, the original catalase spectrum appeared again, indicating that the enzyme had performed its duty and was ready for more work.

What is the purpose of the brief combination of an enzyme with its substrate? The answer to the question hinges on a fact mentioned earlier: that not all collisions between molecules produce chemical reactions. In a 100-cubic-centimeter solution of ethyl bromide and diethyl sulfide, for instance, there are 1.6×10^{34} (16 million billion billion billion) collisions a second, but fewer than one out of every billion billion collisions results in a chemical reaction. The reason for this low proportion of successful hits is that molecules are relatively stable structures, and most of them bounce off each other a bit jarred but essentially unscathed.

Now enzymes do not increase the speed of molecules in solution, nor do they increase the frequency of collisions. Instead they increase the number of fruitful collisions by weakening the structure of substrate molecules so that they react more readily. In combining briefly with its substrate (in the case of catalase the combination lasts less than one 85,000th of a second), an enzyme somehow distorts the architecture of the substrate molecule, converting it from a relatively stable to a highly reactive state. There is evidence that in some cases this effect is achieved by removing electrons and transforming the substrate molecule into a charged ion.

Chemical changes, however, do not necessarily take place simply because a substrate molecule has fitted itself into its enzyme mold and has been activated. With certain enzymes like pepsin, to be sure, this two-molecule union seems to be sufficient for completion of the substrate-converting process. But more often than not a third substance that is not a protein is an added requirement. Catalase, peroxidase and other enzymes seem to have such accessory substances more or less permanently attached to their proteins. These substances are therefore called "prosthetic groups." Many enzymes, however, apparently require only that the accessory substances be available in solution as so-called coenzymes. This field of inquiry is one of the most active and controversial in enzyme research, and the entire question of enzyme auxiliaries needs considerable clarifying. This much is known: in most enzyme processes the protein alone is not enough to speed chemical reactions. Unless coenzyme I is present in the alcohol fermentation system previously described, the process breaks down.

Even more obscure than the mechanism of enzyme action are the factors that control enzyme reactions. What is it that determines when and how quickly enzyme activities shall take place in nerve cells, muscle cells and all of the other specialized units that make up the higher plant and animal organisms?

There is reason to believe that hormones play an important part in controlling and coordinating the workings of enzyme systems. The most significant finding along these lines was made about two years ago by W. H. Price, Carl Cori and S. P. Colowick at Washington University in St. Louis. They discovered that hormones play an essential part in the enzyme system that maintains the balance between sugar in the blood and glycogen in the liver. There is a delicate equilibrium between the hormone insulin, which tends to lower the amount of sugar in the blood by promoting the storage of glycogen in the liver, and a presumed diabetogenic hormone secreted by the pituitary gland, which promotes the metabolism of glycogen and hence tends to raise the concentration of sugar. Diabetes may be the result either of too little insulin or too much of the pituitary hormone. The Washington University group suggested that this upset in hormone balance was directly connected with hexokinase, the same enzyme that initiates alcoholic fermentation.

Hexokinase is utilized in the liver to add a phosphate group to glucose, a preliminary step essential to the storage of sugar. The pituitary hormone, however, inhibits hexokinase activity. Whether it is overproduced or insulin is underproduced, the effect is the same: a relative excess of the diabetogenic hormone, subnormal storage of glycogen in the liver, and rising sugar levels in the blood. The discovery of this process is one of the first connections established between hormones and enzymes. Other hormones, including those that produce dwarfs and giants, probably influence growth and metabolism in a similar manner.

Such speculation brings us again to the notion of enzymes that can be inhibited and activated. Does the diabetogenic hormone inhibit hexokinase by providing substances that occupy strategic parts of the enzyme molecule and thus prevent it from working on its normal substrate? And does insulin counteract the inhibiting effect by removing these substances and "unmasking" the enzyme? Only further investigation can answer these questions, but it is known that the unmasking effect plays a significant role in the control of enzyme action during many biological processes. For example, pepsin, the function of which is to digest proteins, does not enter the stomach ready to act; it is secreted by the stomach walls as the inert substance pepsinogen, which is promptly converted to pepsin by the hydrochloric acid of the gastric juices. The conversion is accompanied by a drop in molecular weight from 42,000 to 38,000, and this may be interpreted as the removal of a protein fragment that masks pepsin action.

An example of mass unmasking is familiar to embryologists. An unfertilized egg cell is fully prepared for the most spectacular burst of biological energy known. It contains structural materials, ATP as a source of energy, and hundreds of enzymes that will engineer the building of a tree or a man from a tiny blob of protoplasm. The enzymes, however, are blocked, probably by specialized coatings, until fertilization takes place. Then, by an unidentified mechanism, the blocking substances are dissolved, hundreds of reactions are set off at once, and the cell begins to grow and divide.

The patient investigation of enzyme action and enzyme control has brought added insight to a whole constellation of biological and medical problems. Enzymes not only speed the vital processes of the developing egg, but play a fascinating part in the act of fertilization. An unfertilized egg is protected by a tough coating of cells cemented with a substance called hyaluronic acid. The sperm carries the enzyme hyaluronidase specifically to break up the barrier and penetrate within.

There is evidence that the single sperm which accomplishes fertilization does not contain enough of the enzyme to break down the barrier by itself, and that the unsuccessful sperm cells must contribute their hyaluronidase. This explanation accounts for the fact that perhaps millions of sperm are necessary for fertilization, although only one penetrates the egg. Working on the theory that some cases of human sterility may be due to a lack of hyaluronidase, some physicians have recently administered extra amounts to a few selected patients and, according to preliminary announcements, normal pregnancy has resulted. Whether or not this simple treatment proves effective in a significant number of cases, any successful treatment for sterility will require an intimate understanding of the enzymes concerned.

Enzymes have also been identified with the toxins of infectious diseases. Thus *Clostridium welchii*, the rod-shaped organism most commonly found in gas gangrene, releases an enzyme called lecithinase. This destroys red blood cells by disintegrating the substance lecithin in their walls. (The same lecithinase is one of the poisons in cobra and rattlesnake venom.) The germ also liberates an enzyme that dissolves the protein connective tissue of muscle, and the "gas" of gas gangrene is produced by a group of enzymes that accelerate a pathological form of fermentation. The effects of many drugs and poisons are similarly tied up with enzyme reactions.

The chemical study of certain coenzymes has had unexpected medical consequences. In 1932 it was found that an essential part of the coenzyme I molecule was nicotinic acid, and three years later

C. A. Elvehjem and his associates at the University of Wisconsin identified the substance as the anti-pellagra vitamin. Other vitamins definitely known to be part of coenzyme molecules include B-1, B-2 and B-6. Whether all vitamins are parts of coenzymes remains to be seen, but the possibility is particularly strong for those factors of nutrition that are needed in "trace" quantities. In amounts of less than one ten millionth of an ounce the new B-12 factor is sufficient to produce measurable rises in the blood counts of anemia patients.

Vitamins are as necessary to some harmful bacteria as they are to human life. This fact has opened the way for putting the competitive inhibition of enzymes to medical use. The possibility was discovered by accident after the introduction of sulfa drugs, though for a long while their effectiveness in curbing germs was a mystery. Then it was observed that the ability of the drugs to inhibit the growth of bacteria was considerably reduced in the presence of para-aminobenzoic acid (PAB), a member of the vitamin B complex and an essential factor in the growth of many organisms. A comparison of the molecular structures of sulfanilamide and the acid soon indicated the reason for the phenomenon.

Germs that need the vitamin presumably incorporate it into their metabolic processes as part of a coenzyme, and things go beautifully until sulfanilamide comes upon the biochemical scene. This sulfa drug is a very close chemical relative of PAB, which is the secret of its medical effectiveness. The resemblance is so close that the bacterium cannot tell the difference and takes up sulfanilamide as if it were a real food factor. By the time the mistake is discovered, the false "vitamin" has been drawn into the enzyme system and jammed the works.

The part played by vitamins and other accessories in enzyme action also throws new light on the importance of trace elements in plant and animal life. In 1895 thousands of sheep on Australian ranches were dying of "bush sickness." Since the disease closely resembled anemia, ranchers tried feeding the animals large doses of iron. The treatment worked in some cases and not in others, the difference depending on the source of the iron.

So the Australian Government imported iron ores from all over the world and compared the samples. After a series of elaborate analyses, it was found that the iron which helped to cure bush sickness contained tiny amounts of cobalt. A sheep's daily requirements were calculated, and it was found that about a millionth of an ounce of cobalt was enough to prevent the malady.

The need for such elements may be connected with the efficient working of enzyme systems, for many enzymes are known to contain or to require the presence of metallic elements in small amounts. One of the key steps in alcoholic fermentation involves an enzyme called enolase, which requires the presence of magnesium ions before it can take effect. Hexokinase, similarly, cannot work without magnesium.

Enzymes have still other practical applications. By breeding improved strains of microorganisms and investigating their enzymes, research workers have increased the efficiency of alcoholic fermentation in the beer, wine and liquor industries.

Enzymes are also used to obtain heating gas, fertilizer and many other valuable materials from sewage and industrial waste. They tenderize meat, tan leather, turn cornstarch into syrups and sugars, and help in the making of dozens of products in the cosmetic, textile and baking industries.

From the lofty perspective of science the fundamental problem is not industrial utility or even the artful crystallization of pure enzymes. The most challenging question is how the human body, or any organism, manufactures enzymes. An enzyme is a protein built up by the body from amino acids and peptide chains. If we assume that an enzyme may be synthesized as simply as possible, it is created when two protein fragments are pieced together by a single enzyme. If that is the process, how is the ultimate enzyme itself synthesized?

The only way of surmounting the difficulty is to assume that certain molecules are capable of forming exact replicas of themselves. This is to say that they act as enzymes for their own synthesis. This is reproduction at the molecular level. Genes, the units of heredity, have been assumed to be such substances. Viruses likewise are autocatalytic. There is some speculation that genes are the ultimate enzyme-makers.

The red bread mold Neurospora is an experimental organism used in the study of this problem. In a whole series of experiments in which mutated Neurospora strains were created by ultraviolet radiation that knocked out a single gene in each case, each missing gene resulted in the organism's inability to synthesize a specific foodstuff, indicating that the heredity-transmitting molecules are directly connected with the production of enzymes. In fact, it has recently been indicated that a strain of Neurospora lacking a particular gene actually lacks an enzyme which can be extracted from strains having a full complement of genes.

To proceed from Neurospora to man, there are certain human diseases that are due to the deficiency of single enzymes. Significantly, these diseases also are inherited according to strict Mendelian laws. The lack of a single enzyme in the metabolism of the amino-acid phenylalanine is sufficient to cause a form of mental defectiveness, while another missing link in the same system is responsible for albinism. Since each of these diseases can ultimately be traced to the hereditary defect of a single gene, and the agency in each case is a missing enzyme, circumstantial evidence argues strongly for the theory that each gene is associated with the making of a single enzyme.

So enzymes bring us finally to the very core of the cell, and to the core of all biological problems. The solution of these problems depends more and more on our understanding of proteins, and particularly of enzymes, self-duplicating and otherwise.

IV. CALCIUM AND LIFE

by L. V. Heilbrunn

THE PROTOPLASM of men and whales, of bedbugs and orchids, of all living things, is composed essentially of the same sort of chemicals. Most of the chemicals are organic; that is to say, they contain carbon. If we burn any living thing, the bulk of it vanishes into thin air: the water goes off as steam and the organic materials are changed for the most part into carbon dioxide and water vapor. There is left a little ash—in the case of a human body, only enough to fill a small urn. Most of this ash comes from parts of the bones and teeth which are not really alive. The tiny remainder comes from the truly living tissues. It makes up only about one part in a hundred of the living material. And yet this ash, which consists of various types of salts, is a very important part of protoplasm, and there can be no life without it.

Living material is of course the most precious of all the materials of the world, but the stuff of which it is made is of the very cheapest. Its salts are the commonest varieties—mainly sodium chloride, or common table salt, and salts of potassium, magnesium and calcium. Each of the salts the body needs has its own importance. But perhaps the most essential of all is calcium.

Without calcium the higher animals of course would have no bones or teeth. These hard parts owe their rigidity to calcium salts; if the salts are dissolved in strong acid solutions, bones lose their stiffness and become quite pliable. But beyond this calcium has other important roles in the body. In the soft tissue as well as in the bones and teeth it is a vital structural material: it is an essential ingredient in the cement that binds cells together in tissues, and it gives firmness to the membranes that enclose the cells themselves.

Its binding property has been demonstrated very clearly in experiments on sea-urchin eggs. Like all animals, the sea urchin begins life as a single cell. After this egg cell has been fertilized, it divides into two, then into four and so on until there is a great mass of cells, all clinging together. But if a sea-urchin egg in the two-cell or four-cell stage is placed in a solution of a substance that removes calcium, the clinging cells tend to fall apart.

Proof of the fact that the firmness of a cell's membrane depends on calcium has come from studies on the amoeba. This wonderful little animal is a single large cell consisting of a droplet of protoplasm surrounded by a relatively rigid cortex. When the cortex contracts, it pushes out fingerlike projections and the animal slowly flows into these and moves forward. The contribution that calcium makes to the stiffness of the cortex can be measured by comparative tests of the cortex's viscosity before and after removal of the calcium. There is a still more conclusive proof that calcium is essential for the formation of a limiting membrane or film around the cell. Ordinarily when one breaks the membrane of an amoeba or a sea-urchin egg, the outflowing proptoplasm soon forms a new membrane. But if the cell is kept in a solution completely lacking in calcium, no new membrane forms to replace the one that has been ruptured; all the protoplasm flows out of the cell and scatters through the surrounding solution.

The reaction whereby a broken cell forms a new membrane in the presence of calcium is called the surface precipitation reaction. It occurs in various types of cells, and indeed it is a characteristic of living substance generally. The reaction of blood to calcium ion is very similar to that of protoplasm. When blood flows from an artery or vein, it clots. But this clotting does not occur if the blood flows into a solution from which calcium is absent. Blood that has been mixed with a little oxalate, which removes calcium from solutions, never does clot, unless a supply of the natural blood-clotting substance called thrombin is added.

Actually the surface precipitation reaction in protoplasm is a form of clotting, and there are many similarities between blood clotting and protoplasm clotting. If a little fluid from clotted blood, which always contains thrombin, is added to a fresh sample of blood, the new blood clots even in the absence of calcium. Similarly a little injured cell material taken from a suspension of smashed or otherwise injured cells can cause a surface precipitation reaction in another suspension of cells in the complete absence of calcium. On the other hand, an overabundance of calcium can prevent clotting, both in blood and in protoplasm. Incidentally, the curdling (clotting) of milk, like that of protoplasm and blood, also is a calcium-influenced process.

The fact that the clotting of protoplasm is similar to the clotting of blood is further indicated by the fact that smashed or injured protoplasm produces substances which can clot blood as well as protoplasm itself. One of the great pioneers in the study of blood clotting, the Esthonian physiologist Alexander Schmidt, demonstrated this fact many years ago. He and his students found that living material of all sorts—muscle, brain, yeast cells, and so on—when damaged gives off substances which greatly hasten the clotting of horse blood. Furthermore, it has been found that heparin, a substance which prevents blood clotting and is widely used in the treatment of thrombosis, can also prevent the clotting of protoplasm.

Protoplasm is strikingly similar to blood in another respect. Like blood, which contains both clotting and anti-clotting substances and is constantly on the verge of clotting, protoplasm is always in a delicate state of equilibrium between a fluid and a clotted state. Living cells are extremely sensitive to outside influences. And the transition from a fluid to a clotted state is one of the most important factors in the activity of the living material.

Consider a muscle cell. A tiny electric shock will make the muscle respond by contracting. So will a sudden impact or a sudden exposure to heat or cold. What happens in a muscle cell when it is thus aroused to activity?

Until very recently no one really knew whether muscle protoplasm was fluid or solid. Nearly 100 years ago the famous German physiologist Willy Kühne saw a small worm swimming in a frog muscle' fiber and suggested that the protoplasm was fluid. But others objected that a fiber with a worm in it was hardly normal. The question was argued back and forth for many years without a conclusion. Centrifuge tests were of no avail. A year or two ago the issue was finally resolved by a young physiologist, Peter Rieser, at the Univerity of Pennsylvania. He inserted small oil drops into the interior of frog muscle fibers. He was able to observe in a number of instances, when the injection did not damage the muscle, that the oil drop rose toward the top of the cell. It rose because it was lighter than the cell material. From the rate of this movement he was able to calculate the exact viscosity of the protoplasm. As a result of this work, we now know that the interior protoplasm in muscle is fluid and that there is a rather thick cortex through which no oil can move. A muscle cell has the same essential physical structure as the cell of an amoeba or a sea-urchin egg: it is composed of fluid protoplasm surrounded by a stiff cortex.

When the interior protoplasm of a muscle fiber clots, the fiber as a whole shortens. What induces the clot, and why does an electric shock or sudden exposure to heat or cold make the muscle shorten? The answer is relatively simple. Muscle protoplasm, like the protoplasm of cells in general, is extremely sensitive to calcium ion. When a muscle fiber is cut, the cut surface immediately produces a plug, very like the new membrane that forms at the surface of an exuding droplet of protoplasm as it emerges from an amoeba or a sea-urchin egg. The more calcium is present in the surrounding fluid, the more rapidly the plug forms. As the calcium enters the cut surface of the muscle fiber, the fiber shortens.

The sensitivity of muscle protoplasm to calcium can also be shown by the injection of small amounts of calcium directly into the interior of a muscle fiber. This is done with an injection needle of very fine bore. The injection of a very dilute solution of calcium salt makes the muscle contract.

We are thus in a position to explain the sensitivity of the muscle to the agents that cause it to contract. When an electric current is sent through a muscle, the muscle contracts at the point where the current emerges, that is to say, at the cathode. Now it is to the cathode that the calcium of a calcium salt migrates. Moreover, we know from the work of Virginia Weimar in our laboratory that heat and cold release calcium from its bond with muscle protein. The outer cortex of an amoeba consists of protein bound with calcium, and almost certainly the cortex of a muscle fiber has the same constitution. Various agents cause the release of calcium from its binding with the protein of the cortex. The agents that arouse the muscle fiber cause calcium to enter the cell. Clotting and shortening then result.

Various other types of cells are affected in the same way. There is some evidence, for example, that the protoplasm of a nerve fiber, like that of a muscle fiber, clots when the fiber is aroused. This is not, of course, the only change that occurs, but it may well be the most significant one.

Calcium is also important for the division of cells. The mechanism of cell division is a particularly important problem because it has a bearing on cancer. Cancer is essentially due to the initiation of division in cells that do not ordinarily divide. The simplest cells in which to investigate the division process are marine egg cells such as sea-urchin eggs. These eggs can be induced to divide by various artificial

264

means, such as strong salt solutions, heat, ultraviolet radiation, acids. All of these agents can be considered to cause a release of calcium from the cortex of the cell. The calcium that is released enters the cell and causes a clotting there. The clot results in the formation of a beautiful spindle-shaped structure—the mitotic spindle. In frog eggs, the formation of the mitotic spindle and cell division can best be induced by the injection into the egg cell of a trace of thrombin, the substance that induces clotting in blood. Similarly, substances extracted from injured cells can cause the appearance of a mitotic spindle in sea-urchin eggs.

Thus the clotting that precedes the appearance of the mitotic spindle (a process I have called mitotic gelation) can be induced either by calcium or by thrombin or thrombin-like substances. And the anticlotting substance heparin can prevent this mitotic gelation and the division of the cell. There are a large number of heparin-like substances, many of them produced by bacteria. At least one of these, a bacterial polysaccharide isolated by M. J. Shear and his associates at the National Institutes of Health, has quite a favorable effect in causing regression of tumors in rats and mice. Unfortunately it is too toxic to use properly on human patients, but there is always a hope that some other naturally occurring polysaccharide may prove more successful for cancer therapy.

Calcium plays many other roles, of course, in the vital process. For example, there is increasing evidence, notably from the work of Albert I. Lansing at Washington University in St. Louis, that the aging of cells and the aging of men and animals is due to accumulation of calcium in cells in an insoluble form. Calcium is known to be important in a great variety of ailments, from acne to hardening of the arteries. We have confined our attention here to the relation of calcium to the physical properties of protoplasm. The major conclusion from our work is that calcium is not only important for the rigidity of animals and cells; it is the prime instigator of vital activity.

PART 6

I. THE GENES OF MEN AND MOLDS
by George W. Beadle
Head of the California Institute of Technology Division of Biology, George Beadle is a geneticist, whose most notable work has been done with the bread mold *Neurospora,* the subject of this chapter. With the great Cal Tech chemist Linus Pauling, he is engaged in a famous research partnership on the fundamental problems of biology and medicine.

II. THE CHEMISTRY OF HEREDITY
by A. E. Mirsky
Since completing his training in biochemistry at Harvard and Cambridge, A. E. Mirsky has been associated with The Rockefeller Institute for Medical Research.

III. THE GENETIC BASIS OF EVOLUTION
by Theodosius Dobzhansky
Born and educated in Russia, Theodosius Dobzhansky came here in 1927 as a research fellow of the Rockefeller Foundation. From then until 1940 he worked with the great American geneticist T. H. Morgan. Today he is professor of zoology at Columbia University. His *Genetics and the Origin of Species* is one of the eminent modern works in the field.

IV. CATACLYSMIC EVOLUTION
by G. Ledyard Stebbins, Jr.
Primarily a plant geneticist, G. L. Stebbins teaches at the University of California's College of Agriculture in Davis, Cal. At present, he is also carrying out there a long research project designed to produce new varieties and species of forage grasses.

V. LETHAL HEREDITY by Willard F. Hollander
Willard Hollander has been a research associate at the Palmetto Pigeon Plant, a center for experiments in heredity at Sumter, S. C.; at the genetics department of the Carnegie Institution at Cold Spring Harbor, N. Y., and at Yale Medical School. At present he is assistant professor of genetics at Iowa State College.

VI. THE MYSTERY OF CORN
by Paul C. Mangelsdorf
Paul Mangelsdorf, director of the Harvard Botanical Museum, began to study the genetic history of corn at Kansas State College 30 years ago. His pursuit of this one plant has taken him into almost every country of North and South America; as a consultant to the Rockefeller Foundation he has helped many of these countries to improve their grain crops.

GENETICS

Introduction

MODERN GENETICS is the vigorous hybrid offspring of Mendel's laws of heredity and Darwin's theory of evolution, two of the towering generalizations of 19th century biology. Reflecting this diverse inheritance, the science today pursues two major lines of investigation. One, the biochemical branch, seeks to discover how Mendel's genes shape each new individual in the image of its predecessors. The other, more mathematical in method, is concerned with whole populations in their interaction with the Darwinian forces of natural selection.

The two approaches, of course, overlap and complement each other in goal and method. Both, for example, seek their primary clues in the same portentous natural accident: the occasional failure of the gene to reproduce itself and its hereditary trait exactly. This makes it possible to match trait to gene and thereby to relate heredity to chemistry. Similarly, each new and different offspring generated by this process of mutation gives the population geneticist another clue in his reconstruction of the origin and diversification of species.

This chapter reflects both aspects of genetics. It starts with microscopic dissection and ends with an adventure in deduction. The account opens with a study of the effect that genes—both normal and defective—have upon the individual organism. Analytical techniques of the most exquisite delicacy and ingenuity have tied specific genes to specific enzymes and thus to the whole metabolic process. From there the discussion goes on to the chemistry of chromosomes, the units within the nucleus of every cell that contain the genes. No one has ever seen a gene, and there is no assurance that we ever will. But one of the principal chemical agents of the genetic process has been located in the chromosomes. It is desoxyribonucleic acid. In the living cell and even in the test tube, it has been shown to be a molecule of prodigious complexity and activity.

The third section of this chapter surveys the theory of evolution in the light of modern genetic knowledge. Evolution in nature is a slow process, but modern geneticists have found ways to speed it up. A million years in man may be genetically equivalent to a few days in bacteria, a few weeks in fruit flies. The rate of change even in these working models of evolution, can now be accelerated by the artificial induction of mutation, according to experimental plan. Such investigations have both refined and extended the original insights of Mendel and Darwin and completely disposed of the racist theories that seek purity in a context where the word is meaningless.

The two discussions that follow this review of hereditary principles take up special aspects of modern theory. Cataclysmic evolution is a process, involving the doubling of chromosomes, whereby new species arise, not by slow change, but suddenly from one generation to the next. Lethal heredity concerns the defective genes which, *sub rosa*, have been found to lurk in every natural genetic strain. They can be uncovered dramatically by inbreeding; they can also be detected in certain cases by the telltale badge of minor abnormalities. The final section—a scientific detective story—is an inquiry into the lineage of domesticated corn. Nothing like modern corn grows today in the wild state and nothing very much like it ever could have grown in the wild state. Corn is completely dependent on man for survival. This improbable but invaluable species of grass has come down to us from a complex of lucky accidents to which only in the latter stages has man contributed a guiding hand.

The usefulness of genetics is too obvious to pause upon; the hybrid marvels of every seed catalogue proclaim it and we should probably have starved to death long ago but for the resourceful pragmatism of the breeders. It is evident from this chapter, however, that practical genetics based on trial and error crossing of strains does not begin to tap the possibilities that emerge from a knowledge of what genes are and how they work. Science justifies itself when it contributes to the innate desire to know, but it is also true that year by year the need to know becomes more imperative for the survival of the proliferating human race.

I. THE GENES OF MEN AND MOLDS

by George W. Beadle

EIGHTY-FIVE YEARS AGO, in the garden of a monastery near the village of Brünn in what is now Czechoslovakia, Gregor Johann Mendel was spending his spare moments studying hybrids between varieties of the edible garden pea. Out of his penetrating analysis of the results of his studies there grew the modern theory of the gene. But like many a pioneer in science, Mendel was a generation ahead of his time; the full significance of his findings was not appreciated until 1900.

In the period following the "rediscovery" of Mendel's work biologists have developed and extended the gene theory to the point where it now seems clear that genes are the basic units of all living things. They are the master molecules that guide the development and direct the vital activities of men and amoebas.

Today the specific functions of genes in plants and animals are being isolated and studied in detail. One of the most useful genetic guinea pigs is the red bread mold *Neurospora crassa*. Its genes can conveniently be changed artificially and the part that they play in the chemical alteration and metabolism of cells can be analyzed with considerable precision. We are learning what sort of material the genes are made of, how they affect living organisms and how the genes themselves, and thereby heredity, are affected by forces in their environment. Indeed, in their study of genes biologists are coming closer to an understanding of the ultimate basis of life itself.

It seems likely that life first appeared on earth in the form of units much like the genes of present-day organisms. Through the processes of mutation in such primitive genes, and through Darwinian natural selection, higher forms of life evolved—first as simple systems with a few genes, then as single-celled forms with many genes, and finally as multicellular plants and animals in which genes are arranged linearly in the threadlike chromosomes of the cell nuclei.

What do we know about these genes that are so all-important in the process of evolution, in the development of complex organisms, and in the direction of those vital processes which distinguish the living from the non-living worlds? Genes are characterized by students of heredity as the units of inheritance. What is meant by this may be illustrated by examples of some inherited traits in man.

269

Blue-eyed people may differ by a single gene from those with brown eyes. This eye-color gene exists in two forms, which for convenience may be designated B (for brown) and b (for blue).

Every person begins as a single cell a few thousandths of an inch in diameter—a cell that comes into being through the fusion of an egg cell from the mother and a sperm cell from the father. This fertilized egg carries two representatives of the eye-color gene, one from each parent. Depending on the parents, there are therefore three types of individuals possible so far as this particular gene is concerned. They start from fertilized eggs represented by the genetic formulas BB, Bb, and bb. The first two types, BB and Bb, will develop into individuals with brown eyes. The third one, bb, will have blue eyes. You will note that when both forms of the gene are present the individual is brown-eyed. This is because the form of the gene for brown eyes is *dominant* over its alternative form for blue eyes. Conversely, the form for blue eyes is said to be *recessive*.

During the division of the fertilized egg cell into many daughter cells, which through growth, division and specialization give rise to a fully developed person, the genes multiply regularly with each cell division. As a result each of the millions of cells of a fully developed individual carries exact copies of the two representatives of the eye-color gene which has been contributed by the parents.

In the formation of egg and sperm cells, the genes are again reduced from two to one per cell. Therefore a mother of the type BB forms egg cells carrying only the B form of the gene. A type bb mother produces only b egg cells. A Bb mother, on the other hand, produces both B and b egg cells, in equal numbers on the average. Exactly corresponding relations hold for the formation of sperm cells.

With these facts in mind it is a simple matter to determine the types of children expected to result from various unions. Some of these are indicated in the following list:

Mother	Father	Children
BB (brown)	BB (brown)	All BB (brown)
Bb (brown)	Bb (brown)	¼ BB (brown)
		½ Bb (brown)
		¼ bb (blue)
BB (brown)	bb (blue)	All Bb (brown)
Bb (brown	bb (blue)	½ Bb (brown)
		½ bb (blue)
bb (blue)	bb (blue)	All bb (blue)

This table shows that while it is expected that some families in which both parents have brown eyes will include blue-eyed children, parents who are both blue-eyed are not expected to have brown-eyed children.

It is important to emphasize conditions that may account for apparent exceptions to the last rule. The first is that eye-color inheritance in man is not completely worked out genetically. Probably other genes besides the one used as an example here are concerned with eye color. It may therefore be possible, when these other genes are taken into account, for parents with true blue eyes to have brown-eyed children. A second factor which accounts for some apparent exceptions is that brown-eyed persons of the *Bb* type may have eyes so light brown that an inexperienced observer may classify them as blue. Two parents of this type may, of course, have a *BB* child with dark brown eyes.

Another example of an inherited trait in man is curly hair. Ordinary curly hair, such as is found frequently in people of European descent, is dominant to straight hair. Therefore parents with curly hair may have straight-haired children but straight-haired parents do not often have children with curly hair. Again there are other genes concerned, and the simple rules based on a one-gene interpretation do not always hold.

Eye-color and hair-form genes have relatively trivial effects in human beings. Other known genes are concerned with traits of deeper significance. One of these involves a rare hereditary disease in which the principal symptom is urine that turns black on exposure to air. This "inborn error of metabolism," as the English physician and biochemist Sir Archibald Garrod referred to it, has been known to medical men for probably 300 years. Its biochemical basis was established in 1859 by the German biochemist C. Bödeker, who showed that darkening of urine is due to a specific chemical substance called alcapton, later identified chemically as 2,5-dihydroxyphenylacetic acid. The disease is known as alcaptonuria, meaning "alcapton in the urine."

Alcaptonuria is known to result from a gene defect. It shows typical Mendelian inheritance, like blue eyes, but the defective form of the gene is much less frequent in the population than is the recessive form of the eye-color gene.

The excretion of alcapton is a result of the body's inability to break it down by oxidation. Normal individuals possess an enzyme which makes possible a reaction by which alcapton is further oxidized. This enzyme is absent in alcaptonurics. As a result alcaptonurics cannot degrade alcapton to carbon dioxide and water as normal individuals do.

Alcaptonuria is of special interest genetically and biochemically because it gives us a clue as to what genes do and how they do it. If the cells of an individual contain only the recessive or inactive form of the gene, no enzyme is formed, alcapton accumulates and is excreted in the urine. The role of the normal form of the gene in enzyme production is thus made clear.

Man, however, is far from an ideal organism in which to study genes. His life cycle is too long, his offspring are too few, his choice of a mate is not often based on a desire to contribute to the knowledge of heredity, and it is inconvenient to subject him to a complete chemical analysis. As a result, most of what we have learned about genes has come from studies of such organisms as garden peas, Indian corn plants and the fruit fly *Drosophila*.

In these and other plants and animals there are many instances in which genes seem to be responsible for specific chemical reactions. It is believed that in most or all of these cases they act as pattern molecules from which enzymes are copied.

Many enzymes have been isolated in a pure crystalline state. All of them have proved to be proteins or to contain proteins as essential parts. Gene-enzymes relations such as those considered above suggest that the primary function of genes may be to serve as models from which specific kinds of enzyme proteins are copied. This hypothesis is strengthened by evidence that some genes control the presence of proteins that are not parts of enzymes.

For example, normal persons have a specific blood protein that is important in blood clotting. Bleeders, known as hemophiliacs, differ from non-bleeders by a single gene. Its normal form is presumed to be essential for the synthesis of the specific blood-clotting protein. Hemophilia, incidentally, is almost completely limited to the male because it is sex-linked; that is, it is carried in the so-called X chromosome, which is concerned with the determination of sex. Many other abnormalities (e.g., color blindness) are sex-linked.

The genes that control blood types in man and other animals direct the production of so-called antigens. These are giant molecules which

272

apparently derive their specificity from gene models, and which are capable of inducing the formation of specific antibodies.

The hypothesis that genes are concerned with the elaboration of giant protein molecules has been tested by experiments with the red mold *Neurospora*. This fungus has many advantages in the study of what genes do. It has a short life cycle—only 10 days from one sexual spore generation to the next. It multiplies profusely by asexual spores. The result is that any strain can be multiplied a millionfold in a few days without any genetic change. Each of the cell nuclei that carry the genes of the bread mold has only a single set of genes instead of the two sets found in the cells of man and other higher organisms. This means that recessive genes are not hidden by their dominant counterparts.

During the sexual stage, in which molds of opposite sex reactions come together, there is a union comparable to that between egg and sperm in man. The fusion nucleus then immediately undergoes two divisions in which genes are reduced again to one per cell. The four products formed from a single fusion nucleus by these divisions are lined up in a spore sac. Each divides again so as to produce pairs of nuclei that are genetically identical. The eight resulting nuclei are included in eight sexual spores, each one-thousandth of an inch long. This life cycle of *Neurospora* is shown in the illustration on page 277.

Using a microscope, a skilled laboratory worker can dissect the sexual spores from the spore sac in orderly sequence. Each of them can be planted separately in a culture tube. If the two parental strains differ by a single gene, four spores always carry descendants of one form of the gene and four carry descendants of the other. Thus if a yellow and a white strain are crossed, there occur in each spore sac four spores that will give white molds and four that will give yellow.

The red bread mold is almost ideally suited for chemical studies. It can be grown in pure culture on a chemically known medium containing only nitrate, sulfate, phosphate, various other inorganic substances, sugar and biotin, a vitamin of the B group. From these relatively simple starting materials, the mold produces all the constituent parts of its protoplasm. These include some 20 amino acid building blocks of proteins, nine water-soluble vitamins of the B group, and many other organic molecules of vital biological significance.

To one interested in what genes do in a human being, it might at first thought seem a very large jump from a man to a mold. Acually it is not. For in its basic metabolic processes, protoplasm—Thomas Huxley's physical stuff of life—is very much the same wherever it is found.

If the many chemical reactions by which a bread mold builds its protoplasm out of the raw materials at its disposal are catalyzed by enzymes, and if the proteins of these enzymes are copied from genes, it should be possible to produce molds with specific metabolic errors by causing genes to mutate. Or to state the problem somewhat differently, one ought to be able to discover how genes function by making some of them defective.

The simplicity of this approach can be illustrated by an analogy. The manufacture of an automobile in a factory is in some respects like the development of an organism. The workmen in the factory are like genes—each has a specific job to do. If one observed the factory only from the outside and in terms of the cars that come out, it would not be easy to determine what each worker does. But if one could replace able workers with defective ones, and then observe what happened to the product, it would be a simple matter to conclude that Jones puts on the radiator grill, Smith adds the carburetor, and so forth. Deducing what genes do by making them defective is analogous and equally simple in principle.

It is known that changes in genes—mutations—occur spontaneously with a low frequency. The probability that a given gene will mutate to a defective form can be increased a hundredfold or more by so-called mutagenic (mutation producing) agents. These include X-radiation, neutrons and other ionizing radiations, ultraviolet radia-

Life cycle of the mold *Neurospora* begins at the top center of this illustration with the hyphal fusion (corresponding to sexual mating in higher forms) of sex a and sex A. The union produces a fertile egg, in which two complete sets of genes are paired. The egg then divides (*center of drawing*), and divides again. This produces four nuclei, each of which has only a single set of genes. Lined up in a spore sac, the four nuclei divide once more to produce four pairs of nuclei, each pair being genetically identical. A group of spore sacs is gathered in a fruiting body. At this point, a skilled worker can dissect each of the sexual spores from the spore sac and plant them separately in culture tubes. Here genetic defects can be exposed by changing the ingredients of the media; here also *Neurospora* will multiply by asexual means, making it possible to grow large quantities of the mold without genetic change.

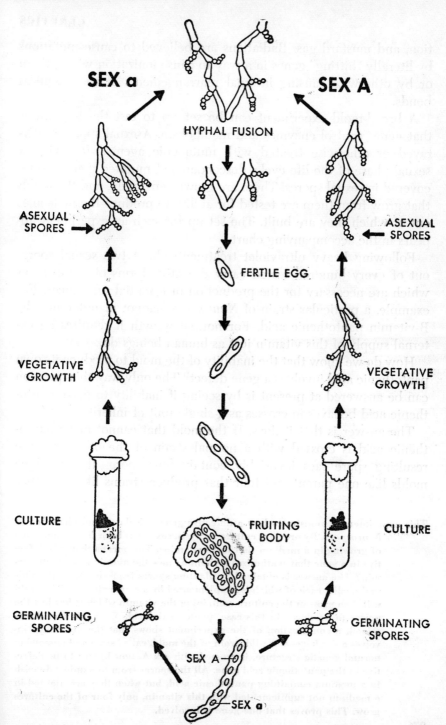

SEX a

SEX A

HYPHAL FUSION

ASEXUAL
SPORES

ASEXUAL
SPORES

FERTILE EGG

VEGETATIVE
GROWTH

VEGETATIVE
GROWTH

CULTURE

FRUITING
BODY

CULTURE

GERMINATING
SPORES

GERMINATING
SPORES

SEX A

SEX a

tion, and mustard gas. Radiations are believed to cause mutations by literally "hitting" genes in a way to cause ionization within them or by otherwise causing internal rearrangements of the chemical bonds.

A bread-mold experiment can be set up to test the hypothesis that genes control enzymes and metabolism. Asexual spores are X-rayed or otherwise treated with mutagenic agents. Following a sexual phase of the life cycle, descendants of mutated genes are recovered in sexual spores. These are grown separately, and the molds that grow from them are tested for ability to produce the molecules out of which they are built. The set up for such an experiment appears in the accompanying chart.

Following heavy ultraviolet treatment, about two sexual spores out of every hundred tested carry defective forms of those genes which are necessary for the production of essential substances. For example, a particular strain of *Neurospora* cannot manufacture the B-vitamin pantothenic acid. For normal growth it requires an external supply of this vitamin just as human beings do.

How do we know that the inability of the mold to produce its own pantothenic acid involves a gene defect? The only way this question can be answered at present is by seeing if inability to make pantothenic acid behaves in crosses as a single unit of inheritance.

The answer is that it does. If the mold that cannot make pantothenic acid is crossed with a normal strain of the other sex, the resulting spore sacs invariably contain four spores that produce molds like one parent and four that produce strains like the other

This experiment demonstrates how a defective gene is isolated and identified in *Neurospora*. By previous culture analysis, sex A has been found incapable of growing in a medium that lacks vitamins. This means that it is defective in a gene that synthesizes a vitamin; now the question is, which vitamin? The answer is obtained by planting spores from the strain in minimal media, each of which is supplemented by a single vitamin. The mold will then grow in the culture containing the vitamin which it has lost the power to produce; in this case, pantothenic acid. With this fact determined, a continuation of the experiment shows that the deficiency involves a single gene. A wild strain of the mold, called sex a and possessing normal genetic structure, is crossed with sex A, now known to be defective in the pantothenic acid gene. All the spores from this union flourish in a medium containing pantothenic acid, but when they are planted in a medium not supplemented with this vitamin, only four of the cultures grow. This proves that one gene is involved.

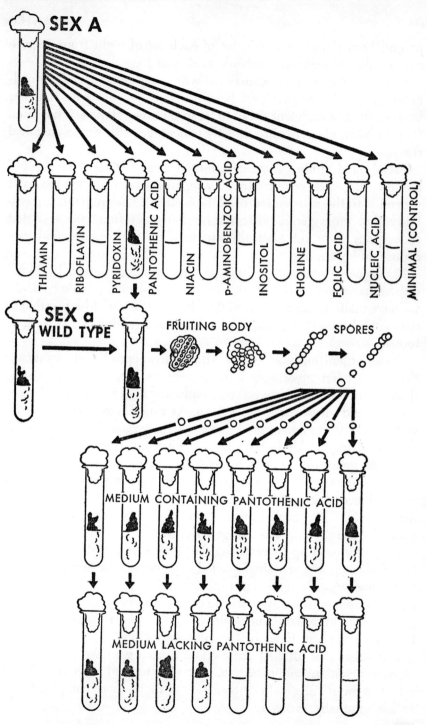

SEX A

THIAMIN
RIBOFLAVIN
PYRIDOXIN
PANTOTHENIC ACID
NIACIN
p-AMINOBENZOIC ACID
INOSITOL
CHOLINE
FOLIC ACID
NUCLEIC ACID
MINIMAL (CONTROL)

SEX a
WILD TYPE

FRUITING BODY

SPORES

MEDIUM CONTAINING PANTOTHENIC ACID

MEDIUM LACKING PANTOTHENIC ACID

parent. Four daughter molds out of each set of eight from a spore sac are able to make pantothenic acid, and four are not.

In a similar way, genes concerned with many other specific bread-mold chemical reactions have been mutated. In each case that has been studied in sufficient detail to be sure of the relation, it has been found that single genes are directly concerned with single chemical reactions.

Bread-mold studies have contributed strong support to the hypothesis that each gene controls a single protein. But they have not proved it to the satisfaction of all biologists. There remains a possibility that some genes possess several distinct functions and that such genes were automatically excluded by the experimental procedure followed.

The general question of how proteins are synthesized by living organisms is one of the great unsolved problems of biology. Until we have made headway toward its solution, it will not be possible to understand growth, normal or abnormal, in anything but superficial terms.

Do all organisms have genes? All sexually reproducing organisms that have been investigated by geneticists demonstrably possess them. Until recently there was no simple way of determining whether bacteria and viruses also have them. As a result of very recent investigations it has been found that some bacteria and some bacterial viruses perform a kind of sexual reproduction in which hereditary units like genes can be quite clearly shown.

By treatment of bacteria with mutagenic agents, mutant types can be produced that parallel in a striking manner those found in the bread mold. This parallelism makes it almost certain that bacterial genes are functionally like the genes of molds.

So we can sum up by asserting that genes are irreducible units of inheritance in viruses, single-celled organisms and in many-celled plants and animals. They are probably nucleoproteins that serve as patterns from which new genes are copied and from which non-genic proteins and other large molecules are produced with configurations that correspond to those of the gene templates.

It is likely that life first arose on earth as a genelike unit capable of multiplication and mutation. Through natural selection of the fittest of these units and combinations of them, more complex forms of life gradually evolved.

II. THE CHEMISTRY OF HEREDITY

by A. E. Mirsky

OUR ANCESTORS are present in our chromosomes, and they reach down to influence the chemistry of every cell in our bodies. That chemistry is the fundamental instrument of heredity there can be no doubt. The genes—the factors of heredity—must depend on the chemical composition of the chromosomes; the growth and division of the chromosomes are chemical processes, and it is by chemical processes that the chromosomes are able to influence the rest of the cell and so make manifest the hereditary factors. When, for example, a chromosome influences the color of a person's eyes, it must do so by taking part in some way in the chemical synthesis of a pigment.

The study of the chemistry of chromosomes is, therefore, a study of life at an elementary level. This article is an account of that fascinating investigation and what has been learned from it.

A living cell, animal or vegetable, has a nucleus within which is a certain number of chromosomes, the number depending on the species of animal or plant. The nucleus may occupy anywhere from less than one hundredth to more than two thirds of the volume of the cell; the rest of the material is the cytoplasm. The chromosomes have individuality: each one differs from all the others in the same set. Within the chromosomes are the genes, arranged in a linear order.

The formation of a new individual begins when the nuclei of an egg cell and a sperm cell fuse to form one cell. The fertilized egg now has two sets of chromosomes, one from each of the parent cells. The two sets are equivalent (except for those chromosomes concerned with sex), and thus there is a pair of each kind of chromosome. The cell then divides to start the building of the body cells of the new organism, and one can see through a microscope that the division proceeds in as orderly and complex a ritual as a courtly 18th-century ballet. Each chromosome grows in size and then splits lengthwise into two identical copies of the original. The couples are promptly separated by two sets of threads that pull them to opposite ends of the cell. The membrane surrounding the nucleus has meanwhile broken down, so that there is no barrier to impede their movement. The two sets of chromosomes become bunched at the opposite points. Once the sets are separated, the threads disappear and a membrane forms around each set. Now there are two nuclei, each containing a complete set of pairs of chromosomes, inherited from the two original parent cells. The two new nuclei are surrounded by cytoplasm, the non-nuclear material of cells. But since cytoplasm divides in a different way from the nucleus, the two daughter cells may have unlike quantities of it.

Thus, by division after division, the new body is built. Every cell of our bodies—each liver cell, brain cell, kidney cell—carries chromosomes and genes from both parents. In the fruit fly, *Drosophila*, we can see the phenomenon distinctly under the microscope. The chromosomes in the salivary gland cells of fruit-fly larvae are so large that we can make out their structural details and see that hereditary factors contributed by both parents are present in each pair of chromosomes.

Knowledge of the relation between chromosomes and heredity has come from two lines of investigation—breeding experiments and microscopic observations. It was the breeding experiments that showed that chromosomes carry hereditary factors. In recent years such experiments on the bread mold neurospora have shown most beautifully that hereditary factors in chromosomes have a precise controlling influence on many chemical processes in the cell.

The simplest and most direct evidence of the nucleus' far-reaching effect on the cell was provided by certain experiments made by the German biologist Joachim Hämmerling in 1931. He performed these experiments on Acetabularia, a little green plant consisting of a single large cell. It has the form of an umbrella, with the exceedingly small nucleus near the tip of the handle. If the hatlike top of the umbrella is cut off, a new one forms, looking just like the one removed. There are many varieties of Acetabularia, each having a distinctively shaped "hat." What Hämmerling found was that if he cut off the "hat" and the nucleus of one of these plants, and then planted in the remaining stump a nucleus taken from another variety of Acetabularia, the new "hat" that formed was like that of the second variety. It was the nucleus that decided what fashion "hat" was made. Hämmerling was even able to graft two different nuclei into the bottom of one "umbrella"; when he did so, the new "hat" was a compromise between the different shapes associated with the two nuclei.

The decisive influence of the nucleus has also been investigated by removing the nucleus from the amoeba, another single-celled organism. It has been known for many years that this operation slows down the metabolism of the cell. Recently Daniel Mazia of the University of California made such an investigation with tracer isotopes of phosphorus—an element well known to play a central role in cell metabolism. The speed with which the "tagged" atoms enter into new combinations in the cell measures the cell's rate of phosphorus metabolism. Mazia cut a number of amoebae in half, one half of the cell containing the nucleus and the other lacking a nucleus. He put the halves containing nuclei in one vessel and the enucleated cell halves in another. Then he supplied tagged phosphate to both lots for the same length of time. He found that the halves with nuclei took tagged phosphorus into their complex phosphorus-containing substances at a normal rate, but the non-nuclear halves took up much less phosphorus. This experiment again showed that the nucleus, though only a minute part of the amoeba, has a decisive influence on the phosphorus metabolism of the whole cell.

Naturally one is led to wonder whether the chemical apparatus by which the chromosomes control all these activities can be related to some special substance in them.

The chromosomes do indeed possess a special substance: desoxyribonucleic acid, called DNA for short. DNA is peculiar to the chromosomes; it is not found in any other part of the nucleus or in the cytoplasm. This was shown many years ago by the well-known "Feulgen reaction." The German biochemist Robert Feulgen had discovered that when DNA is warmed with strong acid and then treated in a certain way with acid fuchsin, it turns a brilliant crimson. He performed this experiment in a test tube and did not apply his test to living cells until some 10 years later. When he did so, he found to his joy that the chemical treatment had not disintegrated the cells and that the network of chromosomes in each nucleus was boldly revealed in brilliant color. The rest of the cell was colorless. Ever since this experiment microscopists have been using the Feulgen reaction on all kinds of plant and animal cells. They have demonstrated that, in general, chromosomes are "Feulgen positive" and nothing else in a cell is. This means that DNA is present in chromosomes and not in other parts of the cell, or, strictly speaking, not in sufficient concentration to be Feulgen positive.

There is another remarkable fact about DNA. The various cells of the body differ greatly in chemical composition: the cells of the liver, kidney, heart, spleen and so on vary in the kind and amount of substances they contain. But every body cell, regardless of type, has the same amount of DNA in its nucleus. And the egg and sperm cells, containing only half as many chromosomes as body cells, have just one half as much DNA.

Measurements have been made for the various tissues of a number of animals. In any one species of animal the quantity of DNA per nucleus is always about the same (within an error of 10 per cent), whether the cells come from the liver, pancreas, spleen or blood. But each species has its own characteristic DNA quota. In the frog, for instance, it is 15 hundred-millionths of a milligram per nucleus; in the shad, 2 hundred-millionths; in the green turtle, 5.3 hundred-millionths.

There are, however, some nuclei in the body which contain two or four sets of chromosome pairs instead of the usual one. They are called polyploid nuclei and we might expect, by analogy with the egg and sperm cells, that they should contain respectively two and four times as much DNA as ordinary nuclei in the same animal. And the analysis of individual cells shows that this is the case. For instance, cells of the rat liver with one-set, two-set and four-set nuclei are found to have DNA contents in the same ratios—1:2:4.

There is, then, a certain quantity of DNA in each set of chromosomes, whether in the egg or sperm, twice this quantity in the sets of chromosome pairs present in most body cells and correspondingly more in the double and quadruple sets of chromosome pairs found in some cells. All this shows that DNA is closely associated with the hereditary factors of chromosomes and most likely forms part of the stuff of which the hereditary factors are made. The chromosomes of course contain other substances besides DNA (e.g., various proteins), but none of those other known constituents is distributed in nuclei in the same regular way.

It should be said that some investigators are convinced that certain cells of a developing embryo contain more than the normal amount of DNA for the organism. But at present it cannot be said whether their measurements are correct or are thrown off by some unconsidered technical point in the procedure. Even if such exceptions to the rule of DNA distribution do exist, they should not occasion surprise, considering how great are the possibilities for cell variation.

To the rule that DNA is restricted to the cell nucleus there is a clearly established exception, but it is an exception which may be said to prove the rule. DNA or something closely related to it has been found in the cytoplasm of egg cells of many organisms; indeed, these cells have far more of the substance in the cytoplasm than in the nucleus. Most egg cells are large with materials required for growth of the embryo. Long before fertilization these materials are fed into the egg from surrounding cells called "nurse cells." In 1936 M. Konopacki, a Polish biologist, showed that in certain animals the nurse cells produce DNA, which is passed into the egg cell along with other nutrient materials. It remains in the cytoplasm until an embryo forms. In other words, the DNA in egg cytoplasm is on its way from one nucleus to another. There is enough of this material in the cytoplasm of one egg cell, according to recent experiments of E. Zeuthen and E. Hoff-Jorgensen in Copenhagen, to supply the nuclei of the thousands of cells that will develop by division from the single fertilized egg.

Now let us consider some experiments which illuminate the dynamic role played by the DNA in the chromosomes. It all began with certain curious observations made by the English bacteriologist Fred Griffith. He was working with pneumococci, the bacteria that cause pneumonia. There are many varieties of pneumococci, and they are classified according to the chemical make-up of the gummy capsule that surrounds the cell, which is different for each type of pneumococcus. Each type reproduces its own kind of capsular gum. When grown under certain conditions, however, pneumococci lose their capsules and reproduce cells that lack capsules. What Griffith did was this: He used two cultures of pneumococci, one of encapsulated cells of type III, and another that had had capsules of type I but had lost the capsules. He killed the encapsulated type III cells by heating them in water. Then he injected these dead cells together with living non-encapsulated cells of type I into a mouse. After a suitable time Griffith examined the mouse and found that pneumococci of encapsulated type III were now growing at a great rate in its tissues! Surely the killed type III cells that had been injected into the mouse could not have multiplied. To make sure of this Griffith injected heat-killed encapsulated type I cells into many mice; in no instance did they multiply. The conclusion to be drawn was clear: It was the living, non-encapsulated type I cells that had multiplied, but they had been converted into encapsulated type III. The dead encapsulated cells had somehow transmitted their hereditary constitution to the living non-encapsulated cells. They must have passed along to the type I cells the ability to make type III capsular gum.

Griffith's results were confirmed by Martin Dawson at the Rockefeller Institute. He found a way of doing the experiment in a test tube: the mixed heat-killed and living cells were placed in a nutrient solution in a test tube and there, as well as in the mouse, the encapsulated, heat-killed cells transmitted their hereditary constitution to the living, non-encapsulated ones. Another important step was made by James L. Alloway, also at the Rockefeller Institute. Pneumococcus cells disintegrate when placed in bile salts; they seem to dissolve, and the fluid in the tube clears. Alloway found that the heat-killed encapsulated cells could transmit their hereditary constitution even after they had been dissolved. In short, it looked very much as if some substance in the cell, rather than the cell as a whole, was responsible for transmitting its hereditary constitution.

The problem now was to hunt for that substance in the disintegrated and dissolved debris of the cell. This problem was undertaken by O. T. Avery, Maclyn McCarty and Colin MacLeod at the Rockefeller Institute. They quickly eliminated the capsular gum itself as a possibility; when they destroyed the gum with an enzyme which decomposes it, the cell debris was still able to transmit the hereditary property of forming a capsule. The investigators then removed from the debris the protein, which makes up the great bulk of material in the cell. They did so by a procedure which leaves the remaining compounds in the solution undamaged. That operation eliminated the protein as a suspect, for what was left of the heat-killed and decimated cell could still transmit its heredity.

With the capsular gum and cell protein out of the way, it became apparent that the effective substance might well be DNA. And further experiments indicated that it was. When the DNA in the remaining material of the cell was decomposed by a purified enzyme known to act specifically on DNA, the debris finally lost its ability to promote the manufacture of capsular gum.

The action of DNA is highly specific. If the DNA derived from another type of pneumococcus is added to non-encapsulated cells, the cells that finally multiply are of the type from which the DNA was derived. There must be a special kind of DNA in each type of pneumococcus.

Transmission of hereditary characteristics in pneumococci by means of DNA provides a beautiful example of one of the fundamental principles of heredity. What is transmitted in us from one generation to the next is not a characteristic eye pigment or blood type or other hereditary trait. Rather, it is a set of factors in chromosomes which are able to influence the activities of the cells so that certain eye pigments and certain substances responsible for blood types are produced. In the pneumococcus the DNA of the pneumococcal chromosomes influences the cell in which it is placed to make a particular kind of capsular gum.

What is the chemical nature of this potent substance? DNA was first discovered in the nucleus by the Swiss biochemist Friedrich Miescher in 1869. He was working with pus cells in the laboratory of Felix Hoppe-Seyler, one of the leading biochemists of the time. At first he had done some experiments on these cells along a line suggested by his teacher. When they did not turn out well, he investigated on the same cells the effect of pepsin, the enzyme of gastric juice which digests proteins. Unlike most biochemists, then or now, Miescher made a practice of examining carefully under a microscope the cells from which he extracted substances. As the pepsin in dilute acid decomposed the proteins of the pus cells, Miescher saw under the microscope that while the structure of the cell as a whole disintegrated, the nucleus remained essentially intact. though it shrank in size. When the peptic digestion was complete. most of the materials of the pus cells had gone into solution. Miescher made a chemical analysis of the residue, consisting of shrunken cell nuclei, and found its composition different from anything else that had previously been prepared from cells. He called the material "nuclein," because it was located in the cell nucleus.

Hoppe-Seyler, at first skeptical about the discovery, soon convinced himself that the work was sound. He himself prepared from yeast cells a substance similar to nuclein. In the meantime Miescher had returned to his native Swiss city, Basel, and there continued his study of nuclein. Basel was fortunately a most suitable place for chemical investigation of the cell nucleus. At that time the Atlantic salmon still swam up the Rhine as far as the falls just above Basel. The fish came up the river to spawn, and when they reached Basel their testes were large with sperm. Nearly every spring Miescher would conduct a sperm-collecting campaign, for salmon sperm are in many respects the most suitable cells for chemical investigation of the nucleus. They are exceedingly rich in DNA; it makes up 50 per cent of their dry weight.

With this favorable material Miescher soon began to make great progress. He found that he could remove the cytoplasm from the sperm cells by immersing them in dilute acid, thereby obtaining clean, well-formed nuclei. It was no difficult matter to extract from the nuclei pure, protein-free DNA. Miescher determined its content of nitrogen, phosphorus, carbon, oxygen and hydrogen. By this time it was clear that "nuclein" was an acid, and another investigator suggested that the protein-free substance be called "nucleic acid."

Miescher was a thorough investigator, not easily satisfied with his own achievements, and when he died in middle age much of his experimental work was found unpublished in his notebooks. In 1897 friends gathered these notes, along with his published papers, into a volume which investigators in this field find well worth poring over today.

Of the other biochemists who entered the field of investigation of the cell nucleus during Miescher's time, two of the most notable were Albrecht Kossel in Heidelberg and P. A. Levene at the Rockefeller Institute. In contrast to Miescher, who worked by himself, both Kossel and Levene had large laboratories and many collaborators—and for the type of problem which now came under investigation, many collaborators were needed.

These investigators, all skillful organic chemists, set themselves the task of unraveling the structure of the nucleic acid molecule. It is a large molecule consisting of half a dozen different, moderate-sized molecules joined together. The first problem was to take apart the large nucleic acid molecule and identify its component molecules. This had to be done gently, for the smaller molecules are themselves quite complex, and rough handling might decompose them. The safest way is to use enzymes, the tools that the organism employs. For some purposes it is possible to use a relatively rough procedure, such as treatment with hot, strong acid.

When the submolecules had been separated, they were examined to see whether they accounted for all the elements found in the large molecule. All the nitrogen could be accounted for by a group of nitrogenous submolecules, of a kind related to uric acid and caffeine. Four of these nitrogenous molecules were found, distinctly different from one another but belonging to the same family of substances. The phosphorus of DNA could be accounted for by the presence of phosphoric acid, which also explains why DNA is an acid. Another submolecule found in DNA is a 5-carbon sugar. From these three types of submolecules DNA is constructed in this way: The 5-carbon sugar molecules are linked together in chains by phosphoric acid links, and to each sugar molecule in the chain is attached one or another of the four nitrogen-containing submolecules. (A fifth nitrogen-containing molecule has recently been found; the amount varies strikingly in different kinds of DNA.)

An intact molecule of DNA is a very large, complicated structure: it may contain as many as 3,000 molecules of the 5-carbon sugar. DNA is an example of what is nowadays called a high polymer. Familiar examples of high polymers are nylon and other substances of which fibers are formed. The characteristic of a high polymer is that some chemical unit is linked together repeatedly to form a big structure. In nylon the unit is relatively simple, there being but one type of submolecule. In DNA the units are far more complex. To learn how they are polymerized to form a giant molecule is a formidable task which has not yet been accomplished. When it is, we shall understand better how DNA functions in the chromosome.

The success of the Feulgen reaction in making chromosomes visible depends upon the fact that DNA is a polymer. When subjected to the procedure of the Feulgen reaction, the polymerized DNA remains insoluble and so becomes stained where it is located in the cell. When DNA that has been depolymerized (*i.e.*, partly decomposed) is treated by the same procedure, it goes into solution; if it were not a polymer it would be washed out of the cell by the Feulgen process.

The effectiveness of DNA as a transmitter of heredity also depends upon its being polymerized. This can be seen in experiments with pneumococci. When the DNA of the heat-killed encapsulated pneumococcus is depolymerized, its ability to transmit the hereditary constitution of the cell is lost.

Besides DNA, every cell possesses another type of nucleic acid.

It is known as ribonucleic acid (RNA). Like DNA, it is composed of phosphoric acid, nitrogen submolecules and a 5-carbon sugar. But RNA's sugar molecule (called ribose) is very different from DNA's (called desoxyribose). It has one more oxygen atom, and this has a big effect on the molecule's properties. Since as much as 48 per cent of a nucleic acid is sugar, and the sugar occupies a central position in the molecule's structure, the difference between the sugars probably is responsible for the many differences between the behavior of RNA and that of DNA.

RNA has a different location in the cell and seems to be concerned with quite different biological functions. It is located largely in the cytoplasm, and it seems to be concerned with synthesis of protein. Protein synthesis is of course one of the central problems in biology, because proteins constitute a large part of living matter, because in the form of enzymes they control nearly all the dynamic processes of the cell, and because they are so complex that they have so far defied all efforts of chemists to solve the riddle of how they are made. Hence RNA is of intense interest to chemists and biologists.

RNA was found in yeast soon after Miescher's discovery of DNA. Since DNA was usually prepared from the cells of animals (fish sperm and the calf thymus gland), while RNA was usually prepared from plant cells (yeast), for many years DNA was known as animal nucleic acid and RNA was known as plant nucleic acid. The first step toward correcting this error came when Feulgen applied his color reaction and found that the nuclei both of plant cells and of animal cells contained DNA. He also stained cells in other ways to try to detect RNA, and he concluded that RNA also probably was present in both plant and animal cells, in this case in the cytoplasm.

To settle more definitely the question of where DNA and RNA are located in cells, Feulgen's pupil Martin Behrens made preparations of isolated nuclei and isolated cytoplasm. From these separated parts of the cell Feulgen and Behrens proceeded to extract nucleic acids and to identify them chemically. They found DNA in the nuclei and RNA in the cytoplasm. It was clear that DNA is in general confined to the nucleus and that at least the bulk of RNA is in the cytoplasm of both animal and plant cells.

For a more precise localization of nucleic acids in the cell further microscopic observations were required, and these were soon made by Jean Brachet in Brussels and Torbjörn Caspersson in Stockholm.

They located the nucleic acids, as Feulgen had, by means of chemical properties which can be detected in a minute amount of material under the microscope. Brachet's method depended upon the presence of phosphoric acid in nucleic acids, Caspersson's on the presence of nitrogen containing submolecules. After investigating many different kinds of cells, they found that there is some RNA in the nucleus, especially in a body called the nucleolus which is attached to a certain chromosome, and that the chromosomes themselves contain some RNA. The amount of RNA in the nucleolus and cytoplasm varies considerably from cell to cell, and in a highly significant way. Certain physiologically active cells, such as those of the heart, skeletal muscles and kidney, contain very little. But cells active in the synthesis of protein, such as those of the glands and those growing rapidly, have a high concentration of RNA in the nucleolus and cytoplasm. Of the large cells lining the stomach those that synthesize pepsin (a protein) have a large amount of RNA, whereas those that form hydrochloric acid have little.

All this certainly indicates very strongly that in some way, not yet understood, RNA plays a part in the synthesis of protein. And this seems especially true of the RNA of the nucleus. Holger Hydén in Sweden has seen under the microscope evidence that in the living cell RNA moves out of the nucleus into the cytoplasm. This may well be one of the mechanisms by which the nucleus influences the surrounding cytoplasm. The clearest sign that the RNA of the nucleus is particularly active comes from an experiment done by R. Jeener in Brussels. He exposed cells to phosphate containing tagged phosphorus and determined that much more phosphorus was incorporated in the RNA of the nucleus than in that of the cytoplasm.

The extreme variation in the amount of RNA contained in an organism's various cells is in striking contrast to the constancy of DNA in the nuclei. The cells of the pancreas of a fowl, for example, have several times as much RNA as do those of the kidney, whereas the nuclei of the two kinds of cells have the same amount of DNA.

Nucleic acids in a cell are not unattached. They are combined with proteins. Very little is known about the protein combinations of RNA. The proteins attached to DNA, on the other hand, were investigated by Miescher and by those who followed him. In the salmon sperm nucleus Miescher discovered, combined with the phosphoric acid of DNA, an unusual protein—far more basic and much simpler in construction than other proteins. It lacked many of the amino acids that are present in most protein molecules. The name of this protein is familiar to diabetics: it is protamine, which is now added to insulin to keep that substance in the blood for a longer period. This is an excellent example of how unpredictable the applications of science frequently are. Who could have anticipated that the strange protein which Miescher discovered in salmon sperm would be combined with a hormone of the pancreas for the treatment of diabetes?

Kossel made an extensive study of the protamines in fish sperm. His work on these simple proteins had a considerable influence on our understanding of proteins in general. Kossel found other basic proteins in the nuclei of red blood cells and of calf thymus, and the author later found them in nuclei of liver, kidney, pancreas and other cells. They are probably present in all cell nuclei. Most of these proteins are somewhat less basic, more complex and larger than those in sperm. During the formation of sperm cells in the testes, the more complex basic proteins in the cells from which they are made are replaced by the simpler protamines. The functions of a sperm cell are to reach the egg and transmit hereditary factors; all the equipment not essential to those functions is trimmed down to a minimum. In fact, the sperm nucleus consists of little more than DNA and its attached basic protein. Even the basic protein is trimmed down to essentials; in the fully formed sperm all that remains of it is the basic part that combines with the phosphoric acid of DNA.

The linear arrangement of the hereditary factors in a chromosome implies that DNA, which is an essential part of these factors, is held in a chromosome in a definite and precise manner. We have found that it is attached to a protein which forms part of the structure of the chromosome. From a mass of isolated chromosomes in a test tube it is possible to extract all the strongly basic protein. This is done with concentrated saline, made slightly acid. Even with the basic protein missing, the chromosomes appear unchanged when examined under the microscope, and they retain their DNA when immersed in a neutral medium. This shows clearly that the DNA is still attached to something in the chromosomes, for in this medium it is freely soluble. If it were not so bound, the DNA would simply float away from the chromosomes. That the material to which the DNA is bound is a protein was proved by treating the chromosomes with pure crystalline trypsin, which digests protein. The polymerized DNA was set free and formed a thick gel. When, on the other hand, we broke down the DNA in chromosomes with an enzyme which digests the nucleic acid, there was left a protein which could be seen as a mass of minute coiled threads quite unlike chromosomes in appearance. It had been deformed and condensed by its separation from DNA. It is the combination of this protein with DNA (to which basic protein also is attached) that forms the chromosome as seen under the microscope. If either DNA or the structural protein is digested, the structure of the chromosome disintegrates.

The amount of structural protein in the chromosomes, unlike that of DNA, is not constant but depends on the over-all activity of the cell. Cells with abundant, metabolically active cytoplasm (e.g., those of the liver or kidney) have a relatively large amount of structural protein in their chromosomes. On the other hand, a cell such as the lymphocyte, with only a scanty layer of cytoplasm around its nucleus, has only one-fifth as much. The metabolically sluggish red blood cell contains less than one-tenth as much.

The fact that the quantity of structural protein in the chromosomes is related to the over-all metabolism of the cell suggests that the structural protein may itself be metabolically active. Experiments show that this is indeed the case. The metabolic activity of the protein was measured by supplying tagged nitrogen, built into an amino acid (glycine). The amino acid can be injected into rats or mice and the fate of the tagged atoms followed in their cells. Einar Hammarsten in Stockholm found that the tagged nitrogen was taken up by the nuclear proteins much more rapidly than by DNA. In our laboratory, carrying the experiment a step further, we showed that structural protein took up nitrogen much more rapidly than did basic protein.

Since the amount of the metabolically active structural protein varies considerably in different nuclei of an organism, it follows that the chromosomes of the different cells must vary in their activity. This also has been confirmed with tagged nitrogen of glycine. Liver and kidney cells of a mouse, for example, have equal quantities of DNA, but the DNA in liver chromosomes takes up nitrogen three times as fast as that in the kidney. Even in the same cell chromosome activity varies during different physiological states. The cells of a digestive gland, such as the pancreas, become far more active when an animal is fed and when the cells are more active, the DNA of their chromosomes takes up 50 per cent more tagged nitrogen. The degree of activity of a chromosome must depend on its surroundings.

The immediate environment of the chromosomes is the cell nucleus. To understand how the chromosomes function we must know the conditions within the nucleus. The effects of chromosome activity are seen in the cytoplasm. The cytoplasm of each type of cell is especially equipped to carry out its specific functions: muscle-cell cytoplasm contains the contractile protein myosin; red-blood-cell cytoplasm has the oxygen-carrying pigment hemoglobin; pancreas-cell cytoplasm has the digestive enzyme trypsin, and so on. In each type of cell the chromosomes are acting on a differentiated cytoplasm. If such a system is to work effectively and harmoniously, the parts must be integrated; there is need for a feedback from the cytoplasm to chromosomes so that the latter's activity is adjusted to the cytoplasm's special requirements. The place to look for evidence of such a feedback is in the nuclear composition.

Does the differentiation of the cytoplasm change the nucleus? There is every indication that it does indeed. The enzymes in nuclei of different cell types of an organism vary, as do those in the cytoplasm. Even when the cytoplasm of a given type of cell has a special enzyme, the same enzyme can occasionally be found in the cell's nucleus as well. For example, the enzyme arginase, which enables liver cells to form urea, is present in both the cytoplasm and the nuclei of those cells. Since changes in the cell environment may produce marked changes in the cytoplasm, it follows that they may also alter the composition of the nucleus, and this has in fact been found to be the case.

Thus the relationship between the nucleus and the cytoplasm is a two-way affair. The hereditary factors in the chromosomes govern the cell as a whole, but the cell in turn influences conditions within the nucleus and so modifies the activity of its chromosomes. Our knowledge of this reciprocal influence, and of the chemistry of the interaction between nucleus and cytoplasm, is still vague, but obviously this is a central problem for our understanding of the cell.

III. THE GENETIC BASIS OF EVOLUTION

by Theodosius Dobzhansky

BIOLOGISTS HAVE IDENTIFIED about a million species of animals and some 267,000 species of plants, and the number of species actually in existence may be more than twice as large as the number known. In addition the earth has been inhabited in the past by huge numbers of other species that are now extinct, though some are preserved as fossils. The organisms of the earth range in size from viruses so minute that they are barely visible in electron microscopes to giants like elephants and sequoia trees. In appearance, body structure and ways of life they exhibit an endlessly fascinating variety.

From the simplest to the most complex, all of these organisms are constructed to function efficiently in the environments in which they live. The diversity and adaptedness of living beings has proved so difficult to explain that during most of his history man has taken the easy way out of assuming that every species was created by God, who contrived the body structures and functions of each kind of organism to fit it to a predestined place in nature. This idea has now been generally replaced by the less easy but intellectually more satisfying explanation that the living things we see around us were not always what they are now, but evolved gradually from very different-looking ancestors; that these ancestors were in general less complex, less perfect and less diversified than the organisms now living; that the evolutionary process is still under way, and that its causes can therefore be studied by observation and experiment in the field and in the laboratory.

Like any historical process, organic evolution may be studied in two ways. One may attempt to infer the general features and the direction of the process from comparative studies of the sequence of events in the past; this is the method of paleontologists, comparative anatomists and others. Or one may attempt to reconstruct the causes of evolution, primarily through a study of the causes and mechanisms that operate in the world at present; this approach, which uses experimental rather than observational methods, is that of the geneticist and the ecologist. This article will consider what has been learned about the causes of organic evolution through the second approach.

Evolution is generally so slow a process that during the few centuries of recorded observations man has been able to detect very few evolutionary changes among animals and plants in their natural habitats. Darwin had to deduce the theory of evolution mostly from indirect evidence, for he had no means of observing the process in action. Today, however, we can study and even produce evolutionary changes at will in the laboratory. The experimental subjects of these studies are bacteria and other low forms of life which come to birth, mature and yield a new generation within a matter of minutes or hours, instead of months or years as in most higher beings. Like a greatly speeded-up motion picture, these observations compress into a few days evolutionary events that would take thousands of years in the higher animals.

One of the most useful bacteria for this study is an organism that grows, usually harmlessly, in the intestines of practically every human being: *Escherichia coli,* or colon bacteria. These organisms can easily be cultured on a nutritive broth or nutritive agar. At about 98 degrees Fahrenheit, bacterial cells placed in a fresh culture medium divide about every 20 minutes. Their numbers increase rapidly until the nutrients in the culture medium are exhausted; a single cell may yield billions of progeny in a day. If a few cells are placed on a plate covered with nutritive agar, each cell by the end of the day produces a whitish speck representing a colony of its offspring.

Now most colon bacteria are easily killed by the antibiotic drug streptomycin. It takes only a tiny amount of streptomycin, 25 milligrams in a liter of a nutrient medium, to stop the growth of the bacteria. Recently, however, the geneticist Milislav Demerec and his collaborators at the Carnegie Institution in Cold Spring Harbor, N. Y., have shown that if several billion colon bacteria are placed on the streptomycin-containing medium, a few cells will survive and form colonies on the plate. The offspring of these hardy survivors are able to multiply freely on a medium containing streptomycin. A mutation has evidently taken place in the bacteria; they have now become resistant to the streptomycin that was poisonous to their sensitive ancestors.

How do the bacteria acquire their resistance? Is the mutation caused by their exposure to streptomycin? Demerec has shown by experimental tests that this is not so; in any large culture a few

294

resistant mutants appear even when the culture has not been exposed to streptomycin. Some cells in the culture undergo mutations from sensitivity to resistance regardless of the presence or absence of streptomycin in the medium. Demerec found that the frequency of mutation was about one per billion; *i.e.*, one cell in a billion becomes resistant in every generation. Streptomycin does not induce the mutations; its role in the production of resistant strains is merely that of a selecting agent. When streptomycin is added to the culture, all the normal sensitive cells are killed, and only the few resistant mutants that happened to be present before the streptomycin was added survive and reproduce. Evolutionary changes are controlled by the environment, but the control is indirect, through the agency of natural or artificial selection. The mechanics of this process is charted on page 296.

What governs the selection? If resistant bacteria arise in the absence of streptomycin, why do sensitive forms predominate in all normal cultures; why has not the whole species of colon bacteria become resistant? The answer is that the mutants resistant to streptomycin are at a disadvantage on media free from this drug. Indeed, Demerec has discovered the remarkable fact that about 60 per cent of the bacterial strains derived from streptomycin-resistant mutants become dependent on streptomycin; they are unable to grow on media free from it!

On the other hand one can reverse the process and obtain strains of bacteria that can live without streptomycin from cultures predominantly dependent on the drug. If some billions of dependent bacteria are plated on nutrient media free of the drug, all dependent cells cease to multiply and only the few mutants independent of the drug reproduce and form colonies. Demerec estimates the frequency of this "reverse" mutation at about 37 per billion cells in each generation.

Evolutionary changes of the type described in colon bacteria have been found in recent years in many other bacterial species. The increasing use of antibiotic drugs in medical practice has made such changes a matter of considerable concern in public health. As penicillin, for example, is used on a large scale against bacterial infections, the strains of bacteria that are resistant to penicillin survive and multiply, and the probability that they will infect new victims is increased. The mass application of antibiotic drugs may lead in

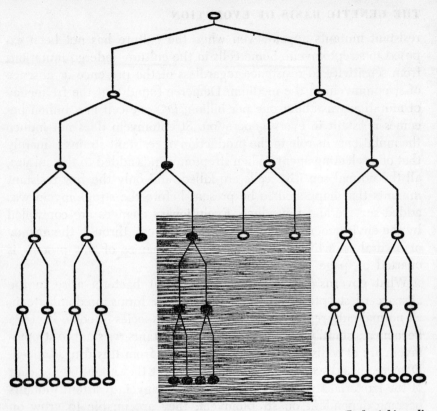

In normal environment the common strain of the bacterium *Escherichia coli* (*white bacteria*) multiplies freely and a mutant strain resistant to streptomycin (*black bacteria*) remains rare because the mutation is not useful. But when streptomycin is added to the environment (*gray area*), the common strain dies and the resistant strain multiplies in its stead.

the long run to increased incidence of cases refractory to treatment. Indications exist that this has already happened in some instances: in certain cities penicillin-resistant gonorrhea has become more frequent than it was.

The same type of evolutionary change has also been noted in some larger organisms. A good example is the case of DDT and the common housefly, *Musca domestica*. DDT was a remarkably effective poison for houseflies when first introduced less than 10 years ago. But already reports have come from places as widely separated as New Hampshire, New York, Florida, Texas, Italy and Sweden that DDT sprays in certain localities have lost their effectiveness. What has happened, of course, is that strains of houseflies relatively resistant to DDT have become established in these localities. Man has unwittingly become an agent of a selection process which has led to evolutionary changes in housefly populations.

Obviously evolutionary selection can take place only if nature provides a supply of mutants to choose from. Thus no bacteria will survive to start a new strain resistant to streptomycin in a culture in which no resistant mutant cells were present in the first place, and housefly races resistant to DDT have not appeared everywhere that DDT is used. Adaptive changes are not mechanically forced upon the organism by the environment. Many species of past geological epochs died out because they did not have a supply of mutants which fitted changing environments. The process of mutation furnishes the raw materials from which evolutionary changes are built.

Mutations arise from time to time in all organisms, from viruses to man. Perhaps the best organism for the study of mutations is the now famous fruit fly, Drosophila. It can be bred easily and rapidly in laboratories, and it has a large number of bodily traits and functions that are easy to observe. Mutations affect the color of its eyes and body, the size and shape of the body and of its parts, its internal anatomical structures, its fecundity, its rate of growth, its behavior, and so on. Some mutations produce differences so minute that they can be detected only by careful measurements; others are easily seen even by beginners; still others produce changes so drastic that death occurs before the development is completed. These lethal mutations are discussed in detail later in this chapter.

In all organisms the majority of mutations are more or less harmful. This may seem a very serious objection against the theory which regards them as the mainspring of evolution. If mutations produce incapacitating changes, how can adaptive evolution be compounded of them? The answer is: a mutation that is harmful in the environment in which the species or race lives may become useful, even essential, if the environment changes. Actually it would be strange if we found mutations that improve the adaptation of the organism in the environment in which it normally lives. Every kind of mutation that we observe has occurred numerous times under natural conditions, and the useful ones have become incorporated into what we call the "normal" constitution of the species. But when the environment changes, some of the previously rejected mutations become advantageous and produce an evolutionary change in the species.

This is not to say that every mutation will be found useful in some environment somewhere. It would be difficult to imagine environments in which such human mutants as hemophilia or the absence of hands and feet might be useful. Most mutants that arise in any species are, in effect, degenerative changes; but some, perhaps a small minority, may be beneficial in some environments. If the environment were forever constant, a species might conceivably reach a summit of adaptedness and ultimately suppress the mutation process. But the environment is never constant; it varies not only from place to place but from time to time. If no mutations occur in a species, it can no longer become adapted to changes and is headed for eventual extinction. Mutation is the price that organisms pay for survival. They do not possess a miraculous ability to produce only useful mutations where and when needed. Mutations arise at random, regardless of whether they will be useful at the moment, or ever; nevertheless, they make the species rich in adaptive possibilities.

To understand the mutation process we must inquire into the nature of heredity. In brief, heredity is self-reproduction, and the units of self-reproduction are the genes. They are considered stable because the copies they make of themselves are true likenesses of the original in the overwhelming majority of cases; but occasionally the copying process is faulty and the new gene that emerges differs from its model. This is a mutation. We can increase the frequency of mutations in experimental animals by treating the genes with X-rays, ultraviolet rays, high temperature or certain chemical substances.

Can a gene be changed by the environment? Assuredly it can. But the important point is the kind of change produced. The change that is easiest to make is to treat the gene with poisons or heat in such a way that it no longer reproduces itself. But a gene that cannot produce a copy of itself from other materials is no longer a gene; it is dead. A mutation is a change of a very special kind: the altered gene can reproduce itself, and the copy produced is like the changed structure, not like the original. Changes of this kind are relatively rare. Their rarity is not due to any imperviousness of

298

the genes to influences of the environment, for genic materials are probably the most active chemical constituents of the body; it is due to the fact that genes are by nature self-reproducing, and only the rare changes that preserve the genes' ability to reproduce can effect a lasting alteration of the organism.

Although the number of genes in a single organism is not known with precision, it is certainly in the thousands, at least in the higher organisms. For Drosophila, 5,000 to 12,000 seems a reasonable estimate, and for man the figure is, if anything, higher. Since most or all genes suffer mutational changes from time to time, populations of virtually every species must contain mutant variants of many genes. For example, in the human species there are variations in the skin, hair and eye colors, in the shape and distribution of hair, in the form of the head, nose and lips, in stature, in body proportions, in the chemical composition of the blood, in psychological traits, and so on. Each of these traits is influenced by several or by many genes. To be conservative, let us assume that the human species has only 1,000 genes and that each gene has only two variants. Even on this conservative basis, Mendelian segregation and recombination would be capable of producing 2^{1000} different gene combinations in human beings.

The number 2^{1000} is easy to write but is utterly beyond comprehension. Compared with it, the total number of electrons and protons estimated by physicists to exist in the universe is negligibly small! It means that except in the case of identical twins no two persons now living, dead, or to live in the future are at all likely to carry the same complement of genes. Dogs, mice and flies are as individual and unrepeatable as men are. The mechanism of sexual reproduction, of which the recombination of genes is a part, creates ever new genetic constitutions on a prodigious scale.

One might object that the number of possible combinations does not greatly matter; after all, they will still be combinations of the same thousand gene variants, and the way they are combined is not significant. Actually it is: the same gene may have different effects in combinations with different genes. Thus N. W. Timofeeff-Ressovsky in Germany showed that two mutants in Drosophila, each of which reduced the viability of the fly when it was present alone, were harmless when combined in the same individual by hybridization. Natural selection tests the fitness in certain environments not of single genes but of constellations of genes present in living individuals.

Sexual reproduction generates, therefore, an immense diversity of genetic constitutions, some of which, perhaps a small minority, may prove well attuned to the demands of certain environments. The biological function of sexual reproduction consists in providing a highly efficient trial-and-error mechanism for the operation of natural selection. It is a reasonable conjecture that sex became established as the prevalent method of reproduction because it gave organisms the greatest potentialities for adaptive and progressive evolution.

Many animal and plant species are polymorphic, *i.e.*, represented in nature by two or more clearly distinguishable kinds of individuals. For example, some individuals of the ladybird beetle *Adalia bipunctata* are red with black spots while others are black with red spots. The color difference is hereditary, the black color behaving as a Mendelian dominant and red as a recessive. The red and black forms live side by side and interbreed freely. Timofeeff-Ressovsky observed that near Berlin the black form predominates from spring to autumn, and the red form is more numerous during the winter. What is the origin of these changes? It is quite improbable that the genes for color are transformed by the seasonal variations in temperature; that would mean epidemics of directed mutations on a scale never observed. A much more plausible view is that the changes are produced by natural selection. The black form is, for some reason, more successful than the red in survival and reproduction during summer, but the red is superior to the black under winter conditions. Since the beetles produce several generations during a single season, the species undergoes cyclic changes in its genetic composition in response to the seasonal alterations in the environment. This hy-

Mendelian segregation is illustrated by four o'clock (*mirabilis jalapa*). The genes of two purebred strains, having red (left) and white flowers respectively, combine in pink hybrid which is furnished with one gene for each strain. When pink hybrid is mated to pink hybrid, however, the next generation will have red and white as well as pink members, each trait being segregated in one of the three possible combinations of the two types of gene. The relative probability of these combinations determines that one fourth will be white (white-white combination), one half will be pink (red-white or white-red combinations) and one fourth will be red (red-red combination).

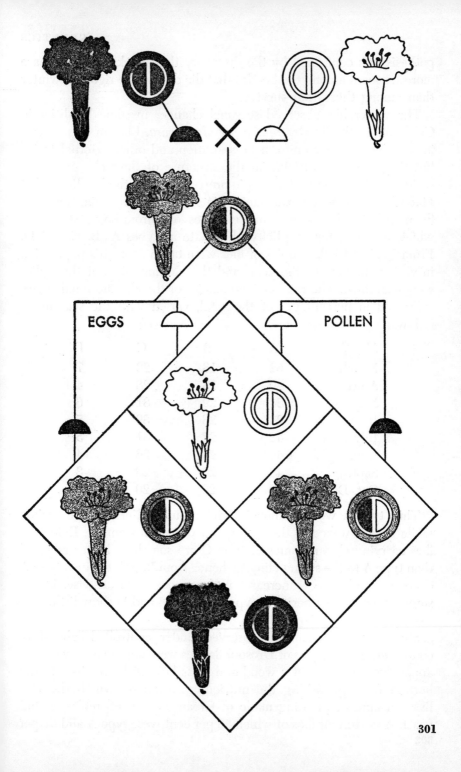

EGGS

POLLEN

pothesis was confirmed by the discovery that black individuals are more frequent among the beetles that die during the rigors of winter than among those that survive.

The writer has observed seasonal changes in some localities in California in the fly *Drosophila pseudoobscura*. Flies of this species in nature are rather uniform in coloration and other external traits, but they are very variable in the structure of their chromosomes, which can be seen in microscopic preparations. In the locality called Piñon Flats, on Mount San Jacinto in southern California, the fruit-fly population has four common types of chromosome structure, which we may, for simplicity, designate as types A, B, C and D. From 1939 to 1946, samples of flies were taken from this population in various months of the year, and the chromosomes of these flies were examined. The relative frequencies of the chromosomal types, expressed in percentages of the total, varied with the seasons as follows:

Month	A	B	C	D
March	52	18	23	7
April	40	28	28	4
May	34	29	31	6
June	28	28	39	5
July	42	22	31	5
Aug.	42	28	26	4
Sept.	48	23	26	3
Oct.-Dec.	50	26	20	4

Thus type A was common in winter but declined in the spring, while type C waxed in the spring and waned in summer. Evidently flies carrying chromosomes of type C are somehow better adapted than type A to the spring climate; hence from March to June, type A decreases and type C increases in frequency. Contrariwise, in the summer type A is superior to type C. Types B and D show little consistent seasonal variation.

Similar changes can be observed under controlled laboratory conditions. Populations of Drosophila flies were kept in a very simple apparatus consisting of a wood and glass box, with openings in the bottom for replenishing the nutrient medium on which the flies lived—a kind of pudding made of Cream of Wheat, molasses and yeast. A mixture of flies of which 33 per cent were type A and 67 per cent type C was introduced into the apparatus and left to multiply freely, up to the limit imposed by the quantity of food given. If one of the types was better adapted to the environment than the other, it was to be expected that the better-adapted type would increase and the other decrease in relative numbers. This is exactly what happened. During the first six months the type A flies rose from 33 to 77 per cent of the population, and type C fell from 67 to 23 per cent. But then came an unexpected leveling off: during the next seven months there was no further change in the relative proportions of the flies, the frequencies of types A and C oscillating around 75 and 25 per cent respectively.

If type A was better than type C under the conditions of the experiment, why were not the flies with C chromosomes crowded out completely by the carriers of A? Sewall Wright of the University of Chicago solved the puzzle by mathematical analysis. The flies of these types interbreed freely, in natural as well as in experimental populations. The populations therefore consist of three kinds of individuals: 1) those that obtained chromosome A from father as well as from mother, and thus carry two A chromosomes (AA); 2) those with two C chromosomes (CC); 3) those that received chromosomes of different types from their parents (AC). The mixed type, AC, possesses the highest adaptive value; it has what is called "hybrid vigor." As for the pure types, under the conditions that obtain in nature AA is superior to CC in the summer. Natural selection then increases the frequency of A chromosomes in the population and diminishes the C chromosomes. In the spring, when CC is better than AA, the reverse is true. But note now that in a population of mixed types neither the A nor the C chromosomes can ever be entirely eliminated from the population, even if the flies are kept in a constant environment where type AA is definitely superior to type CC. This, of course, is highly favorable to the flies as a species, for the loss of one of the chromosome types, though it might be temporarily advantageous, would be prejudicial in the long run, when conditions favoring the lost type would return. Thus a polymorphic population is better able than a uniform one to adjust itself to environmental changes and to exploit a variety of habitats.

Populations of the same species which inhabit different environments become genetically different. This is what a geneticist means when he speaks of races. Races are populations within a species that differ in the frequencies of some genes. According to the old concept of race, which is based on the notion that heredity is transmitted through "blood" and which still prevails among those ignorant of modern biology, the hereditary endowment of an isolated population would become more and more uniform with each generation, provided there was no interbreeding with other tribes or populations. The tribe would eventually become a "pure" race, all members of which would be genetically uniform. Scientists misled by this notion used to think that at some time in the past the human species consisted of an unspecified number of "pure" races, and that intermarriage between them gave rise to the present "mixed" populations.

In reality, "pure" races never existed, nor can they possibly exist in any species, such as man, that reproduces by sexual combination. We have seen that all human beings except identical twins differ in heredity. In widely differing climatic environments the genetic differences may be substantial. Thus populations native in central Africa have much higher frequencies of genes that produce dark skin than do European populations. The frequency of the gene for blue eye color progressively diminishes southward from Scandinavia through central Europe to the Mediterranean and Africa. Nonetheless some blue-eyed individuals occur in the Mediterranean region and even in Africa, and some brown-eyed ones in Norway and Sweden.

It is important to keep in mind that races are populations, not individuals. Race differences are relative and not absolute, since only in very remote races do all members of one population possess a set of genes that is lacking in all members of another population. It is often difficult to tell how many races exist in a species. For example, some anthropologists recognize only two human races while others list more than 100. The difficulty is to know where to draw the line. If, for example, the Norwegians are a "Nordic race" and the southern Italians a "Mediterranean race," to what race do the inhabitants of Denmark, northern Germany, southern Germany, Switzerland and northern Italy belong? The frequencies of most differentiating traits change rather gradually from Norway to southern Italy. Calling the intermediate populations separate races may be technically correct, but this confuses the race classification even more, because nowhere can sharp lines of demarcation between these "races" be drawn. It is
304

quite arbitrary whether we recognize 2, 4, 10, or more than 100 races—or finally refuse to make any rigid racial label at all.

The differences between human races are, after all, rather small, since the geographic separation between them is nowhere very marked. When a species is distributed over diversified territories, the process of adaptation to the different environments leads to the gradual accumulation of more numerous and biologically more and more important differences between races. The races gradually diverge. There is, of course, nothing fatal about this divergence, and under some circumstances the divergence may stop or even be turned into convergence. This is particularly true of the human species. The human races were somewhat more sharply separated in the past than they are today. Although the species inhabits almost every variety of environment on earth, the development of communications and the increase of mobility, especially in modern times, has led to much intermarriage and to some genetic convergence of the human races.

The diverging races become more and more distinct with time, and the process of divergence may finally culminate in transformation of races into species. Although the splitting of species is a gradual process, and it is often impossible to tell exactly when races cease to be races and become species, there exist some important differences between race and species which make the process of species formation one of the most important biological processes. Indeed, Darwin called his chief work *The Origin of Species.*

Races of sexually reproducing organisms are fully capable of intercrossing; they maintain their distinction as races only by geographical isolation. As a rule in most organisms no more than a single race of any one species inhabits the same territory. If representatives of two or more races come to live in the same territory, they interbreed, exchange genes, and eventually become fused into a single population. The human species, however, is an exception. Marriages are influenced by linguistic, religious, social, economic and other cultural factors. Hence cultural isolation may keep populations apart for a time and slow down the exchange of genes even though the populations live in the same country. Nevertheless, the biological relationship proves stronger than cultural isolation, and interbreeding is everywhere in the process of breaking down such barriers. Unrestricted interbreeding would not mean, as often supposed, that all people would become alike. Mankind would continue to include at least as great a diversity of hereditary endowments as it contains today. However, the same types could be found anywhere in the world, and races as more or less discrete populations would cease to exist.

Species, on the contrary, can live in the same territory without losing their identity. F. Lutz of the American Museum of Natural History found 1,402 species of insects in the 75-by-200-foot yard of his home in a New Jersey suburb. This does not mean that representatives of distinct species never cross. Closely related species occasionally do interbreed in nature, especially among plants, but these cases are so rare that the discovery of one usually merits a note in a scientific journal.

The reason distinct species do not interbreed is that they are more or less completely kept apart by isolating mechanisms connected with reproduction, which exist in great variety. For example, the botanist Carl C. Epling of the University of California found that two species of sage which are common in southern California are generally separated by ecological factors, one preferring a dry site, the other a more humid one. When the two sages do grow side by side, they occasionally produce hybrids. The hybrids are quite vigorous, but their seeds set amounts to less than two per cent of normal; *i.e.*, they are partially sterile. Hybrid sterility is a very common and effective isolating mechanism. A classic example is the mule, hybrid of the horse and donkey. Male mules are always sterile, females usually so. There are, however, some species, notably certain ducks, that produce quite fertile hybrids, not in nature but in captivity.

Two species of Drosophila, *pseudoobscura* and *persimilis*, are so close together biologically that they cannot be distinguished by inspection of their external characteristics. They differ, however, in the structure of their chromosomes and in many physiological traits. If a mixed group of females of the two species is exposed to a group of males of one species, copulations occur much more frequently between members of the same species than between those of different species, though some of the latter do take place. Among plants, the flowers of related species may differ so much in structure that they cannot be pollinated by the same insects, or they may have such differences in smell, color and shape that they attract different insects. Finally, even when cross-copulation or cross-pollination can occur, the union may fail to result in fertilization or may produce offspring that cannot live. Often several isolating mechanisms, no one of which is effective separately, combine to prevent inter-breeding.

The fact that distinct species can coexist in the same territory, while races generally cannot, is highly significant. It permits the formation of communities of diversified living beings which exploit the variety of habitats present in a territory more fully than any single species, no matter how polymorphic, could. It is responsible for the richness and colorfulness of life that is so impressive to biologists and non-biologists alike.

Our discussion of the essentials of the modern theory of evolution may be concluded with a consideration of the objections raised against it. The most serious objection is that since mutations occur by "chance" and are undirected, and since selection is a "blind" force, it is difficult to see how mutation and selection can add up to the formation of such complex and beautifully balanced organs as, for example, the human eye. This, say critics of the theory, is like believing that a monkey pounding a typewriter might accidentally type out Dante's *Divine Comedy*. Some biologists have preferred to suppose that evolution is directed by an "inner urge toward perfection," or by a "combining power which amounts to intentionality," or by "telefinalism" or the like. The fatal weakness of these alternative "explanations" of evolution is that they do not explain anything. To say that evolution is directed by an urge, a combining power, or a telefinalism is like saying that a railroad engine is moved by a "locomotive power."

The objection that the modern theory of evolution makes undue demands on chance is based on a failure to appreciate the historical character of the evolutionary process. It would indeed strain credulity to suppose that a lucky sudden combination of chance mutations produced the eye in all its perfection. But the eye did not appear suddenly in the offspring of an eyeless creature; it is the result of an evolutionary development that took many millions of years. Along the way the evolving rudiments of the eye passed through innumerable stages, all of which were useful to their possessors, and therefore all adjusted to the demands of the environment by natural selection. Amphioxus, the primitive fishlike darling of comparative anatomists, has no eyes, but it has certain pigment cells in its brain by means of which it perceives light. Such pigment cells may have been the starting point of the development of eyes in our ancestors.

We have seen that the "combining power" of the sexual process is staggering, that on the most conservative estimate the number of possible gene combinations in the human species alone is far greater than that of the electrons and protons in the universe. When life developed sex, it acquired a trial-and-error mechanism of prodigious efficiency. This mechanism is not called upon to produce a completely new creature in one spectacular burst of creation; it is sufficient that it produces slight changes that improve the organism's chances of survival or reproduction in some habitat. In terms of the monkey-and-typewriter analogy, the theory does not require that the monkey sit down and compose the *Divine Comedy* from beginning to end by a lucky series of hits. All we need is that the monkey occasionally form a single word, or a single line; over the course of eons of time the environment shapes this growing text into the eventual masterpiece. Mutations occur by "chance" only in the sense that they appear regardless of their usefulness at the time and place of their origin. It should be kept in mind that the structure of a gene, like that of the whole organism, is the outcome of a long evolutionary development; the ways in which the genes can mutate are, consequently, by no means indeterminate.

Theories that ascribe evolution to "urges" and "telefinalisms" imply that there is some kind of predestination about the whole business, that evolution has produced nothing more than was potentially present at the beginning of life. The modern evolutionists believe that, on the contrary, evolution is a creative response of the living matter to the challenges of the environment. The role of the environment is to provide opportunities for biological inventions. Evolution is due neither to chance nor to design; it is due to a natural creative process

IV. CATACLYSMIC EVOLUTION

by G. Ledyard Stebbins, Jr.

As a rule the origin of new species is a slow, gradual process. It takes many generations, and sometimes eons of time, for a new kind of organism to arise by the usual evolutionary method of mutation, favorable mating and natural selection. But there is a certain type of evolution in the plant kingdom that telescopes the transformation into one or two steps. In a single generation or two an old species or pair of species suddenly gives rise to a brand-new plant, strikingly different from its parents and capable of breeding true to type in its new form. The occurrence of this cataclysmic type of evolution is by no means uncommon: it was responsible for the origin of cultivated wheat, oats, cotton, tobacco, sugar cane, and probably, in the remote past, of apples, pears, lilacs, willows and many other plants.

The process that causes this abrupt creation of a new species is the accidental doubling of a plant's set of chromosomes. Occasionally, for example, a species of rose that normally has 14 chromosomes may produce an abnormal offspring with 28, or a tobacco species with 24 chromosomes may yield one with 48. Most cases of doubling, however, arise not from such a single species but from hybrids, or crosses between two species. In any case, the double-chromosome descendant is definitely a different plant from the original. It meets the all-important test that marks a truly new species: it can reproduce itself indefinitely, but it cannot mate with the original species, or if it does, the offspring from the mating are sterile.

The story of the discovery and elucidation of this phenomenon, which has given man a tool of great potential usefulness, covers a half-century in time and involves a host of investigators in all parts of the world. Early in this century the Belgian botanists Élias and Émile Marchal, examining certain mosses under the microscope, observed that some of them had double the usual number of chromosomes. At about the same time in England the geneticist R. Ruggles Gates discovered the same curious circumstance in an evening primrose. In Japan Masuto Tahara noticed another interesting fact: in the various species of the chrysanthemum family the chromosome numbers were all multiples of nine—namely, 18, 36, 54, 72 and 90.

What did these facts mean? The first to suggest a plausible answer was Ojvind Winge of Denmark. Using Tahara's series of chrysanthemum numbers he worked out an ingenious hypothesis based on the behavior of chromosomes in the sex cells. In body cells the chromosomes split when the cell divides and each daughter cell receives a full complement of chromosomes. Sex cells are formed by a division of a special kind called meiosis. In meiotic division the chromosomes do not split; hence the sex cells come to have only half as many chromosomes as the body cells from which they originated. (The full number is restored when a pollen cell and an egg cell unite at fertilization.) During the process of meiosis the chromosomes in the sex cells pair off; each seeks a mate with genes like its own and takes a position beside it. Ordinarily it finds its match from the opposite parent; that is, a chromosome derived from the male parent of the plant mates with a chromosome from the female parent of the plant. In the final meiotic division the mates are separated, one set of chromosomes going to one daughter cell and their mates to the other.

These facts explain why a hybrid between two parents of different species is generally sterile. To form functioning sex cells the chromosomes must find mates that are very like themselves in genetic composition. When two different species are crossed, the chromosomes that the hybrid receives from one parent may differ greatly from those it gets from the other, so that pairing of the chromosomes is difficult or impossible. As a result the meiotic cell divisions are much disturbed, and those sex cells that do emerge have abnormal, disharmonious combinations of chromosomes.

Winge analyzed Tahara's series of chrysanthemum chromosome numbers in the light of this knowledge. Suppose the various species of chrysanthemums had arisen from a single original species by successive doublings of the chromosomes. Then the series of numbers should run 18, 36, 72 and so on. But the actual numbers were 18, 36, 54, 72 and 90. How could a species with 54 chromosomes have been formed? Certainly not by doubling 18 or 36. If it was created by chromosome-doubling, it must have come from a plant with 27 chromosomes.

Now it was easy enough to see how a 27-chromosome chrysanthemum might arise by hybridization. Suppose a chrysanthemum species with 36 chromosomes hybridized with one having 18 chromosomes. The combining egg and pollen cells of the two species would have half the full number of chromosomes, or 18 and 9, respectively. When the two combined, the result would be $18 + 9 = 27$, and the hybrid would carry this number of chromosomes in its body cells. This hybrid, as we have seen, would ordinarily be sterile, since its chromosomes would not match. However, Winge suggested, suppose by some accident the hybrid should produce offspring with a doubled set of chromosomes. Then each chromosome would have a duplicate, and they could easily mate. Hence this offspring, with 54 chromosomes, should be fertile. A simplified scheme of this argument is shown in the drawing.

Winge's ingenious theory suffered from the weakness that, so far as he knew, no sterile hybrid had ever actually been converted into a fertile species by doubling its chromosome number. But soon after he published his hypothesis other botanists pointed out that just such a phenomenon had been observed, although Winge apparently was unaware of it. Miss L. Digby of London University had noticed that a certain sterile hybrid primrose growing in Kew Gardens suddenly became fertile when its usual quota of 18 chromosomes was doubled to 36.

Two botanists at the University of California, Roy E. Clausen and T. Harper Goodspeed, proceeded to perform a critical experiment which clearly verified Winge's hypothesis. They crossed the cultivated tobacco species (*Nicotiana tabacum*) with a distantly related wild species of tobacco (*Nicotiana glutinosa*). The resulting hybrid was vigorous but almost completely sterile. By means of careful hand-pollination Clausen and Goodspeed managed to obtain a few seeds, from which they raised several second-generation plants.

These plants turned out to be highly fertile, as Winge's theory predicted. But they also furnished a more convincing confirmation. On the basis of the Mendelian laws of heredity, one would expect the plants to resemble one or the other of the original parents— *N. tabacum* or *N. glutinosa*. This is so because in normal Mendelian heredity the parental characteristics are not changed but simply sorted out to the offspring; chromosomes carrying the father's genes pair with those from the mother, and whether a second-generation offspring shows the traits of one or the other depends on the dominance relations in the particular combination of genes. If Winge's hypothesis was correct, however, when a hybrid doubled its chromosomes the two members of a pair would not come from the father and the mother. Both would be derived from a single original chromosome in the hybrid; they would be the two duplicates of this chromosome created when the hybrid doubled its set. Consequently Winge's theory predicted that the second-generation offspring would have traits not just like those of the paternal or maternal species but somewhere between.

This indeed proved to be the case in Clausen's and Goodspeed's tobacco plants. All of them resembled their hybrid parent much more closely than they did either of the original parental species. They differed from the first-generation hybrid chiefly in their more robust appearance and somewhat larger flowers.

Counts of the chromosomes also confirmed the theory. *N. tabacum* has 48 chromosomes in its vegetative (body) cells and 24 in its pollen and egg cells, while *N. glutinosa* has 24 and 12, respectively. Examination of the first-generation hybrids under the microscope showed that their vegetative cells had 36 chromosomes, that is, 24 plus 12. And the fertile new species produced in the second generation had 72 chromosomes in their vegetative cells, as Winge had predicted they would have.

How did the chromosome-doubling occur? From a careful study of chromosome behavior during pollen formation in the first-generation hybrid, Clausen and Goodspeed concluded that the doubling was caused by accidents during the abnormal cell divisions of the sex cells. In short, this tobacco hybrid behaved in every respect as Winge had predicted.

Meanwhile botanists had accumulated many more facts about chromosome-doubling as a method of plant evolution. Multiple series of chromosome numbers like that in the chrysanthemums had been found in roses, wheats and several other groups of plants. Experimenters had succeeded in doubling the chromosome numbers artificially in tomatoes and nightshades. During the decade from 1925 to 1935, as the phenomenon became widely known, geneticists began to experiment in breeding new plants by this method. The Russian cytologist G. D. Karpechenko synthesized a fertile new species that was a cross between the radish and the cabbage. Unfortunately it combined the worst features of both—it had the harsh leaves of the

Hybrid between two distantly related species (*left*) cannot produce fertile sex cells because the chromosomes cannot form pairs. But if the chromosomes of this hybrid double (*right*), pairing becomes possible and the sex cells are fertile.

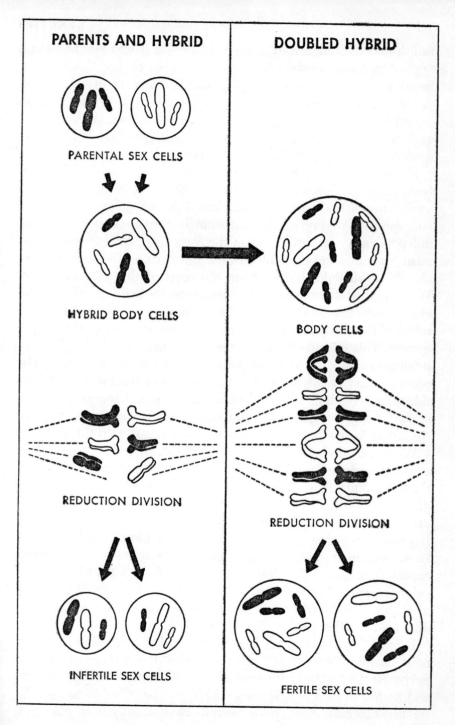

PARENTS AND HYBRID

DOUBLED HYBRID

PARENTAL SEX CELLS

HYBRID BODY CELLS

BODY CELLS

REDUCTION DIVISION

REDUCTION DIVISION

INFERTILE SEX CELLS

FERTILE SEX CELLS

313

radish and the tough root of the cabbage—so it was worthless as a vegetable. Other workers in Germany, the U. S., the U.S.S.R. and Sweden produced a true-breeding doubled hybrid of wheat and rye which may eventually be a valuable crop plant. The Swedish geneticist Arne Müntzing, working along this line, is breeding a grain which, it is hoped, will combine the high milling and baking qualities of wheat with the ability of rye to grow in poor soils.

For a number of years experiments in chromosome-doubling suffered from the difficulty that there was no sure method for doing it artificially. In 1937 this bottleneck was removed. Albert F. Blakeslee and Amos G. Avery of the Carnegie Institution Laboratory and, independently, Bernard Nebel of the New York Agricultural Experiment Station found that colchicine could perform the trick. Colchicine is an alkaloid derived from the roots of the autumn crocus. When applied in very weak concentrations to dividing cells, colchicine permits the division of the chromosomes but prevents the separation of the daughter chromosomes to opposite poles of the mitotic spindle. This effectively blocks the completion of cell division and produces cells with twice the normal number of chromosomes. The concentration of colchicine and the method of treatment vary considerably according to the plant being treated. Through the work of scores of investigators in many countries a great variety of plant species have now been converted into new forms with this drug.

When a nonhybrid plant of some well-established species such as corn, tomatoes or snapdragons has its chromosome number doubled, the new plant differs from its undoubled progenitor in certain definite characteristics. It usually has thicker leaves and stems; its flowers are generally larger and have firmer petals; its seeds are larger. It grows more slowly and therefore usually blooms and sets fruit later. Most important, it is likely to be more or less sterile. For this reason chromosome-doubling of nonhybrid plants is of little practical value for grains, in which the seeds are harvested, though it has given us useful varieties of certain leaf crops such as clover and of snapdragons and other garden flowers.

It is in the treatment of hybrid plants that colchicine has been most useful. Investigators have employed the technique for two purposes: 1) to study the evolutionary ancestry of plant species, and 2) to produce new species which may be of value to man. In the first category some striking results have already been obtained.

One of the most interesting has to do with the origin of wheat.

About 1920 T. Sakamura in Japan and Karl Sax at Harvard University independently discovered that the cultivated wheats include species with three different chromosome numbers. The primitive, small-grained "einkorn" wheat of southeastern Europe and southwestern Asia has only 14 chromosomes in its vegetative cells. The "emmer" wheats, grown in northern Europe and the northern U. S. as stock feed, have 28 chromosomes. The bread wheats have 42 chromosomes. Both botanical and archaeological evidence points to the 14-chromosome wheats as the oldest and most primitive and the 42-chromosome species as the most recent. Furthermore, the behavior of the chromosomes suggests that the 28- and 42-chromosome groups originated as doubled hybrids from species having lower chromosome numbers.

At about the time that Sakamura and Sax made this discovery the British wheat expert John Percival put forward the bold hypothesis that the bread wheats originated from hybridization between emmer wheats and goat grass, a common weed which often grows around the edges of wheat fields in the Mediterranean region. After colchicine became available, several cytologists undertook to test Percival's theory by experiment. From emmer wheat and a particular species of goat grass named *Aegilops squarrosa*, E. S. McFadden and Ernest Sears of the U. S. Department of Agriculture and Hitoshi Kihara of Japan produced doubled hybrids which looked very much like primitive forms of bread wheats. Furthermore, they passed the acid genetic test: hybrids between them and bread wheats were fertile and had normal chromosome behavior. This is proof that the bread wheats actually arose from hybrids between emmer wheats and *Aegilops squarrosa*. This goat grass grows in the region of the ancient Babylonian civilizations where the bread wheats are believed to have originated.

The remarkable fact about the wheat story is that the combination of chromosomes of a moderately useful plant, emmer wheat, and those of a completely useless and noxious weed produced the world's most valuable crop plant. This example should tell us that we cannot always predict in advance whether a particular hybrid will be worthless or a priceless new addition.

The story of the cataclysmic evolution of other plants, notably cotton, has been worked out in a similar manner. Thanks to the experimental method made possible by colchicine, the origin of these doubled hybrids can be established with greater certainty than can any other process of evolution. Moreover, this method enables us to reconstruct the beginnings of plants that are not well represented in the fossil record, and to work out the distribution of plant species in past geological ages when conditions were very different from what they are now.

A striking example of such a deduction concerns the blue-flag iris of the northeastern U. S. (*Iris versicolor*). Edgar Anderson of the Missouri Botanical Garden proved that this species, which has 108 chromosomes, is a doubled hybrid derived from a blue-flag iris of the Mississippi Valley (72 chromosomes) and an arctic iris found in central Alaska (36 chromosomes). This means that the two parent irises, now widely separated, must once have lived close together. During the last ice age, when glaciers covered the northern end of the Mississippi Valley, there must have been many areas directly south of the ice sheet in which conditions were not very different from those that now prevail in central Alaska. The Alaskan iris could have grown together with the Mississippi Valley species in those areas, and the blue-flag iris of the northeastern states probably originated there. When the ice retreated, the new species was better adapted to occupy the newly opened regions than either of its parental species. At present this northeastern blue flag is the only common one in the formerly glaciated areas.

In the case of cotton and some members of the grass family, the original parents or nearest relatives are now so widely separated that they do not even occupy the same continents. There are two species of rice grass in the woodlands of eastern North America which are doubled hybrids that came from parents now separated by thousands of miles of ocean: one parent today is found only in North America and the other in eastern Asia. Evidently certain species of rice grass now confined to Eurasia once existed in North America. Evidence from fossils has shown the same to be true of various species of forest trees.

Up to the present the principal interest in doubled hybrids has been scientific rather than practical. But their practical possibilities have already been recognized in such work as the combination of wheat and rye to produce a hardier grain and in other projects. Beyond this, botanists may be able to combine in one doubled hybrid the valuable qualities of two or more wild plants now useless by themselves, and so create new species of value to man. This is particularly true in the case of forage plants, where vigor of growth and adaptability to a variety of environmental conditions are more important than the quality of the product. Many wild plants that grow in very dry or very cold habitats originated as natural doubled hybrids. Increased knowledge of species of this type should make it possible for plant breeders to synthesize useful plants that can thrive in these severe conditions.

V. LETHAL HEREDITY
by *Willard F. Hollander*

LMOST EVERY SPECIES produces defectives or freaks of one kind or another. Some of these mutations are lethal, some merely disadvantageous. In nature most departures from normal cannot survive long. A pigeon with defective feathers, for example, or a rodent with defective teeth, generally dies soon after its parents cease to care for it. When man intervenes to protect and encourage such freaks, as he often does in the laboratory, they may thrive and multiply.

Mutations are generally inherited as units, according to Mendelian principles. Take for an example albinism in the rabbit. A rabbit with this common color mutation is at a serious disadvantage in the wild state, because the lack of pigment impairs its vision and deprives it of camouflage. When an albino rabbit is crossed with a normal one the albinism seems to disappear—all the first-generation offspring are normal. In other words, albinism is a recessive trait. But if one of these hybrids is mated with an albino, about half of their offspring will have albino genes from both parents and will show albinism. If two hybrids are mated together, we expect to obtain three times as many normal as albino progeny. Finally, if a hybrid is mated with a pure normal, we expect no albino young. By this last type of mating a recessive defect may be transmitted from generation to generation *sub rosa*. Under such circumstances an organism may carry defective genes entirely unsuspected: it is quite impossible to tell whether an animal or plant bears such genes merely by examining it.

Some mutations are dominant, and therefore incapable of hiding. An example is the extra-toed condition found rather commonly in house cats. Extra toes are no handicap to a pet cat, and the trait multiplies merrily. But in the wild this clumsy type of paw would be a disadvantage to a cat.

Sometimes a mutation is so radical that nothing can be done to prolong the animal's life to maturity. This is what is known as a lethal mutation. Often it kills the animal while it is still an embryo. Most lethal mutations are recessive, however, and are carried unsuspected by normal-appearing animals.

Certain lethals are partly recessive but are betrayed by a mild defect or by an apparently irrelevant trait. This identification mark

318

may be called a "badge" of lethality. Sewall Wright of the University of Chicago has described an interesting lethal badge in the guinea pig. On individuals of one family of guinea pigs he found thumbs, which are normally lacking in this animal. When two of the thumbed freaks were mated, about a fourth of their offspring died before birth from hemorrhages. The lethal embryos had many extra toes on each foot.

Among other examples of lethal badges are hairlessness in dogs, deafness with blue eyes in white cats, and crests as well as white color in canaries. In mice, which have been studied extensively, many badge types have been discovered. One of them, recently found by Leonell Strong at Yale University, is a looped tail. When two loop-tailed mice are mated, about one out of four of their young has a cleft vertebral column (*spina bifida*) and an exposed brain; the defective animal dies at birth.

Do lethal mutations exist in man? The answer must be yes. It is hard to get people to give information on such skeletons in the closet, but babies with fatal congenital malformations are known to be born in large numbers—up to 20,000 a year in the U. S. Miscarriages and abortions of defective fetuses are still more common. Though some of these defects may be due to accidents or infections, it would be rash indeed to say that heredity played no part.

The Norwegian investigators Otto Mohr and Christian Wriedt have identified a lethal badge in man in the form of a shortened second phalanx of the index finger. One marriage of two such persons was reported: they had a child who lacked fingers and toes and had other skeletal deformities leading to death at an early age. No doubt other traits that we see now and then are actually lethal badges, but they are so rare, and the marriage of two people with the same defect is so improbable, that we lack evidence of their lethality.

The quickest way to expose lethal traits is by intense and continued inbreeding. In man such matings are generally illegal or taboo; the experience of the race indicates bad results. But brother-sister matings in animals and self-pollination in plants are a standard laboratory practice. The outcome is generally detrimental—unless inbreeding has been customary in the species. When inbreeding begins, the heredity seems to be breaking down. All sorts of defects and weaknesses appear. The average life-span decreases. After a few generations the family often becomes extinct.

But if the family can weather the first few generations (5 in the case of plants, 10 in animals), a leveling-off sets in. Members of the family may show defects or weaknesses, but not new ones, and there is striking uniformity. The type has become fixed. This is the situation already reached in nature by self-pollinated plants such as wheat and peas. No more lethals or defects are hidden—every such trait has been exposed and discarded, if that is possible. Inbreeding therefore is best considered not a destroying agent but a sorting or purifying method. The hybridizing of two purified inbred families or "strains" generally produces a highly vigorous and uniform progeny; the simplest explanation is that the two strains combine the effects of their good genes and are not likely to possess the same recessive defects.

The abundance of hidden lethals and hereditary defects exposed by inbreeding must be seen to be believed. It seems safe to say that very few individuals of an ordinary mixed population fail to harbor one or more. Whence came this multitude of skulking malefactors? A partial answer to this question has been obtained with the little fruit fly, Drosophila. It has been shown that when mutations are induced in fruit flies by X-rays, mustard gas, supersonic vibrations or some other artificial means, most of them are lethal. In nature cosmic rays, radioactive elements and certain chemicals presumably produce mutations in animals and plants at a low but steady rate.

The primary effect of a mutagenic agent appears to be damage to one or more chromosomes in the nuclei of cells. The damage may be anything from an invisible minor change to a complete fracture. Fractures may heal by reuniting; however, if several fractures are induced in the same cell, wrong ends of chromosomes may unite, and serious consequences are then sure to follow. The nucleus may

Doubled chromosomes cause lethal effects when an organism containing them is crossed with a normal organism. Germinal cells, which produce eggs or sperm, normally have two sets of chromosomes (*diploid*). Eggs or sperm derived from them have one set (*haploid*). The germinal cells of an organism whose chromosomes have doubled contain four sets of chromosomes and its eggs or sperm have two sets. In this diagram one such diploid egg is fertilized by a haploid sperm, giving rise to an organism with three sets of chromosomes (*triploid*). The combinations possible in eggs or sperm descended from a cell carrying this uneven number of chromosomes are shown at the bottom of the drawing. Only the first and last chromosome possibilities would produce normal embryos when mated with a normal egg or sperm.

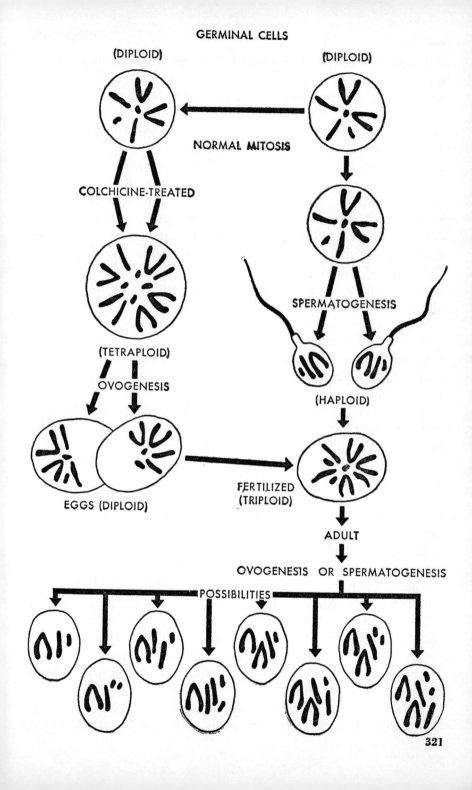

GERMINAL CELLS

(DIPLOID)   (DIPLOID)

NORMAL MITOSIS

COLCHICINE-TREATED

(TETRAPLOID)   SPERMATOGENESIS

OVOGENESIS

(HAPLOID)

EGGS (DIPLOID)   FERTILIZED
                             (TRIPLOID)

                             ADULT

                             OVOGENESIS OR SPERMATOGENESIS

                             POSSIBILITIES

321

lose parts of chromosomes completely, or the chromosomes may be improperly combined in the formation of sperm and egg cells. If both sets of chromosomes do not have identical damage, the offspring of the combination of a sperm and an egg may seem normal, but the potentiality for tragedy in later matings is still there.

When an entire chromosome is lost, the animal or plant may be visibly abnormal but viable. When both the paternal and the maternal members of a chromosome pair are lost, the effects are always lethal. Similarly, the addition of one extra chromosome may not have seriously adverse effects, but two are generally lethal.

On the other hand, when an organism doubles its whole set, acquiring twice the normal chromosome number, it may have little difficulty and reproduce normally. This is what happens when a plant is treated with the drug colchicine. Such a plant may grow exceptionally large flowers with double-size pollen and ovules. A cross between this plant and a normal one, however, produces a hybrid with an odd multiple of the chromosome number. A hybrid of this sort runs into trouble in reproduction, because, as the diagram shows, the dividing cells cannot have equal sets of chromosomes. It seldom produces viable offspring.

Perhaps the greatest problem facing the modern biologist is to learn just how heredity operates. The late Raymond Pearl of Johns Hopkins University likened a living system to a spring-driven clock. He thought that heredity, like the spring, determines the "total energy output" capacity of an individual. It has just so much energy to use up in its lifetime. For example, the fruit fly becomes less active at low temperatures and lives longer. Pearl also drew an analogy between natural death and the wearing out of a machine; he suggested that as the body ages, "organ systems get out of balance and wreck the whole machine."

These analogies are useful in guiding our practical efforts to lengthen life. Breeding, however, remains the most important factor. Disease, that potent ally of the Grim Reaper, may be beaten back by quarantine and by drugs, but without hereditary resistance a population is like tinder waiting for a spark. The American Indians, for example, apparently were far harder hit by the white man's measles than by his bullets.

The effectiveness of selective breeding over many generations may be striking. In agriculture genetics is playing a larger and larger role. New varieties of truck crops are announced every year, and our tremendous harvests of wheat and other grains are vitally dependent on continued breeding of resistance to rusts and other enemies. Poultry breeders are coming to select for "livability" as well as for high egg production.

In human beings applied genetics is a personal matter, seldom willingly turned over to group control. Several medical centers have human heredity clinics which give advice, but in general heredity is ignored as a medical blind alley, leading to fatalism. This is particularly unfortunate because man is now able to damage heredity more efficiently than ever before in history. Research and education in prophylactic measures, particularly concerning radiation hazards, should be increased.

The occurrence of a lethal condition in a family is a tragedy. The consoling aspect of the matter is the tendency for lethals to be self-effacing. We should be concerned, however, with the accumulation of defective genes in the population. Conceivably such recessive defects may eventually accumulate to the point where eliminating them will be virtually impossible. But that will be a long time in the future, and we can hope that before then the advances of science will have solved the problem of detecting the hidden recessives.

VI. THE MYSTERY OF CORN

by Paul C. Mangelsdorf

T HE MOST IMPORTANT PLANT in America is corn. It is grown
in every state and on three-fourths of all the farms of the
U. S. Corn is the backbone of our agriculture. It is the most
efficient plant that we Americans have for trapping the
energy of the sun and converting it into food. True, we consume only
small amounts of corn directly, but transformed into meat, milk, eggs
and other animal products, it is the basic food plant of our civilization.

Yet corn is also a mystery—a botanical mystery as baffling and
intriguing as any in the pages of fiction. The plant has become so
highly domesticated that it is no longer capable of reproducing itself
without man's intervention. A grass, it differs from all other grasses,
wild or cultivated, in the nature of its seed-bearing organ: the ear.
This is a highly specialized inflorescence, or flower cluster, enclosed
in husks, which when mature bears several hundred or more naked
seeds upon a rigid cob. The pollen-bearing inflorescence, the tassel,
occurs separately on the same plant. The ear of corn has no counterpart anywhere else in the plant kingdom, either in nature or among
other cultivated plants. It is superbly constructed for producing grain
under man's protection, but it has a low survival value in nature, for
it lacks a mechanism of seed dispersal. When an ear of corn drops
to the ground, scores of seedlings emerge, creating such fierce competition among themselves for moisture and soil nutrients that usually
all die and none reaches the reproductive stage.

What could have been the nature of the wild or primitive corn
from which this pampered cereal has developed? Where, when and
how was a species, once so hardy that it could survive in the wild,
converted to a cultivated plant so specialized and so dependent
upon man's ministrations that it would soon become extinct if deprived of man's help? These are questions that have puzzled botanists and anthropologists for more than a century. Now, as a result
of research in botany, genetics, archaeology and history, the answers
are a little nearer. The mystery has not been solved, but the web of
circumstantial evidence is drawing tighter and the final solution is
almost in sight.

The first reference to corn in recorded history occurs on Novem-

ber 5, 1492. On that day two Spaniards, whom Christopher Colum-
bus had delegated to explore the interior of Cuba, returned with a
report of "a sort of grain they call maiz which was well tasted, bak'd,
dry'd and made into flour." Later explorers to the New World found
corn being grown by Indians in all parts of America, from Canada
to Chile. Corn proved to be as ubiquitous in the New World as it was
unknown in the Old. There was a great diversity of corn varieties;
all of the principal types we recognize today—dent corn, flint corn,
flour corn, sweet corn and pop corn—were already in existence when
America was discovered.

The evidence that corn originated in America is so overwhelming that it seems sensible to concentrate, if not to confine, our search
for its wild ancestor to the Western Hemisphere. In America corn
has obviously had an ancient history. The seminomadic hunting and
fishing Indians in both North and South America augmented their
diet of fish and game with corn from cultivated fields. The more
advanced Mound Builders of the Mississippi Valley and the Cliff
Dwellers of the Southwest were corn-growing and corn-eating peoples. The highly civilized Mayas of Central America, the warlike and
energetic Aztecs of Mexico and the fabulous Incas of Peru and
Bolivia all looked to corn for their daily bread.

This universal reliance of the pre-Columbian cultures on corn as
the basic food plant, and its great diversity of varieties, greater than
that of any other cereal, bespeak a long period of domestication.
How old is corn as a cultivated plant? Fortunately this investigation
is no longer wholly a matter of guesswork. Radio-carbon dating of
corncobs and kernels found in various ancient sites bears out previous archaeological and geological estimates that the oldest corn yet
found in South America goes back to about 1000 B. C., and the
oldest in North America to not earlier than 2000 B. C. The oldest
prehistoric ears in both North and South America are small and
primitive; they differ decidedly in several characteristics from the
modern varieties of the Corn Belt. Yet almost any American farm
boy would recognize them instantly as corn. So some 4,000 years ago
corn was already well on the road to becoming the unique cereal it
is now.

In what part of America did corn originate? And what kind of
wild grass was it that gave rise to the multitude of present-day
varieties of corn?

One theory has corn originating from a plant called by the Aztecs *teocintle* (now Anglicized to teosinte). Teosinte is undoubtedly the closest wild relative of cultivated corn. Like corn, it has tassels and ears borne separately, although its "ears" contain only five or six seeds, each enclosed in a hard, bony shell—characteristics that make teosinte a most unpromising food plant. Also like corn, it has 10 chromosomes, indicating that it is a closely related species. Teosinte can readily be crossed with corn to produce hybrids that are completely fertile or almost so. If corn came from teosinte, as many botanists have supposed, it must have originated in Guatemala or Mexico, for teosinte is found only in those two areas.

The second principal theory is that corn originated in South America from a peculiar primitive plant called "pod" corn. Primitive pod corn today has virtually vanished; it is no longer found in pure form but as an admixture in modern varieties. As described in early references, and as obtained by inbreeding from present-day mixtures, pod corn has its kernels enclosed in a pod or chaffy shell similar to that found in all other cereals—a condition which almost certainly was characteristic of wild corn.

Which, if either, of these two theories is more likely to be correct? Botanists, in attempting to determine the place of origin of a cultivated plant, place considerable reliance upon two criteria. One is the occurrence of wild relatives of the plant in question; the other is diversity in the cultivated species itself. It is assumed that other things being equal, the region of maximum diversity should coincide with the center of origin, since diversification has progressed longer at the center than at the periphery of the plant's present range. In the case of corn the two clues point in opposite directions: the wild-relative clue points to Guatemala and Mexico, where teosinte, corn's closest relative, grows; the diversity clue points to South America, where, on the eastern slopes of the Andes, occurs the greatest diversity of corn varieties found anywhere in America in a region of comparable size.

Some 20 years ago my colleague Robert G. Reeves and I began working at the Agricultural Experiment Station of Texas A. & M. College on a series of genetic and cytological studies of corn and its relatives to test these two conflicting theories. We hybridized corn with teosinte to determine how the genes that differentiate the two species are inherited and how they are distributed on the chromosomes. We also hybridized corn with tripsacum, a more distant wild relative of corn, which occurs in both North and South America. Our hybrids of corn and teosinte revealed that corn differs from teosinte not by a relatively few genes, as might be expected if the one had been derived from the other as a result of domestication, but by a large number of genes inherited in blocks. Our hybrids of corn and tripsacum, the first such hybrids ever to be made, showed that the chromosomes of tripsacum, 18 in number, differed greatly from those of corn. Microscopic studies of the reproductive cells of the tripsacum-corn hybrids showed little pairing (a criterion of relationship) between the chromosomes of the two species. Nevertheless, there was some chromosome association and consequently some opportunity for exchange of genes. Especially important was the discovery that some of the plants that occurred in later generations of the tripsacum-corn hybrid resembled teosinte in some of their characteristics. This discovery led to the conclusion that teosinte might well be not the ancestor but a descendant of corn—the product of the natural hybridization of corn and tripsacum. Such a possibility had been suggested years earlier by Edgar Anderson of the Missouri Botanical Garden.

Since 1937, when we arrived at this working hypothesis, much additional research has been done on corn, pod corn, teosinte and tripsacum, and upon their hybrids. There is abundant circumstantial evidence, but still no conclusive proof, that teosinte is the product of the hybridization of corn and tripsacum. There is even more evidence to show that teosinte could scarcely have been corn's ancestor. Reeves, who has made an intensive study of the botanical characteristics of corn, teosinte and tripsacum, has found that teosinte is intermediate between corn and tripsacum or is identical with one or the other of these two species in the 50 or more features in which they differ. John S. Rogers, also working at the Texas Experiment Station, has found that numerous genes, many more than previously supposed, are involved in differentiating teosinte from corn. The possibility that these considerable genetic differences could have originated during a few thousand years of domestication seems remote indeed.

So the teosinte theory has become increasingly untenable. Meanwhile the theory that corn originated from pod corn has become more and more plausible. When a modern hybrid form of pod corn is inbred (a process that usually intensifies inherent traits) the result is a plant quite different from ordinary cultivated corn. The ear disappears and the kernels, now borne on the branches of the tassel, are enclosed in glumes, or chaff, as in other cereals. This pure pod corn possesses a means of dispersal, since its seeds are not on a heavy ear but on fragile branches. In the proper environment it could undoubtedly survive in the wild and reproduce itself. It has characteristics like those of many wild grasses; indeed, in its principal botanical features it is quite similar to its wild relative tripsacum. Pure pod corn has virtually all of the characteristics we would expect to find in the ancestral form of corn. Furthermore, it is more than a relative of corn; it *is* corn—a form of corn that differs from cultivated corn in exactly the way a wild species ought to differ from its cultivated counterpart. Finally, all the hereditary differences between pod corn and cultivated corn are traceable to just one gene on one chromosome. Thus a single mutation can change pod corn to the non-podded form, and it has actually done so in many cultures.

The aboriginal wild corn that man began to cultivate undoubtedly had other primitive characteristics in addition to those of the ancestral pod corn. Its kernels, for example, were probably small, hard and pointed. Kernels of this kind are found today in varieties of pop corn. Indeed, the U. S. botanist E. Lewis Sturtevant, one of corn's most astute investigators, concluded more than half a century ago that primitive corn must have been both a pod corn and a pop corn. Evidence is now accumulating to show that Sturtevant was right.

In the remains of prehistoric civilizations unearthed in South America, pop corn predominates over other types. Pottery utensils for popping corn, as well as actual specimens of the popped grains, have been found in prehistoric Peruvian graves. Certainly there is nothing new about the pop corn which modern Americans consume so lavishly as part of the movie-going ritual. Pop corn is an ancient food, and it is quite possible that primitive man first discovered the usefulness of corn as a food plant when a wild corn was accidentally exposed to heat. This would have exploded the small, vitreous, glume-covered kernels, and transformed what to people with no grinding tools other than their own teeth was a very unpromising food into tender, tasty, nutritious morsels.

There is an interesting historical reference which lends support to Sturtevant's conclusion that primitive corn was both a pod corn and a pop corn. A century and a half ago Félix de Azara, the Spanish Commissioner to Paraguay, wrote of a peculiar variety of corn in Paraguay in which small seeds enclosed in "envelopes" were borne in the tassel. When the tassels were heated in hot oil, the kernels exploded to produce "a superb bouquet capable of adorning at night the head of a lady."

By a very simple experiment in our breeding plots, we have succeeded in duplicating exactly the corn Azara described. Pod corn was hybridized with pop corn and was then inbred to produce an earless plant bearing in the branches of the tassel small hard seeds enclosed in glumes. When a tassel of this pod-pop corn was heated in hot oil, it behaved exactly like the corn of Azara. The kernels exploded but remained attached to the tassel to produce the "bouquet" he described.

These recent findings have quite naturally given new impetus to the search for wild corn in South America, since the most convincing and conclusive proof of the pod-corn theory would be the discovery of a primitive pod corn still existing in the wild state. The search for a wild corn has not so far been successful in its primary objective, but it has been quite fruitful in turning up new types of corn, especially less extreme forms of pod corn whose kernels are only partially enclosed in glumes. Perhaps wild corn will still be discovered in some remote protected spot in a region not yet thoroughly explored. The odds are at least even, however, that it no longer exists. Corn in the wild may well have been a plant with low survival value, restricted in its range, and already well on the road to eventual extinction when first used by man.

In the meantime a wholly unexpected discovery, made within the past two years, has furnished direct evidence for the theory that primitive corn was both a pod corn and a pop corn. During the summer of 1948 an expedition sponsored by the Peabody Museum of Harvard University and led by Herbert W. Dick, a graduate student in anthropology, uncovered many cobs and other parts of corn from the accumulated refuse in an abandoned rock shelter in New Mexico known as Bat Cave. This shelter was occupied from about 2000 B. C. to 1000 A. D. Uninhibited by modern concepts of sanitation, its successive generations of occupants allowed refuse and trash to accumulate in the cave to a depth of about six feet. Carefully removed and sifted by the archaeologists, the refuse yielded 766 specimens of shelled cobs, 125 loose kernels and various fragments of husks, leaf sheaths and tassels. The cobs are of particular interest, since they reveal a distinct evolutionary sequence. The oldest, at the bottom of the refuse heap, are the smallest and most primitive. These cobs and loose kernels from the same level prove that the earliest Bat Cave people grew a primitive variety of corn which was both a pop corn and a form of pod corn. The pod corn, however, was not as extreme as the earless synthetic "wild" corn described above. It probably represents a type already partly modified by domestication, more nearly like the weak forms of pod corn still found in South American varieties.

The Bat Cave corn has answered another of our questions: What is the relationship of corn to teosinte? The oldest and most primitive of the Bat Cave corn shows no evidence whatever of having stemmed from teosinte. But beginning about midway in the sequence there is strong evidence of the introduction of a corn that had become contaminated with teosinte. Thus the Bat Cave cobs suggest that early botanical investigators were not completely wrong in believing that teosinte played a role in the evolution of corn. Although teosinte clearly was not the progenitor of corn, it contributed its genes to corn's progress toward its present form.

The Bat Cave remains still leave unanswered the question: Where in America did corn originate? It seems improbable that corn could have been a native of the region where these remains were found, since corn is a moisture-loving plant and the region is now and was then quite dry. Probably it was brought into the Bat Cave region as a cultivated plant from Mexico. Whether corn was native to Mexico or had been introduced there still earlier from South America is an open question.

How did the primitive pod-pop corn that the Bat Cave people grew 4,000 years ago evolve in so short a period, as evolutionary time is measured, into the modern ear of the Corn Belt? Some botanists are inclined to endow the American Indian with unusual abilities as a plant breeder. If the great changes that have occurred in corn in this relatively brief period are the product of his skill, he was indeed remarkably adroit. The corn from Bat Cave does not, however, support this view. On the contrary, there is no evidence that the Bat Cave people were any more concerned with plant improvement than they were with sanitation. If selection was practiced at all, it was probably an unplanned "negative" selection—the good ears were consumed and the leftover nubbins were used for seed. Nevertheless, thanks probably to accidental hybridization with teosinte and with other races of corn, there was a gradual increase in the average size of ears and kernels and an enormous increase in total variation during the 3,000 years of the Bat Cave's history.

The evolutionary sequence in the Bat Cave indicates that four principal factors operated in the evolution of corn during this period: 1) The pressure of natural selection, one of the most important suppressive factors in evolution, was greatly reduced; 2) mutations from the more to the less extreme forms of pod corn occurred; 3) corn was modified by contamination with teosinte; 4) crossing of varieties and races produced new combinations of characters and a high degree of hybridity.

All of these factors contributed to a tremendous increase in variation, so that when man finally did begin to practice selection in corn, he had a rich diversity at his disposal. From this, by accident or design, he chose a combination of characteristics that makes corn the most efficient of all cereals as a producer of foodstuffs.

PART 7

I. **THE VIRUS** *by F. M. Burnet*
One of the world's leading virologists, F. M. Burnet is director of the Walter and Eliza Hill Institute of Medical Research at the Melbourne Royal Hospital, Australia. He has worked principally with the bacterial viruses and the viruses of influenza and Q fever. Dr. Burnet's contributions were recognized in 1951 by a knighthood.

II. **THE MULTIPLICATION OF**
BACTERIAL VIRUSES
by Gunther S. Stent
Gunther Stent is a biologist who holds degrees in chemistry and owes his choice of career to a physicist. Born in Berlin in 1924, he came to this country in 1940 and entered the University of Illinois. There a friend gave him a copy of Erwin Schrödinger's *What Is Life?* Stent was "so impressed by what I read that I decided to have a try some time at becoming a biologist." Today he is with the Virus Laboratory of the University of California.

III. **NATURAL HISTORY OF A VIRUS**
by Philip and Emily Morrison
A theoretical physicist at Cornell University, Philip Morrison has been attracted in recent years to fundamental questions in biology. During the war, Dr. Morrison was a key member of the Manhattan District Research Organization at the Los Alamos Laboratories. Emily Morrison, a freelance journalist, collaborates with her husband on many of his writings.

IV. **A NEW ERA IN POLIO RESEARCH**
by Joseph L. Melnick
Though his career spans only a little more than a decade, Dr. Melnick has had the satisfaction of seeing polio reduced from a remote problem of fundamental investigation to the subject of an immediate and predictable campaign of practical control. His own work—particularly in isolating the several polio virus types *in vitro*—is playing a decisive part in opening the new era in polio research. Dr. Melnick is associate professor of microbiology at the Yale University School of Medicine.

V. **THE COMMON COLD**
by Christopher Howard Andrewes
Christopher Andrewes, one of England's leading virologists, is deputy director of the National Institute for Medical Research and heads the World Influenza Centre (World Health Organization) with headquarters in London. He also heads the Common Cold Research Unit at Salisbury.

THE VIRUS

Introduction

VIRUS RESEARCH, though presented here in a separate chapter, links directly with current inquiries into genetics and the origin of life. It is a striking example of the fact that in modern science there are few islands of investigation; studies that begin along isolated lines and with specific ends soon find themselves enmeshed in the general search for basic laws and governing principles. Virus study was undertaken in the first place for the express purpose of overcoming certain diseases, but it would be continued today if there were no single virus infection left to plague us.

The viruses were the last group of infective agents to succumb to medical research for the reason that they are particularly difficult to investigate. They are so small that they can be seen only in an electron microscope and, until recently, could be grown only in living creatures—mice, rats, rabbits, monkeys and even men—which are expensive, inconvenient, inaccurate and otherwise unsatisfactory testing instruments. Some virus diseases—notably smallpox and yellow fever—were brought under control before the nature of the infective agent was clearly understood, but the systematic campaign against the viruses did not get under way until the 1920's.

The conquest of the viruses is now at its height. Two of the main battles—against influenza and poliomyelitis—are described in detail in this chapter, as is the preliminary skirmishing with the common cold, the one prevalent virus that shows as yet no sign of succumbing to scientific attack. New techniques for growing viruses in culture and for inactivating or modifying them have so accelerated the development of effective countermeasures, that as careful a student of virology as F. M. Burnet can speculate that creative medical research in this field may soon begin to fall off for want of new enemies to conquer.

Meanwhile, however, the virus has become an object of the liveliest interest in its own right. From being a foe to be exterminated, it has been transformed into a valuable ally and tool of modern biology. The nature and habits of this microorganism places it at the focus of the problems

333

that occupy biologists today. Viruses function at an organic level where
the distinction between animate and inanimate matter becomes shadowy.
The study of viruses, like the study of photosynthesis, thus gives us a
main entry to the origin of life. At the same time, they seem to reproduce sexually and to observe laws which parallel those of heredity in
much more complicated organisms. As a kind of "naked gene" they are
invaluable to the geneticist. Finally, viruses multiply their numbers by
attacking, breaking down and assimilating to themselves constituents of
living cells. Accordingly, the viruses are now being put to work as a dissecting tool by investigators like Gunther Stent who are concerned with
the broadest aspects of life within the cell.

The studies in this chapter on bacterial viruses and on that enigmatic
organism, herpes, stem from this more recent interest in the virus, and
they are only remotely connected with medicine. The organisms that
attack bacteria are harmless to higher forms of life. Philip and Emily
Morrison were moved to trace the natural history of the herpes coldsore
virus because this minor nuisance provides a classical case of evolutionary
adaptation of parasite to host. These investigations are perhaps typical
of the virus research of the future—a part of the major plan for discovering the source and demonstrating the relatedness of all life.

I. THE VIRUS

by F. M. Burnet

OLIO, influenza and the common cold—probably the three infectious diseases of most interest to the average person —are all caused by viruses. So are smallpox and yellow fever, most of the "childhood diseases" and a host of rarer maladies. Since the days of Jenner and Pasteur the virus plagues have been studied from every angle that might help toward their understanding and control. It is natural that most of the research in this field should have a strongly medical bias, but quite apart from the problems of human and animal disease, the viruses themselves— their nature, their interaction with the cells they infect, their place in the evolutionary scheme—provide topics of the highest interest. This article is an attempt to give an account of modern experiments and ideas that bear on these matters. It will be based to a considerable extent on the investigations of influenza virus that have gone on in England, America and Australia during the last 20 years. The influenza viruses are my own chief field of interest, and they are also the field in which fundamental study of animal viruses—as opposed to the viruses that attack plants or bacteria —is farthest advanced.

A virus can be defined as a microorganism, considerably smaller than most bacteria, which is capable of multiplication only within the living cells of a susceptible host. This definition immediately indicates the important feature that distinguishes the virologist's problem from that of the classical bacteriologist. A bacterium, say the diphtheria bacillus, can be grown on relatively simple mixtures of sterilized nutrients—the tubes of broth and the plates of nutrient agar that are the bacteriologist's tools of trade. For viruses nothing less than the living cell will serve. An influenza virus can be grown in the nasal passages of a ferret, in the lung of a mouse, in the tissues of a developing chick embryo or in a culture of embryonic cells in a flask, but it will not grow in any nonliving material.

There are two general prerequisites for experimental laboratory work with a man-infecting virus. First, the experimenter must find some convenient animal whose cells can be infected by the virus. If chick embryos or mice, which are cheap and available in virtually unlimited number, will serve, so much the better. Second, the experimental host must show some sign or symptom that will allow the experimenter to know when it is infected.

Any good experimental work must be quantitative. In most experiments with viruses we need to know how much virus is present after such and such a manipulation. Suppose we wish to measure the influenza virus present in an extract from the lungs of a mouse that has just died of the disease. Different quantities of the virus will produce different degrees of consolidation (solidification) of the lungs and we can adopt the convention that one unit of the virus is the amount which on the average produces solidification over 50 per cent of the visible surface of the lungs. To measure the strength of the extract from the fatally stricken mouse, then, we dilute the extract in varying degrees, so that we have samples diluted to one part in 10, one in 100, and so on. Each of these samples is inoculated into the noses of six mice four to five weeks old. A record is kept of deaths in each group and of the aspect of the lungs when the surviving mice are killed seven days after inoculation. If we find that 50 per cent consolidation occurs, on the average, in mice given a 1-to-10,000 dilution of the extract, while other doses produce more or less consolidation, the original extract is reckoned to have a strength (titer) of 10,000 units of virus. This principle of diluting something down in a series of steps until it produces a certain standard degree of action is very cumbersome in practice and not very accurate. But so far no more convenient method of measurement has been found, and for most viruses the dilution technique will probably remain the standard quantitative procedure.

Research on influenza is farther advanced than research on poliomyelitis because until recently (see the discussion of polio later in this section) the technique for measuring the former disease was much the easier. Influenza virus can be measured not only in the mouse and the chick embryo but also in the test tube. When the virus is mixed with a suspension of red blood cells in saline, it causes the cells to clump in easily visible fashion, and various applications of the agglutination technique make it possible to analyze the qualities by which one type of virus differs from another.

The very fact that the influenza virus agglutinates blood cells has provided the most important of all leads to an understanding of how the virus makes effective contact with the cell it is going to infect. Red blood cells themselves are not susceptible to penetration and infection by influenza or any other virus. But the surface of the red cell seems to have essentially the same complex mosaic of chemical components as any other cell from the same species of animal. There is much direct evidence that the action of an influenza virus on the surface of red cells corresponds closely to its action on the susceptible cells that line the air passages of a ferret or a mouse. Experiments with red cells may therefore provide a convenient model of what happens in the more important but less accessible tissues of the lungs and bronchial tubes.

From large numbers of experiments in many laboratories we have a fairly clear picture of the process by which an influenza virus initiates infection of a cell. The virus seems to approach the cell surface through a reaction closely resembling that between an enzyme and the substance it acts upon. The virus particle has on its surface a number of patches which function as enzymes. These enzyme patches attach themselves to and break down certain molecules of a complex carbohydrate that are built into the surface of the cell. The virus can then sink into the substance of the cell and there begin to multiply.

The points on the cell surface to which the virus attaches are spoken of as receptors, and the complex carbohydrate of which they are composed belongs to the class called mucins or mucopolysaccharides. These are sticky substances like those responsible for the stickiness of egg white and saliva or for the slime track of a snail. The receptor mucin is closely related to the mucins that form a protective film over all the moist air and food passages, provide the chemical basis for the blood groups A, B and O and serve as one of the most important of the sex hormones, gonadotrophin. Influenza virus acting as an enzyme, it has been found, will rapidly destroy the activity of the hormone responsible for the sexual development of the immature female rat or mouse—here surely is a most unexpected crossing of paths between two distinct fields of biology.

In the course of work in my laboratory in Melbourne we found that the organism responsible for cholera produces an enzyme of the same type as the influenza virus enzyme. This enzyme is not part of the cholera germ but is set free in soluble form and can be concentrated and purified by chemical methods. We call it RDE (receptor-destroying enzyme). The isolation of this substance provided an opportunity for a very interesting experiment. If RDE destroys the cell receptors, and if influenza virus can enter cells only through such receptors, then an injection of RDE should make an animal immune to influenza. Joyce Stone of our laboratory performed the experiment both on mice and on chick embryos and found that this was indeed the case. The immunity is very short-lasting, however, for the cells regenerate fresh receptors within two or three days.

So far this mechanism of cell entry by viruses has been definitely established only for viruses of the influenza group. But within the last year somewhat similar observations have been made on two groups of viruses closely resembling, but not identical with, the poliomyelitis viruses; mice have been protected against infection by one of these types of virus by prior treatment with RDE. It is too early to say whether developments in this field will have any significant influence on the prevention or treatment of virus diseases of man. But the work suggests that the chemistry of cell entry by viruses may eventually become of great importance to workers in virus diseases.

When a virologist undertakes an investigation of a human disease, his first concern is to find some laboratory host for the virus. His next is usually to "hot up" the virus for the new host so that it will regularly produce whatever symptom or lesion is being used as an index for the presence of the virus. Only rarely does fresh virus from a human patient multiply easily in the laboratory animal. Ordinarily it must be adapted to the animal by a series of transfers, or "passages," from one individual to another in the new host. It follows, therefore, that what the virologist works with is strictly speaking not the human virus he started with but a variant—a laboratory-adapted variant. Sometimes the difference may be very striking indeed. The stock influenza viruses of the laboratory are studied mainly by observing their capacity to produce pneumonia in mice or to agglutinate red blood cells in chick embryos. Influenza virus A as it comes from the human throat is quite incapable of doing either of these things. Similarly the virus used for vaccination against yellow fever, though a live, lineal descendant of a fatal virus taken from a patient who died, produces no illness at all, because it has been changed into a harmless strain by passage through chick embryos.

The capacity of viruses to change their character in nature or in the laboratory is obviously of the greatest practical importance. It is

338

accomplished by the processes of mutation and selective survival. There is much evidence, too, that influenza virus mutations occur in nature and play an important part in determining the timing and extent of epidemics of influenza. For some reason the influenza viruses appear to be especially mutable. This can be positively embarrassing at times. A standard influenza virus is sent to two laboratories which maintain it in slightly different fashions. At the end of 10 or 15 years the descendant viruses in the two laboratories may differ very considerably, and these differences can create confusion or even ill-feeling when investigators, thinking they are working with the same virus, obtain discrepant results.

Most present-day biologists would agree that the most fundamental aspects of living processes are all related in one way or another to the problems of reproduction and variation—the subject matter of genetics. Until recently it was a convenience to believe that viruses lived and reproduced very much like small bacteria; that is, that they multiplied by the same process of enlargement and division and differed from bacteria merely in the fact that they required a more complex nutrition, which only the interior of the living cell could provide. Today it seems that this is almost certainly incorrect. The virus actually multiplying in the cell is something quite different from the virus that passes as the infectious agent from cell to cell or from person to person.

This idea that a virus within the cell is distinct from the infectious particle whose picture is given by the electron microscope came first from studies of the viruses that attack bacteria—the bacterial viruses, or bacteriophages. Like the viruses that cause disease in man or animals, the bacterial viruses are incapable of multiplication except within the cells they infect. The virus particle first makes a chemical union with some component of the bacterium's surface and then by some process penetrates the cell wall and finds itself within the cell substance. After it has made this entry, the virus vanishes for a time; at last, the host cell bursts, liberating the once more identifiable virus many times multiplied.

The process of virus reproduction within the cell is discussed at length later in this section; the point that interests us here is that something akin to genetic exchange takes place when more than one virus infects the same cell. The evidence suggests that the virus on entering the cell liberates or breaks up into a number of subunits, which are sufficiently analogous to the bearers of genetic characters in higher organisms to be called genes. Each gene multiples more or less independently until a large "pool" of genes is created at the expense of the bacterial substance. Then from the pool groups of genes begin to aggregate in such a way that each group contains all the genetic components needed for the construction of the virus particle. These groups may unite components contributed by two parent viruses and these may appear in various combinations in the groups of offspring. Actually not only two but three or even more different viruses entering a bacterium may combine some of their hereditary characteristics in a single virus particle among their progeny! Hybridization is hardly an adequate term for such a process.

There are as yet no studies on animal viruses to match this work on the bacteriophages. Influenza virus, however, has been extensively studied along similar lines, and it is extremely likely that the eventual interpretation of its process of multiplication will be almost identical with that of the bacterial viruses.

In the case of influenza virus, treatment of the particles with ultraviolet light can interfere with the multiplication of living virus in the embryo. It seems that this "killed" virus—killed at least in the sense that it never multiplies in susceptible cells—can often enter a cell and in some way block a component of the cell that is necessary for the reproduction of active virus. This interference effect also occurs when two living viruses infect a cell, if proper experimental methods are used. It plays an important part in the experiments which we shall now have to discuss. These provide more definite evidence as to how influenza viruses multiply.

In experimental biology one can often learn more about the working of an organism by observing its behavior in some alien environment than by watching it in its normal place in nature. We have obtained our most interesting results by injecting influenza virus into the mouse brain, which is even farther than the chick embryo from the virus' natural habitat—the human air passages.

When an ordinary influenza virus is injected into a mouse's brain, even in rather large amount, the mouse may show some evidence of sickness for 24 hours but later recovers completely. The virus is not inert; some sort of abortive multiplication must take place, for the amount of virus often increases slightly in the first 10 or 12 hours and it does not disappear entirely until four or five days later. R. Walter Schlesinger of the New York Public Health Research Institute has obtained strong evidence that when virus enters "alien" cells, instead of multiplying in normal fashion it gives rise to something which may be called "partial virus." His finding was that the blood-agglutination test indicated a much larger amount of virus substance to be present in the cells than did the standard chick-embryo and mouse infection tests. This suggests that the virus off-spring in the alien cells retain the ability to agglutinate blood but have weakened in their power to infect. The conception of "partial virus" is not easy to grasp, and many virologists are chary of offering any detailed interpretation of Schlesinger's facts. But his finding fits in rather neatly with the results of our mouse-brain experiments.

Although no ordinary influenza virus can infect the mouse brain, about 12 years ago a combination of accident and "training" in a laboratory in England did produce a strain of influenza virus that could multiply freely in a mouse's brain and kill the animal. This strain, which remains an influenza virus in every respect except its unusual ability to infect brain cells, we have named "neuro-flu" virus. It can be grown quite normally in chick-embryo cavities, giving highly infectious fluids for experimental use.

When highly diluted "neuro-flu" virus is inoculated into the brains of a group of mice, the animals appear quite normal as soon as they recover from the anesthetic. But after four or five days they begin to sicken, and a day later they die with signs of brain infection. Tests of the brain show that it contains very large amounts of fully active virus.

We found, curiously enough, that when a large amount of ordinary influenza virus is mixed with a little of this neuro-flu, the result of the injection is quite different. It might reasonably be expected that with the double infection the mice would probably die just a little sooner than if they had the neuro-flu alone. In fact they usually show no signs of illness whatever. The explanation is that there occurs a type of "interference," in a rather special technical sense of the word, which is well known to virologists. The effect depends on the relative amounts of the two viruses. A mixture of one part of neuro-flu to 10 or 100 parts of ordinary flu is harmless; when the mixture contains equal amounts of both viruses, there is little inter-ference and the death of the animals is delayed only a short time.

341

An experiment in which mice received a mixture of the two viruses that produced partial interference, with some mice in a group dying and some surviving, yielded another interesting result. Examination of the viruses in their brains showed that there were not two but three types: neuro-flu, ordinary flu and a third type which possessed characteristics of ordinary virus and the most obvious quality of the neuro-flu, namely its capacity to produce fatal brain infections. The most likely, though perhaps not the only, interpretation of these results is that the third type of virus is a "recombinant" in which the qualities of the other two have been combined.

So far there have been no accounts of any other experiments on this "hybridization" of viruses. For technical reasons it may be hard to find other situations in which the process can be shown. It is unjustifiable, therefore, to say that the conclusions derived from the neuro-flu experiments are applicable to other types of virus. Nor, to be quite honest, do I think that other virologists are yet as convinced as I am that the recombination experiments done in my laboratory in Melbourne have all the significance that I have given them. That is only likely to come when the experiments have been repeated and more deeply analyzed in other laboratories.

To the layman the most interesting thing about viruses is their smallness. There is a tendency to feel that until you can see something there is no way of studying it. This of course is a complete fallacy. With the electron microscope we can now produce very detailed pictures of influenza viruses and of the bacterial viruses, and every virologist has been excited and delighted by seeing them. We must know what viruses look like to satisfy our curiosity and to provide background for the refinements in the use of electron microscopy which in the future will make it a really valuable technique. But it is fair to say that what is revealed by these pictures has hardly helped at all in understanding how viruses produce the effects that make them so important. At the present time our pictures are only of the free virus particles; for technical reasons it is not yet possible to see what is happening while the virus is multiplying in the cell. Electron microscopists who are interested in viruses are seeking to devise ways in which clear pictures of what is happening in the early stages of cell infection can be obtained. This is obviously not going to be an easy task, but one can feel reasonably certain that it will be accomplished.

342

So far in this discussion of the virus and the cell, the host has played a purely passive role. The virus is the invader, and the effectiveness of its attack, it would seem, depends only on whether its genetic make-up is appropriate to the host cell concerned. Fortunately life is not like that. There is a rule about infectious disease to which I know of no exceptions: Whenever a parasite and its host species have lived together for many generations, they will have found a *modus vivendi* whereby the parasite species survives without producing more than minor damage to the host species. The dramatic epidemic that kills a high proportion of those it strikes will always on investigation prove to be the result of some new development. In the old days of yellow fever in the West Indies the native population appeared unaffected by the disease, while European armies melted away in a few months under its onslaught. The Europeans were intruders into a virtually stabilized biological equilibrium.

The practical control of a virus disease nearly always depends essentially on obtaining an understanding of the means by which the balance between the virus and the host is maintained in nature and how it can be modified in either direction by biological accident or by human design. In the approach to such an understanding two important related concepts have emerged—"subclinical infection" and "immunization."

A subclinical infection is one in which the infected person gives no sign of any ill effect. In a population attacked by an infectious disease, subclinical infections often greatly outnumber those severe enough to produce unmistakable symptoms of the disease. For example, when a child comes down with a paralyzing attack of poliomyelitis, a careful examination of the rest of the family will commonly reveal that all the other children have the virus in their intestines over a period of a week or two, but they either show no symptoms at all or have only a mild, nondescript illness. Fortunately even a subclinical infection produces heightened resistance or immunity to the virus for a period after the attack. This capacity of mild or subclinical infection to confer immunity is probably the greatest factor in maintaining a tolerable equilibrium between man and the common virus diseases. The trouble is that viruses are labile beings, liable to undergo mutation in various directions, and a virus that causes only mild infection may evolve into one far more deadly.

343

Immunity to virus disease was known long before any virus could be handled in the laboratory. In fact, it was from Jenner's early vaccination attempts against smallpox that the whole science of immunity sprang. But then immunology turned almost wholly to bacterial diseases. The toxin-antitoxin approach, which developed from the late 19th-century discovery of the cause and the means of prevention of diphtheria, for many years dominated the outlook of immunologists. In recent years the study of immunity in virus disease has been renewed, and it has profited from the concepts developed in the bacterial investigations.

One cannot claim that there is full agreement about the nature of immunity to viruses, but it is possible to offer a simplified account which most virologists would accept as at least the most convenient approach to understanding that is available at the present time. This interpretation is that all immunity to viruses is mediated through antibody, and that the effectiveness of the protection depends first on the amount and character of the antibody and second on the availability of the antibody to protect the particular cells that are exposed to infection. Antibodies can be described most simply as modified blood-protein molecules which are capable of attaching themselves firmly to the specific virus or other invading organism that provoked their production by the body. If a sufficient number of antibody molecules can attach themselves to a virus particle, they have a blanketing effect which effectively prevents the virus' attachment to the host cell and its multiplication within the cell.

Antibody appears in the blood a few days after infection and reaches a peak in two to three weeks. The body continues to produce antibody at a slowly diminishing level long after recovery—in some diseases, such as measles and yellow fever, for the whole of life. Immunity is long-lasting only against those diseases in which the virus must pass through the blood at some stage before it produces symptoms. The explanation, in general, is as follows: After any virus infection, the antibody produced in response to it is always concentrated most abundantly in the blood. In a disease such as measles, where the virus must pass through the blood, the large amounts of antibody there waylay any virus in a second invasion and render it inert before it has a chance to create any symptoms of illness. The virus of a disease such as influenza, which does not have to pass through the blood but spreads from cell to cell over the surface of the air passages in the respiratory system, has an easier time. The concentration of antibody here is always less than in the blood, and it soon declines to an amount insufficient for protection. Hence the immunity that follows an influenza infection is less complete and less lasting than that in a disease where the virus must run the gamut of the blood.

To return to the problem of how a tolerable equilibrium between man and a common virus is maintained, the situation can be summarized as follows: The first contact with the virus normally takes place in childhood. How early it will occur depends on how prevalent is the virus and how effective are the social barriers against its spread, such as cleanliness and good housing. The standard virulence of the virus is low, and young children as a rule recover after a mild illness or no illness at all. This induces an immunity not only against virus of normal or low virulence but also against the occasional type that has undergone mutation to higher virulence. The process will never be completely effective. As long as the common virus diseases (measles, influenza, poliomyelitis) persist, there will be epidemics in which some patients will require all the help the physician can provide. But under present-day conditions the great majority of people pass through childhood and middle life with no more than trivial episodes of infectious disease. They have not escaped infection, but by the sequence of subclinical infection and immunization they have been kept from even knowing of its occurrence.

It is against virus diseases not commonly present in their own communities that people most need the protection of artificial immunization. Men having to work or fight in the tropics of Africa or South America must be immunized against yellow fever. Smallpox, still prevalent in many parts of the globe, may enter any country, so vaccination is a necessity for any traveler and desirable for all. In these two instances immunity is produced by procedures which very closely imitate the natural process of subclinical infection. The immunizing agent is a living virus, a variant of the virulent form which can be relied upon to produce no more than trivial symptoms. If its safety can be assured, this is the most effective type of immunization. But in many cases it is not possible to produce a variant that is both effective and safe. The only available method of immunization against such viruses is to inject relatively large amounts of killed virus. On the whole this is not a particularly satis-factory method, and the only human disease against which it has proved reasonably effective is influenza. Provided the proper type of virus is used in preparing the vaccine, and provided the immu-nizing dose is given not too soon before the epidemic, about 80 to 90 per cent protection can be expected.

From its very nature virus research, like bacteriology in general, has tended in the past to concentrate on medical and veterinary problems. It will probably always be carried on against a background of its significance for medicine. But if one looks around the medical scene in North America or Australia, the most important current change he sees is the rapidly diminishing importance of infectious disease. The fever hospitals are vanishing or being turned to other uses. With full use of the knowledge we already possess, the effective control of every important infectious disease is possible. Polio-myelitis is still perhaps an exception to this statement, but its defeat now seems imminent.

Today the most intellectually exciting aspects of virology are not directly concerned with medicine. As I see it, the main interest of the virus to biology now is the possibility of using it as a probe in the study of the structure and functioning of the cell it infects. In many ways the cell is the center of life, the unit from which all but the very smallest organisms are built—ourselves included. All the biological sciences, with biochemistry and genetics in the lead, seem to be converging to attack the central problem of cellular structure and function. In this endeavor the detailed study of the interaction between the virus and the cell promises to be very fruitful.

The very smallness and simplicity of viruses have a special attrac-tion to the biochemist bold enough to look for an answer to the hardest question he can ask: How is the specific pattern of living chemical structure reproduced within the cell? The answer to that will be more likely to come from the study of plant viruses than from any other source. For similar reasons geneticists are attracted to the possibilities arising from the investigations of the recombination of genes in bacterial viruses, now also becoming visible in some of the animal viruses as well.

Microbiology is today the queen of the biological sciences. It can provide biologists with work and pastime and reward for many generations to come.

346

II. THE MULTIPLICATION OF BACTERIAL VIRUSES

Gunther S. Stent

THE process of heredity—how like begets like—is one of the most fascinating mysteries in biology, and all over the world biologists are investigating it with enthusiasm and ingenuity. Of the many angles from which they are attacking the problem, none is more exciting than the experiments on bacterial viruses. Here is an organism that reproduces its own kind in a simple and dramatic way. A virus attaches itself to a bacterium and quickly slips inside. Twenty-four minutes later the bacterium pops open like a burst balloon, and out come about 200 new viruses, each an exact copy of the original invader. What is the trick by which the virus manages to make all these living replicas of itself from the hodgepodge of materials at hand? What happens in the host cell in those critical 24 minutes?

Within the past few years studies with radioactive tracers have made it possible to begin to answer these questions. By labeling with radioactive atoms the substances of the virus or of the medium in which it multiplies, experimenters can follow these materials and trace the events that lead to the construction of a new virus. This article will tell about some of the experiments and the facts learned from them.

The bacterial virus, a tiny organism only seven millionths of an inch long, is a nucleoprotein: that is, a particle made up half of protein and half of nucleic acid. The latter is desoxyribonucleic acid—the well-known DNA which is a basic stuff of all cell nuclei. The role of DNA was considered at some length by A. E. Mirsky in his discussion of chromosome chemistry in the preceding section. We are interested in the respective roles of the two parts of the virus molecule: the protein and the DNA. We are also interested in where the various materials come from when a virus synthesizes replicas of itself inside the bacteria growing in a culture medium.

First let us consider the tracer technique. Suppose we wish to label the DNA part of the virus particles. Since an important constituent of DNA is its phosphate links, we shall label the element phosphorus with the radioactive isotope phosphorus 32. We begin with the medium in which we are growing bacteria that are to be infected by the virus. The culture contains inorganic phosphate as the source of phosphorus for the bacteria. To this medium we add a little radiophosphorus, so that there is one radioactive atom for every billion atoms of ordinary, non-radioactive phosphorus. The bacteria will take up the same proportion of radioactive and ordinary phosphorus. We can tell how much phosphate the bacteria contain simply by counting the radiophosphorus atoms with a Geiger counter: the total amount of phosphorus is a billion times that.

Now if we infect the culture of bacteria with viruses, the virus progeny also will have the same proportion of radiophosphorus. But to measure their phosphorus we must isolate them, for the culture contains a great deal of phosphorus not incorporated in them. We can separate the viruses in three ways: (1) by a series of centrifuging operations that remove the other materials through their differences in weight; (2) by adding non-radioactive bacteria, on which the viruses become fixed and which can then be removed by low-speed centrifugation; or (3) by adding a serum (developed in rabbits) which contains antibodies that combine with the viruses and precipitate them from the culture.

Two radioactive isotopes are used in the bacterial virus work: phosphorus 32 to label phosphate and the DNA part of the virus, sulfur 35 to label the protein part of the virus. Now let us look at the experiments.

All of these experiments were done on bacterial viruses of the strain called T2, which infects the common bacterium *Escherichia coli*. Some years ago two investigators—Thomas F. Anderson of the University of Pennsylvania and Roger M. Herriott of The Johns Hopkins University—observed that something curious happened to bacterial viruses when they were exposed to "osmotic shock," namely, a sudden change in osmotic pressure effected by adding distilled water to the liquid in which they were suspended. These viruses could still attack and kill bacteria. But they had lost their ability to reproduce. Under the electron microscope they looked like sacs that had been emptied of their contents, and a chemical analysis indicated that they had lost all their DNA.

Recently A. D. Hershey and M. W. Chase, working at the Carnegie Institution of Washington genetics laboratory in Cold Spring Harbor, N. Y., repeated and confirmed these experiments with the help of radioactive tracers. The DNA, labeled with radiophosphorus, was indeed removed from the virus by osmotic shock. It remained as DNA in the solution, but it was easily broken down by an enzyme—an indication that it had lost the protection of the protein "coat" of the virus. As for the protein shell of the virus, when separated from the solution and placed in a culture of bacteria it showed all its old power to seize upon and kill the bacteria. It also retained its ability to react with antivirus serum.

This looked very much as if the two parts of the virus had specialized functions. Apparently the virus' ability to attach itself to and kill a bacterium resided in its protein "coat." Did its power to reproduce and build hereditary images of itself reside in its DNA core? Other investigators had found that DNA did control hereditary continuity in bacteria. Hershey and Chase proceeded to investigate the question in their viruses.

They first put viruses in cultures of bacteria that had been killed by heat. The viruses attached themselves to the dead bacteria and apparently poured out their DNA, for the DNA (labeled with radiophosphorus) was easily broken down by the enzyme desoxyribonuclease, just as when it was spilled out from viruses after osmotic shock. Similarly, when bacteria were killed by heat after viruses had infected them, the enzyme again broke down the DNA. The enzyme had no effect, however, on DNA discharged into *living* bacteria. It seems that the living membrane of a bacterium protects DNA from the enzyme, but when the bacterium is killed, its membrane becomes permeable and lets the enzyme through.

What happens to the protein coat of the virus after it has emptied its DNA into the bacterium? Hershey and Chase infected living bacteria with virus, this time labeling the protein with radiosulfur. Then they shook up the suspension of infected bacteria in a Waring blender—the device used for stirring laboratory mixtures and for making milk shakes. The shearing force of the mixer stripped more than 80 per cent of the labeled protein off the bacteria. On the other hand, it did not remove any significant amount of DNA or interfere with the reproduction of viruses within the bacteria. The experiment showed that the virus protein stays outside the bacterium, and its job is finished as soon as it enables the DNA to gain entry into the cell. By the same token, it indicated strongly that the DNA is responsible for reproduction.

Once inside the host, the task of the nucleic acid is to reproduce itself 200-fold. It must also stimulate the production of 200 protein coats exactly like the one it has just shed. Where do the raw materials come from, and how are they put together?

In 1946 Seymour S. Cohen of the University of Pennsylvania, the first investigator to study bacterial virus reproduction with radioactive tracers, conceived an experiment directed to this question. He wished to find out whether the needed raw materials, particularly the phosphorus, came from the bacterial cell itself or from the medium surrounding it. He grew two cultures of bacteria, one in a medium containing radiophosphorus, the other in a non-radioactive medium. Then he removed the bacterial cells from the liquid in the two test tubes and switched them, putting the non-radioactive bacteria in the radioactive medium and *vice versa*. Now he infected both cultures with viruses. When the bacteria burst and the new viruses emerged, he isolated the viruses and measured their radioactivity. The viruses that came out of the non-radioactive bacteria transferred to the radioactive medium were radioactive: they had two thirds as high a concentration of radiophosphorus as the medium in which the bacteria had been immersed. On the other hand, the viruses that came from the radioactive bacteria in the non-radioactive medium had only one third as much radiophosphorus as the bacteria. Cohen therefore concluded that the new generations of viruses had obtained two thirds of their phosphorus from the growth medium while they were being formed and only one third from their host bacteria. This was a great surprise to those bacteriologists who had long supposed that bacterial viruses were formed from ready-made structures already present in the host cell.

At the University of Chicago Frank W. Putnam and Lloyd M. Kozloff, making similar studies with nitrogen 15 as the tracer, have found that the protein of viruses, like their DNA, is derived mostly from substances assimilated from the growth medium.

Cohen's experiment had covered the state of the system at just one stage: the next step was to follow the whole history of the conversion of inorganic phosphorus into virus DNA, from the moment bacteria began to grow in the medium until the newborn viruses finally emerged. At the State Serum Institute of Denmark Ole Maaloe and I extended Cohen's experiment with radiophosphorus, making the switch of radioactive bacteria to a non-radioactive medium and *vice versa* at many different stages in the development of the culture, both

350

before and after infection of the bacteria with virus. In this way we were able to determine just how much of the phosphorus that the bacteria eventually donated to the new viruses was assimilated by them from the medium during the various periods of development. Before they were infected with virus, the bacteria took up that phosphorus at the rate of their own growth, which means that they were using the phosphorus to make their own DNA. But after infection, their assimilation of phosphorus that they were to donate to the viruses increased sharply. Most of the phosphorus the bacteria were now taking up was going directly into the synthesis of new viruses.

We also observed that it takes at least 12 minutes to convert inorganic phosphorus into virus DNA. Hence any phosphorus that is to go into the making of the new viruses must have been assimilated by the bacteria by the end of the first 12 minutes of the 24-minute period during which the viruses are synthesized in the cell. As a matter of fact, A. H. Doermann has found that the 24-minute latent period divides into two 12-minute phases. In experiments at Cold Spring Harbor he opened infected bacteria at various stages. During the first half of the latent period there were no fully formed viruses with infective power within the bacterial cell; even the original invader had disappeared. Then, after 12 minutes, the first infective particle appeared, and more followed until there were 200 just before the cell burst. The explanation is clear. The original invading virus had shed its protein coat on entering the cell and therefore was no longer an infective unit. No virus could appear in the cell until at least one new protein coat had been manufactured and coupled with a unit of DNA. Apparently this proceeding takes some 12 minutes.

It seems that the manufacture of protein and of DNA goes on side by side within the cell. In experiments with radiosulfur as the label, Maaloe and Neville Symonds of the California Institute of Technology have recently shown that by the time the first new infective virus appears, there is already enough virus protein in the cell to form about 60 viruses. On the other hand, in similar experiments with radiophosphorus as the label we have found indications that completed units of DNA do not unite with protein units until the last moment; the particle then becomes infective.

The DNA of the original invading virus is responsible, as we have seen, for reproduction within the cell, both of DNA itself and of protein. How does it go about its job? Putnam and Kozloff labeled viruses with radiophosphorus and followed the radioactivity to see what happened to the phosphorus after the viruses infected bacteria. They found that about 40 per cent of the labeled phosphorus showed up in the viruses' progeny, the rest being discarded in the debris. Experiments with radiocarbon have shown that the same is true of other constituents of the DNA. In other words, about 40 per cent of the DNA of the parent viruses is passed on to the descendants.

How is the old DNA passed on? Is the parent's DNA handed on intact to a single individual virus offspring in each bacterial cell in a random 40 per cent of the cases, or is it distributed generally among the descendants? At Washington University of St. Louis Hershey, Martin D. Kamen, Howard Gest and J. W. Kennedy examined this question. They infected bacteria in a highly radioactive medium (one in every 1,000 phosphorus atoms was radioactive) with non-radioactive viruses. The DNA of the parent viruses was stable; not containing any radiophosphorus, it would not decay by radioactivity. Hence if it was passed on intact, a recognizable number of the viruses' descendants also should have stable DNA. But this was not the case. The descendant population steadily lost its infectivity, due to radioactive decay of its phophorus atoms, until fewer than one tenth of 1 per cent of the descendants were infective.

Is it possible that the hereditary continuity of the virus resides in a fraction consisting of 40 per cent of the DNA, and that the rest of the DNA does not participate in reproduction at all? To answer this ques-

Reproduction of bacterial viruses is shown in four stages; clock indicates progress and approximate time of the reaction. The large stippled object is the bacterium. Virus is shown in two parts: its outer layer (*fine stipple*) is the protein which has the ability to attach itself to the surface of the bacterium and to react with antivirus serum; its core (*black*) is the nucleic acid. First stage of reproduction is infection. In it the virus particle attaches itself, probably by the tail, to the bacterium and the nucleic acid core empties into bacterial cell, the protein coat remaining outside. In the second stage, the "dark" period, the virus nucleic acid within the bacterial cell has begun to induce the formation of new protein coats. However, the coats contain no nucleic acid and there are as yet no infective particles, not even the particle that caused the original infection. In the third phase, called the "rise period", some of the protein coats contain nucleic acid; they are the first infective particles of the new generation. The final stage occurs about 30 minutes after the first. The infected bacterium bursts and releases the new generation of virus particles into the surrounding medium. Only a few of the 200 particles in the new generation of bacterial viruses are depicted.

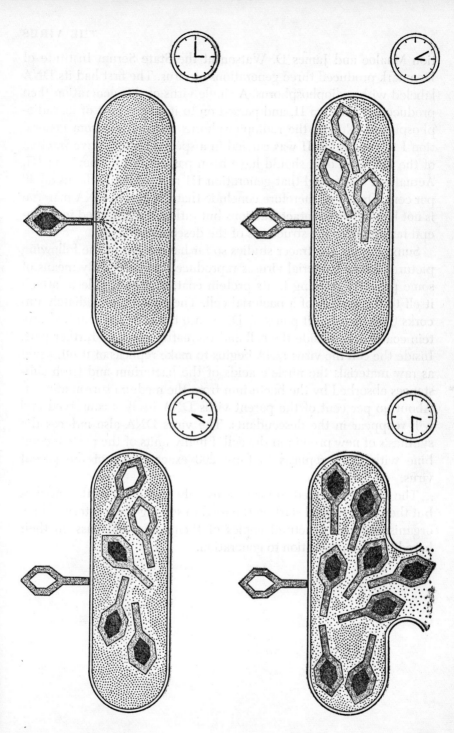

tion Maaloe and James D. Watson at the State Serum Institute of Denmark produced three generations of virus. The first had its DNA labeled with radiophosphorus. A single virus of this generation then produced generation II, and passed on to it 40 per cent of its radiophosphorus. Now if the radiophosphorus transmitted from generation I to generation II was carried in a special reproductive fraction of the DNA, all of it should have been passed on to generation III. Actually it was found that generation III received only the usual 40 per cent. One must therefore conclude that the parent DNA material is not handed on in intact fractions but rather is distributed in a general fashion over the structures of the descendants.

Summing up, the tracer studies so far have given us the following picture of how bacterial viruses reproduce themselves. By means of some property residing in its protein coat, a virus is able to attach itself to the surface of a bacterial cell. The contact immediately uncorks the virus, and it pours its DNA into the cell. The emptied protein coat is left outside the cell and thereafter plays no further part. Inside the cell the virus DNA begins to make replicas of itself, using as raw materials the nucleic acids of the bacterium and fresh substances absorbed by the bacterium from the medium surrounding it. About 40 per cent of the parent virus DNA itself is conserved and will reappear in the descendants. The virus DNA also induces the synthesis of new protein in the cell. Finally units of the protein combine with DNA replicas to form 200 exact copies of the parent virus.

The facts discovered so far give us only an outline of the process, but they seem a good start on the road to solving the mystery of how organisms build structural copies of themselves and pass on their heredity from generation to generation.

III. NATURAL HISTORY OF A VIRUS

by *Philip and Emily Morrison*

IN THE 19th century natural history meant adventure and travel to distant lands. Scientific explorers sought the secrets of birds, flowers and beasts in green jungles, on snow-topped mountains and desert islands. In the 20th century natural history still means adventure, but not always travel. While men continue to pursue fauna and flora to remote parts of the world, there are many who do all their exploring within the walls of a laboratory. Among the microorganisms whose lives and habits have been traced with curiosity, ardor and profit during the past decade is a particularly intriguing virus, the virus of herpes simplex.

One does not need to go on a journey to look for the herpes virus; almost everyone has seen its tracks on his own body. It is the organism that causes the fever blister, or cold sore, on the lip—one of the most common afflictions of man. Only a few Eskimo tribes in arctic Greenland seem to be free of it. The story of how the herpes virus was discovered is of interest not only for itself but also because the methods used and the results obtained in its study are basic to an understanding of the biological nature of all disease. The virus was tracked down only after a long series of investigations, principally by the Australian virologist F. M. Burnet.

Although nearly everyone has fever blisters occasionally, the cause of this affliction had mystified physicians for many years. It does not behave at all like an infectious disease. The blisters recur throughout the lifetime of an individual, usually in exactly the same spot; they sometimes accompany colds, fevers or emotional upsets, as if they were symptoms of a more general disorder; they give no evidence of spreading by contact from one person to another. In contrast, a typical common virus disease like measles begins with considerable fever, catarrhal symptoms and spots in the mouth; a rash appears all over the body; the disease spreads very readily, so that only about one per cent of susceptible individuals fail to get it upon their first close contact with an infected person; usually it attacks an individual only once during his lifetime.

As early as 1912 some investigators discovered that a virus was present in typical herpes blisters. But this information was dismissed by most research men as not significant. How could herpes be caused by a virus, they argued, when it did not conform to the pattern of other known virus infections? Nonetheless a few workers persisted in following this apparently blind alley. An encouraging but by no means conclusive experiment in 1919 showed that some of the liquid obtained from a human fever blister would produce lesions on the cornea of a rabbit. Again the evidence was rejected. A number of other theories were offered to explain the recurring blisters. One of the most widely supported was that herpes was simply a peculiar external irritation which affected certain sensitive cells.

In 1939, however, the virus theory received a tremendous boost from findings by Katharine Dodd, L. M. Johnston and G. J. Buddingh at the Vanderbilt University School of Medicine. They learned that an infectious organism could be recovered from the mouths of infants suffering from a common form of sore mouth known as aphthous stomatitis. They did not see the typical herpes blisters on these children, nor, of course, could they see the virus itself. Their chief evidence that herpes virus was present was the fact that they found herpes antibodies in the children's blood. In other words, they detected the virus not by its usual blister track but by a kind of chemical scent. The antibodies they found had previously been observed in herpes patients.

Spurred by these experiments, Dr. Burnet and other investigators undertook a long, painstaking series of measurements of herpes antibodies, hoping to prove conclusively that herpes was a virus infection and to study the virus in all its peculiar habits.

Most normal adults have antibodies against herpes. These antibodies are a very peculiar breed. Usually an individual who has antibodies against a specific infection is immune from a recurrence of that disease; a notable example is measles. But the investigators found that in the case of herpes the antibodies do not immunize the patient; he may still have repeated attacks of fever blisters. On the other hand, in spite of the wide distribution of the virus some people who have no herpes antibodies never get fever blisters. The herpes antibodies show another peculiarity. In many diseases, notably diphtheria and influenza, the production of antibodies varies with time or with the virulence of the attack. A mild infection produces a moderate concentration of antibodies; after a severe infection the convalescing patient at first has many antibodies, then the concentration diminishes and finally disappears. In herpes, however, the investigators found that it is all or none—there are either a great number of antibodies or none at all. Infants with stomatitis who had no antibodies at the time they became ill had a great many within three weeks, and continued to have them. No one who had ever had a fever blister failed to show a large number of antibodies, and a comparable concentration was found in some people who could not remember having had such blisters. People who have herpes antibodies seem to keep them in their blood for life, and even for some time after death.

With these facts at hand, the picture of the herpes virus began to take definite shape. The investigators determined by filtration that the virus is about 1,500 Angstroms in diameter—roughly the same size as the influenza and rabies viruses. Like other viruses it is essentially a nucleoprotein. It first appears in young children, usually in the form of a mild sore mouth. Once it attacks an individual, it remains with him until death, though it makes its presence known only by occasional blisters or not at all. If you escape herpes as a child, it is somewhat more difficult to acquire it as an adult, but it can be done. A woman worker in Burnet's group acquired the telltale antibodies during their investigation. Upon inquiry, Burnet learned that she had become engaged to a man who had had herpes for some time, and she had evidently acquired it from him!

Infection with the herpes virus seems to vary considerably with socio-economic status, according to investigations made in Australia and England. Among university graduates, who come generally from comfortable homes, only about 37 per cent showed herpes antibody; among groups in lower income brackets, as many as 93 per cent showed antibody. This statistic has been interpreted as evidence that herpes is probably spread by close physical contact, perhaps between the mouth of the child and the saliva of the parent. The greater crowding and poorer hygiene in lower income groups would tend to help the spread. If herpes were spread by airborne droplets rather than by contact, it is difficult to see how anyone could escape, regardless of income.

However that may be, the Burnet group positively identified herpes as an infectious disease, like measles, mumps or any other virus infection except for the unique fact that it persists all through life, almost without symptoms. Only occasionally, when the resistance of the body is lowered by fever, a cold, or some emotional strain, do the infected cells of the lip, where the virus presumably lives, break down into the typical blisters.

So far we have been concerned with the description of herpes virus from the point of view of the explorer and the naturalist. We have identified it and studied its habits. But it is valuable also to assume the point of view of the historian and the ecologist, to trace the virus' origin and development and its relationship to other organisms. Like Darwin, the modern virologists have not stopped at description but have carried their researches into this broad and rather speculative field as well.

The natural habitat of the herpes virus is man; it seems to be as much at home in man as is the mistletoe on the oak. The creature multiplies and gains its food and shelter from the human body, and from its own point of view it is extremely successful, for it invades almost every human being, is never dislodged and is spread constantly to new hosts. But an essential element of this parasite's successful adaptation to its environment is that it does not do much damage to man. This is not because the virus is totally incapable of doing damage; on the contrary, when introduced into rabbits or mice it often infects the central nervous systems of these little animals and produces a true and usually fatal encephalitis. There are a few rare cases known in which herpes virus has caused encephalitis even in human beings.

The generally harmless behavior of this creature in its natural host gives us an exceptionally illuminating insight into the survival value of a biological equilibrium. If you die of encephalitis induced by herpes virus, this is of little advantage to the virus, since it cannot spread in large numbers from the dead. On the other hand, if you carry herpes virus always with you and suffer only an occasional blister, the virus lives happily in equilibrium with you for a relatively long time and has splendid opportunities to grow and spread. Herpes virus is a concrete demonstration of a great generalization made by the late Theobald Smith of the Rockefeller Institute: wherever you find a parasite that has affected a given host over a long period of time, you will find that the infection does not interfere with the survival of the host and is of low virulence. A strange host, such as a mouse or a rabbit, cannot expect such good behavior from the human herpes virus because the host has not had time to establish the basic chemical adjustment necessary for a state of equilibrium. Just as psittacosis is much more virulent in people than in parrots, so herpes is much more virulent in the mouse than in human beings.

The history of herpes in man leads us to speculation on its sociology. We know that virulent diseases like measles and smallpox have not always been world-wide. When measles first strikes in virgin territory (e.g., the Faeroe Islands) it becomes a terrible scourge; so it was with smallpox when it was introduced among the American Indians and among the Samoans. Such diseases are probably diseases of civilization—of cities or large, stable communities of men. Measles is more or less constantly present in all large communities, and there are major epidemics or sudden increases in prevalence at about three-year intervals. It spreads rapidly, but after a comparatively short illness it confers long-time immunity on those who recover. By the age of 20 approximately 85 per cent of the population in a civilized country have had the disease, and second attacks are unusual. The measles virus is able to survive indefinitely in a city because there are always new people to carry and spread it among a constant new crop of children who have no immunity. It would have a hard time surviving in an isolated family or small tribe of cave dwellers on nomads, for if the group were accidentally infected from some outside source, all of its members would be sick, the survivors would be immune and the disease would die out.

The herpes virus, on the other hand, is self-perpetuating. It is the only known virus disease in which a woman who had the disease as a child can infect her own children many years later, and they in turn infect their children. Thus the virus can be forever transmitted even within a single isolated family.

So this newly-found living form, virus of herpes simplex, is not new at all but probably older than the pyramids or the ancient plagues, older than writing or even fire. It is undoubtedly one of the oldest domesticated organisms we know, older even than man's friend the dog. The occasional colony of herpes on your lip is the evidence of a slight disturbance of the nice balance between 10,000 generations of men and countless generations of these little globes of nucleoprotein 1,500 Angstroms in diameter. Only as a result of many mutations and selective adaptations could the herpes virus have achieved so successful and peaceable a relationship and struck so delicate a balance between its own aggressive tendencies and the defensive mechanisms of its human host.

The human race exists in a rather sensitive and unstable relationship with its environment; it can be profoundly disturbed by slight changes in the physiology of even the tiniest parasitic microorganism. It may well be that other diseases of man can profitably be studied in the light of what we have learned about herpes, even though in those diseases the picture is much more complicated. It may be that some of our violent diseases are merely new forms of old ones that have suddenly been modified, perhaps by mutation. It may also be that these virulent diseases are evolving toward the achievement of a new successful equilibrium with their hosts, and will gradually lose their present virulence. Investigators have not failed to observe that old diseases are disappearing just as new ones are emerging. The theory of the age of disease also offers a clue to the understanding of epidemics. It suggests that they represent sudden, violent disturbances in the equilibrium between a parasitic microorganism and the human host.

Thus the investigation of this harmless, seemingly insignificant virus may well have important results in our defenses against more troublesome diseases. The science of immunology becomes ever more important as it becomes evident that the individual living in a large community is unable to protect himself against infections from his neighbors, but must rely upon the continuing exploration and research of today's natural historian, whose daily adventures are just as exciting and perhaps more immediately useful than those of his predecessors who sought the tapir, the okapi, or the high-flying condor.

IV. A NEW ERA IN POLIO RESEARCH

by Joseph L. Melnick

THE DISCOVERY of a way to grow poliomyelitis virus in tissue culture—made three years ago by John F. Enders, Thomas H. Weller and Frederick C. Robbins at the Children's Hospital in Boston—has given a tremendous impetus to the study of this disease. It means the end of the "monkey era" in poliomyelitis research and opens the way to a much wider attack on the problem. Now that experimental work in poliomyelitis is no longer dependent mainly on the infection of monkeys, many more investigators can work on the problem and new kinds of experiments can be undertaken. Advances are already being made on many fronts. The new method stems directly from the discovery that poliomyelitis virus may grow in cultures of human tissues other than those of the brain and spinal cord. Together with other evidence, this suggests that in the body also, poliomyelitis virus can grow in other than nervous-system cells. If this concept is true, it is indeed fortunate, for such cells presumably are not damaged. Actually there is often no sign of disease, and yet the body gains an enduring immunity. Such multiplication of virus without producing disease means that the body, when it meets the virus again at some future date, is able to thwart its attack, and prevent the virus from moving to the nervous system—for it is only in this latter tissue that the damaging lesion is produced. This concept of virus growing in cells outside the nervous system and conferring an immunity to subsequent paralytic infection is opening up new channels for investigation, from which may come a means of preventing the disease.

The materials used in the new culture technique are certain monkey or human tissues, the latter being available after some types of surgical operations. They are cut into tiny fragments about 1/25-inch square. When such a piece is bathed in a suitable nutrient medium, long strands or sheets of cells grow out from it within a few days and continue to grow for several weeks. If virus is added to the test tube, the cells are broken up and destroyed, usually within a few days. That these destructive changes are specifically associated with the growth of virus is shown in two ways. The first is that they fail to occur in uninoculated tubes. The second is that the addition to the culture tube of serum containing specific anti-bodies for poliomyelitis virus prevents this destruction of cells. From this may come a laboratory diagnostic test for poliomyelitis infection in man. The method is already useful for the classification of polio-myelitis viruses and for the quantitative measurement of antibodies to each type.

Three different species or types of poliomyelitis virus exist in nature. This fact was established by a huge program of research, sponsored by the National Foundation for Infantile Paralysis and carried out at several universities, in which vast numbers of monkeys were used and more than $1 million was spent. Each of the types can cause the human disease. It has been found in monkeys that infection with one type confers little or no immunity against the other two. This may explain why paralytic poliomyelitis is sometimes contracted more than once by the same person.

In order to gain a better understanding of poliomyelitis epidemics, it is necessary to know just which types are prevalent during a single outbreak, and the tissue-culture method is ideally suited for such work. For example, in Easton, Pa., during the summer of 1950 tissue-culture analysis conducted by the author, together with John T. Riordan and Nada Ledinko, showed that all three virus types were present in poliomyelitis patients sent to the hospital, but Type 1 accounted for most of the cases. Whether all three types occur in every epidemic, with one type in ascendancy, is a question for future investigation to answer. At least it shows that as far as the polio-myelitis family of viruses is concerned, "birds of a feather flock together."

In addition to the three types of poliomyelitis virus, three other viruses were isolated in tissue culture in Easton. They were found in patients diagnosed as having mild poliomyelitis. They did not belong to the poliomyelitis virus family, for none of the three polio-myelitis antisera neutralized their capacity to destroy cells in tissue culture. Further study has shown that two of the agents can be grouped with the poliomyelitis fellow-travelers—the Coxsackie or C viruses. The remaining agent is as yet unidentified. The following summer there were isolated three similar viruses with the capacity to produce cellular damage in monkey-tissue cultures, but not, as far as could be determined, with the capacity to cause disease in monkeys or other laboratory animals. Thus a whole new field is opening up in which heretofore unrecognized human viruses may be detected by the tissue-culture method.

For many infectious diseases the presence of specific antibodies is generally believed to mean that the individual is immune to that disease. If this assumption is correct for poliomyelitis, tissue culture may provide a simple tool to tell who is immune to poliomyelitis and who is not. This may have great practical significance. By sampling the population periodically to determine the prevalence of antibodies to the three poliomyelitis viruses, health officers may be able to predict the occurrence of epidemics.

The new test-tube method for identifying poliomyelitis viruses and measuring antibodies is not yet available to practicing physicians and health officers. At this writing its use is confined to research laboratories, but it is our hope that in the near future the test will become a practical tool and available on a large scale.

Immunity to poliomyelitis cannot be left to chance, and two approaches to possible immunity and control of epidemics are currently being made. One is based upon the fact that the virus can be grown in cultures of non-nervous tissue, and the other relies on the use of a human blood fraction called gamma globulin, obtained from large pools of human plasma and rich in antibodies to all three poliomyelitis viruses (as well as to other infectious agents).

When some viruses are cultivated in a medium different from the one to which they are accustomed in nature, they sometimes change in their ability to cause the kinds of diseases they once produced. The new virus is still a living agent and causes infection, but this infection does no damage to tissue. And it will produce antibodies which render the subject resistant to subsequent infections by a wild, disease-producing related virus. In this way such diseases as smallpox and yellow fever have been conquered and it appears that rabies will be next on that list. Even though we do not yet possess any attenuated strains of poliomyelitis virus which can be considered safe for human trial, there is good evidence that the viruses being propagated generation after generation in tissue culture already differ from the viruses grown in the spinal cord of laboratory animals. Some of the cultivated strains confer immunity in monkeys even after their capacity to produce paralysis in laboratory animals has almost disappeared.

As it may be a long time before investigators find strains of live virus which can be safely administered to children, work is going forward with vaccines containing killed virus. Monkey spinal cords could never have been entirely safe for this purpose, even if adequate supplies had been available, because the inoculation of nervous-tissue material all too often produces allergic encephalitis, a very serious disease. Test tube-grown viruses which are killed before being incorporated into the vaccine obviate this danger.

Killed viruses have been shown to be effective in producing immunity in animals. Virus particles have an exquisite architecture, characteristic of large nucleoprotein molecules. If this architecture is altered by chemical or physical means, certain properties of the virus are also altered. Several investigators are studying the most effective ways to destroy only that part of the virus responsible for the production of disease. As the rest of the particle, now harmless, would be similar in architecture to a virulent virus, it might be expected to stimulate the production of antibodies and thus confer immunity on a susceptible individual.

Field tests of gamma globulin conducted under the direction of William McD. Hammon of the University of Pittsburgh have proved that this material obtained from normal adult human blood does confer immunity. Hammon himself has pointed out the limitations of this passive type of immunization in which the body receives antibodies from outside but produces none of its own. The period of protection is short and, since the moment of exposure cannot usually be determined, the globulin is often given too soon or too late to confer maximum immunity. In the long run, gamma globulin is unlikely to prove a satisfactory means of preventing poliomyelitis, but it may be a valuable expedient, particularly in epidemic areas, until a safe and reliable vaccine is available. The gamma globulin field trial suggests that the concentration of antibodies needed to confer immunity is lower than had been supposed. If this is confirmed, it means that vaccines capable of stimulating only a low level of antibody development should provide a significant degree of immunity, that a given amount of attenuated or inactivated virus will provide more doses and that these weaker doses will be that much safer for human use.

Tissue-culture methods have provided virologists with a simple *in vitro* method for testing a multitude of chemical and antibiotic agents for their effect on the multiplication of viruses in living cells. Some workers in the poliomyelitis field have already found that certain antimetabolites suppress the growth of the virus in tissue culture. These organic chemicals are structural analogues of compounds found within normal cells, and for this reason they interfere with the normal pathways of metabolism. It is thought that if these pathways can be temporarily blocked, the parasitic virus within the cell may find itself in an environment unfavorable for its propagation.

Altogether the arrival of the tissue-culture technique has greatly encouraged investigators of poliomyelitis. The day when the disease will be brought under control now seems closer.

V. THE COMMON COLD

by *Christopher Howard Andrewes*

WHY HAS the solution of the common cold problem so long eluded us? Perhaps the fact that it is not a dangerous disease has detracted from the pressure to solve the problem. There are, however, more important reasons. Virus diseases are studied mainly by observing the effects produced in experimental animals or plants, for viruses cannot be grown on artificial culture media. Unfortunately there is no convenient experimental animal for investigating the cold virus. Neither the mouse, the guinea pig nor any other readily available species can be infected with colds. The only animal, besides man, that will catch a true cold is the chimpanzee, and chimpanzees are so hard to come by and to handle and so expensive as to be almost useless.

Furthermore, the whole subject of colds is overlaid by stratum upon stratum of folklore, superstition and pseudo-science. Colds touch each of us personally, and it is a human failing that where our own afflictions are concerned our scientific judgment becomes faulty. We are apt, when we unexpectedly catch a cold—or avoid catching one—to attribute this to some unwise or wise act on our part. He who would solve the cold problem would do well to consider only scientifically checked facts and to keep a very critical attitude toward the folklore and toward what his friends tell him about how they catch or avoid colds.

What do we know about colds? There is ample evidence that colds are "catching": we know that an infected person can pass on infection to another person, and that chimpanzees can catch colds from human beings. Yet it is also true that often an individual intimately associated with a cold-sufferer fails to catch a cold at all. Another very well attested fact is that among a group of people isolated on a remote island, particularly if the community is small, colds tend to die out. But when such a cold-free community re-establishes contact with civilization, as by the visit of a ship from outside, its inhabitants are found to be abnormally susceptible and are pretty sure to catch a real "snorter."

In the ordinary way, the immunity acquired after a cold apparently is of brief duration. It is fairly clear that the freedom from colds of isolated groups is dependent not on increased resistance but on disappearance of the cold germ during their isolation. Their fate when they do meet cold germs shows that their resistance has waned with their freedom from attack. The temporary resistance developed by people in ordinary communities seemingly is maintained by frequent and repeated contact with cold virus.

Attempts to transmit colds artificially from one person to another are successful in about 50 per cent of the trials: a number of people at any one time prove resistant. The technique used has been to wash the noses of people who have colds with a salt solution and to drop some of this mucus-containing solution in the noses of normal people. Careful study of these solutions by the standard bacteriological methods has failed to show that any of the cultivable bacteria can be incriminated as the cause of colds. Indeed, during the early stages of colds the nasal secretions often have a subnormal content of bacteria. Several investigators have shown that after nasal secretions from people with colds have been passed through filters so fine that they hold back ordinary bacteria, the filtered material is still infective; it can pass in series from one person to another and continue to produce colds. This indicates that the infective agent is something which can multiply and is smaller than bacteria: *i.e.,* a virus.

I propose to describe here the attack on the cold problem that has been carried out since 1946 in the Common Cold Research Unit of the Medical Research Council at Salisbury, England. Naturally the first objective of a study of this kind is to find a reliable, simple test, if possible, for the presence or absence of the virus. With such a test we could determine whether the virus is present in nasal secretions at various stages of the disease, whether it can be killed by various disinfectants, and so on. The only test we have at present, a very expensive and rather unreliable one, is to drop material in people's noses and see if it produces a cold. As a first step toward attaining our objective of finding a simpler and better test, we had, perforce, to resort to this method, and to set up a very complicated organization to carry it out.

Our "laboratory" is the Harvard Hospital at Salisbury, consisting of a number of prefabricated huts which during the war were used as an American hospital with a staff from the Harvard Medical School and at the end of the war were generously handed over to the British Ministry of Health. As now set up, the unit has six rather luxurious huts, each divided into two separate flats. The 12 flats house 12 pairs of volunteers. We decided to take the subjects in pairs because it seemed obvious that people would be more likely to volunteer if they could have a friend with them. Each pair is kept in isolation and studied for 10 days; a fresh lot of about 24 volunteers comes along every fortnight. During university vacations we have no difficulty in getting students to come; at other times we have managed to keep pretty full with other people coming along in response to appeals on the radio or in the press. Since we started, four years ago, more than 2,000 volunteers have passed through our hands.

What have been the over-all results of our tests with the 2,000 volunteers? The first thing to be said is that those who received harmless control inoculations remained satisfactorily free from colds during their 10-day stay. This is an indication that our quarantine and other precautionary measures are adequate. Of those who received the active secretions taken from people with colds, some 50 per cent, as I have mentioned, caught colds. An interesting point is that many of those who were inoculated with active material seemed to be starting a cold on the second or third day after inoculation but next day had lost all their symptoms: the cold had aborted. Possibly most colds abort naturally. If this is true, it is easy to see why remedies purporting to cure the common cold so often gain a wholly unmerited reputation.

Women, according to our observations, are definitely more susceptible to colds than men; the comparative scores were 55 per cent to 43 per cent. Age has very little effect on vulnerability, within the age group of 18 to 40 from which our volunteers come. The incubation period of a cold is usually two to three days. On the whole the colds have been milder than we expected. Perhaps in ordinary life it is only the severe colds that attract one's attention; we remember the virus' offensive triumphs rather than our own defensive successes. The colds of our volunteers have usually cleared up by the time they go home, several days after the cold's onset. Those departing are given post cards to send back a fortnight later with news of any nasal happenings in the meantime. Not a few relate that the colds, which had cleared up in our sheltered environment, got worse again in the hard outside world.

All this is very interesting, but what about the primary objective, the development of a better technique for studying colds? We cannot yet, unfortunately, report success. Much effort has been devoted to cultivation of the cold virus in fertile hens' eggs, by each of several different techniques that have proved useful with other viruses. No unequivocal success has been attained. Nor have we succeeded in inducing colds in experimental animals. We have tried rabbits, rats, mice, guinea pigs, hamsters, voles, cotton rats, gray squirrels, flying squirrels, hedgehogs, pigs, chickens, kittens, ferrets, baboons, green monkeys, capuchin monkeys, red patas monkeys and a sooty manga-bey. People have told us that they have observed that such-and-such an animal develops colds in captivity when in contact with human colds. Such clues we have followed up, but in vain. The "colds" in these animals have either not been reproducible or have seemed to be due to bacteria rather than to a true cold virus.

Efforts have been made to determine the properties of the cold virus, for knowledge of these should help us toward our objective of the simple laboratory test. We have, for instance, learned something about the size of the virus. Virus-containing fluids can be filtered through a special type of collodion filter having pores of very uniform and accurately graded diameter. We find that the cold virus passes with but little loss in potency through membranes with pores as small as 120 millimicrons, or 120 millionths of a millimeter. In one experiment patients took colds from material passed through a filter with pores of only 57 millimicrons. If we can believe this single positive result, we can deduce that the cold germ is one of the smaller viruses—about as big as that of yellow fever, which has a diameter of some 25 millimicrons. At most the cold virus seems to be no more than about 60 millimicrons across; that is, decidedly smaller than the influenza virus.

Cold virus is very stable when kept frozen at 76 degrees below zero centigrade; some of it has retained its potency for as long as two years at this temperature. This knowledge is very useful to us. At any point we can bottle our "pedigree" strain of virus, put it away in dry ice and forget about it till it is next needed.

We have also established several negative facts about the virus: it is not affected by penicillin or streptomycin, nor is it adsorbed, as is influenza virus, on human or fowl red blood cells.

We need to know not only these rather abstract properties but also something of how the virus behaves in man—the relation of the

369

parasite to its host. We have no way of telling how many virus particles our infectious dose of material contains; it may take thousands of particles to overcome a person's normal resistance and produce a good cold. We have found, however, that nasal secretions are infectious even when diluted up to 1,000 times with salt solution. Apparently the saliva secretions in the front of the mouth also contain much virus. This is important in relation to the means of spread, for most of the liquid expelled when you sneeze comes from the front of the mouth.

Nasal secretions contain plenty of virus during the incubation period of a cold, even before symptoms develop. We found this out by washing out the noses of infected subjects 12, 24, 48 and 72 hours after inoculation and testing the washings for infectivity. The 12-hour specimen was negative, all the rest positive. The subject who provided the washings developed symptoms 48 hours after inoculation. Our impression is that secretions from the early stages of a cold are the most potent, but we have had "takes" with washings taken seven days after inoculation. We have occasionally produced mild colds with secretions from normal, symptom-free people; this suggests that some people may be carriers of cold infection without showing symptoms themselves, as in diphtheria, typhoid fever and many other diseases. There is some evidence that children are especially "efficient" in spreading cold infection.

We have tried many dodges to increase the rate of successful transmission of colds from 50 per cent to 100 per cent. Obviously if we could produce a cold every time we tried, we could work faster and more certainly, and we might find some clues to the reasons for people's varying susceptibility and resistance. But so far we have not been able to increase or decrease the rate of cold "takes" very much.

For instance, we put to the test the practically universal idea that chilling induces colds, or at least increases one's chances of catching a cold. Three groups of six volunteers each were used in this experiment. One lot received a dose of dilute virus, calculated not to produce many colds. The next lot were given no virus but were put through a severe chilling treatment: that is, they had a hot bath and were then made to stand about in a draughty passage in wet bathing suits for half an hour, by which time they felt pretty chilly and miserable. They were further made to wear wet socks for the rest of the morning. A third group received the dilute virus plus the chilling treatment. On one occasion, in a variation of the experiment, the chilling consisted in a walk in the rain, following which the subjects were not allowed to dry themselves for half an hour and were made to stay in unheated flats.

Now this experiment was performed three times. In not one instance did chilling alone produce a cold. And in two out of the three tests chilling plus inoculation with the virus actually produced fewer colds than inoculation alone; in the other the chilled people who also got virus did have more colds than the "virus only" group. So we failed to convince ourselves that chilling either induces or favors colds.

One of the great puzzles about colds is that even repeated attacks do not confer any lasting immunity. We have found some evidence in the blood that the body produces antibodies to cold virus. But in contrast to the antibodies called forth by the virus of a disease such as measles, which guard the body against future attacks, the antibodies that respond to a cold seem almost powerless to help. A person may develop two successive colds within a matter of months. Why are the antibodies so effective against the one disease and so ineffective against the other?

The answer may possibly be that in the latter case the antibodies are not in the right place. In the case of measles the virus always appears in the bloodstream at one stage of the attack, and this is where the antibodies are, ready to intercept and destroy it. On the other hand, the virus of the common cold (and of influenza, in which the antibodies are also relatively ineffectual) attacks the superficial membranes lining the nose and other respiratory passages, without having to pass through the bloodstream and encounter the antibodies. True, some influenzal antibodies apparently can pass from the blood into the mucus covering these membranes, and this doubtless helps to keep infection under control. But there is usually much less antibody there than in the blood.

Consequently a promising approach to the prevention of colds is to try to determine what conditions control the amount of antibody in the mucus. Perhaps frequently repeated contacts with small doses of virus, doses insufficient to produce a manifest cold, stimulate the body to provide the mucus with enough antibody to protect it. Very probably it is the lack of such stimuli that renders small isolated communities so susceptible to colds. It may be that in the future colds may be kept at bay by repeated doses of an attenuated virus taken as a snuff, rather than by vaccines given in the orthodox ways used in other diseases.

As explained earlier, it is not always easy to decide just when a person has or has not got a cold; often this can be determined only by carefully designed techniques. It is still more difficult to decide when a cold has been prevented or cured. A cold's natural duration is extremely variable, and it is likely that at least as many colds abort in their early stages as go on to be full-blown. Hence people often draw rash conclusions about "cures" on slender evidence, as testified by the correspondence sent to us by over 200 persons with the most helpful intentions. Not only can we place little or no faith in conclusions based on individual experience, but we can be readily deceived by trials carried out on a fairly extensive basis.

Unfortunately for the poor public, a "cold-cure" is news and is well publicized. Pricking of a bubble is not news and word of it gets around slowly. Antihistamines, for example, are worthless for curing colds, but they are still sold in quantity for that purpose. Claims to prevent colds by means of oral or other vaccines rest on just as shaky foundations as those of the antihistamines: adequately controlled trials have failed to demonstrate their value.

There are things that can be done to relieve the unpleasantness of colds, but up to the present it still remains true that the untreated cold will last about seven days, while with careful treatment it can be cured in a week!

PART  8

I. ACTH AND CORTISONE *by George W. Gray*

George Gray appears more often in this book than any other author; he was also the most steady contributor to SCIENTIFIC AMERICAN in its first five years under new editorship. In that period the magazine published 10 Gray articles in the fields of astronomy, chemistry, physics, biology, medicine and scientific biography. Quite aside from his other work, Gray's writing for this magazine (which happily continues) would establish him as one of the outstanding science journalists of the day.

II. THE PITUITARY *by Choh Hao Li*

Choh Hao Li came from China to the University of California, Berkeley, as a graduate student of chemistry in 1938. He worked there with the great biologist, Herbert Evans, and is today a professor in the university's department of biochemistry. Dr. Li is especially noted for his study of the growth hormones, which are synthesized in the pituitary.

III. THE ALARM REACTION
by P. C. Constantinides and Niall Carey

P. C. Constantinides was born in Greece; Niall Carey is Irish. In 1949, when they wrote this article, they were both young biologists with medical training working under the celebrated Hans Selye at the Institute of Experimental Medicine and Surgery, Montreal University. Dr. Constantinides is now an associate professor of anatomy at the University of British Columbia in Vancouver. Dr. Carey, following a prolonged illness, is on vacation in England.

IV. SCHIZOPHRENIA AND STRESS
by Hudson Hoagland

In 1944, Hudson Hoagland, with Gregory Pincus as co-director, established the Worcester Foundation for Experimental Biology and Research. This institution, of which he is today executive director, employs a staff of some 100 persons engaged in basic biological and medical research. Professor Hoagland himself works principally in the fields of neurophysiology and endocrinology; the study of schizophrenia described in this essay is typical of his interests.

STRESS

Introduction

THE QUESTIONS about the nature of life raised thus far in this book have been at the micro-level of cell, molecule and atom. In the present section the scale shifts: We step back a pace or two from the laboratory bench to observe the whole organism in interaction with its environment. In particular, we are concerned with its reaction to the heavier demands which the environment makes on it from time to time.

The word "stress," as used in this research, is an all-inclusive term. It covers tuberculosis, pregnancy, a broken leg, exposure to heat or cold, fatigue or bad news from home. The body meets any and every stress laid upon it with the same systemic response. The existence of this generalized response may prove to be the revolutionary medical discovery of our day. It may turn the practice of medicine away from specialization and toward a view of sickness or health as states of the body as a whole. One can feel running through these essays a mood of excitement and enthusiasm such as scientists reserve for the really grand insights into nature.

As told here, the story begins with one of the medical sensations of the past half decade—the arrival of ACTH and cortisone in the clinics of the nation. It was these two hormones that first confronted the practice of medicine with the stress response and its tremendous implications. They were heralded at the outset as specifics for arthritis and related afflictions of the aging human, but experience soon showed that they would alleviate a large number of diseases, entirely unrelated to one another. Their unpredictable and powerful versatility, it soon became clear, derives from the fact that they do not act upon the affliction or the afflicted part, but upon the whole organism.

In the body, in fact, ACTH and cortisone provide the triggering mechanism for the stress response. Investigators have found that the output of cortisone from the adrenals is stimulated by increased output of ACTH from the pituitary gland. This discovery, adding control of the stress response to the other known functions of the pituitary, which include control of growth, clearly established it as the master gland.

For the concept of the stress response itself, medicine is indebted to the radical pioneering of a group of investigators headed by Hans Selye at the University of Montreal. During the ten years preceding the ACTH-cortisone sensations they had been subjecting hundreds of animals to severe and prolonged experimental stress. The animals showed upon dissection clear pathological evidence for what Selye called the "alarm reaction" and the "adaptation syndrome." Their symptoms duplicated most of the conditions found at autopsy in the "old age" diseases of man.

These studies are therefore especially pertinent to the practice of medicine today. Heart, circulatory and kidney diseases have long since displaced the infectious diseases as the principal cause of death in our population. It has been said that the pressure and tension of modern life are the price we pay for the civilization that insulates us so well from the ancient plagues.

The pathology of stress may help to us to understand and cope with that other major public health problem of our time, mental illness. This section closes with a report of an investigation that shows a correlation of ACTH output to schizophrenia. It has been found that, as a class, schizophrenics do not respond normally to stress and that their pituitaries fail to sound the alarm.

I. ACTH AND CORTISONE

by George W. Gray

THE ADRENAL GLANDS, two wrinkled pads of yellowish-brown tissue that lie on top of the human kidney like miniature pancakes, rarely weigh as much as a quarter of an ounce— but on that glandular quarter-ounce depends the health of the whole body. Animals from which the adrenals are removed die in a few days, and persons whose adrenals become involved in the condition that brings on Addison's disease rarely can be kept alive except by injections of adrenal hormones.

Each adrenal consists of an inner core, the medulla, and an outer bark, the cortex. The medulla manufactures only one hormone: adrenalin. The cortex, however, turns out 20 to 30 different compounds. Although these cortical hormones have long been studied, until recently chemists were able to isolate only a few of them, and those few were obtained in such microscopic amounts as to be useless for medical purposes.

World War II brought a great impetus to the research on cortical hormones. During the war it was rumored that pilots of the *Luftwaffe* were being injected with hormones of the adrenal cortex and that these injections enabled them to fly at ease at altitudes of 40,000 feet. When this rumor (eventually found to be false) was first received in Washington in the fall of 1941, it galvanized the medical departments of the Army and Navy into action. They called on the National Research Council for large supplies of cortical hormones. The Council made a quick survey and found that 22 laboratories had been engaged in investigations of the adrenal cortex. The resources of these institutions were immediately enlisted, and a coordinated team effort was organized. Under this accelerated program, several methods of obtaining cortical hormones in pure form were explored. Eventually one of the investigations yielded a marvelous result: the production of cortisone.

The father of cortisone was Edward C. Kendall, chief of the biochemical laboratories of the Mayo Foundation for Medical Education and Research. As early as 1935 he had isolated the first few granules of the substance. It was the fifth in a series of hormones that he had separated from cortical extract, and so he called it compound E. Tests of these hormones on rats and mice showed that three of them—A, B and E—apparently played some part in metabolism. Compound E had a marked effect on muscular activity, and apparently it increased an animal's ability to resist exposure to cold, poisons and other physiological stresses. But to find out what compound E and the other hormones could do for human disease required much larger supplies of the substances than could then be procured.

Chemists have a technique, known as partial synthesis, by which they remodel one molecular structure to produce another of different form. There is a well-known bile acid whose structure, it was believed, might be reshaped into the configuration of either compound A or E. But many steps would be involved, and whether partial synthesis could be applied to convert the acid into one of the hormones was pure speculation. Kendall and his associates were studying the possibilities of this molecular reconstruction when the German rumor of 1941 suddenly made cortical hormones an emergency problem. The Mayo laboratory at once became a center in the government's accelerated research program, and it soon joined forces with a laboratory of Merck and Company, Inc., another important outpost in this research.

Early in 1946 Lewis H. Sarett of the Merck laboratory delivered the results of the first partial synthesis of compound E. This method, however, yielded too minute a quantity to be of clinical use, so the chemists tried again. Two years were spent working out the 37 steps of a new synthesis. Finally, in May, 1948, this effort was crowned with success. By September of that year compound E was being manufactured by the gram instead of the milligram. For the first time there was enough compound E, now named cortisone, to test it on human illness.

The first tests were made at the Mayo Clinic. For many years Philip S. Hench, chief of the Clinic's department of rheumatic diseases, had been investigating two curious phenomena connected with arthritic disease. He had noticed that when an arthritic woman became pregnant, the arthritis usually ceased to trouble her, and that jaundice also seemed to cause arthritic symptoms to fade away. The remissions were only temporary, however, for after the patient had given birth or had recovered from jaundice, the old swellings, stiffness and pain returned. Other doctors had remarked on these coincidences, but apparently Hench was the first seriously to seek an answer. Beginning in 1929 he spent several years in careful observation of arthritic patients during pregnancy and in attacks of jaundice. He came to the conclusion that rheumatoid arthritis might not be of microbial origin, as many authorities held, but might be caused by some basic disturbance of the body's chemistry. He conjectured that the anti-rheumatic factor was probably a substance which the body produced normally at all times but poured into the bloodstream in greater quantities during jaundice and pregnancy. This suggested that the adrenal glands might be the source, for it was already known that under other conditions of stress, such as those imposed by anesthesia, surgical operations and certain bacterial invasions, these glands rapidly increase their secretions.

If it was indeed true that jaundice and pregnancy stimulated the adrenals to secret a hormone that neutralized rheumatism, then the injection of the hormone into arthritic patients ought to have a similar effect. Hench and Kendall had often discussed this hypothesis. They had actually made a trial in a few rheumatoid volunteers with a cortical extract, but this mixed extract produced no conclusive results. Then came the great day when cortisone was available. Dr. Hench selected for his initial test a 29-year-old woman who had had severe rheumatoid arthritis for more than four years. She had undergone many different treatments with no success. Destructive changes in her right hip and in other joints had made them stiff, swollen, tender and painful when moved; her symptoms had progressively increased in intensity, and by the morning of September 21, 1948, when the cortisone test was undertaken, she was hardly able to get out of bed.

The treatment was begun with the daily injection of 100 milligrams of cortisone. No change appeared in the patient's condition the first day or the second, but when she awoke on the third day she moved in bed with ease. By the fourth day stiffness was gone, appetite was good, life was a joy instead of a painful endurance test. Walking became so easy that on the eighth day the patient went downtown on a three-hour shopping tour, and returned with no feeling of stiffness or soreness. Thereafter the daily dose was reduced to 50 milligrams for 4 days and then to 25 for 10 days. These amounts soon demonstrated their insufficiency, for rheumatic symptoms reappeared and it was necessary to return to the larger dose.

Following this trial, Hench and his associates, Charles H. Slocumb and Howard F. Polley, administered cortisone to 13 other arthritic patients. In each instance improvement began within a few days, continued as long as the full dose was given, and was followed by relapse as soon as the dosage was reduced or discontinued. Thus the evidence was unanimous. There could be no doubt that cortisone had the property of opposing rheumatism. And this suggested an explanation for the curious ups and downs that Hench had observed in pregnancy and jaundice: apparently the extra burdens imposed on the body somehow increased the supply or utilization of cortisone.

But what was the mechanism of this stimulation? How did the adrenals know that the body was overburdened and needed extra cortisone to reinforce its resistance? Such questions to an endocrinologist pointed a finger in one direction, toward the master gland: the pituitary.

The very position of the pituitary gland marks it as an organ of superior importance. It is suspended within a bony cavity beneath the brain, and not only enjoys the protection of the surrounding skull but the advantage of close association with the central organ of the nervous system.

The pituitary is even smaller than the adrenal body, being about the size of a grain of corn. It consists of two distinct parts: a bulbous front section known as the anterior lobe, and a smaller rear, the posterior lobe. The anterior lobe is the executive office that controls the internal secretions of many other glands, including the thyroid, gonads and adrenals. It exercises this control by means of specific hormones which it releases into the circulation. The pituitary messenger to the adrenal cortex is a substance known as adreno-corticotropic hormone, or, more simply, ACTH. It is ACTH that commands the cortex to release cortisone. Presumably, therefore, pregnancy and jaundice suppress arthritis through the pituitary. By some means these states of the body cause the pituitary to secrete ACTH, the ACTH in turn causes the adrenal cortex to secrete cortisone, and the cortisone counteracts the rheumatism.

This sequence of cause and effect suggested an alternative method

The endocrine glands all secrete their hormones into the common pool of the blood and the lymph. They are thus able to work delicately orchestrated effects upon one another. An example that has special importance in the topics with which this section of the book is concerned is the interaction of the adrenal and pituitary glands.

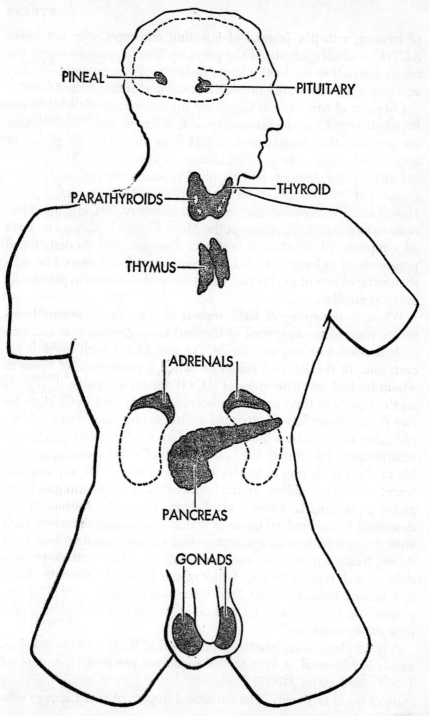

PINEAL

PITUITARY

PARATHYROIDS

THYROID

THYMUS

ADRENALS

PANCREAS

GONADS

STRESS

of treating arthritis. Instead of injecting cortisone, why not inject
ACTH? A small quantity of the pituitary hormone might excite the
adrenal glands to send out a larger quantity of the cortical hormone,
and thus the body would provide its own anti-rheumatism factor.

Chemists of Armour and Company, applying methods worked out
by Choh Hao Li at the University of California, had been working
for years on the extraction of ACTH from the pituitary glands of
hogs slaughtered in the packing houses. In 1946 John R. Mote, medical director for Armour, had supplied research workers with a few
grams, but it had not been tested on arthritis. On February 8, 1949,
Hench and his associates began to administer ACTH in daily injections to two arthritic patients at the Mayo Clinic. Within a few days
all symptoms of the disease began to diminish, and this continued
progressively as long as ACTH was given at its full dosage. The anti-
rheumatic effects of ACTH paralleled those of cortisone in practically
every particular.

When, in the spring of 1949, reports of Hench's pioneering treatments were made to groups of medical men, Armour was immediately flooded with requests for ACTH, and Merck with requests for
cortisone. In the fall of 1949 Mote called a conference of those to
whom he had given supplies of ACTH for an exchange of reports
on the results of their use of the hormone. They met in Chicago for
two days. "Never have I attended such a conference," reported one
physician on his return home. "It was like a religious meeting, with
men popping up all over the house to tell of some seeming miracle.
No medical gathering in history ever heard reports of so many different diseases yielding to treatment with a single drug." Acute
asthma, pneumonia, chronic alcoholism, rheumatic fever—diseases
described in the medical books as widely contrasting disorders, each
with its own pattern of symptoms—had all been mastered, at least
during treatment, by daily injections of ACTH. After listening to two
days of such reports, Walter Bauer of the Harvard University Medical School remarked that "the astonishing ability of ACTH apparently to turn diseases off and on at will marks the opening of a
new era in medicine."

A good place to explore the background of these exciting developments is Montreal. A visit to the endocrine research laboratory of
J. S. L. Browne at McGill University found him in much the same
state of mind as Bauer. "The emotional impact of that Chicago con-

ference was terrific," he said. "As disease after disease was reported on, I sat enthralled, feeling that we were witnessing the beginning of a revolution." The revolution, if revolution it be, was foreshadowed in certain research results that have been accumulating in Montreal over the last dozen years.

Dr. Browne is an authority on the adrenal glands. A colleague in this same field is Hans Selye, who carried on research on the adrenals at McGill from 1932 to 1945 and then transferred to the neighboring University of Montreal, where he now heads the Institute of Experimental Medicine and Surgery.

It was at McGill that Selye and his group began their now-famous studies of how the body reacts to damage, of which more will be said later in this chapter. The main point of their findings is that almost any kind of injury or stress—exposure to cold, burns, fractures, infection, poison, terror or other emotional trauma—calls forth the same general response: a pattern of defensive reactions called the "general adaptation syndrome." Working with animals, Selye found that the adrenal cortex plays a critical role in these defenses. When it is defective, the defenses are enfeebled and the animal passes rapidly to the stage of exhaustion. When the adrenals are entirely removed, resistance to stress practically disappears and death comes quickly. Selye later showed that if the adrenal response to stress goes wrong, this response itself can lead to diseases, such as arthritis and high blood pressure. Two hormones from the cortex are responsible for these abnormal effects. But these hormones can be counteracted by administering cortisone, which also comes from the adrenal cortex. Thus the adrenal cortex can both cause and cure the same disease.

Selye went on to demonstrate that stress or injury always causes the pituitary to produce more ACTH, and that this hormone plays a master part in the body's defense mechanism.

"We were greatly stimulated by these discoveries," related Browne, "and took up the general adaptation syndrome as a major subject for research in human beings as well as in animals. By 1939 Selye and Victor Schenker had devised a method of measuring the amount of cortical hormones in a solution; using this, Paul Weil found that patients suffering from Cushing's Disease had high quantities of the hormones in their urine. Cushing's Disease is a disorder in which the adrenal cortex is chronically overactive; therefore this result was to be expected. Next it was shown that the cortical hormones were in excess quantity in the urine of patients suffering from burns, infections and surgical operations. These studies demonstrated that Selye's theory, originally derived from studies of animals, was also true of man, and we were able to transfer findings from lower animals to the human."

In 1942 Eleanor Venning, an associate of Browne, devised a more sensitive method of measuring cortical hormones and applied it to pregnant women. She found that the cortical output progressively rose during gestation, and in the eighth month was equal to that of a victim of Cushing's Disease. Dr. Venning also found that in the first three to five days of a newborn infant's life the infant has an extremely low supply of these hormones. It had been known before that the adrenal gland is not fully developed in unborn babies but matures within a few days after birth. Does this account for the poor resistance of newborn infants? J. A. F. Stevenson, a medical student on duty in the Montreal Foundling Hospital, had occasion to test this theory. A girl infant in the hospital who had been born prematurely at seven months and weighed only three pounds became desperately ill of pneumonia at the age of five days. Even in oxygen she was blue with cyanosis. The death rate from this combination of prematurity and pneumonia is close to 100 per cent. Stevenson asked if it would be safe to administer an extract of cortical hormones to the dying baby, and Browne told him to give 50 cubic centimeters per day—a dose considered to be enormous at that time. Half an hour after the first injection the baby ceased to be blue, and within five days she was well. "An infant, especially one born at seven months, does not have adequate adrenal protection against the stresses of the outside world," observed Browne.

Studies of emotional stress gave additional evidence of the adrenal cortex's function as the body's defender and repairer. Dr. Venning, examining a woman directly after an outburst of hysteria, found that her cortical hormones, which had normally been 40 to 60 units, rose within 24 hours to 300 units. The cortical hormones of another patient shot up to 250 units upon her receipt of news that her sister was dangerously ill. "These measurements show that mental or emotional upset is just as truly an injury to the body as a bone fracture, a burn or a bacterial infection," remarked Browne.

Since the adrenal cortex acts only under orders from the pituitary, the complete picture of how the body responds to damage awaits a better understanding of the relationships of the master gland. We know that when ACTH is administered to subjects in whom the adrenals themselves are atrophied or missing, the hormone has no effect. In short, ACTH acts only through the adrenal cortex. For patients with defective adrenals, only cortisone will do. But if the adrenals are normal, either cortisone or ACTH will be effective. In either instance, the effect is to raise the level of cortisone in the bloodstream and thus increase the body's resistance to stresses of all kinds.

Some kinds of stress require more cortisone than others. For example, the amount of cortisone called into circulation by pneumonia is sufficient to overcome the damage imposed by rheumatism, but not necessarily sufficient to overcome that of pneumonia. However, as was shown at the Chicago conference, it is possible by means of ACTH injections to call sufficient reserves of cortisone into the circulation to overcome even pneumonia. Maxwell Finland of the Boston City Hospital reported the case of a boy afflicted with lobar pneumonia. Twenty-four hours after administration of ACTH all the pneumonia symptoms had disappeared, and the boy felt fine even though large numbers of pneumococci appeared in his blood. With the extra molecules of ACTH-induced cortisone circulating through the boy's system, it seemingly paid no attention to the bacteria.

Smith Freeman of Northwestern University Medical School reported similar results in a case of tuberculosis. Seventy-two hours after treatment with ACTH began, the telltale sedimentation rate was down, fever was down, appetite was up. With daily injections the patient's condition steadily improved. All signs of tuberculosis disappeared except one—his sputum still swarmed with bacilli. This meant, of course, that he was still a source of infection to others, and more dangerous because the familiar signs of the disease were missing. On the 21st day administration of the hormone was stopped. At once, as if the lid had been lifted from a cauldron, the symptoms erupted again.

Even cancer, in some forms, seems to yield temporarily to the influence of ACTH. O. H. Pearson and a group at the Memorial Hospital reported to the New York Academy of Medicine in January that five patients with leukemia showed improvement within a few days after daily injections began.

"The picture of disease which these recent developments suggest is that of an iceberg," said Browne. "Seven-eighths of an iceberg is submerged, and so is the greater part of the processes of a disease. This invisible area represents the body's basic response to stress of all kinds. It is described in Selye's general adaptation syndrome. But just as certain parts of an iceberg protrude above the water, so certain specialized responses of the body become visible as manifestations of specific stresses. For example, the symptoms of tuberculosis are the manifestations of the injury inflicted by the tubercle bacilli. But underlying them is the general response of the body to damage, any damage, and it has just as great a significance as the special response to the bacilli. Without both the special and the general responses the disease does not exist.

"Each disease is made up of this special pattern outcropping above the general response of the body. Cortisone melts the iceberg so that the symptoms fall below the surface. But if you stop administering the hormones, the iceberg freezes again, and then the tuberculous fever returns, the arthritic stiffness and pain reappear, the asthmatic wheezes and gasps recur, and all other symptoms become visible.

"This philosophical point of view greatly alters our concept of disease," concluded Dr. Browne. "It presents the picture of a basic pathological process at work which when it mounts to a certain magnitude is the disease. And this idea, I may add, is completely at variance with the older views of scientific medicine. It is at variance with the idea of compartmentalized disease, which is the central dogma of modern medical practice. Medical men who recognize the revolutionary and shattering nature of these developments realize that a great adjustment in our thinking has to be made. Here is the pool of Bethesda."

The Bible relates that the pool of Bethesda stood by the sheep gate to Jerusalem, and hundreds of the sick, the crippled, the withered of body and mind waited at its edge. They waited for the moment of healing. According to tradition, an angel descended into the pool at certain seasons and troubled the water. Whoever then first stepped in was healed, no matter what his affliction.

The idea that there exists a universal remedy which is sovereign over all diseases has persisted through the centuries. The medieval search for the elixir of life has been succeeded in more recent times by the familiar examples of homeopathy, osteopathy and chiropractic, each with its one cause and one cure. Scientific medicine has consistently frowned on all unitary theories of the healing art. But the demonstration of what hormones can do brings this whole subject under a new scrutiny. To be sure, there are many blanks in the hormone picture, many questions yet to be answered, many byways to be explored. For example, ACTH may stop acute attacks of gout; but when administered in the quiet period between attacks it may also bring on an upflare of the disease. Like a valve, the hormone seems to work in either direction, depending on which side exerts the greater pressure.

Nor can one overlook the various side effects of the use of these hormones: heightened blood pressure, severe headaches, edema, excessive hair growth, skin eruptions and occasional confused mental states which some patients have had to endure at the cost of being freed from rheumatism. A 10-year-old child who had severe arthritis was quickly relieved by the hormones, but immediately developed diabetes; when the cortisone treatment was stopped, the diabetes disappeared but the arthritis reappeared. At last accounts her physicians were trying to moderate the dosage to a level that would keep her arthritis mild without raising the diabetes to an acute stage. Apparently until science learns more about the treatments, this child will have to endure some symptoms of both diseases.

A fuller understanding is needed of the other hormones produced by the adrenal cortex, for cortisone is only one among many manufactured by this prolific tissue. Some of these other hormones are remarkably similar in structure to cortisone, and they also produce or suppress disease symptoms.

Despite the fragmentary nature of our knowledge of how cortisone works, it is clear that the pages of history have turned a new chapter in man's long search for the mechanism of disease. Thus far we have been able to read only a few disconnected sentences of that new chapter. But they are so amazing in what they tell and so revolutionary in what they imply that the medical world today is watching avidly, one might almost say breathlessly, for the next development.

II. THE PITUITARY

by Choh Hao Li

THE VERY NAME of the pituitary is expressive of the misunderstanding that has shrouded its career. This master gland was discovered by Andreas Vesalius, the 16th-century founder of modern anatomy, in his dissections of the human body. Because of its position just above the nasal passages, Vesalius deduced that its function was to secrete phlegm into the nose, and he therefore named it the pituitary from the Latin word for phlegm.

The beginning of our modern understanding of the pituitary was made in 1886 by the great French neurologist Pierre Marie. Marie observed that people with acromegaly—enlargement of the face, hands and feet due to a resumption of growth after maturity—had always suffered destruction of the pituitary by tumor tissue. He erroneously inferred that the function of the pituitary was to inhibit growth, and that the resumption of growth was caused by the removal of this metabolic brake. Despite his error in interpretation, Marie's observation of a causal relationship between acromegaly and destruction of the pituitary served to put investigators on the right track.

They soon recognized that other metabolic disasters were associated with destruction or damage to the pituitary. Some of these diseases reflect hyperfunction (overactivity) and others hypofunction (underactivity) of the pituitary. True acromegaly occurs after the epiphyseal cartilages, which separate the bones in the growing child, have closed up upon maturity. At this stage any abnormal resumption of growth, having no outlet in the cartilage, takes place in the acra (extremities), including the hands, feet, chin, nose and the soft tissue such as lips and tongue. Another disorder caused by overactivity of the pituitary gland is pituitary gigantism. In this case—not the same as acromegaly—the whole body grows to giant size, with all the parts in normal proportion to one another. It results from a continuous growth of the individual, without the closing of the bone-joint structures, beyond the normal age of maturity. A third major expression of hyperfunction is Cushing's Disease, a generalized disturbance of the pituitary-adrenal system.

Underactivity of the pituitary is best exemplified by Simmonds'

Disease, a disorder that sometimes follows damage to the pituitary by anemia during pregnancy and occurs four times as often in women as in men. This classical syndrome effectively reveals the boundaries of the endocrine dominion ruled by the pituitary. Simmonds' Disease is precipitated by total or near-total atrophy or destruction of the pituitary. The result is an endocrine disaster unmatched by any other disorder of internal secretions. Life gradually comes to a halt. There is a loss of body hair and teeth; genital atrophy; growing muscle weakness; lowering of pulse rate, blood pressure, body temperature and basal metabolic rate. The victim becomes hypersensitive to heat or cold. The hungers of the body, including sexual desire and even the craving for food and drink, are lost. The individual has the appearance of accentuated senility. Toward the end the totally wasted patient appears to exist in a limbo, dedicated to death and indifferent to life.

Less ubiquitous but nonetheless disastrous are two other major diseases of hypofunction linked with pituitary destruction: pituitary dwarfism or infantilism, and Frölich's Syndrome, a eunuchoidlike condition.

The discovery that these endocrine catastrophes were related to failure of the pituitary alerted biologists to the realization that the pituitary must be the source of substances of critical importance to the normal functioning of the human body. To find out precisely what the pituitary's responsibilities were, biologists began to investigate the effects of removal of the gland in experimental animals. Harvey Cushing, S. J. Crowe and J. Homans performed such operations on dogs in 1910. But surgical removal of the pituitary, called hypophysectomy, is an extraordinarily delicate operation. The operation itself was often fatal; it was not easy to keep an animal alive after it was deprived of its pituitary, and even when the animal survived the experimenter could not be sure that its symptoms were not caused by brain damage rather than by removal of the gland.

In 1916 Philip E. Smith, then at the University of California (he is now professor of anatomy at Columbia University), succeeded in removing the pituitary satisfactorily from tadpoles, and by 1926 he had developed a successful operation for the rat. Smith's classical experiments, and those of others, demonstrated that loss of the pituitary in growing or mature rats was followed by sweeping changes: cessation of general body and skeletal growth; a decrease in the size of the liver, spleen and kidneys; atrophy of the gonads, thyroid and adrenal cortex and cessation of estrus in the female; atrophy of the mammary glands and suppression of milk production; lowering of the metabolic rate and disturbance of the carbohydrate metabolism; loss of the ability to adapt. Young hypophysectomized rats remained sexually and otherwise immature, and in these animals life could be maintained only with difficulty.

But it was still not conclusively proved that these changes could be attributed to the removal of the pituitary alone. To prove it beyond doubt, it was necessary to show that the animals' lost functions could be restored by providing a substitute for the excised organ. This proof was not long in coming. In 1930 Smith implanted pituitary tissue under the skin of hypophysectomized rats: the animals thereupon resumed their growth and recovered their other lost functions. Meanwhile other investigators began studies to determine the specific roles of the various parts of the pituitary—the anterior, intermediate and posterior lobes. They found that all of the major body functions described above were controlled by the secretions of the anterior lobe.

By this time it was clear that the substances responsible for the control exercised by the pituitary were hormones. What were the hormones of the anterior pituitary, and what was their relationship to the already familiar hormones of the adrenals, the thyroids and the sexual organs?

For a number of years Herbert M. Evans and Joseph A. Long of the University of California had been investigating the effects of cattle pituitary extracts on rats. In 1921 they had performed an historic experiment which demonstrated that injection of such an extract could cause normal adult female rats to resume growth. Rats given the extract for a long period grew to abnormal size. These investigators concluded that the pituitary might contain a growth-promoting factor. After Smith's successful experiments in reawakening growth in hypophysectomized animals by means of implants of pituitary tissue, Evans and Long found that they could also achieve this result by injecting their growth-promoting extract.

This work stimulated efforts to find other pituitary factors. Smith found evidence for four, two of them involved in the sexual cycle: 1) the follicle-stimulating hormone (FSH); 2) the interstitial cell-stimulating hormone (ICSH); 3) the adrenal-stimulating hormones

(ACTH); and 4) the thyroid-stimulating hormone. P. Stricker and F. Grueter of the University of Paris proposed the existence of a lactogenic, or mammary gland-stimulating hormone. Thus a total of six hormones from the pituitary was postulated. Eventually, by methods developed in the laboratory of the writer and in other laboratories, five of the six hormones were isolated in pure form, and their chemical nature was studied. The thyroid-stimulating hormone, though not yet isolated in a pure form, is available in extracts for experimentation.

The demonstration of the stimulating action of pituitary hormones added a radically new concept to the study of internal secretions. It became clear that with one exception the known hormones of the pituitary do not participate directly in physiological reactions. The pituitary acts like a generator or primer. Its secretions are called tropic hormones, meaning that they "turn," *i.e.*, change, something else. Each tropic hormone has a specific target organ. For example, the target of the adrenocorticotropic hormone ACTH is the adrenal glands; that of the thyrotropic hormone is the thyroid. The tropic hormone spurs the target organ to action. Awakened by the activating hormone, the target organ produces a second hormone which carries out specialized tasks. The one non-tropic exception among the pituitary secretions is the growth hormone, which appears to participate directly in general growth processes.

Painstaking research has determined the functions of each pituitary hormone and shown how disturbances in its production cause disease. The interrelationships of the various hormones are, of course, subtle and complex. Yet the general responsibility of each can be described. The six hormones may be divided into two groups: 1) the gonadotropic hormones, which have to do with the sex cycle, and 2) the metabolic hormones, which tend the nutrition and regulate the chemistry of the body.

The gonadotropic hormones are FSH, ICSH and the lactogenic hormone. In collaboration with their target organs these hormones nurture the reproductive processes.

The metabolic hormones are the thyrotropic, adrenocorticotropic and growth hormones. The thyrotropic hormone, of course, stimulates the thyroid gland to secrete and produce the thyroid hormone thyroxin. The metabolic well-being of the body pivots on the precarious chemical balance among these hormones.

We are concerned here mainly with the growth hormone and ACTH. In rats the injection of the growth hormone produces remarkable results. For example, the pituitaries were removed from a group of young rats. Their growth stopped. Two weeks later the experimenters began to inject these dwarfs periodically with the growth hormone. After a year of this treatment, the rats had grown to about six times their original weight. Although they were giants in over-all appearance, their endocrine glands—the adrenals, the thyroid and the sex glands—were immature. On the other hand, all the bony structures of these infantile giants showed intensive development, and their soft tissues had a pattern of biochemical changes similar to those considered to be indicative of true growth.

For reasons which are not clear, it has not yet been possible to stimulate growth in human subjects, as it has been in rats, with injections of growth hormone. Yet it seems apparent that some day experimental conditions will be found for using this hormone to treat human dwarfism and to combat diseases characterized by wasting of the body's reserves of proteins and other organic compounds.

It is now known that the biological activity of the growth hormone can be modified by simultaneous injections of other hormones. For instance, daily injections of a small dose of thyroxin reinforce the effect of growth hormone in hypophysectomized rats. On the other hand, ACTH counteracts the growth hormone in such rats; it even inhibits growth in normal growing rats.

The mechanisms whereby these hormones influence the processes of growth are not known, but it seems clear that they must act by affecting the metabolism of proteins, since true growth is generally interpreted as the accumulation of proteins. The growth hormone and ACTH also play a part in the metabolism of fats and carbohydrates. Injections of high dosages of ACTH into normal rats produce an excess of sugar in the blood and urine—the familiar symptoms of diabetes. It has repeatedly been demonstrated that ACTH produces diabetic symptoms in man also.

Plainly all this has great implications for medicine. Hormone

The master gland of the body is the anterior lobe of pituitary (the more darkly stippled part in the cross section, opposite); it produces hormones that act directly on the body and that stimulate secretion of hormones by other glands.

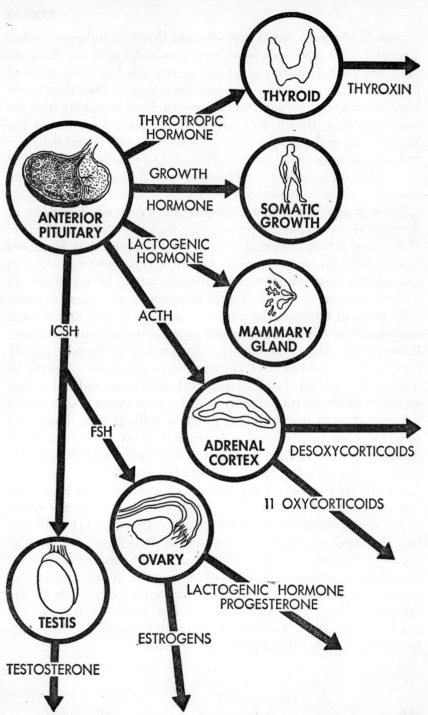

THYROID → THYROXIN

THYROTROPIC
HORMONE

GROWTH
HORMONE → SOMATIC
GROWTH

ANTERIOR
PITUITARY

LACTOGENIC
HORMONE

ACTH

MAMMARY
GLAND

ICSH

ADRENAL
CORTEX → DESOXYCORTICOIDS

11 OXYCORTICOIDS

FSH

OVARY

TESTIS

LACTOGENIC HORMONE
PROGESTERONE

TESTOSTERONE

ESTROGENS

therapy is, of course, an old story—insulin, thyroxin, androgens, estrogens and other hormones have been in wide clinical use for some time. But until 1949 almost no therapeutic use had been made of the hormones from the master gland—the pituitary. Since then a very large amount of exploratory work has been done, beginning with the success of P. S. Hench, E. C. Kendall and their associates at the Mayo Clinic in the treatment of arthritis with ACTH and cortisone and continuing with the investigation of their use for a multitude of other ailments, including some types of mental illness and rheumatic fever.

There is a good reason for pressing the expansion of sources of the pituitary hormones and for pursuing their use in clinical practice. This reason is that treatment with some of the secondary hormones is attended by certain dangers. Animal experiments indicate that administration of target-organ hormones often brings about an adverse reaction in the pituitary. For example, after rats are given large doses of estrogen, the pituitary increases in size, and there is evidence that the enlarged gland manufactures more lactogenic hormone. In other experiments, estrogen has produced pituitary tumors. Large injections of cortisone may cause the pituitary to produce less ACTH. In other words, there is some indication that an excess of the target hormone in the system causes a shift in the dynamic state of the producing pituitary cells. It is as though the pituitary must be active; if it is relieved of its normal duties, it begins to free-lance. The use of the primary hormones such as ACTH would tend to keep the system in better balance.

But ACTH is scarce; the pituitaries of half a million hogs yield but one pound of the hormone. Recent experiments by the writer and others indicate that the biological activity of ACTH resides in peptide fragments obtained when the parent molecule is split. The peptide fragments proved highly active in the treatment of rheumatoid arthritic patients. A number of laboratories in this country and abroad are now actively engaged in the search for methods for the isolation of the ACTH peptide and its final synthesis. It is not too much to hope that a synthetic process may be found for the manufacture of this and related pituitary hormones.

III. THE ALARM REACTION

by P. C. Constantinides and Niall Carey

IN BIOLOGY and medicine it is becoming increasingly difficult to see the forest for the trees. The specialization of modern research leads into ever-narrowing paths. One man spends an entire lifetime studying a single hormone, another an enzyme, another the circulation of the kidneys. Year by year the data pile up; yet in some respects this vast accumulation of facts is leading us no nearer to an understanding of the living organism as a whole. Obviously we have reached a point where it is highly desirable to widen the view, to conduct researches in breadth as well as in depth.

At the Institute of Experimental Medicine and Surgery, the University of Montreal, Dr. Hans Selye and his team of biologists have been pursuing such an investigation for more than a decade, with stimulating results. They have been studying the generalized reactions of a whole animal to the stresses produced by its environment. A living organism consists of salts, enzymes, hormones, energy and a host of other elements, each of which may react in a specific way to some assault from outside; but the response of the organism as a whole is more than the sum of all these reactions. Life, and even death, is a chain reaction, and it is this linked process that Selye's group has been examining.

The particular focus of the investigation was the adaptation of animals to various types of severe or prolonged injury that affect large sections of the body. From this work came the discovery that the animal organism possesses a general defense mechanism which it automatically mobilizes against any damage, whatever the cause. The principal agent of the mechanism is the endocrine system. As the officers of the defense, the hormones call upon various organs of the body for extraordinary efforts. If the stress becomes too great, the animal is destroyed by its own defenses, for ultimately the strain is conveyed to the heart and the circulatory system. Thus this research led directly to a study of high blood pressure, hardening of the arteries and heart failure—the principal causes of death among human beings today.

Selye started these studies some 16 years ago as the result of unexpected findings during some experiments on rats. He was investigating their specific responses to various drugs, poisons and gland extracts. He injected heavy doses, sufficient to kill the rats in a day or two, and made a careful autopsy of every animal. He was surprised to find that every substance he injected produced exactly the same result in three of the animals' organs: 1) the adrenal glands swelled to twice their usual size and changed in color from yellow to brown; 2) the thymus withered away; and 3) the stomach lining was spotted with bleeding ulcers. The puzzling fact was that these reactions were caused by such widely diverse substances as atropine, strychnine, formalin, crude pituitary extracts—all entirely different in chemical structure and mechanism of action. The only factor that the many agents had in common was that all were injected in quantities dangerous to life. Selye reasoned that the responses he observed must represent a nonspecific reaction to general damage as such, regardless of the specific agent that caused the damage.

If this were true, other types of acute stress ought to provoke the same response. Selye tested this assumption by subjecting animals for some hours to cold, to excessive muscular exercise, to fasting, to emotional excitement and to numerous other kinds of injuries. Sure enough, all these nonchemical types of stress elicited in the animals the same unmistakable "alarm reaction" (AR), as he called it.

It was soon discovered that certain characteristic chemical changes in the tissues and body fluids always accompanied the AR. Among the first to be studied were the sugar and the chloride ions of the blood. During the first few hours of exposure to stress, it developed, both of these fall to subnormal concentrations. After a few more hours, they rise above normal values. The two periods are known respectively as the "shock phase" and the "countershock phase" of the AR.

We now know many of the events that take place in the AR. The messengers that marshal the alarm reaction are hormones from the cortex of the adrenal glands. During the first hours of the AR, the hormones are rapidly discharged from the adrenal cortex into the bloodstream and race to the tissues of the body. There they perform their various functions, of which two are definitely known: 1) They keep the composition of the fluid cell environment constant, mainly by retaining salts, particularly sodium, in the solution be-

396

tween the cells. The most important salt-retaining hormone is desoxy-corticosterone, more commonly known as DCA. 2) They promptly build up sugar, a ready energy donor, from other materials, particularly proteins. The best known sugar-forming hormone is cortisone.

How important these two functions are can easily be judged from the fact that animals whose adrenals are removed die with their blood almost drained of salt and sugar. On the other hand, the injection of salt-retaining and sugar-forming adrenal hormones can prolong the life of such animals considerably; it also raises their resistance to otherwise fatal stress.

Yet the adrenal itself does not act independently. It is merely an executive of higher coordinating centers, from which it receives orders as to when to act, how much to act and what hormones to discharge. The adrenal cortex, like almost all other endocrine glands, is under the direct command of the anterior part of the pituitary gland—the "leader of the endocrine orchestra." If the pituitary is removed, the adrenal cortex shrinks and becomes inactive. It can regain its normal size and function only if a new pituitary gland is transplanted into the animal or if pituitary extracts (*i.e.*, hormones) are injected.

A long series of experiments has clearly outlined the AR mechanism. Acute stress acts on the anterior pituitary through some unknown pathway; the pituitary replies by mobilizing the adrenals, which discharge their hormones, which in turn effect most of the other changes.

After the alarm reaction was established, the next major step in the experiments was to investigate the animals' long-range responses to stress. What would happen if the organism were exposed to continuous, prolonged stress of an intensity below the lethal level, a stress strong enough to strain the defenses almost to the limit, yet not sufficiently overwhelming to silence all defense at once?

Animals were subjected to sublethal daily stress with the same agents for several weeks instead of a few days. During the first few days the organism responded with the usual AR. It showed the typical organic and chemical changes; growth and sex functions ceased, and all the signs of an intense tissue breakdown were present.

As the stress continued unabated, the animals that survived the AR began to recover. The adrenals started to refill their empty stores with lipids and reverted to normal size; the thymus began to regain its mass, and such substances as sugar and chlorides in the blood rose to normal or even higher levels. At the height of that state the organism had in some way accomplished an adaptation to the continuing stress. Its organs and their functions were apparently returning to normal. In some instances it was difficult to distinguish such animals from control animals not under stress. This stage, lasting from a few weeks to a month or more, was called the "stage of resistance."

It should be noted, however, that resistance increased only against the one type of stress employed from the beginning. If, in the middle of this recovery period, the stress against which adaptation developed was replaced by a different one, the animals succumbed immediately. Quantitative experiments with graded amounts of stress showed that while the animal's specific resistance to the initial agent increased, its resistance to any other stress decreased.

The adaptation to the original stress was not permanent. As the strain continued after the recovery period, the animals became progressively weaker; the adrenals enlarged again and discharged their lipids; the thymus lost the mass it had recovered; sugar and chlorides fell to dangerous levels; after a few weeks all defenses collapsed and life ceased. This last "stage of exhaustion" was similar to the initial alarm reaction. The end was like the beginning.

Thus the struggle of life against stress was found to consist of three successive acts, all aiming at a balance which was not quite attained during the AR, was achieved during the stage of resistance but was lost again during the stage of exhaustion. Evidently the war of the organism against damage was waged at the expense of a finite capital of "adaptation energy." The whole battle was named the "general adaptation syndrome" (GAS).

The establishment of the GAS opened a number of fascinating fundamental problems. Life as a whole could be regarded as a GAS that ends when adaptation energy runs out. More immediately, the phenomenon suggested some studies of great medical interest.

Some types of stress are so severe that an animal can develop resistance for only a very short period; others permit a prolonged adaptation before the animal becomes exhausted. Animals can adapt themselves to cold, for example, for periods as long as two or three months. And such animals presented quite unexpected changes. The arteries were enormously thickened and their bore was narrowed almost to obliteration in numerous districts of the body; the heart was abnormally large and filled with nodules very like those appearing in human rheumatic disease; the kidneys were largely destroyed through hardening and closing of their vessels—as in human nephrosclerosis—and the blood pressure rose more than 50 per cent. In other words, long-lasting stress had produced in these animals hypertension and cardiovascular disease.

This was a finding of the highest importance in experimental medicine. It suggested that these diseases might be caused by the pituitary-adrenal mechanism, perhaps by the excessive production of their hormones. If one could reproduce the changes found in the animals by loading the organism with large quantities of pituitary and adrenal hormones, this would indicate that at least degenerative diseases are the consequences of "over-adaptation," i.e., too vigorous a marshalling of the body's defenses.

The experiment was made. A number of animals were dosed with large amounts of these hormones. Extracts would not do for this purpose: one is seldom sure exactly how much hormone they contain or that they include everything produced by the gland in the natural state. Fortunately the previously mentioned adrenal hormone desoxycorticosterone, or DCA, was available in pure, crystalline form. Because chemists had not yet succeeded in synthesizing any pituitary hormones, it was decided to use the whole anterior lobe of this gland, powdered and suspended in water. In laboratory terms the product, "lyophilized anterior pituitary," is referred to as LAP. Continuous injections of large amounts of DCA or LAP were considered to be equivalent to the prolonged and excessive secretion of hormones by the adrenal or the pituitary, respectively.

The results were remarkable. In three weeks the animals that were injected with DCA developed severe hypertension and hardening of the kidneys. Those treated with LAP showed a strikingly similar picture, though after a somewhat longer interval.

In medical research one can never lose sight of the ultimate objective, namely, the cure of patients. The investigator first devotes every effort to reproduce a disease in animals, and when he has succeeded he turns about and tries to destroy that disease. In searching for ways to combat the diseases produced by too much hormone production, one of the most obvious targets would be to try to neutralize the hormonal excess, in other words, to find a chemical antidote. Logical as it seems at first sight, this is too complicated a task at present. In the first place, we do not yet know the chemical mechanism of the hormones' action. Secondly, we must not forget that the organism needs those hormones, even if by overproducing them it poisons itself with its own defense substances.

A more practicable approach was suggested by experience in the treatment of other endocrine diseases. Some of these diseases can be alleviated by control of the diet. A case in point is diabetes, in which the basic trouble is a hormone deficiency. In moderate cases diabetes can be completely controlled by a diet low in sugar.

It was conceivable that the experimental hypertension produced in animals by overdoses of hormones or by stress might flourish on some diets and be suppressed by others. The animals were therefore subjected to a great variety of diets, a process which had to be pursued by trial and error because there was little indication as to what diets might be helpful.

From the many tests, two facts emerged clearly. One was that experimental hypertension produced by DCA was markedly affected by salt in the diet. A high salt intake increased both the frequency and the intensity of the pathological changes caused by that hormone. Contrariwise, when the animal was fed a salt-free diet, it was immune to hypertension, even when considerable amounts of DCA were injected. The second finding was that hypertension caused by stress or LAP was not affected by salt at all but was influenced by protein in the diet. A low-protein diet afforded considerable protection to the animals, while a high-protein intake aggravated the damage.

Thus sodium favored the adrenal hormones, and proteins favored stress or the pituitary hormones in their injurious effect on blood vessels and blood pressure. The why and wherefore of these results is still unknown. It may be that DCA cannot act without the simultaneous presence of sodium. Perhaps the pituitary manufactures adrenal-stimulating hormones from food proteins. Research on these questions is now going on. One of the present objectives is to find out whether it is the total quantity of proteins that counts or a protein constituent, *i.e.*, an amino acid.

In any event, these experiments tend to strengthen the case for the widely held belief that some forms of human cardiovascular disease are due to hormonal derangements. Medical experience has taught doctors that patients with high blood pressure fare best on a low-sodium, low-protein diet. This is exactly what the animals needed to withstand the destruction of their blood vessels by prolonged stress or by hormones.

The research of Selye's group yielded another key fact—that in this whole adaptation process the kidneys are somehow deeply involved. They are early victims of damage in the resistance phase of the GAS or during the inundation of the body with pituitary and adrenal hormones. But they also seem to be something more than passive targets. A great deal of work since the turn of the century has shown that the kidney itself can become the active cause of the most malignant hypertension. There is considerable evidence now that under certain abnormal conditions parts of the kidney tissue may stop their normal function, which consists in filtering blood and producing urine, and start producing hormones that raise the blood pressure. In the rat, this was beautifully demonstrated by what is now known as the "endocrine kidney" of Selye. By a surgical operation that interferes with the blood supply of one kidney, the whole organ is transformed into an endocrine gland, and in a few days the blood pressure rises to fatal levels. It is a particular feature of the endocrine kidney technique that only one kidney is transformed into a gland while the other gets all the damage.

Correlation of the evidence derived from all the numerous experiments on the GAS has led Dr. Selye to formulate the following current hypothesis: Long-lasting stress provokes an excessive production of adrenal-stimulating hormone in the anterior pituitary; this forces the adrenal cortex to an intensive discharge of DCA-like hormones which, among other things, affect the kidney in such a way as to release hypertensive substances.

In a sense the research is only beginning. Its implications are tremendous. In the GAS we seem to see the merest outlines of a great biological chain reaction which can be set off by almost any stress and which may frequently lead to the suicide of the organism. Some of the links in this chain are still missing, but its essential structure has been amply confirmed. As a result, large-scale research in this field is now in progress in many laboratories.

Should further research prove that chronic stress can produce the same disorders in man as in animals, it would appear that the most frequent and fatal diseases of today are due to the "wear and tear" of modern life. One might question whether stress is peculiarly characteristic of our sheltered civilization, with all its comforts and amenities. Yet these very protections—modern labor-saving devices, clothing, heating—have rendered us all the more vulnerable and sensitive to the slightest stress. What was a mild stress to our forebears now frequently represents a minor crisis. Moreover, the frustrations and repressions arising from emotional conflicts in the modern world, economic and political insecurity, the drudgery associated with many modern occupations—all these represent stresses as formidable as the most severe physical injury. We live under a constant strain; we are losing our ability to relax; we seek fresh forms of physical or mental stimulation.

Thus it would not be surprising to find that much of our organic disease derives from psychological trauma, with the general adaptation syndrome as the bridge that links one to the other. If this be true, medicine may eventually find a cure for the consequences of stress; but prevention of the basic causes will remain a task that lies beyond its reach.

IV. SCHIZOPHRENIA AND STRESS

by Hudson Hoagland

NEARLY HALF of the hospital beds in the United States are occupied by patients suffering from mental illness, and about a third of these patients have a psychosis known as schizophrenia or dementia praecox. This is an all too common and serious form of insanity, affecting nearly one per cent of the population. Brilliant people often develop it and are lost to society.

A quotation from *The Biology of Schizophrenia,* by R. G. Hoskins, describes the behavior of these patients:

"The psychosis represents a bizarre mélange of behavioral normality and abnormality. The core disturbance, the so-called process schizophrenia, is perhaps still best expressed in [the German psychiatrist E.] Kraepelin's definition of the psychosis as 'a peculiar disorganization of the inward coherence of the psychic personality with predominating damage to the affective life and will.' In other words, the patient cannot think straight, feel straight, or will straight. His logic limps woefully. He often substitutes phantasy for reasoning and to him argument from analogy is singularly convincing. . . .

"It is these latter peculiarities that led [the German psychiatrist Eugen] Bleuler to ascribe primary importance to 'disorders of the association processes.' Memory and orientation are frequently well preserved. The patient often shows little disturbance of comprehension. Despite frequent appearances to the contrary, he is usually rather well aware of what goes on about him. The fundamental disorder of the thinking processes leads ultimately to such deviations from normality as hallucinations, delusions, poor judgment, incongruity of emotions—often with apparent neutrality or indifference—incoherence in train of thought and displacement of normal volitional responses by automatic or impulsive reactions."

While it is true that everyone probably has a threshold beyond which he breaks down under the batterings of life, it is not true that the mental breakdown need take the form of schizophrenia. And why some people develop this crippling psychosis under very little apparent stress and others do not under great stress is a challenging mystery.

It is an article of faith among most students of the biological sciences that human conduct, from the simplest reflex to the mental achievements of a Shakespeare, depends upon the dynamic functioning of bodily processes and particularly of those of the nervous system, which, of course, includes the brain. Just as physiological events determine mental processes, so we believe that mental processes can determine some physiological events.

In the following discussion we shall not be concerned with the very important techniques of the psychiatrists, who are primarily concerned with the treatment of mental disease at the psychological level. Rather we shall review a specific aspect of the physiology of the adrenal gland which appears to be relevant to understanding the nature of schizophrenia. This work has been carried out during recent years by a group at the Worcester Foundation for Experimental Biology working in collaboration with the research staff of the Worcester State Hospital. The Foundation group has consisted primarily of Gregory Pincus, Harry Freeman, Fred Elmadjian, Louise Romanoff, James Carlo. David Stone and the author.

Hans Selye's study of the "alarm reaction," discussed in detail earlier in this chapter, indicates that the adrenal cortex, stimulated by the pituitary's release of ACTH, plays a significant role when the organism is under stress. During the war Gregory Pincus and the author studied fatigue in aviators and found that individual differences in resistance to stress were correlated with secretion of hormones from the adrenal cortex. In subsequent studies we found that a wide variety of common workaday stresses produced enhanced activity of this gland, and later we decided to extend our studies to the large group of mental patients who have notably failed to meet the stresses of daily life. We asked ourselves whether these psychotic patients might show inadequacies in this very general stress-response system. Just as the engineer tests the strength of structural materials by stressing them and measuring the strains, it seemed reasonable to stress psychotic patients and normal people by the same standard procedures and compare their adrenal responses. A number of methods are available for measuring the functioning of the adrenal-cortex hormones in man. Because of their far-reaching effects on metabolic processes, the concentrations of various substances in the blood and urine indicate adrenal activity.

Since 1941 we have been studying the effects of various stresses on adrenal-cortex function among normal .men and women and

404

among mental patients. Our various studies have involved approximately 200 normal men and women and 100 mental patients, for the most part chronic male schizophrenics.

In our experiments we have used such clearly physiological stresses as exposure to heat or cold and the ingestion of large doses of sugar. All these stresses increase the activity of the adrenal cortex in normal people. We have measured adrenal function before and after the stress and fatigue resulting from prolonged operation of a pursuit meter, a flight-simulating device that has airplane-type controls and is used to test coordinating ability. We have studied activity before, during and after 152 training flights of 16 Army instructor pilots; similar tests were made of 56 flights of seven civilian test pilots. On the purely psychological side, we have also recorded adrenal-cortex function during interviews, during especially designed frustration tests, and during examinations given to college students in regular courses. All of these tests, with the exception of the last one and the actual airplane flights, have been administered both to schizophrenic patients and to normal control groups.

Among normal individuals we find that, within limits, the greater the stress the greater the hormonal output; and we have been able to correlate measurements of the degree of fatigue in the pursuit-meter test with adrenal-cortex secretions. Patients suffering from psychoneuroses (which for the purposes of this discussion may be regarded as less severe forms of mental breakdown) generally exhibited adrenal responses similar to those of normal persons, although often in somewhat exaggerated form.

The schizophrenic group showed a subnormal ability to respond to these tests with enhanced adrenal output as measured by certain of our urinary indexes, despite the fact that their normal secretion, as measured after 24 hours of rest, was little different from that of the general population. A schizophrenic does not have an underproductive adrenal cortex, as does the sufferer from Addison's Disease, but the organ is generally less responsive to stress and to situational demands.

One might think that the relative unresponsiveness was due to lack of interest in the tests and an other-worldly detachment from the experimental procedure, but this notion does not account for the facts. Both patients and normal controls were exposed to heat or cold at the same time and both groups sweated or shivered alike. The controls showed enhanced adrenal activity over that of the patients. Moreover, the purely internal stress of assimilating large amounts of sugar administered in our sugar-tolerance tests revealed significant differences in the adrenal-cortex responses between patient and control groups. In psychological tests, including performance on the pursuit meter, the patients were cooperative and their interest in the situation seemed to be as great or greater than that of the control subjects.

The basis of response depression among the patients was next investigated. Perhaps their adrenal-cortex secretions were normal but their body tissues were less reactive to the hormones. This possibility was excluded by injecting patients and controls with standard quantities of cortical extract. Both groups showed the same reactivity to the injected hormone.

The depression must occur then either because the brain fails to excite the pituitary to its normal ACTH discharge or because the ACTH inadequately stimulates the adrenal cortex. To investigate these alternatives, we injected patients and controls with 25-milligram doses of ACTH and tested our eight indexes before injection and at several standard intervals of time afterwards. The results were striking. The ACTH produced vigorous responses in all the controls; but most of the patients showed much less response. It was therefore reasonable to believe that the patients' stress defects resulted from the inability of ACTH from their own pituitaries to excite their adrenals adequately.

We next selected a group of schizophrenic patients who by our urinary indexes failed to react to 25 milligrams of ACTH and injected them with 75 to 100 milligrams. These extra-large doses produced some response. To assure ourselves that smaller doses had not failed because of possible inadequacies of the hospital diet, we fed a group of nonresponsive patients for two weeks on an ample protein and vitamin diet and again injected them with 25 milligrams of ACTH. But they were still as unresponsive as before.

The statement that schizophrenic patients do not give normal adrenal-cortex responses to stress or to ACTH is based on statistical group comparisons obtained by using specific measures of adrenal function. Several indexes commonly used show group differences, others do not. Ability to respond must be quantitatively defined in terms of the specific measures applied.

SCHIZOPHRENIA AND STRESS

What then do these findings mean in terms of schizophrenia? They certainly do not imply that adrenal stress-response depression is the one and only "cause" of the psychosis. It may be an important factor but it may also be a result of the disorder and not a cause. A psychosis is primarily a failure in interpersonal relations determined by a breakdown of higher mental processes which, in turn, depend upon the dynamic patterning and conduction of impulses in billions of nerve pathways in the brain. One of the most conspicuous indexes of stress depression among schizophrenic patients is provided by adrenal regulators of salt balance, and our data on potassium excretion have been particularly consistent in separating patients from controls. Potassium is of great importance in the generation and propagation of nerve messages which are waves of electrical action and constitute the physical basis for thought processes and behavior. Work from our laboratory has shown that the content of potassium in rat brain is altered following stress and that the adrenal cortex regulates brain potassium. Furthermore, neurophysiologists have demonstrated that important electrical properties of nerve depend on its potassium content. It is possible that faulty potassium metabolism in the face of the repeated stresses of daily life may be a factor in the development of a psychosis. This is only one of several physiological hypotheses implied by the findings.

It may be that chemical deficiencies of the kind we have been discussing, perhaps genetically determined, make some persons more vulnerable than others to the stresses of living. They may never become psychotic, especially if their lives present few problems, but under more severe environmental and personalized stresses their physiological defects may result in brain malfunction with consequent psychotic disturbances. This hypothesis warrants further investigation.*

* Since this article was published in 1949, further work with improved methods has indicated that there are certain qualitative and quantitative divergences from normal in the adrenal physiology of psychotic patients at rest as well as when stressed. Adrenal abnormalities are also found in psychotic patients other than those diagnosed as schizophrenic. No new therapy has as yet resulted from these findings and it is still uncertain whether the abnormalities are a result or a cause of the psychosis.

407

ANIMAL BEHAVIOR

Introduction

THERE IS no province of nature so full of pitfalls for the unwary explorer as the field of animal behavior. Adaptation for survival has yielded as many varieties of behavior as of species. The instances of good sense and foolishness, of charm and grim rapacity, of artless simplicity and high complexity to be found on all sides engage the susceptibilities of our imagination and invite quick jumps to conclusion.

The creatures assembled for examination here suggest the range of natural invention. The army ant is dumb compared to the honey bee, which may be said to have a language; it manages nonetheless to conduct a comparable order of social existence. An otherwise normally endowed spider cooperates in its own execution by a particular but ordinary sort of wasp. The male stickleback stages an elaborate ritual of courtship and then assumes complete care of the offspring. At the higher level of mammalian psychology, we encounter the drama of inner-conflict.

How can we explain what goes on inside the big and little heads of these and other creatures?

The pathetic fallacy, which ascribes purpose to the hive, offers not only the easiest explanation but, by all odds, the most engaging. Anthropomorphism is the stock and trade of the nature faker. But some of the most naïve examples are to be found in the literature of science.

Nor is this the only booby trap to be scouted by the animal psychologist. One that is cleverly camouflaged, especially to the pedant and the sophisticate, is the meaningless tautology. The researcher isolates a unit or an aspect of the animal's behavior. For this he invents a name, sometimes imaginative, often merely polysyllabic. Thereafter the behavior is explained by invocation of the word. Two of the more familiar examples of this blind alley or circular process are given by the common coins of "instinct" and "complex."

As might be expected, therefore, the science of animal behavior is rife with controversy. In truth, there are disagreements, which the careful reader may detect, among the authors represented in this section. In the privacy of their professional channels of communication, the conflicts come more plainly to the surface and sometimes develop measurable heat. Grown men spend their lives investigating the behavior of ants, bees, spiders, sticklebacks and neurotic cats because they are concerned, in the end, with questions of human behavior.

There is nothing yet in the findings on animal behavior to challenge the towering superiority of man's capacity to reason. Comparative psychology discovers wisdom in the differences, not the similarities, between human and animal behavior. Our little brothers organize societies and use tools, but their activities in every case can be reduced to sub-rational explanations. Animal behavior shows us that the same must be said for much of the behavior of man. The compass of reason is narrowed. But, by the same token, the cultivation and the extension of reason are shown to be essential to the survival of the species.

I. THE ARMY ANT

by T. C. Schneirla and Gerard Piel

*Wherever they pass, all the rest of the animal world
is thrown into a state of alarm. They stream along the
ground and climb to the summit of all the lower trees
searching every leaf to its apex. Where booty is plenti-
ful, they concentrate all their forces upon it, the dense
phalanx of shining and quickly moving bodies, as it
spreads over the surface, looking like a flood of dark-
red liquid. All soft-bodied and inactive insects fall an
easy prey to them, and they tear their victims in pieces
for facility in carriage. Then, gathering together again
in marching order, onward they move, the margins of
the phalanx spread out at times like a cloud of skirmishers from the flanks of an army.*

THAT IS HOW Henry Walter Bates, a Victoran naturalist, described the characteristic field maneuvers of a tribe of army
ants. His language is charged with martial metaphor, but
it presents with restraint a spectacle which other eyewit-
nesses have compared to the predatory expeditions of Genghis Khan
and Attila the Hun.

Army ants abound in the tropical rain forests of Hispanic America,
Africa and Asia. They are classified taxonomically into more than
200 species and distinguished as a group chiefly by their peculiar
mode of operation. Organized in colonies 100,000 to 150,000 strong,
they live off their environment by systematic plunder and pillage.
They are true nomads, having no fixed abode. Their nest is a seeth-
ing cylindrical cluster of themselves, ant hooked to ant, with queen
and brood sequestered in a labyrinth of corridors and chambers
within the ant mass. From these bivouacs they stream forth at dawn
in tightly organized columns and swarms to raid the surrounding
terrain. Their columns often advance as much as 35 meters an hour
and may finally reach out 300 meters or more in an unbroken stream.
For days at a time, they may keep their bivouacs fixed in a hollow
tree or some other equally protected shelter. Then, for a restless
period, they move on with every dusk. They swarm forth in a solemn,
plodding procession, each ant holding to its place in line, its forward-
directed antennae beating a hypnotic rhythm. At the rear come
throngs of larvae-carriers and, at the very last, the big, wingless

411

queen, buried under a melee of frenzied workers. Late at night they hang their new bivouac under a low branch or vine.

The army ant, observers are agreed, presents the most complex instance of organized mass behavior occurring regularly outside the homesite in any insect or, for that matter, in any subhuman animal. As such, it offers the student of animal psychology a subject rich in interest for itself. But it also provides an opportunity for original attack on some basic problems of psychology in general. The study here reported, covering the behavior of two of the Eciton species of army ants, was conducted by Schneirla over a 20-year period with extended field trips to the Biological Reservation on Barro Colorado Island in the Panama Canal Zone and to other ant haunts in Central America. In undertaking it, he had certain questions in mind. The central question, of course, was how such an essentially primitive creature as the ant manages such a highly organized and complex social existence. This bears on the more general consideration of organized group behavior as an adaptive device in natural selection. There was, finally, the neglected question of the nature of social organization. This is primarily a psychological problem because it concerns the contribution of individual behavior and relationships between individuals to the pattern of the group as a whole. It was expected that reliable data on these questions in the instance of the army ant might throw light on similar questions about human societies.

The ant commends itself to study by man. Measured by the dispassionate standard of survival, it stands as one of the most successful of nature's inventions. It is the most numerous of all land animals both in number of individuals and number of species (more than 3,500 at present count). It has occupied the whole surface of the globe between the margins of eternal frost.

The oldest of living families, the ant dates back more than 65 million years to the early Jurassic period. More significant, the societies of ants probably evolved to their present state of perfection no less than 50 million years ago. Man, by contrast, is a dubious experiment in evolution that has barely got under way.

Lord Avebury, a British myrmecologist, marveled at "the habits of ants, their large communities and elaborate habitations, their roadways, possession of domestic animals and, even, in some cases, of slaves!" He might have added that ants also cultivate agricultural

crops and carry parasols. It is the social institutions of ants, however, that engender the greatest astonishment. The sight of an army
ant bivouac put the British naturalist Thomas Belt in mind of Sir
Thomas More's *Utopia*. The Swiss naturalist Auguste Forel urged
the League of Nations to adopt the ant polity as the model for the
world community.

The marvels of ant life have led some thinkers into giddy speculation on the nature of ant intelligence. Few have put themselves so
quaintly on record as Lord Avebury, who declared: "The mental
powers of ants differ from those of men not so much in kind as in
degree." He ranked them ahead of the anthropoid apes. Maeterlinck
was more cautious: "After all, we have not been present at the delib-
erations of the workers and we know hardly anything of what happens in the depths of the formicary." Others have categorically
explained ant behavior as if the creatures could reason, exchange
information, take purposeful action and feel tender emotion.

Obviously anthropomorphism can explain little about ants, and it
has largely disappeared from the current serious literature about ant
behavior. Its place has been taken, however, by errors of a more
sophisticated sort. One such is the concept of the "superorganism."
This derives from a notion entertained by Plato and Aquinas that a
social organization exhibits the attributes of a superior type of individual. Extended by certain modern biologists, the concept assumes
that the biological organism, a society of cells, is the model for social
organizations, whether ant or human. Plausible analogies are drawn
between organisms and societies: division of function, internal communication, rhythmic periodicity of life processes and the common
cycle of birth, growth, senescence and death. Pursuit of these analogies, according to the protagonists of the superorganism, will disclose
that the same forces of natural selection have shaped the evolution
of both organism and superorganism, and that the same fundamental
laws govern their present existence.

This is a thoroughly attractive idea, but it possesses a weakness
common to all Platonistic thinking. It erects a vague concept, "organism" or "organization," as an ultimate reality which defies explanation. The danger inherent in this arbitrary procedure is the bias it
encourages in the investigator's approach to his problem. The social
scientist must impose on his work the same rules of repetition, systematic variation and control that prevail in the experimental sciences.

413

Wherever possible he should subject his observations to experimental tests in the field and laboratory. In the area we are discussing this kind of work may at times seem more like a study of ants than an investigation of problems. But it yields dependable data.

The individual ant is not equipped for mammalian types of learning. By comparison with the sensitive perceptions of a human being, it is deaf and blind. Its hearing consists primarily in the perception of vibrations physically transmitted to it through the ground. In most species, its vision is limited to the discrimination of light and shadow. These deficiencies are partially compensated by the chemotactual perceptions of the ant, centered in its flitting antennae. Chiefly by means of its antennae, the army ant tells friend from foe, locates its booty, and, thanks to its habit of blazing its trail with organic products such as droplets from its anal gland, finds its way home to the nest. In any case, the ant has little need of learning when it crawls out of the cocoon. By far the greater part of its behavior pattern is already written in its genes.

How the essentially uncomplicated repertory of the individual ant contrives, when ants act in concert, to yield the exceedingly complex behavior of the tribe is one of the most intricate paradoxes in nature. This riddle has been fruitfully explored during the past generation under the guidance of the concept of "trophallaxis," originated by the late William Morton Wheeler of Harvard University, who ranks as the greatest of U. S. myrmecologists. Trophallaxis (from the Greek *trophe,* meaning food, and *allaxis,* exchange) is based upon the familiar observation that ants live in biological thrall to their nestmates. Their powerful mutual attraction can be seen in the constant turning of one ant toward another, the endless antennal caresses, the licking and nuzzling. In these exchanges they can be seen trading intimate substances—regurgitated food and glandular secretions. Most ants are dependent for their lives upon this biosocial intercourse with their fellows. There is strong evidence that an interchange of co-enzymes among larvae, workers and queen is necessary to the survival of all three. Army ant queens unfailingly sicken and die after a few days of isolation.

The well-established concept of trophallaxis naturally suggests that clues to the complex behavior of the ant armies should be sought in the relationships among individuals within the tribe. Most investigators have looked elsewhere, with invariably mistaken results. In attempting to explain, for example, why an ant army alternates between periods of fixed bivouac and nomadic wandering, a half-dozen reputable scientists have jumped to the simplest and most disarmingly logical conclusion: food supply. The ants, they declared, stay in one place until they exhaust the local larder and then move on to new hunting grounds. Schneirla has shown, however, that the true explanation is quite different.

The migratory habits of the ant armies follow a rhythmically punctual cycle. The *Eciton hamatum* species, for example, wanders nomadically for a period of 17 days, then spends 19 or 20 days in fixed bivouac. This cycle coincides precisely with the reproductive cycle of the tribe. The army goes into bivouac when the larvae, hatched from the last clutch of eggs, spin their cocoons and, now quiescent, approach the pupal stage. At the end of the first week in permanent camp, the queen, whose abdomen has swollen to more than five times its normal volume, begins a stupendous five- to seven-day labor and delivers the 20,000 to 30,000 eggs of the next generation. The daily foraging raids, which meanwhile have dwindled to a minimum, pick up again as the eggs hatch into a great mass of larvae. Then, on about the 20th day, the cocoons yield a new complement of callow workers, and the army sets off once more on its evening marches. The rise and fall of this rhythm is shown in the accompanying sketch.

In determining this pattern of events Schneirla logged a dozen ant armies through one or more complete cycles, and upwards of 100 through partial cycles. Observations were set down in shorthand in the field. In the course of the last field trip, from February to July, 1953, broods of more than 80 colonies were sampled, most of them repeatedly at intervals of a few days.

A sentimentalist presented with this new picture of the army ant's domestic habits will perhaps decide that the ants stay in fixed bivouac to protect the queen and her helpless young through the time when they are most vulnerable. Doubtless this is the adaptive significance of the process. But the motivation which carries 100,000 to 150,000 individual ants through this precisely timed cycle of group behavior is not familial love and duty but the trophallactic relationship among the members of the tribe. A cocooned and slumberous pupa, for example, exerts a quieting influence upon the worker that clutches it in its mandible—somewhat as a thumb in the mouth pacifies an infant. But as it approaches maturity and quickens within its cocoon, the pupa produces precisely the reverse effect. Its stirring and twitching excite the workers to pick up the cocoon and snatch it from one another. As an incidental result, this manhandling effects the delivery of the cocoon's occupant.

The stimulus of the emerging brood is evident in a rising crescendo of excitement that seizes the whole community. Raiding operations increase in tempo as the hyperactive, newly delivered workers swarm out into the marching columns. After a day or two, the colony stages an exceptionally vigorous raid which ends in a night march. The bivouac site is left littered with empty cocoons. Later in the nomadic phase, as the stimulus of the callow workers wanes, the larvae of the next generation become the source of colony "drive." Fat and squirming, as big as an average worker, they establish an active trophallactic relationship with the rest of the tribe. Workers constantly stroke them with their antennae, lick them with their mouth parts and carry them bodily from place to place. Since the larvae at this stage are usually well distributed throughout the corridors and the chambers of the overnight bivouac, their stimulus reaches directly a large number of the workers. This is reflected in the sustained vigor of the daily raids, which continue until the larvae spin their cocoons.

These observations are supported by a variety of experimental findings in the field and laboratory. The role of the callow workers

Life cycle of army ant tribe is governed by closely interlocked cycles of behavior (above) and reproduction (below). Rhythm is established in large part by punctuality of queen's five-week cycle of ovulation. She goes into labor at end of first week of the statary phase of the behavior cycle and lays 20,000 to 30,000 eggs in next few days. At this time, tribe is in fixed bivouac, and workers are conducting minimal daily foraging raids. Statary phase continues for another 10 days, long enough for eggs to develop into larvae and for preceding generation of larvae to come through the pupal stage and yield a crop of callow workers. Chemical and physical stimuli of wriggling larvae and hyperactive callows now help to energize the nomadic phase of the behavior cycle, in which vigorous raiding induces daily change of bivouac. During this 17 day period, callows merge into the ranks of mature workers, the larvae attain their growth and start spinning cocoons, and the queen's gaster begins to swell with eggs. Raiding diminishes, and the tribe settles again in a fixed bivouac. This synchronism of behavior and reproductive cycles, providing maximum security at the successive biological crises of reproduction, plays an obviously critical role in the survival of the species.

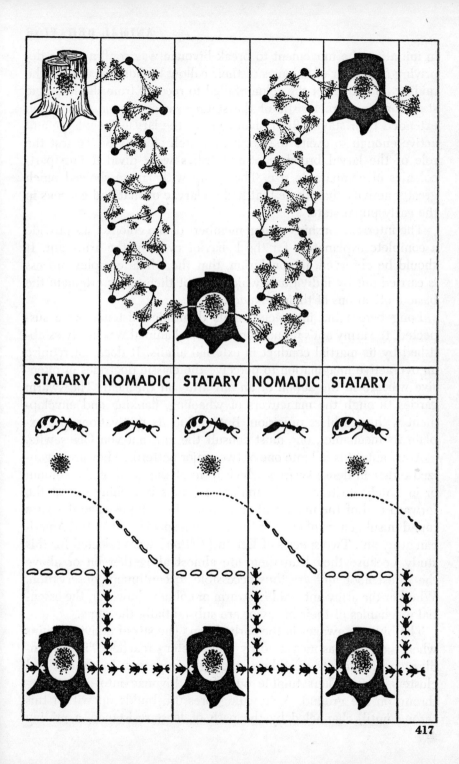

STATARY	NOMADIC	STATARY	NOMADIC	STATARY	

in initiating the movement to break bivouac was confirmed by depriving a number of colonies of their callow broods. Invariably, the raiding operations of the colony failed to recover from the lethargic state that is characteristic of the statary phases. Some tribes even extended their stay in fixed bivouac until the larvae grew large and active enough to excite the necessary pitch of activity. To test the role of the larval brood, captured tribes were divided into part-colonies of comparable size. The group with larvae showed much greater activity than those that had no larvae or that had cocoons in the early pupal state.

The interrelationships among members of the colony thus provide a complete explanation for the behavior cycle of the army ant. It should be observed, in conclusion, that the whole complex process is carried out by individuals which do not themselves originate the basic motivations of their behavior.

Long before the intricacies of its domestic existence were suspected, the army ant's reputation as a social animal was firmly established by its martial conduct in external affairs. It does not require an overactive imagination to perceive the classic doctrines of offensive warfare in the action of an ant army in the field. The swarm carries through the maneuvers of wheeling, flanking and envelopment with a ruthless precision. But to find its motivations and explain its mechanics, one must consult the ant, not von Clausewitz.

Army ant raids fall into one of two major patterns. They are organized either in dense swarms which form at the head of the column or in a delicate tracery of capillary columns branching out at the forward end of the main raiding column. Both types of raiding are found in subgenera of each of the common species of Central American army ant. Two species of Eciton (*Eciton*) were selected for this study because they lead their life almost altogether on or above the forest floor and are thus accessible to continuous observation. Whether the army ants raid in swarm or column, however, the essential mechanics of their behavior are substantially the same.

The bivouac awakes in the early dawn. The stir of activity begins when the light (as measured by photometer) reaches .05 foot candles, and it mounts steadily as the light increases. In strands and clusters, the workers tumble out of the bivouac into a churning throng on the ground. A crowding pressure builds up within this throng until, channeled by the path of least resistance, a raiding

column suddenly bursts forth. The ants in the column are oriented
rigidly along the line of travel blazed by the chemical trail of the
leaders. The minims and medium-sized workers move in tight files
in the center. The "workers major," displaced by the unstable footing
afforded by the backs of their smaller fellows, travel along each side.
This arrangement no doubt lends suggestive support to the major's
legendary role of command. It has an adaptive significance in that it
places the biggest and most formidable of the workers on the flanks.
Unless disturbed, however, the majors hug the column as slavishly
as the rest. The critical role of the tribal chemical in creating this
drill sergeant's picture of order may be demonstrated by a simple
field experiment. Removal of the chemically saturated litter from the
trail brings the column to an abrupt halt. A traffic jam of ants piles
up on the bivouac side of the break and is not relieved until enough
ants have been pushed forward to re-establish the chemical trail.

Appearances are less ordered at the front of the column, where
the "scouts" and "skirmishers" are most frequently observed. The
timid individual behavior of the forward ants scarcely justifies such
titles. The Eciton is a far from enterprising forager. It never ventures
more than a few inches into the chemically-free area ahead. Even
this modest pioneering is stimulated principally by physical impact
from the rear. At the end of its brief sally, the Eciton rebounds
quickly into the column. It is here that the critical difference between column and swarm raiding arises. The column-raiding ants
are somewhat freer in their pioneering behavior and so open new
pathways more readily. In the swarm raiders the comparatively
reluctant progress of the forward elements creates a counterpressure
against the progress of the column. This forces the head of the column into a broad elliptical swarm which arrays itself at right angles
to the line of march. With ants pouring in from behind, the swarm
grows steadily in size as it moves forward, often achieving a width
of more than 15 meters.

The path of an ant army, whether in swarms or columns, shows no
evidence of leadership. On the contrary, each individual makes substantially the same contribution to the group behavior pattern. The
army's course is directed by such wholly chance factors as the stimulus of booty and the character of the terrain. On close inspection,
therefore, it appears that the field operations of ant armies approximate the principles of hydraulics even more closely than those of

419

military tactics. This impression is confirmed by analysis of the flanking maneuver as executed by the swarm raiders. A shimmering pattern of whirls, eddies and momentarily milling vortices of ants, the swarm advances with a peculiar rocking motion. First one and then the other end of the elliptical swarm surges forward. This action results in the outflanking of quarry, which is swiftly engulfed in the overriding horde of ants. It arises primarily, however, from an interplay of forces within the swarm. One of these forces is generated by the inrush of ants from the rear. Opposed by the hesitant progress of the swarm, the new arrivals are deflected laterally to the wing which offers least resistance. This wing moves forward in a wheeling motion until pressure from the slow advance of its frontal margins counterbalances the pressure from the rear. Pressure on the opposite wing has meanwhile been relieved by drainage of the ants into the flanking action. The cycle is therewith reversed, and a new flanking action gets under way from the other end. External factors, too, play a role in this cycle. The stimulus of booty will accelerate the advance of a flank. The capture of booty will halt it and bring ants stampeding in for a large-scale mopping-up party. But raiding activity as such is only incidental to the process. Its essential character is determined by the stereotyped behavior of the individual ant with its limited repertory of responses to external stimuli.

The profoundly simple nature of the beast is betrayed by an ironic catastrophe which occasionally overtakes a troop of army ants. It can happen only under certain very special conditions. But, when these are present, army ants are literally fated to organize themselves in a circular column and march themselves to death. Post-mortem evidence of this phenomenon has been found in nature; it may be arranged at will in the laboratory. Schneirla has had the good fortune to observe one such spectacle in nature almost from its inception to the bitter end.

The ants, numbering about 1,000, were discovered at 7:30 a.m. on a broad concrete sidewalk on the grounds of the Barro Colorado laboratories. They had apparently been caught by a cloudburst which washed away all traces of their colony trail. When first observed, most of the ants were gathered in a central cluster, with only a company or two plodding, counterclockwise, in a circle around the periphery. By noon all of the ants had joined the mill, which had now attained the diameter of a phonograph record and was rotating

THE ARMY ANT

somewhat eccentrically at fair speed. By 10:00 p.m. the mill had
divided into two smaller counterclockwise spinning discs. At dawn
the next day the scene of action was strewn with dead and dying
Ecitons. A scant three dozen survivors were still trekking in a ragged
circle. By 7:30, 24 hours after the mill was first observed, the various
small myremicine and dolichoderine ants of the neighborhood were
busy carting away the corpses.

This peculiarly Eciton calamity may be described as tragic in the
classic meaning of the Greek drama. It arises, like Nemesis, out of
the very aspects of the ant's nature which most plainly characterize
its otherwise successful behavior. The general mechanics of the
mill are fairly obvious. The circular track represents the vector of
the individual ant's centrifugal impulse to resume the march and the
centripetal force of trophallaxis which binds it to its group. Where
no obstructions disturb the geometry of these forces, the organization of a suicide mill is almost inevitable. Fortunately for the army
ant, the jungle terrain, with its random layout of roots and vines,
leaves and stones, disarrays the symmetry of forces and liberates the
ant from its propensity to destroy itself.

The army ant suicide mill provides an excellent occasion for considering the comparative nature of social behavior and organization
at the various levels from ants to men. Other animals occasionally
give themselves over to analogous types of mass action. Circular
mills are common among schools of herring. Stampeding cattle,
sheep jumping fences blindly in column and other instances of pell-
mell surging by a horde of animals are familiar phenomena. Experience tells us that men, too, can act as a mob. These analogies are the
stock-in-trade of the "herd instinct" schools of sociology and politics.

We are required, however, to look beyond the analogy and study
the relationship of the pattern to other factors of individual and
group behavior in the same species. In the case of the army ant,
the circular column really typifies the animal. Among mammals, such
simplified mass behavior occupies a clearly subordinate role. Their
group activity patterns are chiefly characterized by great plasticity
and capacity to adjust to new situations. This observation applies
with special force to the social potentialities of man. When human
societies begin to march in circular columns, the cause is to be found
in the strait-jacket influence of the man-made social institutions
which foster such behavior.

421

As for "specialization of functions," that is determined in insect societies by specialization in the biological make-up of individuals. Mankind, in contrast, is biologically uniform and homogeneous. Class and caste distinctions among men are drawn on a psychological basis. They break down constantly before the energies and talents of particular individuals.

Finally, the concept of "organization" itself, as it is used by the superorganism theorists, obscures a critical distinction between the societies of ants and men. The social organizations of insects are fixed and transmitted by heredity. But members of each generation of men may, by exercise of the cerebral cortex, increase, change and even displace given aspects of their social heritage. This is a distinction which has high ethical value for men when they are moved to examine the conditions of their existence.

II. THE LANGUAGE OF THE BEES

by August Krogh

KARL VON FRISCH, the Austrian naturalist, began working with bees about 40 years ago when he showed that, contrary to prevalent opinion, these insects are not entirely color-blind. From that beginning, he went on to a lifelong study of the other senses of bees and of many lower animals, especially fish. The experiments to be described here were almost all made after the last war in a small private laboratory that von Frisch maintains at Brunnwinkl in the Austrian Alps.

Von Frisch's early experiments showed that bees must possess some means of communication, because when a rich source of food (he used concentrated sugar solution) is found by one bee, the food is soon visited by numerous other bees from the same hive. To find out how they communicated with one another, von Frisch constructed special hives containing only one honeycomb, which could be exposed to view through a glass plate. Watching through the glass, he discovered that bees returning from a rich source of food perform special movements, which he called dancing, on the vertical surface of the honeycomb. Von Frisch early distinguished between two types of dance: the circling dance (*Rundtanz*) and the wagging dance (*Schwänzeltanz*). In the latter a bee runs a certain distance in a straight line, wagging its abdomen very swiftly from side to side, and then makes a turn. Von Frisch concluded from his early experiments that the circling dance meant nectar and the wagging dance pollen, but this turned out to be an erroneous translation as will presently appear.

In any case, the dance excites the bees. Some of them follow the dancer closely, imitating the movements, and then go out in search of the food indicated. They know what kind of food to seek from the odor of the nectar or pollen, some of which sticks to the body of the bee. By means of some ingenious experiments, von Frisch determined that the odor of the nectar collected by bees, as well as that adhering to their bodies, is important. He designed an arrangement for feeding bees odoriferous nectar so that their body surfaces were kept from contact with it. This kind of feeding was perfectly adequate to guide the other bees. In another experiment, nectar having the odor of phlox was fed to bees as they sat on cyclamen flowers.

423

When the bees had only a short distance to fly back to the hive, some of their fellows would go for cyclamen, but in a long flight the cyclamen odor usually was lost completely, and the bees were guided only by the phlox odor.

The vigor of the dance which guides the bees is determined by the ease with which the nectar is obtained. When the supply of nectar in a certain kind of flower begins to give out, the bees visiting it slow down or stop their dance. The result of this precisely regulated system of communication is that the bees form groups just large enough to keep up with the supply of food furnished by a given kind of flower. Von Frisch proved this by marking with a colored stain a group of bees frequenting a certain feeding place. The group was fed a sugar solution impregnated with a specific odor. When the supply of food at this place gave out, the members of the group sat idle in the hive. At intervals one of them investigated the feeding place, and if a fresh supply was provided, it would fill itself, dance on returning and rouse the group. Continued energetic dancing roused other bees sitting idle and associated them to the group.

But what was the meaning of the circling and wagging dances? Von Frisch eventually conceived the idea that the type of dance did not signify the kind of food, as he had first thought, but had something to do with the distance of the feeding place. This hypothesis led to the following crucial experiment. He trained two groups of bees from the same hive to feed at separate places. One group, marked with a blue stain, was taught to visit a feeding place only a few meters from the hive; the other, marked red, was fed at a distance of 300 meters. To the experimenter's delight, it developed that all the blue bees made circling dances; the red, wagging dances. Then, in a series of steps, von Frisch moved the nearer feeding place farther and farther from the hive. At a distance between 50 and 100 meters away, the blue bees switched from a circling dance to wagging. Conversely, when the red group's food was brought gradually nearer to the hive, the dance changed from wagging to circling in the 50 to 100 meter range.

Thus it was clear that the dance at least told the bees whether the distance exceeded a certain value. It appeared unlikely, however, that the information conveyed was actually quite so vague, for bees often feed at distances up to two miles and presumably need more

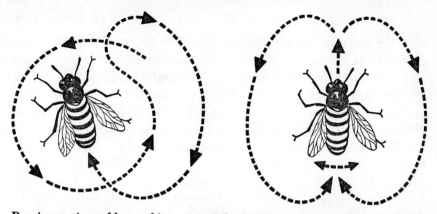

Dancing motions of bee on hive convey information to other members of hive. In "Rundtanz," or circling dance, at left the bee turns in alternately clockwise and counter-clockwise circles. In "Schwänzeltanz," or wagging dance, at right the bee runs forward in straight line, wagging its abdomen from side to side, then circles around to retrace her steps again. Number of wags and number of turns has been shown to be related to distance of feeding place from hive.

precise guidance. The wagging dance was therefore studied more closely. The rate of wagging is probably significant, but it is too rapid to follow. It was found, however, that the frequency of turns would give a fairly good indication of the distance. When the feeding place was 100 meters away, the bee made about 10 short turns in 15 seconds. To indicate a distance of 3,000 meters, it made only three long ones in the same time. A curve plotted from the average of performances by a number of bees shows that the number of turns varies regularly with the distance, although the correspondence is not very precise in individual cases.

How accurately do the bees respond to what is told them? Von Frisch studied this problem with several experiments, of which the following is the most conclusive. The feeding table was placed in a certain direction and at four different distances in four trials. At each trial plates containing the same odor but no food were placed in the three other directions and in each case at nearly the same distance as the food source. At short distances (about 10 meters) the bees searched almost equally in all directions. But beginning at about 25 meters they evidently had some indication of the right direction, for the plate with food was visited by much larger numbers than the plates at the other points of the compass.

How did the returning bees indicate to the other bees in the hive the direction of the feeding place? A key to the answer was given by the known fact that bees use the sun for orientation during flight. A bee caught far from the hive and liberated after a few minutes will fly straight back. But if it is kept in a dark box for a period, say an hour, it will go astray, because it continues to fly at the same angle to the sun's direction as when it was caught. Von Frisch deduced that the bee dance must signal direction in relation to the position of the sun. Obviously a horizontal direction can not actually be shown on the vertical surface of a honeycomb, but von Frisch discovered that the bees transpose direction and designate the top of the comb as the horizontal position of the sun. When the sun, as seen from the beehive, is just above the feeding place, the straight part of the dance is vertical with the head up. When the feeding place is in the opposite direction, the straight part again is vertical, but with the head down. And when the food is not in line with the sun, the bee shows the horizontal angle between the sun and the feeding place by pointing at the same angle from the vertical on the honeycomb.

This indication of direction changes continuously throughout the day with the changing position of the sun, which is always represented on the vertical. The dance is normally performed in complete darkness within the hive, yet the bees, roused by, following and imitating the dancer, correctly interpret the signals to an accuracy within a few degrees. It can be observed without disturbing the bees in photographic red light, which is invisible to them.

It is a very curious fact, for which no explanation has been found so far, that the position of the sun in the heavens is correctly used by the bees even when it is hidden behind an unbroken layer of clouds, and when in addition the hive is placed in surroundings totally unknown to the bees. This precaution is necessary because in territory that the bees know well they are experts in using landmarks. It appears possible that infrared rays from the sun, penetrating the clouds, may guide the bees. Experiments have shown that bees are not stimulated by heat rays as such, but the possibility cannot be excluded that the eyes of bees could be sensitive to near infrared although insensitive to visible red. This point has not so far been investigated for lack of a suitable light filter. Von Frisch has also undertaken some experiments to determine how the bees would cope with the problem of a mountain ridge or tall building which forced.

them to make a detour. He found that they would indicate the air-line direction from the hive to the feeding place, but would give the distance that they actually had to fly.

Von Frisch tells me that he himself considered some of these results so fantastic that he had to make sure that ordinary bees which had not been experimentally trained could also do the tricks. They could, and moreover they continued to work efficiently on honeycombs removed from the hive. While studying these "wild" bees, von Frisch became curious to see what would happen if the honeycomb was put in a horizontal position instead of the vertical. To his surprise the bees responded by pointing directly toward the feeding place, and they kept on doing this even when the honeycomb was slowly rotated like a turntable. It looked as if the bees had a magnet in them and responded like a compass needle, but experiments showed them to be not the least affected by magnetic force. This method of pointing also takes place under natural conditions, the bees often performing horizontal dances in front of the entrance to the hive.

On the other hand, experiments showed that on the underside of a horizontal surface the bees were unable to indicate any direction, and it turned out that their signals were disturbed when the horizontal surface was placed in the shade. Von Frisch therefore decided to test directly their power of indicating direction on a horizontal surface in the dark. A movable chamber was built to enclose the observer and the observation hive. By photographic red light or even by diffuse white light in a tent, the bees proved unable to indicate any direction on a horizontal surface (although they can work with precision in the dark on a vertical one). They were not restrained from dancing, and the stimulated bees, thoroughly confused, searched for food equally in all directions. The sun can be replaced in these experiments by any artificial light source of sufficient strength. But only if such a light is placed in the direction, corresponding to that of the sun at the time, are the bees led toward the feeding place. Placed in any other position, the light will lead them astray.

At this point, some contradictory evidence turned up. On several occasions the bees had proved able to give correct instructions on a horizontal surface even when the sun was not directly visible. Therefore, the experiment was made of removing the north wall of the observation chamber, which allowed the bees to see only the sunless sky. In clear weather this proved sufficient to give them the correct orientation. Indeed, it was eventually found that when light from a blue sky came into the chamber through a tube 40 centimeters long and only 15 centimeters in diameter, this bare glimpse of the sky sufficed to orient the bees toward the sun's position. Light from a cloud, however, was without effect when seen through the tube, and sky light reflected by a mirror was misleading. The most probable explanation is that the bees are able to observe the direction of the polarized light from the sky and thereby infer the sun's position.

When the honeycomb is tipped to an inclination between vertical and horizontal, the bees respond by giving information that combines direct pointing with use of the vertical to indicate the sun's position. This of course results in a deviation from the true course. Analysis of earlier experiments, in which light from the sky complicated the reactions of bees on a vertical honeycomb, showed that the perturbations could all be quantitatively explained on the same basis.

I have tried to give a very condensed account of the principal results which von Frisch has so far obtained. This series of experiments constitutes a most beautiful example of what the human mind can accomplish by tireless effort on a very high level of intelligence. But I would ask you to give some thought also to the mind of the bees. I have no doubt that some will attempt to "explain" the performances of the bees as the result of reflexes and instincts. Such attempts will certainly contribute to our understanding, but for my part I find it difficult to assume that such perfection and flexibility in behavior can be reached without some kind of mental processes— I do not venture to proclaim them as "thoughts"—going on in the small heads of the bees.

III. THE SPIDER AND THE WASP

by Alexander Petrunkevitch

IN THE FEEDING and safeguarding of their progeny insects and spiders exhibit some interesting analogies to reasoning and some crass examples of blind instinct. The case I propose to describe here is that of the tarantula spiders and their arch-enemy, the digger wasps of the genus Pepsis. It is a classic example of what looks like intelligence pitted against instinct—a strange situation in which the victim, though fully able to defend itself, submits unwittingly to its destruction.

Most tarantulas live in the tropics, but several species occur in the temperate zone and a few are common in the southern U. S. Some varieties are large and have powerful fangs with which they can inflict a deep wound. These formidable looking spiders do not, however, attack man; you can hold one in your hand, if you are gentle, without being bitten. Their bite is dangerous only to insects and small mammals such as mice; for a man it is no worse than a hornet's sting.

Tarantulas customarily live in deep cylindrical burrows, from which they emerge at dusk and into which they retire at dawn. Mature males wander about after dark in search of females and occasionally stray into houses. After mating, the male dies in a few weeks, but a female lives much longer and can mate several years in succession. In a Paris museum is a tropical specimen which is said to have been living in captivity for 25 years.

A fertilized female tarantula lays from 200 to 400 eggs at a time; thus it is possible for a single tarantula to produce several thousand young. She takes no care of them beyond weaving a cocoon of silk to enclose the eggs. After they hatch, the young walk away, find convenient places in which to dig their burrows and spend the rest of their lives in solitude. The eyesight of tarantulas is poor, being limited to a sensing of change in the intensity of light and to the perception of moving objects. They apparently have little or no sense of hearing, for a hungry tarantula will pay no attention to a loudly chirping cricket placed in its cage unless the insect happens to touch one of its legs.

But all spiders, and especially hairy ones, have an extremely delicate sense of touch. Laboratory experiments prove that tarantulas can distinguish three types of touch: pressure against the body wall, stroking of the body hair and riffling of certain very fine hairs on the legs called trichobothria. Pressure against the body, by a finger or the end of a pencil, causes the tarantula to move off slowly for a short distance. The touch excites no defensive response unless the approach is from above where the spider can see the motion, in which case it rises on its hind legs, lifts its front legs, opens its fangs and holds this threatening posture as long as the object continues to move.

The entire body of a tarantula, especially its legs, is thickly clothed with hair. Some of it is short and woolly, some long and stiff. Touching this body hair produces one of two distinct reactions. When the spider is hungry, it responds with an immediate and swift attack. At the touch of a cricket's antennae the tarantula seizes the insect so swiftly that a motion picture taken at the rate of 64 frames per second shows only the result and not the process of capture. But when the spider is not hungry, the stimulation of its hairs merely causes it to shake the touched limb. An insect can walk under its hairy belly unharmed.

The trichobothria, very fine hairs growing from disklike membranes on the legs, are sensitive only to air movement. A light breeze makes them vibrate slowly without disturbing the common hair. When one blows gently on the trichobothria, the tarantula reacts with a quick jerk of its four front legs. If the front and hind legs are stimulated at the same time, the spider makes a sudden jump. This reaction is quite independent of the state of its appetite.

These three tactile responses—to pressure on the body wall, to moving of the common hair and to flexing of the trichobothria—are so different from one another that there is no possibility of confusing them. They serve the tarantula adequately for most of its needs and enable it to avoid most annoyances and dangers. But they fail the spider completely when it meets its deadly enemy, the digger wasp Pepsis.

These solitary wasps are beautiful and formidable creatures. Most species are either a deep shiny blue all over, or deep blue with rusty wings. The largest have a wing span of about four inches. They live on nectar. When excited, they give off a pungent odor—a warning that they are ready to attack. The sting is much worse than that of a bee or common wasp, and the pain and swelling last longer. In the adult stage the wasp lives only a few months. The female produces but a few eggs, one at a time at intervals of two or three days. For each egg the mother must provide one adult tarantula, alive but paralyzed. The mother wasp attaches the egg to the paralyzed spider's abdomen. Upon hatching from the egg, the larva is many hundreds of times smaller than its living but helpless victim. It eats no other food and drinks no water. By the time it has finished its single gargantuan meal and become ready for wasphood, nothing remains of the tarantula but its indigestible chitinous skeleton.

The mother wasp goes tarantula-hunting when the egg in her ovary is almost ready to be laid. Flying low over the ground late on a sunny afternoon, the wasp looks for its victim or for the mouth of a tarantula burrow, a round hole edged by a bit of silk. The sex of the spider makes no difference, but the mother is highly discriminating as to species. Each species of Pepsis requires a certain species of tarantula, and the wasp will not attack the wrong species. In a cage with a tarantula which is not its normal prey the wasp avoids the spider, and is usually killed by it in the night.

Yet when a wasp finds the correct species, it is the other way about. To identify the species the wasp apparently must explore the spider with her antennae. The tarantula shows an amazing tolerance to this exploration. The wasp crawls under it and walks over it without evoking any hostile response. The molestation is so great and so persistent that the tarantula often rises on all eight legs, as if it were on stilts. It may stand this way for several minutes. Meanwhile the wasp, having satisfied itself that the victim is of the right species, moves off a few inches to dig the spider's grave. Working vigorously with legs and jaws, it excavates a hole 8 to 10 inches deep with a diameter slightly larger than the spider's girth. Now and again the wasp pops out of the hole to make sure that the spider is still there.

When the grave is finished, the wasp returns to the tarantula to complete her ghastly enterprise. First she feels it all over once more with her antennae. Then her behavior becomes more aggressive. She bends her abdomen, protruding her sting, and searches for the soft membrane at the point where the spider's leg joins its body—the only spot where she can penetrate the horny skeleton. From time to time, as the exasperated spider slowly shifts ground, the wasp turns on her back and slides along with the aid of her wings, trying to get under the tarantula for a shot at the vital spot. During all this maneuvering, which can last for several minutes, the tarantula makes no move to save itself. Finally the wasp corners it against some obstruction and grasps one of its legs in her powerful jaws. Now at last the harassed spider tries a desperate but vain defense. The two contestants roll over and over on the ground. It is a terrifying sight and the outcome is always the same. The wasp finally manages to thrust her sting into the soft spot and holds it there for a few seconds while she pumps in the poison. Almost immediately the tarantula falls paralyzed on its back. Its legs stop twitching; its heart stops beating. Yet it is not dead, as is shown by the fact that if taken from the wasp it can be restored to some sensitivity by being kept in a moist chamber for several months.

After paralyzing the tarantula, the wasp cleans herself by dragging her body along the ground and rubbing her feet, sucks the drop of blood oozing from the wound in the spider's abdomen, then grabs a leg of the flabby, helpless animal in her jaws and drags it down to the bottom of the grave. She stays there for many minutes, sometimes for several hours, and what she does all that time in the dark we do not know. Eventually she lays her egg and attaches it to the side of the spider's abdomen with a sticky secretion. Then she emerges, fills the grave with soil carried bit by bit in her jaws, and finally tramples the ground all around to hide any trace of the grave from prowlers. Then she flies away, leaving her descendant safely started in life.

In all this the behavior of the wasp evidently is qualitatively different from that of the spider. The wasp acts like an intelligent animal. This is not to say that instinct plays no part or that she reasons as man does. But her actions are to the point; they are not automatic and can be modified to fit the situation. We do not know for certain how she identifies the tarantula—probably it is by some olfactory or chemo-tactile sense—but she does it purposefully and does not blindly tackle a wrong species.

On the other hand, the tarantula's behavior shows only confusion. Evidently the wasp's pawing gives it no pleasure, for it tries to move away. That the wasp is not simulating sexual stimulation is certain, because male and female tarantulas react in the same way to its advances. That the spider is not anesthetized by some odorless secretion is easily shown by blowing lightly at the tarantula and making it jump suddenly. What, then, makes the tarantula behave as stupidly as it does?

No clear, simple answer is available. Possibly the stimulation by the wasp's antennae is masked by a heavier pressure on the spider's body, so that it reacts as when prodded by a pencil. But the explanation may be much more complex. Initiative in attack is not in the nature of tarantulas; most species fight only when cornered so that escape is impossible. Their inherited patterns of behavior apparently prompt them to avoid problems rather than attack them. For example, spiders always weave their webs in three dimensions, and when a spider finds that there is insufficient space to attach certain threads in the third dimension, it leaves the place and seeks another, instead of finishing the web in a single plane. This urge to escape seems to arise under all circumstances, in all phases of life and to take the place of reasoning. For a spider to change the pattern of its web is as impossible as for an inexperienced man to build a bridge across a chasm obstructing his way.

In a way the instinctive urge to escape is not only easier but often more efficient than reasoning. The tarantula does exactly what is most efficient in all cases except in an encounter with a ruthless and determined attacker dependent for the existence of her own species on killing as many tarantulas as she can lay eggs. Perhaps in this case the spider follows its usual pattern of trying to escape, instead of seizing and killing the wasp, because it is not aware of its danger. In any case, the survival of the tarantula species as a whole is protected by the fact that the spider is much more fertile than the wasp.

IV. THE CURIOUS BEHAVIOR
OF THE STICKLEBACK

by N. Tinbergen

WHEN I was a young lecturer in zoology at the University of Leyden 20 years ago, I was asked to organize a laboratory course in animal behavior for undergraduates. In my quest for animals that could be used for such a purpose, I remembered the sticklebacks I had been accustomed as a boy to catch in the ditches near my home and to raise in a backyard aquarium. These former pets soon proved to be ideal laboratory animals. They are so tame that they submit unfrightened to laboratory experiments, for the stickleback, like the hedgehog, depends on its spines for protection and is little disturbed by handling. Furthermore, the stickleback turned out to be an excellent subject for studying innate behavior, which it displays in some remarkably dramatic and intriguing ways. We found it to be the most reliable of various experimental animals that we worked with (including newts, bees, water insects and birds), and it became the focus of a program of research in which we now use hundreds of sticklebacks each year. The stickleback today is also a popular subject in various other zoological laboratories in Europe, notably at the universities in Groningen and Oxford. To us this little fish is what the rat is to many American psychologists.

My collaborator J. van Iersel and I have concentrated on the stickleback's courtship and reproductive behavior. The sex life of the three-spined stickleback (*Gasterosteus aculeatus*) is a complicated pattern, purely instinctive and automatic, which can be observed and manipulated almost at will.

In nature sticklebacks mate in early spring in shallow fresh waters. The mating cycle follows an unvarying ritual, which can be seen equally well in the natural habitat or in our tanks. First each male leaves the school of fish and stakes out a territory for itself, from which it will drive any intruder, male or female. Then it builds a nest. It digs a shallow pit in the sand bottom, carrying the sand away mouthful by mouthful. When the depression is about two inches square, it piles in a heap of weeds, preferably thread algae, coats the material with a sticky substance from its kidneys and shapes the weedy mass into a mound with its snout. It then bores a tunnel in the mound by wriggling through it. The tunnel, slightly shorter than an adult fish, is the nest.

Having finished the nest, the male suddenly changes color. Its normally inconspicuous gray coloring had already begun to show a faint pink blush on the chin and a greenish gloss on the back and in the eyes. Now the pink becomes a bright red and the back turns a bluish white.

In this colorful, conspicuous dress the male at once begins to court females. They, in the meantime, have also become ready to mate: their bodies have grown shiny and bulky with 50 to 100 large eggs. Whenever a female enters the male's territory, he swims toward her in a series of zigzags—first a sideways turn away from her, then a quick movement toward her. After each advance the male stops for an instant and then performs another zigzag. This dance continues until the female takes notice and swims toward the male in a curious head-up posture. He then turns and swims rapidly toward the nest, and she follows. At the nest the male makes a series of rapid thrusts with his snout into the entrance. He turns on his side as he does so and raises his dorsal spines toward his mate. Thereupon, with a few strong tail beats, she enters the nest and rests there, her head sticking out from one end and her tail from the other. The male now prods her tail base with rhythmic thrusts, and this causes her to lay her eggs. The whole courtship and egg-laying ritual takes only about one minute. As soon as she has laid her eggs, the female slips out of the nest. The male then glides in quickly to fertilize the clutch. After that he chases the female away and goes looking for another partner.

One male may escort three, four or even five females through the nest, fertilizing each patch of eggs in turn. Then his mating impulse subsides, his color darkens and he grows increasingly hostile to females. Now he guards the nest from predators and "fans" water over the eggs with his breast fins to enrich their supply of oxygen and help them to hatch. For a day or so after the young emerge the father keeps the brood together, pursuing each straggler and bringing it back in his mouth. Soon the young sticklebacks become independent and associate with the young of other broods.

To get light on the behavior of man, particularly his innate drives and conflicts, it is often helpful to study the elements of behavior in a simple animal. Here is a little fish that exhibits a complicated pattern of activities, all dependent on simple stimuli and drives. We have studied and analyzed its behavior by a large number of experiments, and have learned a good deal about why the stickleback behaves as it does.

Let us begin with the stimulus that causes one stickleback to attack another. Early in our work we noticed that a male patrolling its territory would attack a red-colored intruder much more aggressively than a fish of some other color. Even a red mail van passing our windows at a distance of 100 yards could make the males in the tank charge its glass side in that direction. To investigate the reactions to colors we made a number of rough models of sticklebacks and painted some of the dummies red, some pale silver, some green. We rigged them up on thin wires and presented them one by one to the males in the tank. We found that the red models were always more provoking than the others, though even the silvery or green intruders caused some hostility.

In much the same way we tested the influence of shape, size, type of body movement and other stimuli, relating them to specific behavior in nest building, courting, attack, zigzag, fanning and so on. We discovered, for example, that a male swollen with food was courted as if it were a female.

As our work proceeded, we saw that the effective stimuli differed from one reaction to another, even when two reactions were caused by the same object. Thus a female will follow a red model wherever it leads; she will even make frantic efforts to enter a nonexistent nest wherever the model is poked into the sand. Once she is in a real nest, she can be induced to spawn merely by prodding the base of her tail with a glass rod, even after she has seen the red fish that led her there removed. At one moment the male must give the visual signal of red; at the next, this stimulus is of no importance and only the tactile sensation counts. This observation led us to conclude that the stickleback responds simply to "sign stimuli," *i.e.*, to a few characteristics of an object rather than to the object as a whole. A red fish or a red mail truck, a thrusting snout or a glass rod—it is the signal, not the object, that counts. A similar dependence on sign stimuli, which indicates the existence of special central nervous mechanisms, has been found in other species. It seems to be typical of innate behavior, and many social relationships in animals apparently are based on a system of signs.

Sticklebacks will respond to our stimuli only when they are in breeding condition. At other seasons they ignore the signs. This fact led us to investigate the internal factors that govern the fish. The obvious way to study such fluctuations is to measure the frequency and intensity of a response under standard stimulation. For some of these tests we used either uniform models or live fish confined in glass tubes so that we could control their movement. To measure the parental drive we adopted the standard of the number of seconds spent in fanning a given number of eggs per time unit.

The stickleback's drives in the breeding sequence wax and wane in a series of cycles. Each drive runs its course in regular succession: first the male gets the urge to fight, then to build a nest, then to court a female, then to develop the brood. He will not start to build, even though material is available, until he has defended his territory for a while. Nor will he court until he has built the nest; females that approach him before the nest is finished are driven off or at best are greeted with a few zigzags. Within each cycle also there is a fixed rhythm and sequence; for example, if you fill up the pit the male has dug, he will dig one again before collecting nest material. After the pit has been filled several times, however, the fish will build the nest without completing the pit. The development of his inner drive overcomes outside interference.

It seems likely that the rise and fall of inner drives is controlled by hormonal changes, and we are now studying the effects on these drives of castrating and giving hormones to the males. One interesting finding so far is that castration abolishes the first phases of mating, but has no effect on the parental drive. A eunuch stickleback, when given a nest of eggs, ventilates it with abandon.

In any animal the innate drives themselves are only the elementary forces of behavior. It is the interaction among those drives, giving rise to conflicts, that shapes the animal's actual behavior, and we have devoted a major part of our work with the stickleback to this subject. It struck us, as it has often struck observers of other animals, that the belligerent male sticklebacks spent little time in actual fighting. Much of their hostility consists of display. The threat display of male sticklebacks is of two types. When two males meet at the border of their territories, they begin a series of attacks and retreats. Each takes the offensive in his own territory, and the duel seesaws back and forth across the border. Neither fish touches the other; the two dart back and forth as though attached by an invisible thread. This behavior demonstrates that the tendency to attack and the tendency to retreat are both aroused in each fish.

When the fight grows in vigor, however, the seesaw maneuver may suddenly change into something quite different. Each fish adopts an almost vertical head-down posture, turns its side to its opponent, raises its ventral spines and makes jerky movements with the whole body. Under crowded conditions, when territories are small and the fighting tendency is intense, both fish begin to dig into the sand, as if they were starting to build a nest! This observation at first astonished us. Digging is so irrelevant to the fighting stimulus that it seemed to overthrow all our ideas about the specific connection between sign and response. But it became less mysterious when we considered similar instances of incongruous behavior by other animals. Fighting starlings always preen themselves between bouts; in the midst of a fight roosters often peck at the ground as though feeding, and wading-birds assume a sleeping posture. Even a man, in situations of embarrassment, conflict or stress, will scratch himself behind the ear.

So it appears that the stickleback does not start digging because its nest-building drive is suddenly activated. Rather, the fish is engaging in what has been called a "displacement activity." Alternating between the urge to attack and to escape, neither of which it can carry out, it finally is driven by its tension to find an outlet in an irrelevant action.

A similar interaction of drives seems to motivate the male when he is courting. In the zigzag dance the movement away from the female is the purely sexual movement of leading; the movement toward her is an incipient attack. This duality can be proved by measuring the comparative intensity of the two drives in an individual male and relating it to his dance. Thus when the sex drive is strong (as measured by willingness to lead a standard female model) the zig component of the dance is pronounced and may shift to complete leading. When the fighting drive is strong (as measured by the number of bites aimed at a standard male model) the zag is more emphatic and may become a straightforward attack. A female evokes the double response because she provides sign stimuli for both aggression and sexuality. Every fish entering a male's territory evokes some degree of attack, and therefore even a big-bellied female must produce a hostile as well as a sexual response.

This complexity of drives continues when the fish have arrived at the nest. A close study of the movement by which the male indicates the entrance shows that it is very similar to fanning, at that moment an entirely irrelevant response. This fanning motion, we conclude, must be a displacement activity, caused by the fact that the male is not yet able to release his sex drive; he can ejaculate his sperm only after the female has laid her eggs. Even when the female has entered the nest, the male's drive is still frustrated. Before he can release it, he must stimulate her to spawn. The "quivering" motion with which he prods her is much like fanning. It, too, is a displacement activity and stops at the moment when the eggs are laid and the male can fertilize them. It is probable that the male's sex drive is frustrated not only by the absence of eggs but also by a strong conflict with the attack drive, which must be intense when a strange fish is so near the nest. This hostility is evident from the fact that the male raises his dorsal spines while exhibiting the nest to the female.

The ideas briefly outlined here seem to throw considerable light on the complicated and "irrelevant" activities typical of innate behavior in various animals. Of course these ideas have to be checked in more cases. This is now being done, particularly with fish and birds, and the results are encouraging.

I am often asked whether it is worth while to stick to one animal species for so long a time as we have been studying the stickleback. The question has two answers. I believe that one should not confine one's work entirely to a single species. No one who does can wholly avoid thinking that his animal is The Animal, the perfect representative of the whole animal kingdom. Yet the many years of work on the stickleback, tedious as much of it has been, has been highly rewarding. Without such prolonged study we could not have gained a general understanding of its entire behavior pattern. That, in turn, is essential for an insight into a number of important problems. For instance, the aggressive component in courtship could never have been detected by a study of courtship alone, but only by the simultaneous study of fighting and courtship. Displacement activities are important for an understanding of an animal's motivation. To recognize them, one must have studied the parts of the behavior from which they are "borrowed" as well as the drives which, when blocked, use them as outlets.

Concentration on the stickleback has also been instructive to us because it meant turning away for a while from the traditional laboratory animals. A stickleback is different from a rat. Its behavior is much more purely innate and much more rigid. Because of its relative simplicity, it shows some phenomena more clearly than the behavior of any mammal can. The dependence on sign stimuli, the specificity of motivation, the interaction between two types of motivation with the resulting displacement activities are some of these phenomena.

Yet we also study other animals, because only by comparison can we find out what is of general significance and what is a special case. One result that is now beginning to emerge from the stickleback experiments is the realization that mammals are in many ways a rather exceptional group, specializing in "plastic" behavior. The simpler and more rigid behavior found in our fish seems to be the rule in most of the animal kingdom. Once one is aware of this, and aware also of the affinity of mammals to the lower vertebrates, one expects to find an innate base beneath the plastic behavior of mammals.

Thus the study of conflicting drives in so low an animal as the stickleback may throw light on human conflicts and the nature of neuroses. The part played by hostility in courtship, a phenomenon found not only in sticklebacks but in several birds, may well have a real bearing on human sex life. Even those who measure the value of a science by its immediate application to human affairs can learn some important lessons from the study of this insignificant little fish.

V. EXPERIMENTAL NEUROSES

by *Jules H. Masserman*

As Auguste Comte pointed out a century ago, a science generally develops through three phases of evolution: mystic, taxonomic and dynamic. Psychiatry, the branch of medicine devoted to the study and treatment of disorders of behavior, admirably illustrates Comte's generalization. Its first phase —the mystic, ritualistic approach—lasted well beyond the Middle Ages: as late as 1783 an insane woman in Switzerland was judged to be an emissary of the devil and burned as a witch. About two centuries ago psychiatry entered the second phase: that of recording and classifying behavior. Man's first observations of the complexities of his own conduct were understandably biased and inaccurate, and his classifications arbitrary and dogmatic; indeed, we are even yet prone to appraise one another with clinical stares and smug appraisals, such as "compared to me, you are an introvert," a "schizoid," a "cyclothyme"—or some other deviate with a resoundingly meaningless appellation. Most psychiatrists admit that there are still residues of mysticism and irrational dogmatisms in our field.

Nevertheless, we must in all fairness note that through psychosomatics modern psychiatry has recently achieved a reunion with clinical medicine, that modern psychoanalysis is steadily becoming more scientific and less doctrinaire, and that social psychiatry is establishing productive relationships with anthropology, sociology and other humanistic disciplines. Moreover, psychiatry has begun to re-explore its data, hypotheses and methods by experimental research.

The laboratory and clinical studies of Ivan Pavlov, Horsley Gantt, H. A. Liddell, J. Hunt, David Levy, O. H. Mowrer, Curt Richter and many others have indicated that certain basic tenets on which much of modern dynamic psychiatry implicitly rests are demonstrable in nearly all behavior—animal as well as human, "normal" as well as "abnormal." This article will describe how these tenets, incorporated into a more comprehensive system of biodynamics, have been developed and elaborated in various experiments conducted during the last 15 years in the Division of Psychiatry of the University of Chicago and, more recently, in the Department of Nervous and Mental Diseases of Northwestern University.

The principles of biodynamics may be condensed into four relatively simple statements:

1. *All behavior is actuated by the current physical needs of the
organism in the processes of survival, growth and procreation.* Thus
a simple want for calcium or for warmth or even for relief from
bladder tension, if sufficiently urgent, will take precedence over more
complex physiological "instincts" which are considered basic in
some systems of psychology.

2. *Behavior is adaptive to the "external" environment not in any
objective sense, but according to the organism's special interpretation of its milieu, which depends upon its own capacities ("intelligence") and its unique association of experiences.* Thus two crossed
pieces of burning wood may signify only a marshmallow-toast to one
human being, self-congratulatory "white supremacy" to another, and
abject terror of death to a third.

3. *When accustomed methods of achieving a goal are frustrated,
behavior turns to substitute techniques or becomes oriented toward
alternate goals.* Thus if a man's methods of wooing a girl meet with
rebuff, he tries a) other methods, b) another girl or c) another
goal, such as success as a religious prophet, as a jazz drummer or
perhaps as a psychologist.

4. *When two or more accustomed modes of response become
mutually incompatible, physiologic tension, or "psychosomatic anxiety," becomes manifest and behavior becomes vacillating, inefficient
and unadaptive ("neurotic") or excessively substitutive, erratic and
regressive ("psychotic").*

To study these general principles of biodynamics experimentally,
one might utilize any animal with sufficiently high capacities for
perception, integration and reaction—the rat, the dog, the cat or the
monkey. In most of the experiments to be described we employed the
cat, because it has fairly simple motivations and relatively high
intelligence. To actuate the animal's behavior we might have chosen
any one of several stimuli—thirst, cold, pressure or pain, erotic excitement or the like. We found that hunger for food, though a relatively
complex need, is the most convenient: it is easily renewable, is satia-
ble in easy stages, and is neither as climactic nor potentially as
traumatic as are sexuality, cold, pain or other physiologic tensions.

In a typical experiment a cat was deprived of food for a day, then
placed in a glass-enclosed experimental cage at one end of which

was a food-box with a partly open hinged lid. The animal readily learned to obtain pellets of food from this box by prying the lid farther open so it could reach them. The animal was then taught (a) to wait for various combinations of sound and light signals before attempting to feed, (b) to manipulate various electrical switches so as to set off these signals for itself, and (c) to close two or more switches a given number of times in definite sequence or in response to cues. If the training of the animal was too rapid for its age and capacities—and cats seemed to vary in intelligence as much as human beings do—the animal sometimes became recalcitrant, inept and resistive. If, however, the training process was adjusted to the individual cat, its behavior was efficient, well integrated and successful; indeed, pussy presented the appearance of a "happy" animal, as indicated by her eagerness to enter the laboratory, her avidity for the experimenter and the food-switch, and her *legato sostenuto* purring while she worked for her reward.

The animals were then subjected to various frustrations. For example, after a cat had been trained to depress a disc-switch to obtain food, the switch was so rearranged that its manipulation produced little or no reward. The animal would then develop a marked tendency to push down upon other objects in its environment, such as saucers, loops, boxes or other cats. This obsessive manipulative activity took many forms: sitting on the switch or on similar small platforms rather than in more comfortable places, prying into the experimenter's clothes instead of into the food-box, and so on.

Under other provocations the animals even exhibited conduct patterns which, when seen in human beings, have been called, misleadingly, "masochistic." Thus a cat was trained to accept a mild electric shock as a signal for feeding, and then taught to press a switch and administer the shock to itself in order to obtain the food. The intensity of the shock was then gradually increased to as much as 5,000 volts of a pulsating 15-milliampere condenser discharge; yet the animal continued to work the switch avidly for the food. Even when the reward was discontinued for long periods, the animal persisted in its accustomed pattern of depressing the switch, apparently solely for the substitutive experience of a "painful" electric shock. The observations suggest, however, that, contrary to Freud's paradoxical postulate of a death instinct, "masochistic" behavior is not basically "self-punitive" but rather a seeking for survival by patterns of response that seem awry only to an observer unacquainted with the unique experiences of the organism. In the light of the reactions revealed by these experiments, and by clinical investigations, we can understand why a woman may enjoy only certain "painful" forms of sexual intercourse when we learn that she reached her first orgasm while being beaten or raped; she may thereafter value all aspects of this erotic experience, including those considered by others as "painful." Similarly, we can cease to wonder why a man marries a succession of shrewish wives if we determine under deeper analysis that what appears to others to be nagging and persecution simply represents to him the security he had once experienced with his overattentive but devoted mother.

More complex frustrations, arising from social interactions, can also be demonstrated in animal groups with revealing clarity. In one type of experiment two trained cats, after a given feeding signal, are faced with a single food reward. At first they may skirmish a bit at the food-box. Soon, however, all external evidences of competition abate and only one of the animals—usually the more alert and intelligent—responds to the signal while its partner, though hungry, waits patiently until the "dominant" animal is either satiated or removed from the cage. Stable hierarchies of "privilege" can be produced in groups of four or more animals. The same animals may, however, range themselves in different orders of hierarchy for different activities. In short, there evolves a stratified "society" with fixed rankings in various activities.

One particularly enlightening variant of these experiments seemed to reproduce in cats "worker-parasite relationships" that are usually seen only in more elaborate forms of social organization. Two cats, each of which had been trained to manipulate a switch to obtain food, were placed in a single cage. The cage was equipped with a barrier between the switch and the food-box, so that the animal which essayed to work the switch could not reach the food-box until after its less enterprising partner had eaten the pellet. Under these circumstances some pairs of cats evolved a form of cooperative effort; they alternately worked the switch to feed each other. This cooperation, however, lasted no longer among cats than it does among men. One animal sooner or later showed tendencies toward "parasitism"; it ate the pellets produced by its partner's efforts but refused to leave the food to manipulate the switch. The worker

444

animal, finding its own "cooperative" behavior completely unrewarding, in turn ceased to produce food. Both animals, the parasite usually near the food-box and the worker near the switch, lolled about the cage for hours in a travesty of a sit-down strike. But as hunger increased, the relatively undernourished cat that had worked the switch usually would discover that if the switch were depressed six or eight times in rapid succession to release as many food pellets, he could scramble back to the box in time to get the last pellet or two before the parasitic partner gulped them all. In these experiments the end result was that the "worker" animal labored hard for a meager living while supporting its parasitic partner in leisure—a form of relationship apparently accepted by both animals. A few "workers" learned to jam the switch so that food was provided without further effort. They thus found a "technological" solution to a "socio-economic" problem.

Now it is a noteworthy fact that even in circumstances of direct rivalry these animals seldom became hostile or combative toward one another. Indeed, overtly aggressive behavior occurred so infrequently that special experiments had to be devised to determine the specific circumstances under which such behavior could be elicited. These studies demonstrated that animals are likely to become overtly belligerent only under two sets of conditions: 1) when they are displaced from a position of social dominance to which they have become thoroughly accustomed, or 2) when their goal-seeking activities are internally inhibited by neurotic conflicts.

The first situation is illustrated by this series of experiments: Let four cats, designated as Group A, compete for food under controlled conditions until Cat A1 emerges dominant, with A2, A3 and A4 in order below him. Let another group, B, range themselves correspondingly as B1, B2, B3 and B4. If A1 and B1 are now paired, a new contest of speed and skill occurs. As before, each animal at first strives for the food directly and diverts none of its energies into physical attacks on the other. Once again, of course, one animal emerges dominant—say B1. A1 now gives up its efforts to obtain the food reward as long as B1 is in the cage. But between signals A1 may sit on the food-box menacing B1 with tooth and claw, or it may even attack B1 viciously, although it makes no effort to follow up such attacks with sallies at the food.

The second type of situation that leads to aggression—the production of a neurotic conflict in an animal—can also be demonstrated experimentally. If, for example, the dominant animal in a group is made fearful of feeding on signal he will abandon this learned response and permit a subdominant animal to feed instead —yet attack the latter between feedings.

We shall consider briefly how these experimental neuroses are produced in animals and the methods by which the behavior of such animals may be restored to "normal." This portion of our work is perhaps the most relevant to clinical psychiatry, in its older, limited sense as the study of the "abnormalities" of behavior.

The concept of "conflict" has been central to many theories about the causes of neurotic aberrations. In biodynamics this concept is somewhat clarified by postulating that patterns of behavior come into conflict either because they arise from incompatible needs, or because they cannot coexist in space and time. This general statement can be exemplified by a relatively simple method of producing an experimental neurosis in animals:

A cat was trained to manipulate an electric device which first flashed a light, then rang a bell and finally deposited a pellet of breaded salmon in a food-box. The animal was permitted over a period of months to become thoroughly accustomed to this routine of working for the food. One day, however, just as the animal was about to consume its reward for honest labor it was subjected to a physically harmless but "psychically traumatic" stimulus, e.g., a mild air-blast across its snout or a pulsating condenser shock through its paws. The animal dropped the food, beat a startled retreat from the food-box and began to show hesitation and indecision about again manipulating the switch or approaching the food-box. When it did try again, it was permitted to feed several times but then subjected once more to the disruptive blast or shock. After from two to seven repetitions in as many days of such conflict-inducing experiences, the animal began to develop aberrant patterns of conduct so markedly like those in human neuroses that the two may be described in the same terms.

Neurotic animals exhibited a rapid heart, full pulse, catchy breathing, raised blood pressure, sweating, trembling, erection of hair and other evidences of pervasive physiologic tension. They showed extreme startle reactions to minor stimuli and became "irrationally" fearful not only of physically harmless light or sounds but also of

446

closed spaces, air currents, vibrations, caged mice and food itself. The animals developed gastro-intestinal disorders, recurrent asthma, persistent salivation or diuresis, sexual impotence, epileptiform seizures or muscular rigidities resembling those in human hysteria or catatonia. Peculiar "compulsions" emerged, such as restless, elliptical pacing or repetitive gestures and mannerisms. One neurotic dog could never approach his food until he had circled it three times to the left and bowed his head before it. Neurotic animals lost their group dominance and became reactively aggressive under frustration. In other relationships they regressed to excessive dependence or various forms of kittenish helplessness. In short, the animals displayed the same stereotypes of anxiety, phobias, hypersensitivity, regression and psychosomatic dysfunctions observed in human patients.

In nearly every case these neurotic patterns rapidly permeated the entire life of the animals and persisted indefinitely unless "treated" by special procedures. Some of the therapies are strikingly similar to those used in the treatment of human neuroses.

A neurotic animal given a prolonged rest of three to 12 months in a favorable home environment nearly always showed a diminution in anxiety, tension, and in phobic-compulsive and regressive behavior. The neurotic patterns were prone to reappear, however, when the animal was returned to the laboratory, even though it was not subjected to a direct repetition of the conflictual experiences. To draw a human analogy, a soldier with severe "combat fatigue" may appear recovered after a rest in a base hospital, but unless his unconscious attitudes are altered his reactions to latent anxiety recur cumulatively when he is returned to the locale of his conflicts.

If a neurotically self-starved animal which had refused food for two days was forcibly tube-fed, the mitigation of its hunger reduced its neurotic manifestations. In another experiment a hungry neurotic cat was prevented from escaping from the apparatus and was pushed mechanically closer and closer to the feeder until its head was almost in contact with a profusion of delectable pellets. Under such circumstances some animals, despite their fears, suddenly lunged for the food; thereafter they needed less mechanical "persuasion," and finally their feeding-inhibition disappeared altogether, carrying other neurotic symptoms with it. This method is in some ways akin to pushing a boy afraid of water into a shallow pool. Depending on what his capacities are for reintegrating his experiences, he may either find that there was no reason for fear or go into a state of diffuse panic. Because of the latter possibility, ruthless force is generally considered a dangerous method in dealing with neurotic anxieties.

The example of normal behavior sometimes has favorable results. An inhibited, phobic animal, after being paired for several weeks with one that responds normally in the experimental situation, will show some diminution in its neurotic patterns, although never complete recovery. It is well known, of course, that problem children improve in behavior when they have an opportunity to live with and emulate the more successful behavior of normal youngsters—although more specific individual therapy is nearly always necessary to complete the "cure."

A neurotic animal becomes exceedingly dependent upon the experimenter for protection and care. If this trust is not violated, the latter may retrain the animal by gentle steps: first, to take food from his hand, next to accept food in the apparatus, then to open the box while the experimenter merely hovers protectively, and finally to work the switch and feed without special encouragement from the "therapist." During its rehabilitation the animal masters not only its immediate conflicts but also its generalized inhibitions, phobias, compulsions and other neurotic reactions. This process may be likened to the familiar phenomenon of "transference" in clinical psychotherapy. The neurotic patient transfers his dependent relationship to the therapist, who then utilizes this dependence to guide and support the patient as the latter re-examines his conflictful desires and fears, recognizes his previous misinterpretations of reality and essays new ways of living until he is sufficiently successful and confident to proceed on his own.

We have also tested on these animals the effects of drugs, electroshock and other physical methods used in the treatment of behavior disorders. Sedative and narcotic drugs were first tried on normal animals. In one series of experiments an animal was taught 1) to open a food-box, 2) to respond to food-signals, 3) to operate the signal-switch, 4) to work two switches in a given order, and finally 5) to traverse a difficult maze to reach one of the switches. If the animal was then drugged with a small dose of barbital, morphine or alcohol, it became incapable of solving the maze but could still work the food-switches properly. With larger doses, it could "remember" how to work only one switch; with still larger doses, earlier stages of learning also were disintegrated, until finally the animal lost even the simple skill required to open the food-box. In other words, in moderate doses a drug disorganizes complex behavior patterns first while leaving the relatively simple ones intact.

Now if an animal is made neurotic and then is given barbital or morphine, its anxiety reactions and inhibitions are significantly relieved. Instead of crouching tense and immobile in a far corner or showing fear of the feeding signals, it opens the food-box and feeds (albeit in a somewhat groggy manner), as though for the time being its doubts and conflicts are forgotten. Obviously the recently formed, intricate neurotic reactions are relatively more vulnerable to disintegration by the sedative drugs than the animal's preneurotic patterns.

In one variant of these studies, animals which were drugged with alcohol and experienced relief from neurotic tensions while partly intoxicated were later given an opportunity to choose between alcoholic and nonalcoholic drinks. Significantly, about half the neurotic animals in these experiments began to develop a quite unfeline preference for alcohol; moreover in most cases the preference was sufficiently insistent and prolonged to warrant the term "addiction." This induced-dipsomania generally lasted until the animal's underlying neurosis was relieved by nonalcoholic methods of therapy. In still another series of experiments we observed that the administration of hypnotic drugs, including alcohol, so dulled the perceptive and memory capacities of animals that while thus inebriated they were relatively immune to emotionally traumatic experiences. It hardly needs pointing out, in this connection, that many a human being has been known to take a "bracer" before bearding the boss, flying a combat mission or getting married, and that temporary escapes of this nature from persistent anxieties often lead to chronic alcoholism.

We also investigated the effects of cerebral electroshock on neurotic animals. The shock produced by the 60-cycle current usually employed in this treatment acted upon animals like an intoxicant drug, disintegrating complex and recently acquired patterns of behavior in both "normal" and "neurotic" animals. Unlike most drugs, however, electroshock produced permanent impairment of behavioral efficiency and learning capacity. Weaker or modified currents such as are now being tested clinically (*i.e.*, the direct square-wave Leduc type) produced lesser degrees of deterioration in our animals, but also had less effect on their neurotic behavior. All in all, these experiments supported the growing conviction among psychiatrists that electroshock and other drastic procedures, though possibly useful in certain relatively recent and acute psychoses, produce cerebral damage which charges the indiscriminate use of such "therapies" with potential tragedy.

All this is only a condensed summary of a long series of experiments designed to analyze the biodynamics of behavior and to discern principles that may apply to human behavior and to psychotherapy. To be sure, the gap between the responses of cats, dogs or monkeys in cages and the conduct of man in society is undeniably wide; certainly man, of all creatures, has developed the most elaborate repertoire of "normal," "neurotic" and "psychotic" behavior patterns. Yet, as elsewhere in medicine, the best way to unravel an especially complex problem is to take it into the laboratory as well as the clinic, to investigate it by specially designed experiments, to check the results with a rigid self-discipline that eliminates subtle errors and cherished preconceptions, and so to advance bit by bit toward clearer formulations of general principles and more pertinent applications of them. Such experimental and operational approaches, when correlated with clinical practice, may dissolve the verbal barriers among the various schools of medical psychology and foster a needed rapprochement between psychiatry on the one hand and scientific medicine and the humanities on the other.

Beyond this, the work in biodynamics presents some fundamental social implications. Our observations of the causes of aggressive behavior among animals support the clinical and sociological conclusions of Karen Horney, John Dollard and others (including the author) that hostilities among human beings also spring from the frustrations and the anxiety-ridden inhibitions of their persistently barbaric culture—not, as Sigmund Freud believed, from an inborn, suicidal "death instinct."

PART 10

I. THE ANTIQUITY OF MODERN MAN
by Loren C. Eiseley

Loren Eiseley's professional interests range from general biology to cosmology, but he is primarily an anthropologist and is chairman of that department at the University of Pennsylvania. His field work has been focused on early man in the New World; he has done extensive digging in the western U. S. and Mexico. In addition to writing profusely in his professional field, Professor Eiseley contributes short stories and verse to the popular magazines.

II. THE APE-MEN by Robert Broom

At the time of his death in 1951, Robert Broom was curator of the Transvaal Museum in Pretoria, South Africa. Born in Paisley, Scotland, he was educated as a physician. His profession, however, served merely as a means of livelihood during the many years in which his fossil hunting led him into odd corners of the world. As early as 1903 Dr. Broom began to gravitate toward South Africa, believing that he would there eventually find the fossil precursors of man. In addition to his work on the ancestry of man, he is generally credited with having located the origin of mammals.

III. THE SCARS OF HUMAN EVOLUTION
by Wilton M. Krogman

Wilton Krogman divides his work between studies of the very old and the very young. He teaches physical anthropology at the Graduate School of Medicine, University of Pennsylvania, and is also director of the Philadelphia Center of Research in Child Growth. His investigations vary, therefore, from the teeth of prehistoric man to the weight and height of Philadelphia school children. He is also occasionally consulted by the coroner's office as an expert on skeletal identification.

IV. MAN'S GENETIC FUTURE by Curt Stern

As an experimental geneticist, Curt Stern has worked almost exclusively with fruit flies, but as a teacher of genetics he has in recent years devoted more attention to the direct study of inheritance in man. Born and educated in Germany, Professor Stern came to this country in 1932 as a fellow of the Rockefeller Foundation and, like Theodosius Dobzhansky, studied under T. H. Morgan. Today he is professor of zoology at the University of California, Berkeley.

V. IS MAN HERE TO STAY? by Loren C. Eiseley

In demonstration of his versatility, Dr. Eiseley opens this section of the book with an essay on anthropology and closes it with one on genetics.

ORIGIN OF MAN

Introduction

A<small>N EARLIER</small> section has treated of evolution from the viewpoint of biology; here the writers deal with human evolution historically. Granted that men and molds both sprang from the same primordial protoplasm, what is the path by which the human line descends? And where does that path tend in the future? There are those who see us climbing forever through hierarchies of excellence, but that is a dogma and inaccessible to proof. The geneticists and evolutionists who try to look at least a page or two ahead do not offer us the satisfaction of knowing that we are the rough forebears of a better race. Wherever man stands is by his definition the center of time and space; the discussions which follow try to place him more objectively in the evolutionary pattern. For many readers, this will prove to be the most fascinating and sobering section of the book.

The subject of man's forebears has always been clouded by prejudice and preconceptions. It is difficult, as Loren Eiseley points out here in his history of 10 skulls, even for scientists to avoid the error of believing that there was something special about the creation of the human species, that we are the goal toward which evolution has been striving from the beginning. Dean Swift shakes us for a moment with his Houyhnhnms, but we soon regain our place at the apex of the great plan. We sense a difference of kind between man and the beasts and the gulf strikes us as too wide for the selective processes of evolution to bridge. Hence the romantic appeal of that still unidentified hero of the folklore of evolution, the missing link.

It is perfectly possible that we have already held a skull of the missing link in our hands. Robert Broom's account of his diggings in South Africa suggests that this may well be so. His man-apes take us farther back down our branch toward the trunk of the family tree. But how is one to distinguish ape-man from man-ape? At what point do you say that the race of man has been born? The first man, perhaps, was the one who knew himself as "I," but you cannot tell from an empty skull whether or not the brain it formerly held was burdened with the consciousness of self.

What we can do is fill in the gaps in the hope that eventually the chain will be complete, even though the link himself cannot be found. The progress in that work has been great in our century and the results have been surprising. It now seems that we are not the only race of men the world has known, though we may be the only one meriting the title of *sapiens*. *Homo sapiens,* it appears, is not the heir but the contemporary and successful rival of those beetle-browed, huge and lumbering men who died off in the successive ice ages. Our own line goes back much farther than anyone had suspected when the search began; it is no wonder, perhaps, that ancient legend tells us of titans on the earth.

The thread of descent is traced by comparative measurement of the bones turned up from time to time in excavations, in caves and in cliffsides. Wilton Krogman shows another way of tracing it, in the bones and organs of our own bodies. The anatomical result of evolutionary adaptation is far from satisfactory; the architecture of our bodies is characterized by makeshift solution, rather than purposive design. Perhaps no price was too great to pay for the advantage of walking erect. We have evolved into creatures that do some things exceptionally well, even though we are scarcely commendable as all-around animals.

And what of the future? Curt Stern, speaking for the geneticists, estimates that the human race has some 80 trillion genes in its hereditary storehouse; the literally countless combinations in which these can appear argues a tremendous potential flexibility in the race. A factor that must be weighed against these riches in any guess as to man's destiny is the high degree of his specialization. Generally speaking, highly specialized animals do not endure; a change of the environment finds them with their genetic resources exhausted, incapable of new adaptation. But man's specialization—intelligence—is a novelty in the world. It permits him to shape environment largely to his needs. The survival value of intelligence may seem less obvious today than it did in the time of Darwin. Short-range pessimism, however, has no place in the long vistas of evolution.

I. THE ANTIQUITY OF MODERN MAN

by Loren C. Eiseley

TEN SKULLS of Ice Age Europe, covering roughly the period from 50,000 to one million years ago, have baffled a generation or more of geologists and anthropologists. The mystery concerns the fossils' age, and this, in turn, affects the entire question of how the human line has evolved. The key questions in the mystery are whether the beetle-browed Neanderthal man was really our ancestor or an unhappy cousin doomed to extinction, whether *Homo sapiens* is a recent arrival or a hardy species that has stood the test of evolution for several hundred thousand years. In short, how old is modern man? How far back can his characteristic features—our features, the friendly faces that greet us at the club—be traced?

During the summer of 1947 a portion of the mystery was solved. In the cave of Fontèchevade in the French Department of Charente, a few fragments of an old skull were brushed carefully out of the ancient clays. The strange fact about this skull was that it did not look strange; it was a skull very much like your own.

To an anthropologist, that was astounding enough. The great French prehistorian Henri Vallois came and marveled. A few letters were exchanged among scientists. One whose theories had been blasted protested harshly that there must be some mistake. Before there had always been some reason to dismiss such findings as the fumblings of amateurs or an accident of nature that had misplaced the fossils. But this time there could be no mistake, and the doubters grew angrily silent. It was the end of an era, and a new interpretation of human history was now in order.

At Fontèchevade Mademoiselle Henri-Martin, a quiet, amiable French scientist, daughter of a famous archaeologist, continued to busy herself with the restoration of the skull she had discovered. No inquiring reporters intruded, and it was just as well. After six years of laborious effort in the earth one did not want to be hasty; one should establish one's evidence beyond doubt.

Like all true stories the tale of the skull is difficult to tell because the threads are many and lead to strange places and even stranger characters. You can say it began with Darwin or the priest Mac-Enery, or with the eccentric American doctor, Robert Collyer. It is all of this and more, because it concerns man's infinite yearning to know the truth about himself, and that truth he will never possess until he has trekked backward into time far enough to see his own footprints merge humbly into those of the lesser beasts.

Archaeology is just a little over 100 years old, and in that century, man's notions about his history have altered tremendously. Looking back, we can discern two periods of firmly held preconceptions about human origins and we can see also their successive rejection. Three episodes sketch the stages through which this controversy has passed.

A little over 100 years ago, a Catholic priest, Father J. MacEnery, began to carry on some excavations in Kent's Cavern, a famous old cave in the south of England. The time was one in which the Biblical conception of creation still reigned. Mankind, it was thought, could be no older than 6,000 years. Georges Cuvier, the great French paleontologist of the time, is reputed to have tossed out of the window in disgust a human jaw brought to him by someone who thought it associated with fossil animals of the distant past. Scientists and laity alike slapped their thighs and roared with laughter at the ideas of lunatics who talked about tools and bones older than the world itself. Nevertheless that world was changing. Strange things had been found in caves in Germany and France—unbelievable things, of course—but Father MacEnery was curious. He left his contemporaries chortling in their taverns and set out with a shovel to investigate.

In the echoing galleries of the cavern, behind the town of Torquay, the priest found his answer. From the cavern floor he unearthed implements of stone and bone lying in the same stratum with the bones of extinct animals—the great cave bear, the mammoth, the rhinoceros. Father MacEnery, Roman priest, had stepped across an invisible threshold; he had entered the Pleistocene.

It is true he was not quite the first to dig in the English caves. Dean Buckland, then reader of geology at Oxford, had dug at Paviland. Then in his *Reliquae Diluvianae* of 1823 he had given the lie to all he had seen by maintaining that the strange associations of men and beasts he had found were deceptive; the human remains had been accidentally carried into the caves after the time of the universal Deluge. Dean Buckland was an authority. He had reconciled theology and science.

But MacEnery shook his head. No, he maintained. The evidence pointed otherwise. Men had lived here long ago. Men had lit their fires here and cooked their food. Men far away in time, contemporaries of the great gray mammoths.

Father MacEnery spoke, but the Dean thundered. He was the leading authority on caves and the priest evidently was moved to reconsider. He laid aside his book in manuscript; *Cavern Researches* was not published until long after his death. He dug no more. It was 30 years before science accepted what Father MacEnery had seen when he lifted up his torch and looked full into the world of ice. And it took the rest of the century and the long thought of a biological genius, Charles Darwin, to convince the world that human time must be measured in eons and that on the trail back into the past the bodies of men and animals melt and flow and change from age to age like the hills they move upon.

Even then, perhaps, the vision was still beyond us. The human mind always tends to erect new dogmas, to shelter itself in hastily erected systems against what is not known or what proves at last to be unknowable. The forms of paleoanthropic, big-browed fossil men began to be discovered. Though their numbers were few, scientists fitted them into a system—a single line of ascent leading to modern man. A form like Pithecanthropus, for example, led on in the following age to Neanderthal man, and the latter was regarded as our own direct ancestor. At the other end of the succession, the beginning, was an age generally conceived of as differing little from a modern chimpanzee.

The sequence was thought of as short and very direct. The time scale was still being underestimated, and western Europe, actually marginal to the Asiatic land mass, was unconsciously overemphasized as an evolutionary center for mankind. In addition, certain preconceptions were making it difficult to survey the problem of the origin of modern man in an unprejudiced light.

The most obvious of these preconceptions was, of course, the idea that since the remains of Neanderthal man had been found in European deposits immediately underlying our own species, we must be a later breed. Thus there could be no valid remains of *Homo sapiens* that were as old as Neanderthal man in Europe. Aleš Hrdlička, for example, in his Huxley Memorial Lecture of 1927 at London, scoffed at the idea that modern man might have developed before Neanderthal. In his mind there was no doubt that Neanderthal man, placed in the Mousterian period some 100,000 years ago, had slowly been transformed into a creature like ourselves sometime during the middle of the last great ice sheet. The final transformation he attributed, rather crudely, to the selective effect of a rigorous glacial climate.

Curiously enough, however, almost from the beginning there were faint clues that pointed in another direction. For illustration the case of Robert Collyer might be cited. He was an American physician residing in London and actively interested in everything from hypnotism to bones. Intrigued, perhaps, by the Darwinian controversy, he purchased a human jawbone and published a paper about it in 1867. The fossil was submitted to T. H. Huxley and other famous authorties of the day. None seems to have been particularly impressed.

Collyer's claim for the antiquity of his specimen lay in its fossilized state and the fact that it had been obtained from a gravel pit near Foxhall at a depth considerably below the surface. Perhaps the fact that it had once changed hands for a glass of beer did not inspire confidence in its origin. At all events, after it had passed under many eyes, interest waned, *largely because the jaw was modern in appearance.* The disappointed doctor is believed to have turned homeward to America. With him went the Foxhall mandible. Together they vanish from the sight of science. An engraving of the jaw which has come down to us, however, suggests that it did indeed look like modern man's.

The irony of the tale lies in the fact that long, long afterward, in 1922, the English archaeologist Reid Moir relocated the old Foxhall quarry and established an early Pleistocene cultural horizon within it. If the jaw actually came from this level, as Moir believed, we would have undoubted evidence that a form of man like ourselves was wandering on the European continent long before the time of Neanderthal man.

The quarry in which the discovery was made should have been investigated immediately. But unfortunately attention centered on the mandible itself and, since there was nothing about it that the anatomist could surely regard as primitive, interest quickly faded. It cannot be too often emphasized that if the type of man that now exists should prove to be very old, only geology and the study of man's associated tools and implements will have established the fact.

The finds accumulated. Sir Arthur Keith, the great English scholar, catalogued many of them in his work *The Antiquity of Man* in 1925. There were other discoveries in France, in Italy and again in England. Always the doubt remained. Nor was it all mere prejudice. Our digging luck had been bad. When one finds a Neanderthal man, one knows one is handling ancient material. With our own human type, the bones may tell nothing or may speak in riddles. We must have other evidence of an irrefutable character.

Sir Arthur recognized this when he wrote, a little wearily, of the Galley Hill specimen*: "The anatomist turns away from this discovery because it reveals no new type of man, overlooking the much greater revelation—the high antiquity of the modern type of man, the extraordinary and unexpected conservatism of the type. The geologist regards the remains with suspicion for two reasons—first, he has grown up with a belief in the recent origin, not only of modern civilization, but of modern man himself. He expects a real anatomical change to mark the passage of a long period of time. . . . Moreover, a very primitive type of man survived in Europe. . . . Hence the rejection of all remains . . . which do not conform to this standard."

There the argument stood. The Peking men were discovered— low-browed, small-brained, more primitive than Neanderthal. Their datings were not much older than the time suggested for certain of the sapiens specimens. Yet to imagine these two forms as standing so close to each other on the time scale with the one directly ancestral to the other strained credulity. The authenticity of the Galley Hill cranium seemed even less plausible than before. Then in 1935 the fragments of another skull were found, 24 feet below the surface of a gravel pit at Swanscombe, England.

The details of that discovery need not detain us. Here we are concerned only to note that these fragments, which unfortunately did not include the face or forehead, suggested very strongly a true *Homo sapiens* type. And this was associated with the Acheulean culture in geological deposits dated in the Second Interglacial! By comparison, Neanderthal man was alive just yesterday. There was no reasonable doubt of the skull's position, no reasonable doubt as to its geology or the sort of tools found with it. The anatomist W. E.

* The Galley Hill skull is no longer regarded as of early Ice Age antiquity. Fluorine dating methods suggest that, while pre-historic, it is probably post-glacial.

Le Gros Clark allowed that the skull gave evidence that already in early Paleolithic times the human brain had "acquired a status typical of *Homo sapiens*."

Nevertheless, the evidence was not complete. The face was missing. Were we sure, after all, that the face was like our own? Might it not have carried the heavy brow ridges of at least an advanced Neanderthal type? To confuse us further, finds in Palestine, at the much later date of the Third Interglacial, something over 100,000 years ago, suggested a Neanderthal type evolving in the direction of *Homo sapiens*. It was either that or a hybrid mixture between an already existing modern type of man and his heavy-browed relative. Once more argument raged. Even Sir Arthur Keith seemed to waver in his espousal of the antiquity of modern man. It is, then, by this involved and twisted pathway that we come to Mademoiselle Henri-Martin and the deposits in Charente.

The cave lies at the side of a small valley near the village of Montbrun. It had long been known to students of prehistory as having yielded a succession of stone industries extending from Mousterian times to the much later Magdalenian period of the post-glacial era. At the base of the Mousterian cultural layer—everywhere associated in Europe with Neanderthal man—earlier workers had struck a solid floor of stalagmite. There they had stopped.

Mademoiselle Henri-Martin was not so easily deterred. Near the mouth of the cave she broke through the stalagmitic floor and found, in the red, sandy clay underneath it, an older, cruder flint industry marked by large flakes which the French prehistorian Henry Breuil has termed Tayacian. It is regarded as a flake culture transitional between the Mousterian and an earlier period.

Many cultural horizons contain no human remains, but here, abandoned among flint chips and the bones of animals, lay a human skull. One can imagine the eager brushing away of earth, the careful manipulation of tools. Here, certainly, must lie an ancestral Neanderthal. This is the Third Interglacial time. The long, cold night of the Fourth Glacial is still far away in the future.

The skull is too worn, too delicate to free quickly from the encasing earth. The hours go on. It is seen not to be complete; finds of such great antiquity rarely are. Nevertheless, the two parietal bones forming the major part of the sides of the head appear. Part of the occipital bone at the back becomes visible, and a fragment of the

460

frontal. It is not, however, the part of the frontal that can tell us about the brow ridges. But for all that, this skull has an oddly familiar look.

In the bony debris painstakingly gathered by the workers, another human fragment was discovered—a very odd fragment that might easily be tossed aside by the inexperienced. Apparently belonging to a second individual, it is the final key to a story that might otherwise have ended like the debate over the Swanscombe skull. This is a glabellar fragment—a little piece from just over the root of the nose and including a little part of the orbit of the eye. There is no trace of a brow ridge. The orbital edge has the delicate sharpness of a modern woman's. This is *Homo sapiens!* This fossil woman saw with living eyes the warmth-loving fauna of the Third Interglacial. In the trench with her lie scattered the remains of *Rhinoceros mercki* and a Mediterranean turtle. The woolly mammoth and the woolly rhino of the last glaciation have not yet come. In the opinion of Henry Vallois, the fossil seems to validate the authenticity of the Swanscombe discovery and, over and beyond, to confirm the existence of a non-Neanderthaloid type on the continent prior to the Mousterian period.

He states after an exhaustive analysis of the fragment of frontal bone that "it falls into agreement only with a straight and almost vertical forehead such as that of the most fully evolved *Homo sapiens.*" After reconstruction, the entire skull cap appears "to have dimensions comparable to those of living Europeans." Vallois italicizes in his report the remark that *"this is the first time that man certainly not Neanderthal although earlier than the Neanderthal has been found in Europe under such conditions."*

Again and again, in the case of previous discoveries, the question of intrusive burial had arisen—the possibility, in other words, that the bones were younger than the cultural stratum in which they were found. But Dr. Hallam Movius, a leading authority on the Old World Paleolithic, says: "There can be no question concerning the fact that these finds were *in situ* [in their original site] when discovered by Mademoiselle Henri-Martin: they come from an undisturbed horizon sealed below a thick, unbroken and continuous layer of stalagmite that underlies the Mousterian level at this locality. Furthermore, the fauna demonstrate that these deposits were accumulated under conditions of the warm temperate climate of Third

461

Interglacial times. And the archaeological material is definitely older than the Mousterian from a typological point of view."

Despite a few vaguely "primitive" traits, which can often be duplicated in modern skeletons, the existing remains of this fossil cannot be clearly differentiated from equivalent structures in modern man. At the time of the discovery there was thus a strong feeling among archaeologists that proof of the greatly extended antiquity of our own species was at hand. The writer shared that view and is still inclined to accept it as the most satisfactory explanation of the skulls. Nevertheless, an unexpected hindrance has arisen. Dr. Kenneth Oakley of the British Museum, using the fluorine dating method (flourine content of a fossil bone increases with age), has shown that the Piltdown skull, instead of being an early "dawn man," probably belongs to the third Interglacial.

Now Piltdown itself is a puzzle upon which innumerable papers have been written. In brief, the specimen involves a thick skull vault of *sapiens*-like appearance associated with a very primitive anthropoidal jaw. No other specimen like it is known, and the validity of associating the jaw and skull has often been challenged. Dr. Oakley's dating does, however, place the Piltdown specimen from England on a time level with the Fontéchevade fragments. Since these lack the lower jaw, numerous anthropologists have been struck by the possibility that the Fontéchevade individuals may have had similarly primitive mandibles with a "modern" upper face. Therefore, until a *sapiens* lower jaw is found in deposits of the third Interglacial, many scientists will be loath to commit themselves finally on the precise classification of the Swanscombe and Fontéchevade crania.

I am inclined, by the fact that a *sapiens* lower jaw is already present among the Palestine "hybrids" of somewhat similar age, to believe that we are perhaps being over-cautious about accepting the French specimens as of true *sapiens* character. It seems unlikely, for example, that the chimpanzee-like jaw of the Piltdown skull was transformed into a typical modern mandible within the short time

Brain pans of skulls found at Charente and Swanscombe suggest that *Homo sapiens* far predates Neanderthal man. In drawings they are shown as true "modern" types which the author considers a likely possibility, but final classification of these second- and third-interglacial individuals awaits discovery of lower jaws. In any case, it is now certain that beings with essentially modern brain cases lived long before Neanderthal man.

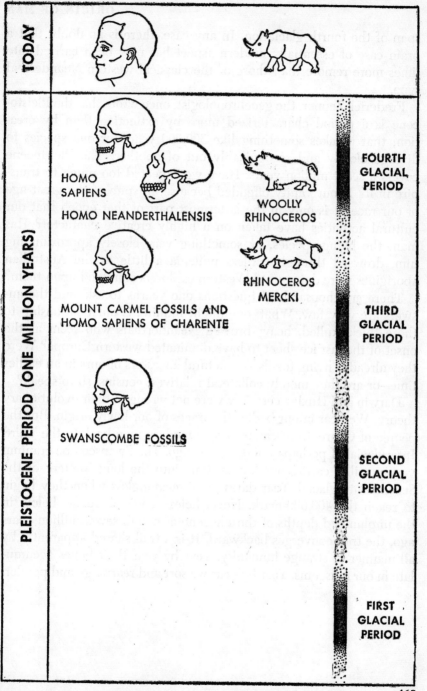

TODAY

PLEISTOCENE PERIOD (ONE MILLION YEARS)

HOMO
SAPIENS

HOMO NEANDERTHALENSIS

WOOLLY
RHINOCEROS

RHINOCEROS
MERCKI

MOUNT CARMEL FOSSILS AND
HOMO SAPIENS OF CHARENTE

SWANSCOMBE FOSSILS

FOURTH
GLACIAL
PERIOD

THIRD
GLACIAL
PERIOD

SECOND
GLACIAL
PERIOD

FIRST
GLACIAL
PERIOD

span of the fourth glaciation. In any case, there is no doubt that a brain case of essentially modern aspect has now been carried into times more remote than those of the classical western Neanderthal type.

Frederick Zeuner, the geochronologist, once wrote that the Pleistocene is a period characterized more by extinction than by creation; that it takes something like 500,000 years for one species to diverge clearly and recognizably out of another. The Pleistocene covers a scant million years. Have we expected too much to transpire in it? Is our vanity offended because, in spite of the great age of our race, it is only in the lattermost part of that epoch that our cultural activities have taken on a highly creative character? Has man, the living species, or something very closely approximating him, drowsed through endless millenia a little as the Australian aborigines were doing until Western explorers stumbled upon them?

There are thousands of questions one yearns to ask, and the answers are very few. What, one might inquire, is our relationship to those thick-skulled, heavy-browed Neanderthals who seem at the onset of the last ice sheet to have dominated western Europe? Were they already "living fossils"—structural ancestors of ours in an earlier time—or are they merely collateral relatives; cousins, so to speak?

Darwin and Huxley certainly were not wrong in their evolutionary theory. We bear in our bodies the traces of our lowly origin. But the people of Charente threaten to take modern man back to the Third Interglacial of perhaps 100,000 years ago. The Swanscombe cranium very possibly carries our human type into the long summer of the Second Interglacial. Year dates grow meaningless when they begin to reach the 400,000-mark. Nevertheless, somewhere far below in the unplumbed depths of the Pliocene of one to seven million years ago, the trail converges backward. It is a trail shared apparently by all manner of strange humanity. Year by year their bones accumulate in our museums. Year by year we sort and rearrange and ponder.

by Robert Broom

T HOUGH THERE WERE many eminent evolutionists before Charles Darwin, it was he who in 1859 first convinced the majority of scientists that man had evolved from some ape or apelike being. Of course there were some who were unconvinced, and even today there are still a few; but no scientist of any eminence, so far as I know, would hold now that man is a special creation. Some deny that he evolved by "natural selection" as Darwin suggested, but every scientist probably is satisfied that whether man developed gradually or arose by a sudden mutation, he certainly came from parents who were apes or apelike.

Yet who were his ancestors? In 1892 Eugene Dubois found the famous Pithecanthropus skull in the Dutch East Indies and after that a number of fossil skulls were discovered in various parts of the world. But they were all identified as belonging to various types of early man, and threw little or no light on man's origin. In 1924, however, came a discovery that opened a new chapter in the story of man.

Near Taungs in South Africa is a large lime deposit with a few caves where fossil bones have been preserved. Numerous skulls of small apes or baboons had been found in these caves. Toward the end of 1924 a quarryman named M. de Bruyn one day blasted out a small skull which he immediately recognized as that of a being not unlike man. It was believed that he had discovered a fossil Bushman, and the skull was sent to the anatomist Raymond Dart in Johannesburg for his opinion. Dart, after cleaning and studying the skull, promptly sent off a paper to London claiming that it was that of a being intermediate between a higher ape and man. He called it *Australopithecus africanus*.

When the paper appeared, on February 7, 1925, all English and American scientists who expressed an opinion were unanimous in declaring that Dart had made a serious blunder; his little Taungs skull, they held, was only a variety of chimpanzee. Immediately after the discovery was announced I went to Johannesburg to see it and made a very careful examination, especially of the teeth. I was at once convinced that Dart was essentially correct in his conclusion, and that this was practically the "missing link"—the most important fossil find ever made. I wrote a paper supporting Dart. The noted paleontologist William J. Sollas of Oxford University, to whom I sent a median section of the skull, also was converted to our view, and became our strongest ally.

The discussion went on for years. Many in England and America eventually came around to the opinion that Dart was right, but as the skull was that of a young child, most anthropologists remained unconvinced.

In 1936, having taken a post in the Transvaal Museum in Pretoria at the suggestion of General J. C. Smuts, I resolved to look for a skull of an adult Australopithecus. I felt that even if I did not get what I sought, I was sure at the least to find other interesting fossil forms in the rich cave deposits. I started to work at caves near Pretoria, and immediately found a considerable number of new fossil mammals. Early in August, 1936, two of Dart's students visited me and told me of caves at Sterkfontein, near Krugersdorp, where they had found some small fossil baboon skulls. This seemed so promising a locality that we all visited the cave on Sunday, August 9.

The caves had been known for over 40 years. Mining operations were being carried on for impure lime in the deposits. G. W. Barlow was the manager of the quarrying operations and the caretaker of the caves. He told me that he had once worked at Taungs, and that he knew something about the skull. I asked him if he had ever found anything like it at Sterkfontein, and he said he rather thought he had. Any nice bones or skulls he found he sold to visitors, and had not worried about what they were. I asked him to keep a sharp lookout for anything like an ape-man skull, and he said he would.

When I visited the caves again some days later, Barlow handed me two-thirds of a beautiful fossil brain cast. It had been blasted out that morning, and it was manifestly the brain cast of a fossil ape-man. I hunted among the blasted debris for some hours, but could get no more of the skull except the cast of the top of it, which I cut out of the side wall of the cave. Next day, after some three more hours' hunting, I found all the base of the skull, both upper jaws, badly displaced, and some fragments of the brain case. When the bones were all cleaned and assembled, we found we had most of the skull, except for the lower jaw, of a creature which we eventually called *Plesianthropus transvaalensis*.

During 1936, 1937 and part of 1938 we found many other remains of this Sterkfontein ape-man—bits of skulls, isolated teeth and parts of limb bones. (We now have many skulls and about 130 teeth of Plesianthropus.) One June day in 1938, when I arrived at the workings, Barlow said, "I have something nice for you this morning," and handed me a beautiful palate of a large ape-man with one molar tooth in position. I said, "Yes, that is a nice specimen. I'll give you a couple of pounds for it." He was quite pleased, but did not seem inclined to tell me where he had got the specimen. The matrix was different from that in the Sterkfontein cave, and I was sure it came from some other locality. When I insisted on knowing where it had come from, Barlow told me a schoolboy named Gert Terblanche had brought it to him from somewhere at Kromdraai, a farm about two miles away.

I traced Gert to his school and the boy drew from his trouser pocket and handed to me four of the most beautiful fossil teeth ever found in the world's history. Two of the four fitted on the palate Barlow had given me. I promptly bought the teeth from Gert, and he told me he had another nice piece hidden away. After school he took me to the spot—an outcrop of bone breccia deposit—and uncovered a very fine jaw with several beautiful teeth.

In the next few days we sifted all the ground in the close neighborhood and recovered nearly every scrap of tooth or bone in the place. When all the bits were cleaned and joined, it was found that we had the greater part of the left side and of the right lower jaw of a very fine skull, with many of the teeth well preserved. The skull differed in a number of characters from that found at Sterkfontein, and it had a larger brain. In some respects it was more human; in a few, less human. We described it as a new genus named *Paranthropus robustus.*

Some English critics considered that I was too daring in identifying two entirely new genera on the basis of skulls which they thought were probably only adult skulls of the Taungs ape. When later we found the jaw of a baby Paranthropus, we discovered that not only are the Taungs, Sterkfontein and Kromdraai ape-men different genera; they perhaps belong to different subfamilies.

Although work at the caves was almost entirely stopped during the war, we had plenty of material collected to keep us busy. At the beginning of 1946 a book was published giving a full account of all the ape-man remains that had been found up to that date.

In August, 1947, General Smuts phoned to ask me to see him. He recognized that we were discovering the origin of man, and that the work at the caves must be carried on. He told me that whatever money I required would be provided by the Government.

We therefore resumed work and found at Sterkfontein a crushed palate of a young ape-man and part of the upper jaw of a baby. On April 18, 1948, a lucky blast revealed a perfect skull of an adult female. This is the finest fossil skull ever discovered—more important than the Pithecanthropus skull of Java, the jaw of Heidelberg man or the skulls of Peking man. Those were all remains of early man. This was the skull of a being not yet man but nearly man. The skull is practically human in all respects, except that the brain is small—only 480 cubic centimeters.

Our next discovery was an almost perfect male jaw, the most notable feature of which is that though the canine tooth is larger than in man it has been ground down in line with the other teeth exactly as in man. This never happens in males of the anthropoid apes. Then we made an even more important find—a nearly perfect pelvis. This structure, human in all essentials, proves that the ape-men walked on their hind legs.

In October, 1948, Wendell Phillips of the University of California Expedition suggested that I might start work at a new cave deposit; he would finance the work and we could share the results. We started at a promising deposit at Swartkrans, only a mile from the main Sterkfontein cave. Immediately we were amazingly successful. Within a few days we discovered a lower jaw of a new type of ape-man much larger than any we had known before. I named it *Paranthropus crassidens*.

My assistant, J. T. Robinson, discovered at Swartkrans a beautiful, nearly perfect lower jaw. It is really huge, possibly larger than the giant jaw from Java that has been called Meganthropus by the Dutch anthropologist G. H. R. von Koenigswald. It almost seems to confirm the view of the noted anthropologist Franz Weidenreich that "there were giants in the earth in those days," as stated in Genesis.

Primate family tree, as proposed by Dr. Broom, places the ape-men Australopithecus, Paranthropus and Plesianthropus on the main branch of human descent. Modern apes, on the contrary, came from an earlier offshoot of the trunk.

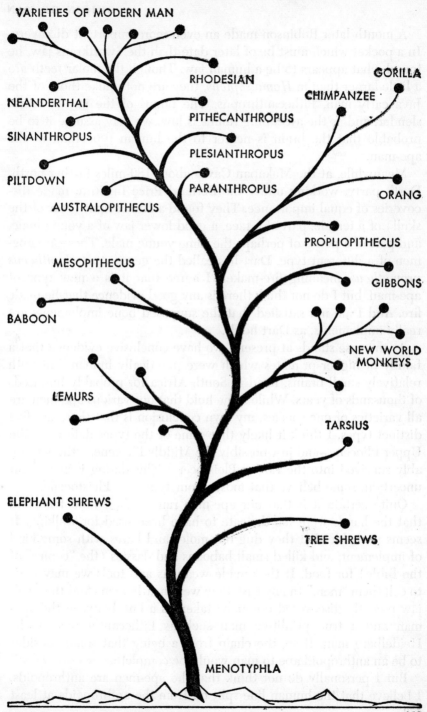

VARIETIES OF MODERN MAN

RHODESIAN

GORILLA

CHIMPANZEE

NEANDERTHAL

PITHECANTHROPUS

SINANTHROPUS

PLESIANTHROPUS

PILTDOWN

PARANTHROPUS

ORANG

AUSTRALOPITHECUS

PROPLIOPITHECUS

MESOPITHECUS

GIBBONS

BABOON

NEW WORLD
MONKEYS

LEMURS

TARSIUS

ELEPHANT SHREWS

TREE SHREWS

MENOTYPHLA

A month later Robinson made an even more important discovery. In a pocket which must be of later date than the Swartkrans jaw, he found what appears to be a human jaw. Though the molar teeth are a little larger than in *Homo sapiens*, they are not unlike those of the Java early man, Pithecanthropus. The ramus of the jaw is rather slender, and as the ascending ramus is low, we may assume it to be probable that the brain is nearer to the human type than in the ape-man.

Meanwhile, at the Makapan Caves about 180 miles farther north, Dart's party, working under the Bernard Price Institute made discoveries of equal importance. They found a fine occiput (rear of the skull) of a female, part of a face, a good lower jaw of a young male, and part of a pelvis of perhaps the same young male. These are ape-men of a different type. Dart has called the genus *Australopithecus prometheus*, meaning fire-maker, I agree that it is a new type of ape-man, but I do not think there is any good evidence that he made fire. And I am not satisfied that the supposed bone implements are really implements, as Dart holds.

As the case stands at present, we have conclusive evidence that a family of higher primates which were practically human, but with relatively small brains, lived in South Africa for probably hundreds of thousands of years. While some hold that our various ape-men are all varieties of one species, my own conclusion is that there are five distinct types. I think it likely that some of the types date from the Upper Pliocene—and just possibly the Middle Pliocene. Others probably survived into the Lower Pleistocene. This dating is at present uncertain; some believe that most of our types are Pleistocene.

Quite certain it is that our ape-men ran on their hind feet, and that the hands were too delicate to have been used for walking. It seems probable that they dug out moles and hares with some kind of implement, and killed small baboons and dassies (the "conies" of the Bible) for food. If they made weapons and tools we may have to call them "men." In any case they were nearly men. And the small jaw recently discovered might be taken as a link between the ape-man and a true primitive man such as Pithecanthropus or the Heidelberg man. If so, the chain from a being that some consider to be an anthropoid ape to man would be complete.

But I personally do not think that the ape-men are anthropoids. I believe that the human line split off from the anthropoids at least

470

as early as the Lower Oligocene, perhaps 25 million years ago, and that the nearest known type to man's remote ancestor is not a chimpanzeelike ape but the little fossil ape Propliopithecus of Egypt. I suggest that the ape-men of South Africa came on a different line from the higher apes, and that one of them became the ancestor of *Homo sapiens.*

III. THE SCARS OF HUMAN EVOLUTION

by *Wilton M. Krogman*

I
T HAS BEEN SAID that man is "fearfully and wonderfully made." I am inclined to agree with that statement especially the "fearfully" part of it. As a piece of machinery we are such a hodgepodge and makeshift that the real wonder is we get along as well as we do. Part for part our bodies, particularly our skeletons, show many scars of Nature's operations as she tried to perfect us.

I am not referring to our so-called vestiges—those tag-ends of structures which once were functional, such as the remnant of a tail at the base of the spine, the appendix, the pineal or "third" eye, the misplaced heart openings of "blue babies," or the like. Nor do I mean the freak variations that crop up in individuals. I am discussing the imperfect adaptations the human race has made in getting up from all fours.

We have inherited our "basic patents," as W. K. Gregory of the American Museum of Natural History calls them, from a long line of vertebrate (backboned) ancestors: from fish to amphibian to reptile to mammal and finally from monkey to ape to anthropoid to *Homo sapiens.* In all this evolution the most profound skeletal changes occurred when we went from a four-legged to a two-legged mode of locomotion.

Gregory has very aptly called a four-legged animal "the bridge that walks." Its skeleton is built like a cantilever bridge: the backbone is the arched cantilever; the vertebrae of the forward part of the backbone are slanted backward and those of the rear forward, so that the "thrust" is all to the apex of the arch; the four limbs are the piers or supports; the trunk and abdomen are the load suspended from the weight-balanced arch; in front the main bridge has a drawbridge or jointed crane (the neck) and with it a grappling device (the jaws).

When all this was up-ended on the hind limbs in man, the result was a terrific mechanical imbalance. Most of the advantages of the cantilever system were lost, and the backbone had to accommodate itself somehow to the new vertical weight-bearing stresses. It did so by breaking up the single-curved arch into an S-curve. We are born, interestingly enough, with a backbone in the form of the simple ancestral arch, but during infancy it bends into the human shape. When we begin to hold our head erect, at about the age of four months, we get a forward curve in the backbone's neck region; when we stand up, at about a year, we get a forward curve in the lower trunk; in the upper trunk and pelvic regions the backbone keeps its old backward curve.

But we achieve this at a price. To permit all this twisting and bending, Nature changed the shape of the vertebrae to that of a wedge, with the thicker edge in front and the thinner in back. This allows the vertebrae to pivot on their front ends as on hinges, like the segments of a toy snake. On the other hand, it also weakens the backbone, particularly in the lower back region, where the wedge shape is most pronounced. Heavy lifting or any other sudden stress may cause the lowermost lumbar vertebra to slip backward along the slope of the next vertebra. The phrase "Oh, my aching back" has an evolutionary significance!

There are other ways in which the backbone may literally let us down. The human backbone usually has 32 to 34 vertebrae, each separated from its neighbor by a disc of cartilage which acts as a cushion. Of these vertebrae 7 are cervical (in the neck), 12 thoracic (upper trunk), 5 lumbar (lower trunk), 5 sacral (at the pelvis) and 3 to 5 caudal (the tail). Most people have 12 pairs of ribs, borne by the 12 thoracic vertebrae, but occasionally the transverse processes of the next lower segment, the first lumbar vertebra, are so exaggerated that they form a 13th pair of ribs. In some people the lowest (fifth) lumbar vertebra is fused with the sacral vertebrae. The latter are usually united into one bone, called the sacrum, but sometimes the first sacral vertebra fails to join with its mates. All these idiosyncrasies can cause trouble.

The "Achilles' heel" of our backbone is the unstable lower end of the vertebral column. This is where we reap most of the evil consequences of standing up on our hind legs. It is a crucial zone of the body—the pathway for reproduction and the junction point where the backbone, the hind end of the trunk and the legs come together. The skeletal Grand Central Station where all this happens is a rather complicated structure consisting of the sacrum and the pelvis. The pelvis is not only a part of the general skeletal framework of the body but also a channel for the digestive and urogenital systems and the coupling to which the muscles of the hind legs are attached. When we stood up on our hind legs, we burdened the pelvis with still another function, namely, bearing the weight of the upper part of the body. How have we changed our pelvis to adapt it to its new position and burdens?

The pelvic structure is made up of three sets of paired bones, the ilium, the ischium and the pubis. The three bones meet at each side in the hip socket, where the head of the thighbone articulates. In standing erect man tilted the whole structure upward, so that the pelvis is at an angle to the backbone instead of parallel to it. The relative position of the three pelvic bones changed, with the pubis now in front instead of below. The bones also were altered in shape. The iliac bones, formerly elongate and bladelike (in the anthropoids), are now shortened and broadened. They form the crests of our hips, and they help support the sagging viscera, especially the large intestine.

The greatest change is in the zone of contact between the iliac bones and the wedgelike sacrum—the so-called sacroiliac articulation. Here are focused the weight-bearing stresses set up by the erect posture. Two things have happened to adapt the pelvic structure for "thrusting" the weight of the trunk to the legs. The area of contact between the sacrum and the iliac bones has increased, stengthening the articulation. In the process the sacrum has been pushed down, so that its lower end is now well below the hip socket and also below the upper level of the pubic articulation. This has brought trouble, for the sacrum now encroaches upon the pelvic cavity and narrows the birth canal that must pass the fetus along to life. Furthermore, the changes have created an area of instability which far too often results in obscure "low back pain" and in "slipped sacroiliacs."

The shortening of the iliac bones has increased the distance between the 12th (lowest) rib and the top or crest of the ilium. This has given us our waist, but it has also materially weakened the abdominal wall, which now, for about a palm's breadth, has only muscle to support it. The greatest weakness of the upright posture is the lower abdominal wall. In four-legged animals the gut is suspended by a broad ligament from the mechanically efficient convex vertebral arch. The burden of carrying the weight of the viscera is distributed evenly along the backbone. Up-end all this and what

474

happens? First of all, the gut no longer hangs straight down from the backbone but sags parallel to it. Secondly, the supporting ligament has a smaller and less secure hold on the backbone. One result of the shift in the weight-bearing thrust of the abdominal viscera is that we are prone to hernia.

Nature has made a valiant effort to protect our lower belly wall. She invented the first "plywood," and made it of muscle. Three sheets of muscle make up the wall, and their fibers crisscross at right and oblique angles. This is all right as far as it goes, but it has not gone far enough: there is a triangular area in the wall which was left virtually without muscular support—a major scar of our imperfect evolution.

The upright posture required a major shift in the body's center of gravity, but here Nature seems to have done a pretty good job. The hip sockets have turned to face slightly forward instead of straight to the sides; the sockets and the heads of the thighbones have increased in size, and the neck of the thighbone is angled a bit upward. As a result of this complex of adjustments the bodily center of gravity is just about on a level with a transverse line through the middle of the hip sockets, and the weight of the trunk upon the pelvis is efficiently distributed on the two legs.

Though it does not directly involve the skeleton, I might mention here that the blood circulation is another factor that is not helped by our upright position. Since the heart is now about four feet above the ground, the blood returned to the heart from the veins of the legs must overcome about four feet of gravitational pull. Often our pumping system and veins find the job too much, and the result is varicose veins. The lower end of the large intestine also is affected, for its veins, when up-ended in a vertical position, become congested more easily, so we get hemorrhoids.

Even more serious is the danger to the circulation along the vertebral column. Two great vessels, an artery and a vein, run down this column. At the level where these vessels divide into two branches, one for each leg, the right-sided artery crosses over the left-sided vein. In a quadruped this presents no problems, but in the erect position the two vessels must cross a sharp promontory of bone at the junction of two vertebrae, and the viscera piled up in the pelvis press down and back on them. During pregnancy the pressure may increase so much that the vein is nearly pressed shut, making for very poor venous drainage of the left leg. This is the so-called "milk leg" of pregnancy.

Going back to the skeleton, it is clear that the two-legged posture places a much bigger burden on our feet. They have adapted themselves to this by becoming less of a grasping tool (as in the monkeys) and more of a load-distributing mechanism. We have lost the opposability of the big toe, shortened the other toes and increased the length of the rest of the foot. The main tarsal bones, which form the heel, ankle joint and most of the instep, now account for half the total length of the foot, instead of only a fifth as in the chimpanzee. We have also achieved a more solid footing by developing two crosswise axes, one through the tarsals and the other through the main bones of the toes. The little-toe side of our foot is relatively neglected—the toe is little because it is not so useful. Our fallen-arch troubles, our bunions, our calluses and our foot miseries generally hark back to the fact that our feet are not yet healed by adaptation and evolutionary selection into really efficient units.

Now let us go to the other extreme—to the head. A lot has gone on there, too. We have expanded our brain case tremendously, and there can be no doubt that many of the obstetrical problems of Mrs. H. Sapiens are due to the combination of a narrower pelvis and a bigger head in the species. How long it will take to balance that ratio we have no idea. It seems reasonable to assume that the human head will not materially shrink in size, so the adjustment will have to be in the pelvis; *i.e.*, evolution should favor women with a broad, roomy pelvis.

If the head has increased in size, the reverse is true of the facial skeleton. Bone for bone the face has decreased in size as we proceed from anthropoid to man. To put it succinctly, we have a face instead of a snout.

What about the teeth in that face of ours? All mammals have four kinds of teeth: incisors in front, canines at the corner, premolars and molars along the sides. With but few exceptions the mammals have both a milk set and a permanent set of teeth. About 100 million years ago, or maybe a bit more, the first mammals had 66 permanent teeth, of which 44 were molars or premolars. Most mammals today have 44 teeth, including 28 molars and premolars. But man, and the anthropoid, has only 32 teeth—8 incisors (upper and lower), 4 canines, 8 premolars and 12 molars. The loss has been greatest in molars, next in incisors, then in premolars, with the canine a veritable Rock of Gibraltar.

While the face bones have decreased in size, our teeth have remained relatively large. Many orthodontists believe that this uneven evolutionary development may be partly responsible for the malocclusion of teeth in children. Certain it is that some human teeth are apparently on the way out: the third molars ("wisdom teeth") are likely to be impacted or come in at a bad angle, and many people never have them at all. Perhaps in another million years or so we shall be reduced to no more than 20 teeth.

IV. MAN'S GENETIC FUTURE

by Curt Stern

Assuming the human species is here to stay, what is likely to happen to us genetically? Will the human stock improve, deteriorate or remain the same? Is the future predestined, or can we direct it?

To answer such questions we must consider mankind's hereditary endowment as a whole and the distribution of this endowment among individuals. Let us assume, as we may for the purpose of this discussion, that the human germ cell has exactly 20,000 genes. That means that every one of the more than two billion people on earth today has acquired a set of 20,000 genes from the father and a similar set from the mother. These are shuffled like two decks of cards to produce new sets of 20,000 genes in the individual's own germ cells. Everyone has the same 20,000 kinds of genes but some genes appear in more than one form. Many probably occur in only one variety and are the same for everyone; others may show two, three, four and up to 100 varieties. In any case, the total pool of genes in the earth's population at present is some 80 trillion (two billion people times 20,000 pairs of genes each). This is the storehouse from which the genetic future of man will be furnished.

The number of different possible combinations of the varieties of genes is huge, so huge indeed that of the hundreds of billions of sperms one man produces during his lifetime, no two are likely to be identical in the combination of genic varieties. The shuffling of the genic cards makes it unlikely that any person on earth (with the exception of identical twins) has ever exactly duplicated any other person in genic make-up, or ever will in the future.

This does not mean, however, that our inheritance is an entirely random affair. If men and women were completely promiscuous in mating—socially, racially and geographically—then one genic combination would be as likely as any other, and people might vary individually much more than they actually do. There are times and places where man does approach such random mating, for example, during great migrations and large military occupations, when one group may sow its genic varieties among those of another group. As a rule, however, a potential child within a given group does not draw on the whole storehouse of mankind's genes. Usually his genes

478

will come from a socially, nationally and racially segregated part of the store.

Yet for thousands of years the barriers separating the store of human genes into compartments have been progressively lowered, and with the increase of human mobility in our era of world-wide transportation many barriers will undoubtedly disappear. Tribes, minor races and other subgroups will vanish. A diffusion of genes from one group to another is bound to occur, however slowly and gradually, and in time it will tend to eliminate all partitions in our storehouse.

Will this be good, bad or immaterial for mankind? We cannot answer this question without evaluating the racial differences of the present. Have the present combinations of genic varieties originated in a haphazard way or are they the result of selective forces in the earliest prehistory of mankind which adapted the different races to specific environments? It is probable that both chance and design have played a role. Thus the racial differences in blood types (Rh and so on) seem to be just accidental and of no adaptive significance. On the other hand, it is likely that the differences in pigmentation and breadth of nose between the Africans and the Caucasians were evolved to fit the differing climates in which these peoples lived. Does this mean that the leveling of the genic partitions will make the world's peoples less fit to cope with their respective environments? Such a conclusion might be justified if we could assume that the originally adaptive traits have the same significance today as they had 100,000 years ago. But has not man created new influences which effectively alter his environment in such a fashion that the external physical factors continuously decline in importance? Housing and clothing, food and medicine, occupation and training have changed radically, and it may well be that these new factors have superseded the old ones.

What of mental differences among races? Whether or not such differences exist has not been established; exact knowledge of the genic distinctions between groups is most lacking where it most matters. This is not only because psychologists have found it difficult to invent standard tests to measure the inborn capacities of different races but also because there is great variability in mental traits within any one group.

The 20,000 pairs of genes in the fertilized egg control a multitude 'of interactions whose full complexity far transcends our understanding. In every trait of the individual numerous genically induced reactions are involved. There is no absolute, one-to-one relation between a specific gene and a specific trait—no gene for dark hair or for height or for "mental endowment." Each gene is only a link in the development of a trait; it is necessary for the process that results in the specific trait, but it does not invariably produce the trait in question. A gene for clubfoot, for instance, makes for an inclination, a potentiality, toward the appearance of clubfoot, but whether this potentiality will become reality depends on the interplay of life processes. A slight variation in timing or in the environment may decide one way or the other. The clubfoot defect may appear in one foot, in both feet or in neither.

The amount of variation in some life processes is small, in others large. A man's blood type, for example, remains the same throughout his life, but the color of his hair changes. Are the traits that distinguish different races variable in expression or invariable products of their genic endowments? It seems as a first approximation that genes for physical traits are more rigid in expression than those for mental traits. The Caucasian's hair remains straight or wavy and the Negro's kinky, regardless of any change in environment or training. It is otherwise with mental traits. A normal man's genetic endowment provides him with a wide potential for mental performance, from very low to very high. As with a rubber balloon, the state of expansion of his mind at any given time is hardly a measure of its expansibility. In human evolution those genes that allow the greatest mental adaptability, that possess the greatest plasticity of expression, seem to have undergone preferential selection in all races. If this is actually so, then the different genic varieties for mental traits may be comparably evenly distributed among all human groups, and the disappearance of the present barriers subdividing man's genic storehouse would not greatly affect mankind's mental potentialities.

What role will differences in reproduction play among the various socio-economic groups within populations? It is well known that the lower socio-economic layers of Western societies have higher birth rates than the upper ones. Do these layers differ in their stocks of genes? We cannot say with any certainty. The difficulties of research in this important field are great. We do not know, for instance, to what extent intelligence scores reflect true genetic factors in addition to education and environment, which they certainly reflect to a large degree. Nevertheless, the evidence strongly suggests that hereditary mental differences between socio-economic groups do exist. The mean intelligence scores of children at the higher socio-economic levels are consistently higher than those of lower groups, whether the tests are made in the U. S., in the U.S.S.R. or in any other country. That environment is not the sole reason for such differences is indicated by comparative studies on comparable groups of children, particularly twins reared together and separately. It is hard to avoid the conclusion that there are mean differences in the genetic endowment of different socio-economic groups, although the individual endowments within each group cover the whole range from very low to very high. Since the groups that seem less well endowed intellectually produce the most children, a deterioration of the genetic endowment of the population should result.

This large-scale difference in reproduction rates is a rather recent phenomenon. It is primarily the result of birth control, which did not become an important social practice until the second half of the 19th century. So far the upper and middle groups of Western countries have adopted birth control much more widely than the lower ones. But there is reason to believe that the use of contraceptive measures will spread through the whole population, and that the group differentials in fertility will be diminished, although perhaps not obliterated.

Before we become too alarmed over the possibility that the genetic stocks of Western peoples may deteriorate, it would be well to obtain an estimate of the rate of this suspected deterioration. Such analyses as have been made suggest that the decrease of valuable genic varieties is probably much smaller than a naive consideration would suggest. High intelligence undoubtedly is based not on single varieties of genes but on the cooperation of many genes. The valuable varieties must be present, singly or in partial combinations, even in the great mass of individuals who score low in intelligence. From there they can, in the course of a single generation reconstitute an appreciable number of the "best" combinations. In other words, the population at large constitutes a great reservoir, and the possible loss of valuable genic varieties possessed by the small upper layers of the population tells only a part of the story.

Those who consider man's genetic future usually lay emphasis on the possibility that the human race is preserving and accumulating genes that cause physical and mental defects. In former times, it is argued, most of the unfortunates who inherited these defects died an early death, whereas now medical and social care keeps these people alive and permits them to pass on the bad genes to the next generation. There is some truth in this argument, but the case is overstated. The most severely handicapped individuals, such as idiots and complete cripples, usually do not reproduce, either because they cannot or because they are kept in confinement. Furthermore, civilization has neutralized some of the so-called defects; the survival of "bad" genes in modern times is often possible because the ingenuity of modern man provides for environments in which the "bad" genes lose their adjective.

Nevertheless, there *are* many defective genes in the human storehouse that are loading man's genetic future unfavorably. They are the anomalies, such as the inherited juvenile cataract, the cleft palate, the predisposition to schizophrenia, which are not serious enough to prevent reproduction but cannot be remedied readily or completely by medical treatment. The changes in mankind's genic substance now proceed more or less automatically. The eugenics movement proposes to make the process less haphazard in the future by applying nationally designed population policies, looking toward the elimination of genetic misfits and an increase in the number of those with superior genetic endowments. The difficulties of such a program are great, and the enthusiasm of its proponents, who have often been motivated by race or class prejudices, has been a drawback rather than a help to the success of the movement. It is now clear that we need very much more knowledge before a far-reaching blueprint can be designed. Fortunately, it is also clear that the problem is not as urgent as it seemed to past generations. The storehouse of human genes is so immense that neither mankind as a whole nor any of its groups is likely to undergo any serious genetic deterioration within the next century or two. If the threatened breakdown of our civilization on the cultural level can be avoided, we shall have time to work out the tasks of the genetic future.

V. IS MAN HERE TO STAY?

by Loren C. Eiseley

THERE IS a widespread tendency to conceive of the course of evolution as an undeviating upward march from the level of very simple organisms to much more complex ones. We are inclined to think of man as the crown and culmination of this movement and the natural point of origin for any further progress. The syllogism runs something like this: Evolution is an upward movement. Man is the most intelligent form of life on the planet. Therefore he will continue to dominate the earth throughout future time, or he will himself give rise to some more perfect and intellectual species as far superior to us as we are superior to our heavy-browed, lumbering forerunners of the Pleistocene.

This last statement is very significant. In it lies the major source of the confusion we manifest about human destiny. We know that man has moved along a particular line that has led to greater and greater intellectual triumphs. We know his brain has grown and his body has altered. We call this process evolution, and we tend not to understand why it cannot go on through an indefinite future. The confusion lies in the fact that we fail to distinguish adequately between progressive evolution in a single family line and those greater movements which adjust life to the rise and fall of continents or the chill winds of geological climate.

There is a pulse in the earth to which life in the long sense adjusts, but it is a rhythm so slow that it is imperceptible in short-line evolution. We can grasp its significance and its indifference to the aspirations of individual life forms only when we call the roll of the ages and note the number of the vanished. Even if we concentrate only upon the Age of Mammals, ignoring the strange departed amphibians of the Paleozoic or the stalking giants of the Age of Reptiles— even then we discover that whole orders and families have passed out of existence. Many of these creatures were highly successful in their day. Yet as one compares the durability of the simpler creatures with that of the more efficient, one may be led to comment cynically that to evolve is to perish.

The subject is a very complex one, of course, because obviously the completely inadaptive organism cannot master a shifting environment. Life must evolve to live. Why, then, are we confronted

483

with the paradox that he who evolves perishes? Are we not the highest animal? And what, among all things that fly or creep or crawl, is more apt to inherit the future than we are?

The one great biological principle that seems to deny man's hopes for continued dominance of the planet is known as the "law of the unspecialized." It is one of the curious ironies of scientific history that the discoverer, or at least the formalizer, of this law was a devout Quaker scientist who put forth his views during the full flush of 19th-century enthusiasm for evolutionary progress. He was Edward Drinker Cope, undoubtedly one of the greatest naturalists America has ever produced.

The gist of Cope's brilliant generalization is that the leaps ahead in evolution generally take off from comparatively unspecialized forms of life, rather than from the most highly developed. In Cope's words: "The highly developed, or specialized types of one geological period have not been the parents of the types of succeeding periods but . . . the descent has been derived from the less specialized of preceding ages." It is the more adaptable and generally the smaller forms that are best able to meet the onset of new conditions which destroy the already dominant and successful types. The first amphibians arose not from a highly successful fish but from a slow-swimming, foul-water form which had to be peculiarly adaptable. Similarly, the first mammals came not from one of the specialized dragon reptiles but from a smaller and much less specialized reptile which was learning to control its blood temperature. Another climbing, jumping reptile became a bird. All were small, all were the fortunate possessors of traits that offered the potentialities of successful adaptation to new climates or new media.

Every one of these insignificant, stumbling but remarkably endowed creatures founded explosive dynasties. Climbing out of the marshes to the uninhabited air or land, entering into some still region whose previous occupants were dead, they radiated with amazing rapidity into a diversity of forms. The new forms grew ever more specialized as they adapted to the particular niches in the environment that they came to occupy. Many of these specializations are of quite remarkable character. Yet in the long course of evolution they threaten to reduce the adaptability of the form in case it should ever find its particular evolutionary corridor blocked or destroyed. The problem is a little like that presented to an elderly glass-blower, let us say, when glass-blowing becomes mechanized. His environmental zone has changed, yet he is old; he is no longer able to master a new field. He is through—and so it may be, in even more brutal ways, in the world of animal life.

Thus the evolutionary paradox becomes plain: The highly and narrowly adapted flourish, but they move in a path which becomes ever more difficult to retrace or break away from as their adaptation becomes perfected. Their proficiency may increase, their numbers may grow. But their perfect adaptation, so necessary for survival, can become a euphemism for death. Climates change, vegetation changes, enemies perfect their weapons, continental ice-fields advance, indirect competitors may smother the corridor. Sooner or later an impasse develops, an impasse which a small, omnivorous creature that has "specialized" in generalized adaptability and inconspicuousness may escape, but which the perfected evolutionary instrument never can. Consider, for example, the disaster that would overtake an animal like the tubular-mouthed anteater if extinction were to overtake the social insects. The anteater could never readjust. He would starve in the midst of food everywhere available to the less specialized.

The question we are mainly interested in is: Is man a specialized or a generalized creature? Are we the refined end-product of an evolutionary line whose genetic plasticity has about reached its limit, or are we the departure point for an undreamed of future? An answer of sorts can be given, but it has to be given with great care and with much attention to precise definition. We will want to ask, first of all, whether there really is such a thing as an "unspecialized" animal. My answer is that there is, and that, furthermore, it robbed our kitty's food dish last night. It is no mere intellectual abstraction.

Probably the major confusion that has developed about Cope's principle of the unspecialized animal is the tendency to imagine it as some kind of inchoate "archetype" creature capable of galloping off in several evolutionary directions at once. No such animal ever existed, and Cope never intended to suggest that it had. The term "unspecialized" is used only in a comparative sense, and in this sense we need not look far, even today, to find examples. The opossum that stole up our back stairway last night and turned the cat's tail into a frightened bush has marched unchanged through 80 million years of geological upheaval. As George Gaylord Simpson observes in his widely read recent book, *The Meaning of Evolution*: "It has been suggested that all animals are now specialized and that the generalized forms on which major evolutionary developments depend are absent. In fact all animals have been more or less specialized, and a really generalized living form is merely a myth or an abstraction. It happens that there are still in existence some of the less specialized—that is, less narrowly adapted and more adaptable —forms from which radiations have occurred and could, as far as we can see, occur again. Opossums are not notably more specialized now than in the Cretaceous and could almost certainly radiate again markedly if available space were to occur again." Nature, in other words, seems to keep available creatures whose life zones are broad enough for renewed experimentation if the need arises.

It is evident, then, that there are two currents in the evolutionary flow, although neither is completely separate from the other. There are times when only a person gifted with foreknowledge of the future could indicate which of these streams means progress. But in general one of the streams is not really progressive. It is simply the perfection of specialization: the creation of the ideally adjusted parasite, of the glowing monster of the abyssal seas, the saber-toothed tiger, the fish with batteries in its belly. All of this is remarkable beyond words, but beneath these superficial diversions a deeper flow has carried life up from the waters, perfected its chemical adjustments, conquered the land, stabilized bodily temperature, developed nervous systems of growing complexity and brought into being the mind, whereby the universe examines itself.

So far this broad upward movement has never retraced its steps. Not that it has not wandered or specialized, or lost itself in peculiar and constricted niches, but once a new level of organization has been attained, it has not been lost, and the old has dwindled in importance. The Crossoptyrigian fishes gave rise to the amphibians and vanished almost totally. The amphibians, making further lung and limb adjustments, then gave rise to the reptiles. The latter then contributed the two great living groups—the mammals and birds. In all these cases it was not the largest or most highly specialized of the new classes that produced the succeeding forms. It was instead the smaller, less spectacular and more adaptable types. Man, who derives from a comparatively generalized and ancient order of mammals, has opened a strange new corridor of existence—the cultural corridor. With the appearance of culture the biologist is confronted with a true innovation.

There exist in various obscure parts of the globe certain ancient and remarkable forms of life. They are, one might say, the immovable immortals. Is that quick-witted, volatile and short-lived parvenu, man, destined to join their company? Has his mastery over environment, the greatest yet achieved by any animal, created the first highly specialized but truly adaptable organism? Does his one great specialization—his brain—mean escape at last from the disasters that have stalked all other forms?

There is a creature something like man that may provide a hint, though it crawls in another shape. Like man, it is an agriculturist and a city builder. It numbers, like man, in the millions, and like man it has mastered the problem of food storage and distribution. In its dark cities it knows something of the common warmth and security, the thrusting back of the harsh natural environment, that man has so recently achieved. This creature is the ant.

You will object at once that ants are physically and mentally remote from men. So they are, but in their tremendous, if minute, activities they have achieved a remarkably humanlike adaptability. The important point is that the ants have led their present lives for more than 80 million years, while man's civilization is scarcely more than 7,000 years old. They are the oldest cosmopolites; they have sheltered longest, grown food, escaped many of the violences of the mammalian world. We shall want to ask just one question: "Have they changed?" It would seem they have changed very little, if at all. They are one of the small "immortals."

The reason for this long life without noticeable change would seem to lie in a perfect environmental balance. Even the creatures' parasites are old. The remarkable instinct-built cities, playing a part roughly equivalent to our own metropolises, have provided shelter, food and protection. The stability of perfect adjustment has set in.

It can justifiably be contended, of course, that man, by reason of his cultural malleability, his ability to invent, to progress, to introduce changes into his environment, is in a much more dynamic and unstable relationship with nature than the social insects. But it is also true that man's cultural proclivities are directed toward making life easier for himself. He prepares food which makes an elaborate dentition superfluous, and which actually encourages its disappearance. His machines transport him with little effort on his own part. His clothing, his air-conditioned houses, his medical devices all protect him from the harsh natural environment that controls the survival and directs the evolution of other animals. As a living organism man is still susceptible in some degree to environmental influences and genetic drift, but natural selection has ceased to operate intensively upon him, except in so far as it may perfect his urban adjustment. To be sure, competition in implements and methods of warfare may well determine the increase or relative decline in significance of particular racial types in given moments of human history. There is nothing in the present life of man, however, to suggest the likelihood of striking increases in brain development or other remarkable innovations in human structure. We may expect at most a few mild changes toward a reduced dentition and other small adjustments if civilization and its luxuries continue.

Man, in other words, gives every sign of having reached by a different road from that of the social insects, an equivalent environmental mastery. It would take a formidable and unforeseen world cataclysm to thrust him once more naked into the wilderness out of which he emerged. It is conceivable that his propensities for destruction may bring about his self-extinction, but because of his worldwide distribution and enormous expansion in numbers this is extremely unlikely.

The 19th century drew from the century before it a concept of human progress which the evidence of the earth's history does not entirely justify. Evolutionists do not see at work any inner perfecting principle that would automatically improve a given organism after it has achieved a certain stability of relationship with its environment. Rather the pace of evolution steadily becomes slower, until the vicissitudes of time demand new adjustments or force the now-specialized organism toward extinction.

Darwin, like his 18th-century forerunners, believed in progressive change and predicted that "we may look with some confidence to a future of great length. . . . All corporeal and mental endowments will tend to progress toward perfection." Yet curiously this quotation lies at the close of a paragraph in which he said: "Of the species now living very few will transmit progeny of any kind to a far distant futurity."

IS MAN HERE TO STAY?

The primate order is old. Man is a comparatively young branch of that order, but his great brain marks him as specialized in a way peculiarly apt to bring an end very soon to his physical modifications and advancement. Indeed, there is evidence that *Homo sapiens* has not altered markedly for perhaps a hundred thousand years. Yet man's strange specialization has introduced a new kind of life into the universe—one capable within limits of ordering its own environment and transmitting that order through social rather than biological heredity.

If man can master quickly his individualistic propensities for destruction, he may be able to become another of the small immortals. Even to this, however, judging by the records of the geological past, there will come an end some day. Sooner or later Cope's law of the unspecialized will have its chance once more.

THE BRAIN AND
THE MACHINE

Introduction

T HE HUMAN BODY is an unspecialized machine; it can thread a needle,
shape a plank or dig an excavation. What we normally call a machine
is a model of the human body specialized to do a particular task exceedingly well. A steam shovel, which is really a forearm with cupped hand
and unbreakable fingernails, will outdig any 20 men, but it has no other
proficiency. Being the heirs of the industrial revolution, we take for
granted the mechanical fragments of ourselves that have assumed most
of the heavy, repetitive and precise jobs once performed by ingenious
but puny and fallible human beings. It was not always so; machines were
once an amazement and they were both admired and feared. Reformers
preached the dawn of a new freedom, but the workmen threw their
wooden shoes into the cogwheels.

Now it appears that we are entering the first stage of another machine
age; this time we are building models of the human brain. The brain is
also an unspecialized instrument, and we do not yet know how far we
can go with our models in aping its versatility. One of the papers
that follow describes a machine that can play a game of chess
and, like as not, defeat the man who built it. That feat is surely an
occasion for wonder and a degree of uneasiness. The machine that sup-
planted human muscle was regarded as a threat to man's livelihood, but
a machine that supplants his nervous system may be seen as a threat to
his prestige and power as the unique thinking organism.

This section presents an account of work going forward on both kinds
of brain, human and electronic. It opens with one of the most illuminating
general surveys of present knowledge about the human brain to be found
anywhere in print. The priority given this article should not, however,
suggest that study of the human brain has helped us to design machines.
Even though the brain has been mapped to the considerable extent indicated by George Gray, the subtlety, variety and possibility that are vested

with such economy in the living substance of the brain beggar our present comprehension. The machine, on the other hand, has proved helpful to our investigations in this marvelous realm of the unknown. For example, communication theorists have devised radical economies of notation for the transmission of information over their electric circuits, and neurologists are now experimenting to determine whether or not the organic nervous system uses a similar shorthand. By the same token, the powerful concept of feedback, which is central to all mechanical control design, is being applied to the study of biological control systems.

The computers described by Harry Davis and the feedback circuits explained by Arnold Tustin provide evidence that the new revolution in technology is already far advanced. The machines far outstrip human capacities in certain jobs that have seemed previously to belong to us alone. They are to be found at work in research laboratories and administrative offices as well as in factories, conducting operations that are too complex, too fast or too dangerous for control by the human brain except through the mediation of its mechanical analogs.

During these early years of the new revolution, automatic control systems are for the most part still working under close human supervision. They are endowed with no more freedom of action than a linotype machine. But Shannon's chess player and Walter's tortoises—both admittedly electronic playthings—have been conceived for the serious purpose of forecasting a later stage in control technology. As time goes on, the machines will undoubtedly show an increasing ability to make judgments and to issue orders independently of their builders. This means that men will have to take an increasingly longer view of the goals toward which their machines are tending. Eventually, they may have to decide which areas can be safely turned over to automatic control and which they must reserve for themselves. That presumably is one decision which the human brain can never delegate to its electronic model.

I. THE GREAT RAVELLED KNOT

by George W. Gray

THOUSANDS OF MILLIONS of nerve cells are woven into the texture of the human brain, and each can communicate with near or distant neighbors. Judson Herrick, the University of Chicago neurologist, has calculated that if only a million of these nerve cells were joined two by two in every possible way, the number of combinations would total $10^{2,783,000}$. This is a figure so tremendous that if it were written out and set up in the type you are reading, it would more than fill two volumes of this size. And we may be sure that the brain has many times a million nerve cells, each capable of groupings of far more than two cells per hookup.

Life has created innumerable patterns in its long climb from the Archeozoic ooze, but none can compare in intricacy of design and virtuosity of function with "The great ravelled knot," as the famous English physiologist Sir Charles Sherrington described it, by which we feel, see, hear, think and decide. One can trace the evolution of this "master tissue of the human body" from fish to man, and observe brain part after brain part originate as each succeeding species becomes better adapted to the complex conditions of life on land, more versatile in its capacity for survival—and more intelligent. Similarly, in the developing human embryo the brain forms by the dual process of multiplying the number of cells and increasing their specialization. In the beginning, a few days after conception, certain skin cells are selected as tissue for nerve function. From this microscopic neural tube the spinal cord forms, and simultaneously the hindbrain, midbrain and forebrain develop from the same germinal structure.

It is the forebrain that attains the crowning organization and integration of the nervous system—the cerebral cortex. Beginning as an insignificant segment of the embryonic brain, this gray mantle eventually grows so large that it must fold in on itself in wrinkles to accommodate its expanding surface to the walls of the skull. When fully grown, the cerebral cortex completely covers the brain structures from which it developed. It overshadows and dominates them, taking control of many of their functions. From every nerve cell, or neurone, fibers pass to other neurones, both of the cortex and of the other brain parts. Millions of lines of communication connect one region of gray matter with another, and these in turn with distant

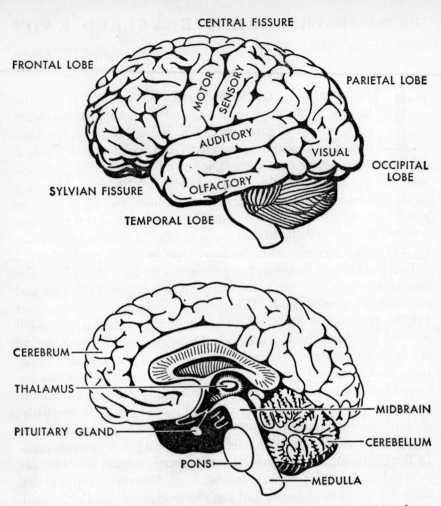

CENTRAL FISSURE

FRONTAL LOBE

PARIETAL LOBE

MOTOR
SENSORY
AUDITORY
VISUAL
OLFACTORY

OCCIPITAL
LOBE

SYLVIAN FISSURE

TEMPORAL LOBE

CEREBRUM

THALAMUS

MIDBRAIN

PITUITARY GLAND

CEREBELLUM

PONS

MEDULLA

Cerebral cortex is the organ which gives mankind its primary distinction from lower animals. Folding and convolutions which fissure its surface permit large surface area within confines of skull. Most important anatomical features are identified by labels surrounding the drawing; functional areas, to which impulses come from sensory organs, are indicated by labels on surface. In lower drawing, near hemisphere of the cerebrum has been removed to show inner surface of opposite hemisphere in relationship to lower brain, which is shown in cross section.

organs. By such means the brain is in communication with the lungs, the heart and other organs; with the specialized cells which serve as the receptors of touch, taste, smell, vision, hearing and other sensations; and with the muscles which produce action.

The cortex may be compared to a holding corporation formed to integrate and extend the services of a number of older companies which are housed in the stem of the brain. Under the consolidation the older companies are not abolished. They are continued as useful adjuncts of the more modern organization: to take care of routine activities such as breathing and digestion, to serve as channels of communication, perhaps to be held in reserve as stand-by agencies capable of resuming their former higher functions in emergencies. But the offices of inquiry and foresight, of planning, initiative, the creating of new ideas, the venturing into new projects, are executed by the holding corporation upstairs, and control of the consolidated system is administered there.

This roof brain is the supremely distinctive organ of the human species. What goes on within its network of cells makes the fundamental difference between man and brute. The functioning of the cerebral cortex not only distinguishes man from the animals, but more than any other faculty it distinguishes man from man.

No cortex is an exact duplicate of another, either in the number or size of its convolutions. Indeed, it seems likely that each roof brain is as individual to its possessor as his face, but certain surface landmarks are characteristic of all. The most conspicuous is the longitudinal division into two approximately equal hemispheres. Then there is the large fissure that cuts laterally across each hemisphere, originating in the longitudinal division, traveling over the cerebral crest, and continuing down the side of the brain in a direction which if continued would bring it about opposite the ear. This great central fissure (also called the Rolandic fissure) is found in the brain of men, apes and monkeys. Another standard feature which appears also in the primates is the Sylvian fissure. It is the cerebral Grand Canyon, a deep gorge which emerges from the bottom of each hemisphere and curves upward and back along the side. The Sylvian fissure perpetuates the name of the 17th-century French anatomist, Franciscus de le Boe Sylvius.

These two prominent depressions provide natural boundaries for subdividing the hemispheres into regions, and almost the first efforts of brain anatomists were directed toward mapping these sections. All that part of each hemisphere which lies in front of the central and Sylvian fissures was designated the frontal lobe. The bulbous rear, which lies under the occipital bone of the skull, was named the occipital lobe. And between these front and rear lobes two others

495

were early laid out: an upper intermediate zone (the parietal lobe, so-called because it lies under the parietal bone of the skull) and a lower intermediate zone (the temporal lobe, below the Sylvian fissure). There is also a limbic lobe, mapped in the cleft between the hemispheres, around the root of the cerebral cortex, where the gray convolutions face one another.

In subdividing the cerebrum into these lobes, the early anatomists apparently had no thought of identifying special functions with each. The accepted idea was that the brain acted as a unit; if by accident or disease one part of the cortex became incapacitated, its faculties were taken over by other parts. This was the almost undisputed view up to the early years of the 19th century.

Then, beginning in the first decade of that century, vigorous attempts were made to equate specific regions of the brain with specific aptitudes and moral qualities. This pseudo-science of phrenology located bumps for such characteristics as musical ability, manual dexterity and misanthropy; it flourished among the gullible. The mapping of the true sensory and motor areas of the brain began at about the same time but it was a slow and cautious business. The science of the day had no direct access to the living brain, but there were certain conditions that provided indirect evidence. One of these was the paralytic stroke affecting only one side of the body.

Marc Dax of Paris noticed that whenever a stroke paralyzed the right side of a right-handed person, the patient usually suffered some loss in his faculty of speech. It was known from anatomical dissection that the nervous pathways of the left hemisphere cross over in the brain stem and pass on to the muscles in the right side of the body. From this Dax reasoned that the speech center must lie in the left hemisphere, and that the same injury which paralyzed the control of the muscles also damaged the speech center.

Localization was carried a step further in 1861 by another French doctor, Paul Broca. He asserted, on the basis of autopsies performed on two paralytics, not only that the brain's control of the vocal cords was confined to the left frontal lobe, but that it was localized in a small area at the base of the third frontal convolution. Broca's fissure and Broca's area have been recognized landmarks of the frontal lobe ever since.

Closely following Broca came the English neurologist Hughlings Jackson, who began to apply the idea of localization to his study of epilepsy, chorea and other brain diseases. By 1869 Jackson had arrived at a clarifying generalization, the astuteness of which is especially impressive because he had no means of checking his speculation by surgery or animal experimentation. Guided entirely by what he observed in patients and found in post-mortem examinations of brains, Jackson announced that there was a primary functional division of the cerebral cortex which cut across both hemispheres. All the *sensory* functions of the brain—its reception of sights, sounds, touches and other signals from sensory organs—were confined to the lobes back of the central fissures, he suggested, while all of its *motor* functions were located in front of them.

The first verification of this bold hypothesis came a year later in Berlin when G. Fritsch and E. Hitzig began experimenting with the brains of dogs. By applying weak electrical currents to the frontal region of a dog's right hemisphere, they obtained movements of the left legs. Similarly, by stimulating frontal areas of the left hemisphere, they obtained muscular responses from the right side of the animal. But when the same electrical currents were applied to the back of the brain, no muscle gave the slightest response although the stimulus was many times repeated. At last, it seemed, a clue had been found to the general organization of the cortex.

But, it was reasoned, if sensory reception is confined to the back lobes of the brain and motor control to the frontal lobe, should there not be some difference in the cellular structure of the contrasting regions? This led to a large-scale search of brain tissue. It was not until methods of staining tissues for microscopic study were devised that real progress could be made. Then histologists were enabled to see that the gray matter of the cerebral cortex is built of several kinds of cells arranged in six layers: 1) the surface or molecular layer, a paving of small structures called horizontal cells; 2) a layer of granular cells, small and roundish; 3) pyramidal cells; 4) a closely packed section of granular cells; 5) a layer of more numerous and larger pyramidal cells; 6) a bottom layer of smaller, spindlelike cells.

This six-layer pattern is typical, but in examining segments taken from different cortical regions the investigators found variations in it. The most striking contrast appeared when the area immediately in front of the central fissure was compared with areas to the rear, notably those which have been identified with seeing, hearing and other sensory functions. Microscopic surveys of these sensory areas

497

show that the pyramidal cells of the third and fifth layers are much reduced in size, while the two granular layers are thickly populated with their characteristic small globular cells. By contrast, just the reverse was found in the motor area. Thus not only brain areas but brain cells appear to be consecrated to specific functions. Granular cells function in the reception of sensory messages while pyramidal cells play a corresponding role in transmitting motor signals.

Can we go further? We yearn to know the nature of the nerve impulses that can bring about such an exquisite orchestration of activity. Where and how do they operate so that one impulse is interpreted as a touch, another as a sound, a third as a sight? And where does memory dwell, where are judgment, the imagination, all those higher faculties that we call intellectual, artistic, moral?

No one would claim that more than a beginning has been made toward answering these questions. Some of the answers are more fragmentary than others, but efforts at completing them never slacken, for the incentives to the search are compelling. Fundamentally the incentives are two: intellectual curiosity and the desire to alleviate disease. Not only laboratory investigators but clinicians, following in the train of Broca and Jackson, have been prolific contributors of new knowledge. It has taken many minds, many stratagems, the use of many tools and techniques to chart our present map of the brain.

Today the principal tools for prospecting the brain are electrical. This is only natural, for the gray matter itself is a generator of electric impulses, and the messages that it receives from the sense organs and the directives that it issues to the muscles are all electrical in nature. Electric currents can therefore be used to stimulate the brain in a way that is entirely normal to its function. Conversely, electric impulses generated by the brain and its tributary system of nerves can be picked up and measured to determine the degree of activity in any selected area.

Of the two general kinds of electrical prospecting techniques, one works from the brain outward to the body responses, and the other from the sense organs inward to the brain. In the first method a delicate electrode carrying an alternating current at low voltage is applied to a selected area of the exposed brain. This technique has been most successful in exploring the motor areas of the cortex. The body-to-brain method, on the other hand, introduces no electric cur-

498

rent, but merely picks up the currents that the brain itself generates. In this kind of research the investigator applies an appropriate stimulus to a sense organ, and electrodes moved over the brain determine the destination at which the sensory message arrives. The stimulus may be a slight touch on the bottom of the foot, a flashing light or a sound. The skin, eye or ear then starts a nerve impulse which moves to the cortex, and the area which receives the message announces the arrival by increasing its electrical output. These discharges are so delicate that it was not until the development of the vacuum-tube amplifier that researchers were able to build receivers sufficiently sensitive to measure them.

The vacuum-tube amplifier has been harnessed to this task in two ways: by the cathode-ray oscillograph and by the electroencephalograph. In the electroencephalograph, the feeble electrical discharges picked up from the brain may be amplified millions of times to produce voltages which when relayed through an electromagnet cause a pen to write the pattern of electrical pulsations on a moving paper tape. These brain waves provide an exact record of the fluctuating electrical activity of the brain area upon which the electrode rests.

The principle of the cathode-ray oscillograph is the same, but the manner of applying it is somewhat different. Here the brain currents deflect a beam of electrons moving in a cathode-ray tube. The moving beam is projected on a fluorescent screen and appears as a quivering luminous line. Whenever the brain area under investigation flares with increased electrical activity, the line pulses in unison with the accelerated discharge. The frequency and amplitude of peaks and valleys in line are measures of the electrical output of the discharging brain cells.

Surgery has also contributed prodigiously to the localization of brain function. Certain parts of the brain have been removed from animals, and the subsequent behavior of the animals has provided direct evidence of the functions that were related to the lost areas. Performing surgery on human beings is another fruitful source of information. Of course human brains are not deliberately exposed for experimental studies, but when the skull must be opened to remove a tumor, to excise a portion of diseased cerebral tissue or for any other clinical reason, it is often possible to make experimental observations of localization, sometimes to confirm in the human brain what has been discovered in lower animals.

The meaning that a nerve message conveys does not depend on its source. Whether it is sent by eye, ear, nose, taste buds or organs of touch will make no difference unless the message reaches the appropriate nerve endings in the brain. Any nerve impulse arriving at the auditory area and carrying sufficient voltage to discharge its neurones is received and interpreted as a sound, no matter what its origin. Presumably one might hear a smell, if an impulse starting from the olfactory organs should get switched in transit and arrive at the brain's center for hearing.

Certain varieties of focal epilepsy provide dramatic evidence that the brain can generate its own sensations. One victim of this disease reported that just before he was seized with convulsions he always saw rings of light. Another patient's preliminary sensation was sound; he heard discordant noises. There have been cases in which the first sign of an epileptic seizure was a foul smell or a curious taste. These abnormal sensations are produced within the brain by the spontaneous discharge of certain hyperactive cells, and it happened that in one patient the hyperactive cells were connected with the visual area, in another with the auditory area, and so on.

The parts of the cortex that have become specialized for the reception of sensory messages and for the dispatch of motor directives are known as projection areas. For all the motor functions, the projection area is where Fritsch and Hitzig found it by electrical stimulation 78 years ago. In man this means the frontal lobe, just forward of the central fissure.

Facing this motor region, stretching along the rear slope of the central fissure and occupying the adjacent plateau of the parietal lobe, is the somatic sensory projection area, the region where sensations of touch are received from all parts of the body. Far back in the brain, at the very rear of the occipital lobe, is a whitish patch known as the striate cortex, the visual projection area. The upper rear of the temporal lobe, on the lower bank of the Sylvian fissure, is the auditory projection area. Impulses generated by odors pass from the nerve endings in the nose to the olfactory bulb, on the underside of the cortex, and from this diminutive area are distributed to a number of ill-defined areas. Actually the neurologists know very little of the topography and physiology of smell. They know even less of the projection area for taste.

The sense organs that send messages literally project images of themselves upon the brain. The cochlea, that spiral harp of the inner ear with its coiled membrane of nerve tissue attuned to vibrate over the entire scale of audible frequencies, is the critical organ of hearing—and an image of the cochlea is projected on the auditory area of the temporal lobe. The part of the eye upon which the lens focuses an image of what we see is a tiny section in the center of the retina. This microscopic patch is projected precisely on the visual area of the striate cortex in the occipital lobe, though in an enlarged replica. The brain actually magnifies the picture which illuminates the rods and cones of the retina by several thousand times. Just what sort of images the taste buds and smell organs project would be difficult to imagine. But when one reaches the somatic sensory area and the motor area, there remains no serious doubt or speculation. The image projected here is that of a little man—a grotesque and somewhat dismembered miniature of the human body.

A number of distinguished neurologists, most of them surgeons, have explored the motor and somatic functions of the human brain, and in terms of localization we know more of these two areas than of any other parts of the cerebrum. Horsley, Bidwell and Sherrington in Britain, Keen, Cushing and Ransom in the United States, Foerster in Germany, and others pioneered this field. The most extensive studies have been made by Wilder Penfield and his associates at the Montreal Neurological Institute. With the assent of patients upon whom he performed surgery, Dr. Penfield used electrical prospecting methods to survey the cerebral cortex. Now he has data from several hundred persons.

These accumulated results show that the amount of brain surface related to a specific part of the body is not proportional to the size of the part but to the extent of its use. The area concerned with the hands and fingers looms larger than those related to the feet and toes because we make more use of our hands. The projection of the lips occupies more of the somatic sensory area than all the rest of the head. The brain represents the somatic and motor functions as a kind of dismemberment of the body, with arms and legs joined, torso almost nonexistent, head separated from body, and tongue separated from head.

This distortion of the body as the brain projects it becomes less exaggerated in animals farther down the scale of evolution. A research group at the Johns Hopkins Medical School has studied somatic representations in a succession of animals—monkeys, dogs, cats, sheep, pigs, rabbits and rats. Clinton N. Woolsey, one of the directors of this work, says that it was not until the group studied the rat that it found a brain whose projection gave a reasonable facsimile of the animal's body. Though grotesque, with enormous head and exaggerated lips, it is approximately ratlike.

These distortions tell us that each cortex reflects the pattern of the body's daily life. In a pig's brain most of the somatic projection area is devoted to snout; in a spider monkey, with its prehensile tail, there is an enormous tail area; in some dogs it is the olfactory area that holds a position of prime importance. Several times in recent experiments secondary projection areas have been found quite distinct from the previously identified dominant areas. E. D. Adrian of England's Cambridge University noticed this phenomenon first in connection with the somatic areas of the cat. Secondary projection areas for sight and hearing have also been located in the cat brain and Dr. Penfield at Montreal has discovered a second somatic area in man. Almost nothing is known yet about these sub-areas of projection in the cortex. Experiments with the hearing of dogs suggest that the secondary auditory area comes into action only when loud sounds are heard and that the image projected is less detailed than the pattern imprinted on the primary hearing centers.

The sights, sounds, touches and other signals that the projection areas receive are a miscellany of random information, every item of which would be endlessly new, bewildering and useless were it not for the functioning of other areas of the roof brain. The burned child avoids the fire, but not because of the sharp signal of pain received in the somatic sensory area. It is the association that teaches the lesson—the association of pain with the sight of fire, perhaps with the sound of a warning scream, with the muscular action of drawing away, and so on.

The brain must have memory to relate the information of the moment with that of the past and to recognize its significance. This means millions of functional correlations, countless hookups of sensory centers with one another and with motor centers, repeated exchanges of data for analysis, comparison and synthesis. These elaborative functions of the cortex are performed by the association areas.

It would be meaningless to say that association areas are more important than projection areas, for without the latter, if we can imagine such a thing, the cortex would have no information about the outside world and no means of voluntarily controlling the body's muscular action. But even though it be indispensable, projection is the lowest level of cortical activity. As we go down the scale of animal intelligence we find the proportion of brain devoted to projection areas growing greater, and that occupied by association declining. In the rat almost the entire cortex is given over to projection, says Dr. Woolsey, "and it is difficult to see where there is any room for association areas."

In man more than three-fourths of the roof brain is occupied by association areas. For example, that patch of striate cortex at the rear of the occipital lobe, the sensory area upon which the retina projects its images, is surrounded by an association area known as the parastriate cortex. Encircling this, and so closely interwoven that the boundary is obscure, lies a second visual association area, the peristriate cortex. It is possible to trace fibers connecting these three. In addition, fibers from the peristriate area run beneath the central fissure to connect with parts of the frontal lobe. Thus the seeing department of the brain, though housed in a small area in the back of the occipital lobe, has connections which link it with much of the roof brain. Even if we consider only the parastriate and peristriate surfaces, they have many times the area of the visual projection center.

An injury that destroys the visual projection area in both hemispheres causes total organic blindness—cortical blindness, the neurologists call it. If the injury is confined to the parastriate (first association) area, the victim can see but is unable to recognize or identify what he sees. This is mind blindness, a form of agnosia (loss of recognition). If it is the peristriate (second association) area that suffers the injury, the mind may have no difficulty recognizing objects but cannot recall their appearance when they are not in view. This kind of disability was early recognized in disturbances of the use of language, when the patient was unable to associate the printed or written word with any meaning. So the loss of function of the second visual association area is commonly called word blindness, one of the so-called sensory aphasias (loss of speech).

During a poliomyelitis epidemic in Los Angeles, a hospital nurse fell victim to the infection. She escaped paralysis, but in about three weeks it became manifest that her visual faculty had been damaged.

"When we asked her to read," relates J. M. Nielsen, who reports the case in his book *Agnosia, Apraxia, Aphasia,* "she claimed she could not see. When an O about ten centimeters in height was written for her, she . . . kept turning the paper about and peering in various directions, but was unable to read it. She then traced it with her finger (proof that she could see) and immediately read it correctly. Other letters were then tried, and it was found that she could see even small letters and could read complete sentences if she was allowed to trace the letters with her fingers. Keys on a ring were exhibited to the nurse; she shook her head, but as soon as the keys were rattled she said, 'Keys.' Looking at an orange meant nothing, but when the nurse smelled it, recognition was immediate."

The agnosia here was of the first order, mind blindness, caused by a disturbance of the functioning of the parastriate cortex. There was no disability of hearing, smell or touch, no disorder of the motor faculties, so the case was relatively simple compared with some complications in which mind blindness is combined, for example, with mind deafness, or, as sometimes happens, when an agnosia of one order of association is combined with an aphasia of another order.

Various brain injuries have disabled the other senses in a manner corresponding to the cortical blindness, mind blindness and word blindness of the visual area. From this we infer that each sense has its successive areas of association, although the actual topography of the areas is not completely known for the visual faculty and is even less definitely mapped for the others.

The motor side of the cortex, in the frontal lobe, also has its association areas. Broca's area, mentioned earlier in this article, is one of these. This association area for the elaboration of speech function is found normally only on one side of the brain: in the left hemisphere of right-handed persons, in the right hemisphere of the left-handed. Nearby is the association area for certain motor functions controlling manual dexterity. Anthropology teaches that the complicated business of developing and using language is closely related to using tools and developing other skills of the master hand. Some suggest that the earliest language may have been a system of signaling with the hands. Broca's area is closely connected with other association areas of the motor side of the cortex and also with areas on the sensory side, *e.g.,* with the visual area (for reading language), the auditory area (for hearing language) and possibly

504

with the somatic area as well (for correlations essential to writing).

Damage to the motor association areas or their connecting fibers may bring two kinds of result: 1) apraxia; 2) motor aphasia. In the first instance the individual is unable to perform purposeful movements—he suddenly finds he cannot tie a shoelace or thread a needle or guide a pen. Told to sign a letter, he cannot do it, although he wrote the letter without difficulty. Apraxia is the motor equivalent of sensory agnosia—a disability of the first order of motor elaboration. In motor aphasia the ability to speak is affected, just as in sensory aphasia the subject no longer understands the significance of what he sees, hears, or touches. In other words, when motor association areas of the first order are damaged, purposeful movement is impaired; when those of the second order are damaged, speech is impaired. Broca's area appears to be an elaborative zone of the second order.

Loss of speech is the most frequently encountered symptom of higher functional impairment, and it may take many forms. Sometimes the disability seems a mere eccentricity, like that of the patient reported by H. A. Teitelbaum of Johns Hopkins who could read the digits 5 and 7 but not 57. Complete speechlessness is the extreme form, though usually the victim can say a few simple words like yes or no. Sometimes the speech is meaningless jargon. It is quite common for aphasic patients to say the same word over and over again, and often it is a word they don't want to say, while at the same time they can't form the word they do want to say. The writer knew one man who could recite poems, quote Shakespeare and sing songs without skipping a syllable, and yet was unable to use the same words in conversation.

These disorders of association rarely occur as uncomplicated conditions, the effects of which point unerringly to specific areas of disturbance. Often there is a mixture of symptoms that confounds all our efforts to portray the great ravelled knot as a compartmented organization. The roof brain is not that simple.

The frontal lobes are the largest segments of the brain. After mapping the extensive motor areas in front of the central fissure and other motor association areas even farther forward, the neurologists are left with considerable territory still to explain. This prefontal region, the prow of the brain, overhanging the eyes like a gray canopy, does not respond to electrical stimulation. For that reason it has been called the silent area. From classical times it was regarded as the dwelling place of memory and of the higher intellectual faculties—the seat of intelligence.

But a hundred years ago a quarryman in Vermont sustained a violent injury to his frontal lobes and did not seem to suffer a serious impairment of intelligence. The quarryman had drilled a hole in a ledge of stone, and, after laying in the gunpowder, was tamping it with a crowbar, when suddenly the charge ignited. The crowbar shot upward into the man's cheek, passed into his skull, and tore an ugly wound in both frontal lobes. Someone rushed to his assistance, pulled the steel out, and by a miracle the wound healed. Months later the quarryman returned to work. Although he was not able to take on his former job as a foreman, he proved to be entirely capable as a worker. His memory was good, his skill as a stoneworker seemed about the same as before the accident, but everyone associated with him noticed a marked change in his behavior. He was profane in speech, indifferent to the interests of others, careless of his obligations—traits which the neighbors were disposed to overlook, remembering that he had passed through a nerve-wracking experience.

Numerous frontal-lobe experiments have been attempted with animals, and one that will go down in history was begun at the Yale Medical School in 1933. In the physiological laboratory there, John F. Fulton and Carlyle Jacobsen were curious to see what effect removal of the prefrontal region would have upon two chimpanzees. From October to the following March the two apes Becky and Lucy were put through intensive training. Then a surgical operation was performed removing the prefrontal region of one hemisphere in each animal. The operation did not change their behavior appreciably. After the wounds healed the chimpanzees were subjected to intelligence tests again and their responses continued as before. In June another operation was performed, removing the remaining prefrontal region from each. When Becky and Lucy were given the tests this time, Fulton and Jacobsen found that a radical change had taken place. The tantrums that used to flare up after the chimpanzees had made the wrong choice and had been denied food or other rewards no longer appeared. "If a wrong choice were made now," said Dr. Fulton, "the animal merely shrugged its shoulders and went on doing something else."

Fulton and Jacobsen reported the experiment to a meeting of medical men in London in 1935. At the close of Fulton's address Egaz Moniz, a neurologist from Lisbon, proposed: "Why wouldn't it be feasible to relieve anxiety states in men by surgical means?" Dr. Fulton admits that the suggestion of so immediately applying the result of an animal experiment to the treatment of human illness rather startled him. But within a year Moniz had enlisted the co-operation of a surgeon, Almeida Lima, and together they had operated on 50 hopeless mental patients in Portugal. Dr. Lima did not remove any part of the cortex, but severed the pathways between the prefrontal region and the thalamic center in the brain stem. Because it was the fibrous white tissue that was cut, he called the operation leucotomy (from the Greek *leukos*, meaning white). The following year, the first leucotomy in the U. S. was performed by Walter Freeman and James W. Watts in Washington. The method has since been taken up by other surgeons and close to 2,000 persons in North America have been operated on by leucotomy or related techniques.

This severing of the connecting fibers apparently releases the "new" brain of the prefrontal region from the emotional dominance of the "old" brain of the cerebral stem—though we have no knowledge of the nature of this emotional dominance. Whatever the mechanism, there have been amazing transformations of violently insane persons into seemingly normal ones. Sufferers from involutional melancholia and other dementias associated with middle age, and even schizophrenics, have benefited. Some leucotomies are unsuccessful, but it is claimed that better than 60 per cent of the subjects have shown improvement following the operation. Adverse personality changes also result, however, and efforts are now being made to determine the total effect of leucotomy, weighing the good —the relief of psychotic symptoms—against the bad—the deterioration in personality. Among these personality changes are intensified selfishness, indifference to moral obligations, failure to foresee the consequences of acts, gauche manners and emotional instability.

Wars and accidents have provided thousands of cases of men with brain injuries, as have surgical operations for removal of tumors and other diseased frontal tissue. While examining battle-wounded men, Kurt Goldstein of Montefiore Hospital in New York was impressed by the lack of imagination and by the defective judgment found in many with frontal-lobe injuries. Dr. Goldstein observed that such a man, so long as he was confronted with concrete situations with which he had had experience, seemed perfectly normal, but when the situation was new and a method of meeting it had to be improvised, the patient's deficiency was strikingly apparent; he was unable to assume an attitude toward the abstract.

A more extensive and prolonged study of effects of prefrontal loss is reported in *Brain and Intelligence* by the psychologist Ward C. Halstead. Halstead has been directing a laboratory for the investigation of neurological patients at the University of Chicago Clinics since 1935. He has examined 237 persons, including brain-injured patients, psychiatric patients and normal individuals used as a control group. Defining biological intelligence in operational terms of four basic factors, the psychologist has carried each of his subjects through an extensive series of tests to measure individual ability in each factor. The findings of his 12-year inquiry may be briefly summarized as follows: that biological intelligence is represented throughout the cerebral cortex; that its representation is not equal throughout; that it reaches its maximum in the cortex of the frontal lobes. Dr. Halstead concludes that "the frontal lobes, long regarded as silent areas, are the portion of the brain most essential to biological intelligence."

But we are still in a realm of speculation so far as a completely consistent picture of brain organization is concerned. Some authorities cling to the idea that learning, intelligence, imagination and the other intellectual faculties are a function of the brain-as-a-whole—and certainly there is evidence for such a view, along with the evidence for localization. Despite the substantial progress that has been made in identifying certain unmistakable areas (of which this article is a brief review), the cortex remains a vast entanglement of interconnecting lines and nodes. It is these interconnections that present the supreme enigma of neural organization. Whether the brain of man is capable of unraveling and comprehending its own complexity is, of course, a question. The very existence of that unresolved complexity constitutes a challenge. Physics, chemistry and mathematics, of which only limited use has been made by neurology up to now, will undoubtedly become major partners in the grand-scale teamwork of research that is ahead.

II. MATHEMATICAL MACHINES

by Harry M. Davis

NEW REVOLUTION is taking place in technology today. It both parallels and completes the Industrial Revolution that started a century ago. The first phase of the Industrial Revolution meant the mechanization, then the electrification, of brawn. The new revolution means the mechanization and electrification of brains.

The 19th-century revolution was based on the transformation and transmission of energy; the 20th-century revolution is based on the transformation and transmission of information. A number, a letter of the alphabet; a dark or light spot in a picture; an "on" or "off" signal; a decision between "yes" and "no"; a judgment as to "more" or "less"; a logical discrimination among "and," "or," and "neither." These are the raw materials and the products of "information-processing systems" that are assuming many of the human functions of calculation, communication and control.

How far has this revolution gone? Much of it is already taken for granted in our daily lives—in the shape of radio, television, telephone dial systems. Other phases affect us less obviously—robot pilots for aircraft; electronic navigation systems; automatic controls in factories; the many kinds of radar. All such devices have functions that are comparable, in one way or another, with the processes of human thinking.

But the machines that above all others deserve the title of "brains" are the electronic computers which easily solve problems so intricate and laborious that they stagger the most patient mathematician. They read, they write, they do arithmetic—all at rates ranging from a thousand to a million times faster than the human eye, mind, and hand.

To the question, "Do these machines really think?" one can get various semantic interpretations of the words "really" and "think," including the rejoinder, "How much do people really think?" Claude E. Shannon, who describes an electronic chess playing machine later in this chapter, aptly answered the "do they think" question when he said on one occasion that the performance of the newest machines "will force us either to admit the possibility of mechanized thinking or to further restrict our concept of thinking." Calculator designers and psychologists seem to gain more respect for the human brain the more they learn about its mechanical competitors. In strictly electrical and chemical terms, the human brain is the most efficient of computing machines, although it is also the slowest. It does not need kilowatts of power to energize its nerves, nor blowers to ventilate it; the electrical brains definitely emit more hot air. Warren McCulloch, the neuropsychiatrist, has pointed out that if a calculator were built fully to simulate the nerve connections of the human brain, it would require a skyscraper to house it, the power of Niagara Falls to run it, and all the water of Niagara to cool it.

"The more I deal with these machines," said one expert, "the more impressed I am with how dumb they are." They do nothing creative. They can only follow instructions, which must be reduced to the simplest terms. If the instructions are wrong, the machines go wrong.

On the other hand, the machines are not subject to distraction. They concentrate all their faculties on the problem at hand. They can do a complicated calculation in less time than it takes a human being, for example, to react to a red traffic light by signaling his right foot to move from gas pedal to brake. Thus they can take over an immense burden of mental labor, handling mathematical and logical data in quantity just as the assembly lines have converted hand labor to mass production.

There are two families of computers—the digital and the analogue. This article is primarily concerned with the digital type but the analogue computers came first. The celebrated "mechanical brain" at M.I.T., for which Vannevar Bush gained fame two decades ago, was an analogue machine. Radar and gun directors leaned heavily upon them, and they will continue to be important in many ways. So let us examine these machines before clarifying the distinction between analogue and digital computers and dissecting the latter.

The speedometer in your automobile is a simple example of the analogue computer. In proportion to the speed of the drive shaft, a centrifugal force is set up which moves a needle to the appropriate place on the "miles per hour" dial. This is an operation of differential calculus—a stationary needle position on the dial displays the rate of change in the position of the car.

The gears of your car, or of your watch, also do arithmetic: they multiply and divide, although, of course, the multiplier is fixed by a built-in gear ratio. The differential, between the rear wheels, is what its name implies: a mechanical subtracting machine, any extra speed gained by one wheel being subtracted from the other. All these ideas are actually used in mechanical computing machines of the analogue type; among their components are gears and cams and differentials.

Bush's "differential analyzer" solved problems in calculus. Different elements of the machine were set for various parts of the equation, all being geared together so that the only answers to come out would be the ones that were true to the equation's requirements.

There are electrical counterparts of the mechanical analogue computer. Instead of a mechanical position, we can have an electrical charge; instead of a velocity, we may have an electrical current, or the magnetic force induced by it. Circuits with resistance, inductance and capacitance are set up to behave in accordance with stipulated equations. An electrical transformer can multiply in the same manner as a pair of gears. Vacuum tube circuits can integrate. M.I.T. has an electrical successor to the mechanical analyzer which looks like a telephone central station.

The analogue computers are likely to be less bulky and expensive than the digital type; they provide quick solutions. But like the slide rule (which is also of the analogue class, because it translates logarithms into physical distances) they have a limit to their possible accuracy. For the higher refinements of calculation, the digital or logical computer is now coming to the fore.

The digital computer is distinguished by the fact that it does not measure; it counts. It never responds to a greater or lesser degree; at every stage of its action, it is an "all or nothing" device, operating with discrete signals that either exist or do not exist.

The simplest digital computer is the human hand, from which, of course, we have our decimal system. Corresponding to such primitive indicators of a numerical unit as a finger, a pebble or a stylus scratch, the new automatic computers represent digits by such methods as:

A round hole in a strip of tape.
A square hole in a piece of cardboard.
A current in an electromagnet.
An armature attracted to the magnet.

511

A closed pair of electrical contacts.

A pulse of current in an electrical transmission line.

An electronic tube in which current is permitted to flow from filament to plate.

A magnetized area on a steel or alloyed wire.

A magnetized area on a coated tape.

A darkened area on a strip of photographic film.

A charged area on the face of a cathode-ray tube.

A moving ripple in a tank of mercury.

In each case there is no measurement of gradations in the signal. There is either a hole or no hole, contact or no contact, current or no current, pulse or no pulse. The designers simply have to make sure that there will be no ambiguity. They have to leave enough room on a tape, for example, so that a magnetized area will not get confused with an unmagnetized one. This sort of consideration, however, only limits the compactness and in some respects the speed of the machine; it does not affect the accuracy or the number of decimal places to which a calculation can be carried out.

Any such imprint or setting is called a "memory." The use of "memory" as a technical term of the computer trade has bolstered their anthropomorphic analogy to "brains." But there really is nothing surprising in it. Every photograph, every printed page, every canceled check, is a form of mechanical memory.

The important thing about the computer devices is not that they can record and remember numbers, but the fact that they are peculiarly adapted to yield up the memory content quickly, and in a form suited for immediate transportation and processing in other parts of the same machine. The "on-off" or "yes-no" kind of contrast makes it easy to transfer a record from one form to another: a pattern of punched holes instantly becomes a pattern of closed switches, or a pattern of conducting electronic tubes; it is the pattern that represents the number.

Of course, the number 1,000,000 is not represented by a million spots, holes, or ripples. Even when transmitted at the rate of a microsecond per unit, that would mean a million steps and take the unconscionable time of a full second. The answer, obviously, is the same one we use in ordinary calculation—position value coding, of which the decimal system is the most familiar example.

512

The first artificial digital computing device was the abacus, a manually operated mechanical memory of great antiquity, yet far in advance of mere finger counting. It is still efficiently used in many parts of the world, including the Chinese laundries of the U. S. The next advance was the adding wheel. Gottfried Wilhelm Leibnitz, who invented the calculus independently of Isaac Newton, also invented the stepped wheel that became the basis for the first commercial calculator. Adding wheels of one sort or another led to the modern desk calculator. Without the various office machines that add, subtract, and multiply at the touch of fingers on a keyboard, it would be difficult to imagine an economy that includes vast insurance companies, closely audited chain stores, and banks competing for personal accounts at 10 cents a check. Were it not for the mechanization of business mathematics, modern industry would need more bookkeepers than factory hands. The scientists, ranging from social statisticians to nuclear physicists, have made full use of these commercial devices. The adding machine is as much a tool of the laboratory as the test tube and the oscilloscope.

A further development, suitable to an electrified era, began in the U. S. with the census expert Herman Hollerith. He developed the first punched-card machine using the position of holes to remember numerical data. In the 1890 census it proved its worth by reducing the labor in half. The Tabulating Machine Company which he organized in 1896 was later consolidated into the International Business Machines Corporation.

The IBM card, standardized in a size of three and a quarter by seven and three-eighths inches, with its 80 columns of 12 punching positions each, has become a kind of mathematical coinage. It is interchangeable among a variety of punching, sorting, tabulating, calculating, and accounting machines, which deal with the cards mechanically, electrically, and electronically. The same kind of card may hold the data of an astronomical orbit, an accountant's audit, a corporation's income tax, or a subscription to a magazine.

The IBM cards, or variations thereof, not only paved the way to the modern electronic brains, but in many cases they constitute a vital part of them. The cards with punched holes speak just the kind of language that an electrical machine understands. From them the machine reads its assignment, upon them it spells out its answers, and with stacks of them it forms its library or memory.

Quite comparable in importance to the development of modern computers, although somewhat overlooked, is the telegraphic printer or teletype. This is the machine that can be seen typing, no hands, in every newspaper office, every news agency bureau, every telegraph office; it is the machine that reads and writes at a distance. The essential elements in this mechanized communication are the relay and the perforated tape. The relay is the lineal descendant of the telegraph relay, switching on new electrical circuits as an armature responds to the electromagnetic pull induced by an incoming signal. The perforated tape has room for five meaningful holes, plus a sixth little sprocket hole which serves to advance it through the reading apparatus.

Note well the mathematical meaning of five holes, or lack of holes, across the width of a paper tape. How many different meanings can be conveyed by this "five-unit code?" The answer is 32, and the way we arrive at it illustrates the binary system of numeration that appears in the most advanced electronic brains.

The first position on the tape may be either perforated or blank; that accounts for two possibilities. Each of these can be associated with a second position that either has a hole or has not; that makes four possibilities. The third, fourth, and fifth positions each in turn doubles the number of alternatives, giving a total of 2^5 or 32. Of these, 26 are used for ordinary purposes of communication to represent the letters of the alphabet, and five for other commands to the mechanical typist: space (between words), carriage return, line feed, shift to letters, shift to figures. (One of the 32 positions is usually not utilized for any signal.) The ability to put such commands in code and to have them carried out by relays is another major ingredient of the electronic brain.

It is possible to build a fairly fast computer using nothing much besides teleprinters and tape punchers, a switchboard and wiring, and a collection of adroitly interconnected relays. It is not only possible—it has been done by the Bell Telephone Laboratories.

The perforator presents the problem in the form of punched tape. The relays (standard items of any telephone dial center) do the calculating. The trick of adding is to wire their contacts and coils in such a way that the closing of any two relays energizes the relay that represents the sum. For instance, if relay No. 1 and relay No. 3 are simultaneously closed, the only possible path presented to an incoming current is the one leading through its closed contacts to the coil of relay No. 4. However, if the incoming current arrives from the "carry in 1" wire (meaning that there was a "1 to carry" from the previous column of the addition), the path leads to relay No. 5.

Compared with the electronic devices we shall discuss below, the relay computers are slow. On the other hand, they are reliable. One machine ran 1,500 hours without a failure. They check themselves, refusing to let an error go through even when they are unattended. When some element in the apparatus fails, the machine simply drops that portion of the problem and proceeds to the next.

What do relays really do? They are nothing but switches operated by electricity, so that one signal opens a valve or gate through which another current can pass. The very same thing can be done by an electronic tube—it is for this reason that the British use the word "valve" where we use "tube." The electronic valve does what the relay does—and does it thousands of times faster. In the relay, a mechanical object must move through space to open and close the gate. In the electronic tube, only electrons move, and their speed can best be indicated by the fact that in a radar set, electronic currents reverse themselves millions and even billions of times per second.

The first mathematical computer to employ the electronic swiftness of radio and radar was built in wartime secrecy for the Army Ordnance Department at the University of Pennsylvania, and later moved, at a cost of about $100,000, to the Ballistic Research Laboratories at the Aberdeen Proving Ground. It is the Electronic Numerical Integrator and Computer, which is known familiarly as Eniac. This phenomenal machine is so complicated that no single person, even among its inventors, knows every part of its wiring or the function of each of its 18,000-odd tubes. However, some of its arrangements are quite simple. The system of electronic number storage, or memory, is obvious almost at first glance.

If you were admitted by Army Ordnance officers to the new air-conditioned quarters of the Eniac, you would observe the walls lined with panels of radio tubes. You are shown one panel called an accumulator, and you are told that it is capable of remembering a 10-digit number. It can register any number from zero to 9,999,999,-999—but only one such number at a time.

THE BRAIN AND THE MACHINE

An accumulator consists essentially of 100 vacuum tubes (one might say 200 tubes, since each is a double triode). They are arranged in 10 columns of 10 tubes each. Reading from right to left, you have the units column, the tens column, the hundreds column, and so on. In each column, the bottom tube represents zero, the second tube represents 1, the third tube 2, and so on to 9.

To make things even easier, there is a neon light in front of each tube which goes on when that tube is in the "indicating" state. Only one tube in each column can be indicating. If the number is 5,384,293,768, tube 5 will be excited in the first column, tube 3 in the second column, and so on. If the number stays put for a few seconds (which in practice it does not, except during demonstrations or checkups) it is quite easy even for the untutored to read the number written on the wall.

The Eniac has 20 such accumulators, which occupy about half of the machine's total space. Thus it can store just 20 numbers (of 10 digits each) in its "electronic memory." At least one of the accumulators will be in use at any given moment in a dynamic way— either sending out the number which it has been holding or receiving a new number. The new number may come into a blank accumulator, on which it registers, or it may come into an accumulator which already holds a number. In the latter case, the new one is automatically added.

And here, at a flash, we see the machine's wonderful possibilities. An accumulator can absorb a 10-digit number (adding it, if necessary, to its existing contents) in just one five-thousandth of a second. Or, to use the kind of time unit which is more suitable to this discussion—in 200 microseconds.

One peculiarity of the high-speed calculators is that it takes no more time to do the most elaborate addition than the simplest. The Eniac needs as long to add 1 and 1 as to add two full 10-digit numbers. In fact, an accumulator gets its maximum workout when it is required to subtract 1 from 1. Since it works in one direction only, it cannot simply go back one step from 1 to zero. Instead, the minus 1 is set in as 9,999,999,999, i.e., 10 billion minus 1. It then adds 1 to 9,999,999,999.

In performing this addition, the machine functions as follows: A single pulse of electricity representing the number 1 goes into the units column. At the first tube position it finds that a free path exists for it to go to the second tube. There, also, the gate is open, and so on through successive tubes to the top of the column. The impulse travels to the tube that represents number 9. The "flip-flop" circuit at 9 position has been in the "on" condition. At this point three things happen: First, the impulse turns the top tube to "off," clearing away the 9. Second, it continues to the next tube position, representing zero, and switches that to the indicating condition. The units column now reads zero, as it should. But if you were doing this mentally you would say "zero and one to carry." To correspond to this, as the third step the activation of zero causes a single pulse to be carried to the ten column. Here the pulse runs through precisely the same routine, finding its way up the column to change the excitation from 9 to zero, and sending a carry into the hundreds column. This is repeated until the entire array has been changed from 9,999,999,999 to 0,000,000,000. The machine, doing it the hard way, has carried out the calculation 1 minus 1 equals zero.

Suppose the number to be added is 1,000,000,011. Then single pulses will enter at the same time into the units column, the tens column, and the billions column. After each runs through its cycle, another signal comes along which instructs each column to yield its "carry" signal, if it has one in store, to the column on its left.

What about multiplying and dividing? These could be done, and in some calculators are done, by repeated addition and subtraction. Multiplying 52 by 7 simply means setting up the number 52 and adding 52 six times. Similarly, division is repeated subtraction. There are, however, some short cuts. Eniac uses a built-in multiplication table. This is wired up to give immediately the product of any two digits. The sums of all the products are then fed into an accumulator and added up. By this method the entire multiplicand can be multiplied by one digit in the time corresponding to one addition. The complete multiplication of a pair of 10-digit numbers can be accomplished in 1/350 of a second.

Any numerical computation, whether it deals with an equation or a table of numbers, can be reduced to a succession of the basic arithmetical operations.

Although Eniac was a tremendous success in its way, opening the path to all-electronic computation, nothing exactly like it will ever be built again. It has serious limitations. Chief among them is a vast discrepancy between the speed with which it can compute and the time it takes for it to become aware of the problem and to spell out its answers.

For a given kind of problem, it must be instructed by plugging in wires and setting switches—both of them manual operations that go at the poor speed of human hands. In addition it has so many components that there are numerous sources of malfunction requiring elaborate maintenance routines. Another limitation, which became evident to its designers even while it was being built, is the limited electronic memory and the large space required for storing numbers in electronic tubes. The tubes require 120 kilowatts of power, and another 20 kilowatts is needed for the blower equipment to take away the heat in the tubes. This led to the conclusion that the use of 100 tubes to spell out a 10-digit number is wasteful of space and power.

There is a way to make a memory composed of electronic tubes more compact. The IBM Selective Sequence Electronic Calculator in New York is an example. The machine deals with 14-digit numbers. On the Eniac principle each number would require 1,400 double tubes. This machine gets by with 560. (In both the Eniac and the IBM machine there are various auxiliary tubes to control the circuits, but these are ignored here.) This is done by modifying the decimal system, using a hybrid called "binary decimal."

Any number from zero to 9 can be represented by only four tubes, provided each tube has a numerical value associated with its position in the group, and provided that more than one of the group can be "on" at the same time. The values chosen are 1, 2, 4 and 8. Thus:

1 is represented by tube 1.
2 is represented by tube 2.
3 is represented by tubes 1 and 2.
4 is represented by tube 4.
5 is represented by tubes 4 and 1.
6 is represented by tubes 4 and 2.
7 is represented by tubes 4, 2 and 1.
8 is represented by tube 8.
9 is represented by tubes 8 and 1.
0 is represented by all tubes off.

This system of binary-decimal digits is applied not only to electronic tubes but also to banks of relays and rolls of paper tape, which provide a vast storage of numbers and sequence instructions. At the input and output connections to the external world, the machine has an automatic device for translating between the decimal system and the binary-decimal. On the other hand, there are fields—engineering, for example, or mathematical physics—where problems can be handled on a pure binary system.

The binary system permits the "on-off" kind of memory to be used with 100 per cent efficiency. As long as we have no special reason for compromising with the decimal notation, we can have our tubes signify 1, 2, 4, 8, 16, 32, 64, 128, 256, 512, 1,024, doubling at each stage, as far as we like. Any intermediate number can always be assembled by appropriate combinations of the tubes.

At this point we find that we no longer need the decimal notation at all. There is another notation which serves the purpose better. The binary system of numbers really requires only two symbols, which are usually written as 0 and 1. Let us first compare it with the above list of doubling numbers:

Decimal notation	Binary notation
0	0
1	1
2	10
4	100
8	1,000
16	10,000
32	100,000
64	1,000,000
128	10,000,000
256	100,000,000

One inference is obvious at a glance. To multiply a number by 2, just add a zero. Or, in a machine, shift the columns one step to the left. Thus multiplying becomes as easy as adding.

Now let us see how the other numbers fit in. We will take just the first few:

Decimal	Binary
3	11
5	101
6	110
7	111
9	1,001

Thus it is clear that every number can be represented by some combination of the two symbols 1 and 0. The presence of one or the other constitutes a binary digit. The phrase binary digit has been abbreviated in a new term—"bit," a rather neat usage because it is essentially a bit of information. In the binary system we need 10 bits to represent a decimal thousand, 20 bits to represent a million, 30 bits to represent a billion, and approximately 33 bits to represent the ten-billion figure of an Eniac accumulator. It takes only one electronic tube to represent a bit. Thus 33 tubes do the work that in the Eniac requires 100 and in the binary-decimal system would require 40. A small section of the binary table is worked out in the chart.

In addition, binary notation thrives on the simple contrast of two alternatives—the ultimate in interchangeability and transportability. The 1 and 0 we have used in the above explanation can be translated physically, and without any further coding, into all sorts of inter-changeable effects—spatial, electrical, magnetic. For example, 1 may be represented by an electric pulse and 0 by the absence of a pulse; 1 by magnetic north and 0 by magnetic south, and so on.

Any such representation may be used as long as it is consistent and interchangeable. As a magnetized tape passes under a coil, the presence or absence of a magnetized spot is converted into the presence or absence of an electrical signal, which in turn can be routed to an electronic tube. If there is a signal (meaning 1) the tube can

Binary number system is used in modern computers because it makes possible the handling of large numbers—and hence large volume and variety of information—in a minimum number of circuits made up of simple "on" and "off" elements. System has only two digits, 1 and 0, which are indicated above by white circles ("on") and black circles ("off"). This chart shows binary code for numbers from one to 16, presenting in each horizontal row a combination of white and black circles that represents the decimal number in column at far left. The position of circle in a row determines its value; the value of each position doubles from right to left as indicated by the figures across the top. To translate binary number coded in each row into the corresponding decimal digit, simply add up value of white or "on" circles. Thus, the four "on" circles in the next to last row add up to 15. With five positions, the capacity goes on up from 16 to 31, and doubles with the addition of each new position. How rapidly the capacity of the system increases is indicated by the fact that with 33 positions (or circuits) it is possible to express any number from one to 8,489,934,592. Using the decimal system, a computer would require 100 circuits to attain an equivalent figure.

DECIMAL NUMBERS	16	8	4	2	1
1	●	●	●	●	○
2	●	●	●	○	●
3	●	●	●	○	○
4	●	●	○	●	●
5	●	●	○	●	○
6	●	●	○	○	●
7	●	●	○	○	○
8	●	○	●	●	●
9	●	○	●	●	○
10	●	○	●	○	●
11	●	○	●	○	○
12	●	○	○	●	●
13	●	○	○	●	○
14	●	○	○	○	●
15	●	○	○	○	○
16	○	●	●	●	●

be flipped from a 0 position to a 1, or from a 1 to a 0. The beauty of the binary system is that addition of a digit always means a simple reversal of the condition of the "memory"—adding a pulse where there is none, or wiping it out if there is one (converting 1 to 0).

The next step is the realization that the binary system does not merely apply to numbers. It applies to logic. For 1 and 0, we can substitute "yes" and "no." Thus, for example, a binary machine may be adjusted to deal with double negatives, making "no" and "no" add up to "yes." The vacuum tube lends itself especially well to acting out such logical concepts as "and," "or" and "neither." To illustrate the idea of "and," a four-element tube has two grids that act as "gates" controlling the current from grid to plate. If both of them are normally held at a strong negative voltage, current can flow only if both are turned positive by an incoming signal. Signals on each are a "necessary" condition for any current to come out of the tube; neither alone provides a "sufficient" condition. On the other hand, external circuits can be so arranged that a signal from either one of two sources will make the tube conduct; such an arrangement carries out the idea of "or," since one source or the other will suffice.

Such electronic gates are at the heart of the traffic and control systems of the high-speed calculators. Each gate may have one, two, or more locks, requiring certain simultaneous conditions to be satisfied before they will permit the machine to proceed to the next step.

The speed of the electronic tube is the key to the speed of the new calculators. We have already seen that the tubes are used in three ways: first, as a memory device which can receive and hold numbers; second, as an arithmetical unit; and third, as a gate for directing the flow of electrical traffic to different parts of the machine in accordance with instruction signals. Vacuum tubes have a fourth important function in making the calculators possible; they are employed, somewhat as in radio and television sets, to amplify worn-out signals that have lost their sharpness and strength, reshaping them as good as new.

Of the four functions, the least efficient is the one described at the beginning of this article—storage. And that is because it takes the length of a wall to store any appreciable quantity of numbers, whether the tubes are used 10 to a decimal digit, or four per decimal digit, or even one per binary digit. For really large-scale computations, machines need a memory that is both more compact and more capacious than any that can be achieved with electronic tubes.

One recource is to have the machine print out partial answers which can then be fed back into it—or other machines—when needed. The IBM Selective Sequence Calculator makes considerable use of this method. As the machine solves some mathematical function, the resulting table of values is perforated in a coded set of punched holes on a wide roll of heavy paper, not unlike an old-fashioned player-piano roll. This is mounted on an arrangement of cylinders and pulleys. Later, when the machine is ready to refer to the table, it "looks up" the proper constant by letting the paper run through it until it finds the appropriate value. On a similar principle, the machines can have a virtually infinite memory capacity in the form of punched cards, but the speed of this method is limited by the slow rate at which the machine reads the cards—100 cards per minute.

To get to another order of magnitude for compactness and speed of memory, designers have turned to methods other than perforated paper. Among them are: magnetic tape, photographic film, charged cathode-ray-tube surfaces and, most remarkable of all, columns of mercury in which numbers are stored in the dynamic form of waves moving at the speed of sound.

Mercury memory came out of wartime radar. One tactical difficulty with radar was the fact that an enemy plane could not well be distinguished against a solid background. It could conceal itself from the radar eye by hiding in front of a mountain as well as behind it. Engineers reasoned they could get around this difficulty if they could find some automatic way of canceling out a fixed echo and showing only those received from a moving object. This was finally accomplished, toward the end of the war, with a device called the delay line. The idea was to make an echo signal travel tardily within the set so that the following echo, arriving perhaps a thousandth of a second later, would overtake it. If the two matched, they would cancel each other out. Thus echoes from fixed terrain would not show. But if the echoes came from different places, as they would from a moving airplane, both of them would appear on the face of the oscilloscope.

One of the best means of delaying such a signal was found to be an "acoustic" line in which the signal would generate a ripple in a tube of liquid mercury. Traveling in the mercury at the speed of sound, the signal would come out at the far end an appreciable interval after its entrance. The exact period of delay could be adjusted by changing the temperature of the mercury or the length of the path through it.

In one of the more recent calculators, an 18-inch column of mercury maintained at 65 degrees provides a delay of 336 microseconds. Since successive pulses in this machine are separated by only one quarter of a microsecond, this delay means that at any given moment the 18-inch length of mercury has in storage four times 336, or 1,344 binary digits. For convenience of executive administration, this is divided into 32 "words," each containing 30 binary digits, a pulse space for plus or minus sign, and 11 more unused spaces (or time intervals) to separate successive words.

The mercury tube is the kind of "brain" in which information is supposed to go in one ear and come out of the other. An electrical signal arriving at the delay line causes a quartz crystal to expand by the well-known piezoelectric effect. The crystal pushes against the mercury, and a ripple runs through faster than the eye could follow. After 1/3,000 of a second it arrives at the far end, presses against another crystal, and generates a new electrical impulse. This is built up in an amplifier and fed back into the front end again. The cycle repeats 3,000 times a second. Thus the digit goes around and around, and would do so just about forever if nothing intervened. At the desired moment, however, an electronic gate opens in the amplifier to switch the signal into some other circuit—the electronic adder, for example. Or the signal can be erased, simply by instructing the amplifier not to amplify at the moment when the signal comes around.

At the present writing the most popular system for calculating machines is based on the mercury ripples, supplemented by electronic computing circuits, magnetic wire or tape for intermediate and erasable memory, punched cards or paper tape for a still more permanent library of memories and accumulated answers, and automatic printers for spelling out the answers.

What does this all mean in practical terms, for today and tomorrow? How will it affect business, government, military affairs, science, and mathematics itself?

It is tempting to make grand generalizations. The writer is indebted to Samuel N. Alexander, chief of the Bureau of Standards' electronic computer section, for a rather hardheaded appraisal of some real prospects—purposes for which various branches of the government are submitting their bids to be among the first to get such machines.

First, there is the matter of enormous numbers of routine substitutions in formulas. An important example is the adjustment of maps. Many nations may be correctly mapped, but two adjoining countries are likely to have a discrepancy of several feet at their mutual boundaries; the corrections require a tremendous number of separate calculations, and both the Army Map Service and the Coast and Geodetic Survey would like electronic assistance.

Secondly, there are elaborate engineering computations which now require enormous effort—for example, a roomful of people spending more than a year and a half checking stress estimates to transfer a design from a model to a full-size airplane. The slowness of ordinary computation was partly responsible for the fact that no U. S. bomber flew in the last war which had not been designed before the war began. This lag should be greatly reduced by the new computers.

Thirdly, there is a group of uses that have to do with "program procedure." What is the best way to distribute available manpower, funds, equipment, and so forth, to maximize a particular effect or to minimize cost? For example, the armed forces can make up various menus which will satisfy the soldier's needs for calories, vitamins and minerals. But each food item also has various such qualities as perishability, compactness and cost. How can you meet the dietary requirements with minimum cost, minimum shipping weight, or minimum time of delivery? Normally the mathematical labor in figuring out the advantages of every possible combination is too great, so only a few combinations are studied thoroughly.

Fourthly, since the digital calculator is essentially a logical machine, it can make all sorts of quick decisions that now require an alert and hard-pressed human being. The Air Navigation Development Board is considering the use of computing machines at airport control towers to relieve the traffic control men of many elementary, stereotyped decisions. And the Research and Development Board of the National Military Establishment has a committee at work considering how electrical computers can be rigged to play out war games.

In the present state of world affairs military applications are still uppermost in the thinking about calculators. The last war undoubtedly stimulated the development of these machines, and the threat of renewed war is a continued stimulant. It seems that electronic computers, associated with radar, will be the main defense against high-speed bombers, tracking the bomber and guiding the anti-aircraft missile to a collision course. And future long-range offensive missiles will most likely radio back what position information they can gather, and have it processed in a home computer, out of which radio-transmitted answers will give the missile its instructions for further navigation.

Should the war clouds dissipate, the cleared air would show the electronic brains contributing mightily to the advance of science and the efficiency of both business and government. They may even offer a technical means of contributing to world peace by helping to make world government practical. The "curse of bigness," which has affected big corporations and big governments alike, is very largely due to the difficulty of any small group of men knowing what is going on in a vast and far-flung enterprise. The trouble with planning is that the planners cannot know all the facts bearing on their plans; much less have they the time and ability to figure out all the consequences that will result from one or another course of action. These considerations have brought about a feeling for an optimum size for any single administration, whether government or business, beyond which the operation gets too unwieldy to compete with smaller, better integrated units. The electronic computers or information processing systems may well move that optimum size upward.

Though they replace other kinds of human mental effort, the mathematical machines will never replace the mathematician. More mathematicians will be needed, but the nature of their work will be changed. As theoretical physicists make ever more general equations expressing the basic forces of the universe with fewer and fewer symbols, their abstractions get ever farther from the numerical values of the laboratory experiment and ordinary life. Even Albert Einstein is not free of this difficulty. His approach toward a unified field theory, a generalization into which relativity fits as a part, may or may not be correct. Einstein does not know. The trouble, he once told the writer, is that "tremendous labors of the most brilliant minds, possibly for several generations, will be needed to translate the general theory into specific cases which can be tested by experiment." Perhaps the new computers, like the one being built by Einstein's colleagues at the Institute for Advanced Study, will help to shorten this labor, and thus allow investigators of the mysteries of the universe to test their own theories to determine whether they are on the right track.

by Arnold Tustin

OR HUNDREDS OF YEARS a few examples of true automatic control systems have been known. A very early one was the arrangement on windmills of a device to keep their sails always facing into the wind. It consisted simply of a miniature windmill which could rotate the whole mill to face in any direction. The small mill's sails were at right angles to the main ones, and whenever the latter faced in the wrong direction, the wind caught the small sails and rotated the mill to the correct position. With steam power came other automatic mechanisms: the engine-governor, and then the steering servo-engine on ships, which operated the rudder in correspondence with movements of the helm. These devices, and a few others such as simple voltage regulators, constituted man's achievement in automatic control up to about 20 years ago.

In the past two decades necessity, in the form of increasingly acute problems arising in our ever more complex technology, has given birth to new families of such devices. Chemical plants needed regulators of temperature and flow; air warfare called for rapid and precise control of searchlights and anti-aircraft guns; radio required circuits which would give accurate amplification of signals.

Thus the modern science of automatic control has been fed by streams from many sources. At first, it now seems surprising to recall, no connection between these various developments was recognized. Yet all control and regulating systems depend on common principles. Indeed, studies of the behavior of automatic control systems give us new insight into a wide variety of happenings in nature and in human affairs. The notions that engineers have evolved from these studies are useful aids in understanding how a man stands upright without toppling over, how the human heart beats, why our economic system suffers from slumps and booms, why the rabbit population in parts of Canada regularly fluctuates between scarcity and abundance.

The chief purpose of this article is to make clear the common pattern that underlies all these and many other varied phenomena. This common pattern is the existence of feedback, or—to express the same thing rather more generally—interdependence.

We should not be able to live at all, still less to design complex control systems, if we did not recognize that there are regularities in the relationship between events—what we call "cause and effect." When the room is warmer, the thermometer on the wall reads higher. We do not expect to make the room warmer by pushing up the mercury in the thermometer. But now consider the case when the instrument on the wall is not a simple thermometer but a thermostat, contrived so that as its reading goes above a chosen setting, the fuel supply to the furnace is progressively reduced, and, conversely, as its reading falls below that setting, the fuel flow is increased. This is an example of a familiar control system. Not only does the reading of the thermometer depend on the warmth of the room, but the warmth of the room also depends on the reading of the thermometer. The two quantities are interdependent. Each is a cause, and each an effect, of the other. In such cases we have a closed chain or sequence—what engineers call a "closed loop."

Not all automatic control systems are of the closed-loop type. For example, one might put the thermometer outside in the open air, and connect it to work the fuel valve through a specially shaped cam, so that the outside temperature regulates the fuel flow. In this open-sequence system the room temperature has no effect; there is no feedback. The control compensates only that disturbance of room temperature caused by variation of the outdoor temperature. Such a system is not necessarily a bad or useless system; it might work very well under some circumstances. But it has two obvious shortcomings. Firstly, it is a "calibrated" system; that is to say, its correct working would require careful preliminary testing and special shaping of the cam to suit each particular application. Secondly, it could not deal with any but standard conditions. A day that was windy as well as cold would not get more fuel on that account.

The feedback type of control avoids these shortcomings. It goes directly to the quantity to be controlled, and it corrects indiscriminately for all kinds of disturbance. Nor does it require calibration for each special condition. Diagrams for the two kinds of control are given on page 535.

Feedback control, unlike open-sequence control, can never work without *some* error, for the error is depended upon to bring about the correction. The objective is to make the error as small as possible. This is subject to certain limitations, which we must now consider.

Any quantity may be subjected to control if three conditions are met. First, the required changes must be controllable by some physical means, a regulating organ. Second, the controlled quantity must be measurable, or at least comparable with some standard; in other words, there must be a measuring device. Third, both regulation and measurement must be rapid enough for the job in hand.

As an example, take one of the simplest and commonest of industrial requirements: to control the rate of flow of liquid along a pipe. As the regulating organ we can use a throttle valve, and as the measuring device, some form of flowmeter. A signal from the flowmeter, telling the actual rate of flow through the pipe, goes to the "controller"; there it is compared with a setting giving the required rate of flow. The amount and direction of "error," *i.e.*, deviation from this setting is then transmitted to the throttle valve as an operating signal to bring about adjustment in the required direction. In flow-control systems the signals are usually in the form of variations in air pressure transmitted through a system of small-bore pipes.

For further light on the principles involved in control systems let us consider the example of the automatic gun-director. In this problem a massive gun must be turned with great precision to angles indicated by a fly-power pointer on a clock-dial some hundreds of feet away. When the pointer moves, the gun must turn correspondingly. The quantity to be controlled is the angle of the gun. The reference quantity is the angle of the clock-dial pointer. What is needed is a feedback loop which constantly compares the gun angle with the pointer angle and arranges matters so that if the gun angle is too small, the gun is driven forward, and if it is too large, the gun is driven back.

The key element in this case is some device which will detect the error of angular alignment between two shafts remote from each other, and which does not require more force than is available at the fly-power transmitter shaft. There are several kinds of electrical elements that will serve such a purpose. The one usually selected is a pair of the miniature alternating-current machines known as selsyns. The two selsyns, connected respectively to the transmitter shaft and the gun, provide an electrical signal proportional to the error of alignment. The signal is amplified and fed to a generator which in turn feeds a motor that drives the gun.

530

This gives the main lines of a practicable scheme, but, if a system were built as just described, it would fail. The gun's inertia would carry it past the position of correct alignment; the new error would then cause the controller to swing it back, and the gun would hunt back and forth without ever settling down.

This oscillatory behavior, maintained by "self-excitation," is one of the principal limitations of feedback control. It is the chief enemy of the control-system designer, and the key to progress has been the finding of various simple means to prevent oscillation. Since oscillation is a very general phenomenon, it is worth while to look at the mechanism in detail, for what we learn about oscillation in man-made control systems may suggest means of inhibiting oscillations of other kinds—such as economic booms and slumps, or periodic swarms of locusts.

Consider any case in which a quantity that we shall call the output depends on another quantity we shall call the input. If the input quantity oscillates in value, then the output quantity also will oscillate, not simultaneously or necessarily in the same way, but with the same frequency. Usually in physical systems the output oscillation lags behind the input. For example, if one is boiling water and turns the gas slowly up and down, the amount of steam increases and decreases the same number of times per minute, but the maximum amount of steam in each cycle must come rather later than the maximum application of heat, because of the time required for heating. If the first output quantity in turn affects some further quantity, the variation of this second quantity in the sequence will usually lag still more, and so on. The lag (as a proportion of one oscillation) also usually increases with frequency—the faster the input is varied, the farther behind the output falls.

Now suppose that in a feedback system some quantity in the closed loop is oscillating. This causes the successive quantities around the loop to oscillate also. But the loop comes around to the original quantity, and we have here the mechanism by which an oscillation may maintain itself. To see how this can happen, we must remember that with the feedback negative, the motion it causes would be opposite to the original motion, if it were not for the lags. It is only when the lags add up to just half a cycle that the feedback maintains the assumed motion. Thus any system with negative feedback will maintain a continuous oscillation when disturbed if a) the

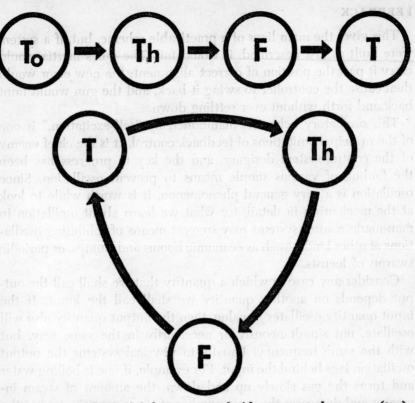

Feedback loop (at bottom) is here contrasted with open control sequence (top). In a hypothetical heating system, the outside temperature (To) might be employed to actuate the thermostat control (Th) on the furnace (F) to adjust room temperature (T). But there is no provision to determine whether the room temperature has attained the level desired. This self-regulating principle is uniquely provided by the feedback circuit. Here the variable which is to be controlled, the room temperature, itself actuates the thermostat. It thus controls the performance of the furnace.

time-delays in response at some frequency add up to half a period of oscillation, and b) the feedback effect is sufficiently large at this frequency.

In a linear system, that is, roughly speaking, a system in which effects are directly proportional to causes, there are three possible results. If the feedback, at the frequency for which the lag is half a period, is equal in strength to the original oscillation, there will be a continuous steady oscillation which just sustains itself. If the feedback is greater than the oscillation at that frequency, the oscillation builds up; if it is smaller, the oscillation will die away. These effects are sketched in the accompanying graphs.

This situation is of critical importance for the designer of control systems. On the one hand, to make the control accurate, one must increase the feedback; on the other, such an increase may accentuate any small oscillation. The control breaks into an increasing oscillation and becomes useless.

To escape from the dilemma the designer can do several things. Firstly, he may minimize the time-lag by using electronic tubes or, at higher power levels, the new varieties of quick-response direct-current machines. By dividing the power amplification among a multiplicity of stages, these special generators have a smaller lag than conventional generators. The lag is by no means negligible, however.

Secondly, and this was a major advance in the development of control systems, the designer can use special elements that introduce a time-lead, anticipating the time-lag. Such devices, called phase-advances, are often based on the properties of electric capacitors, because alternating current in a capacitor circuit leads the voltage applied to it.

Thirdly, the designer can introduce other feedbacks besides the main one, so designed as to reduce time-lag. Modern achievements in automatic control are based on the use of combinations of such devices to obtain both accuracy and stability.

Long before man existed, evolution hit upon the need for anti-oscillating features in feedback control and incorporated them in the body mechanisms of the animal world. Signals in the animal body are transmitted by trains of pulses along nerve fibers. When a sensory organ is stimulated, the stimulus will produce pulses at a greater rate if it is increasing than if it is decreasing. The maximum response, or output signal, occurs before the maximum of the stimulus. This is just the anticipatory type of effect (the time-lead) that is required for high-accuracy control. Physiologists now believe that the anticipatory response has evolved in the nervous system for, at least in part, the same reason that man wants it in his control mechanisms—to avoid overshooting and oscillation. Precisely what feature of the structure of the nerve mechanism gives this remarkable property is not yet fully understood.

Fascinating examples of the consequences of interdependence arise in the fluctuations of animal populations in a given territory. The interdependence of animal species sometimes produces a periodic oscillation. Just to show how this can happen, and leaving out complications that are always present in an actual situation, consider a territory inhabited by rabbits and lynxes, the rabbits being the chief food of the lynxes. When rabbits are abundant, the lynx population will increase. But as the lynxes become abundant, the rabbit population falls, because more rabbits are caught. Then as the rabbits diminish, the lynxes go hungry and decline. The result is a self-maintaining oscillation, sustained by negative feedback with a time-delay.

The periodic booms and slumps in economic activity stand out as a major example of oscillatory behavior due to feedback. In 1936 the economist John Maynard Keynes gave the first adequate and satisfying account of the essential mechanisms on which the general level of economic activity depends. Although Keynes did not use the terminology of control-system theory, his account fits precisely the same now-familiar pattern.

Keynes' starting point was the simple notion that the level of economic activity depends on the rate at which goods are bought. He took the essential further step of distinguishing two kinds of buying—of consumption goods and of capital goods. The latter is the same thing as the rate of investment. The money available to buy all these goods is not automatically provided by the wages and profits disbursed in making them, because normally some of this money is saved. The system would therefore run down and stop if it were not for the constant injection of extra demand in the form of new investment. Therefore the level of economic activity and employment depends on the rate of investment. This is the first dependence. The rate of investment itself, however, depends on

Oscillation is inherent in all feedback systems. The drawing at top shows that when a regular oscillation is introduced into the input of a system (*lighter line*), it is followed somewhat later by a corresponding variation in the output of the system. The dotted rectangle indicates the lag that will prevail between equivalent phases of the input and the output curves. In the three drawings below, the input is assumed to be a feedback from the output. The first of the three shows a state of stable oscillation, which results when the feedback signal (*thinner line*) is opposite in phase to the disturbance of a system and calls for corrective action equal in amplitude. The oscillation is damped and may be made to disappear when, as in the next drawing, the feedback is less than the output. Unstable oscillation is caused by a feedback signal that induces corrective action greater than the error and thus amplifies the original disturbance.

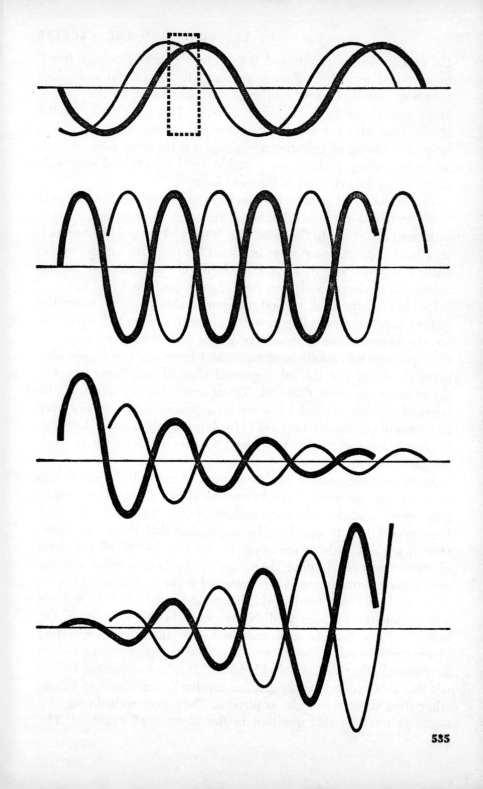

the expectation of profit, and this in turn depends on the trend, present and expected, of economic activity. Thus not only does economic activity depend on the rate of investment, but the rate of investment depends on economic activity. This clarification of the mechanisms at work immediately suggests many ways in which, by proper timing of investment expenditure, by more rational business forecasting, and so on, a stable level of optimal economic activity may be achieved in the near future.

The examples of feedback given here are merely a few selected to illustrate general principles. Many more will be described in other articles in this issue. In this article on "theory" I should like to touch on a further point: some ways in which the properties of automatic control systems or other complex feedback systems may be investigated in detail, and their performance perfected.

Purely mathematical methods are remarkably powerful when the system happens to be linear. Sets of linear differential equations are the happy hunting ground of mathematicians. They can turn the equations into a variety of equivalent forms, and generally play tunes on them. For the more general class of non-linear systems, the situation is quite different. There exact determination of the types of motion implied by a set of dependences is usually very laborious or practically impossible for human calculators. However, electronic computers are now used to predict behavior in such complex systems.

At the commencement of this account of control systems it was necessary to assume that the human mind can distinguish "cause" and "effect" and describe the regularities of nature in these terms. It may be fitting to conclude by suggesting that the concepts reviewed are not without relevance to the grandest of all problems of science and philosophy: the nature of the human mind and the significance of our forms of perception of what we call reality.

In much of the animal world, behavior is controlled by reflexes and instinct-mechanisms in direct response to the stimulus of the immediate situation. In man and the higher animals the operation of what we are subjectively aware of as the "mind" provides a more flexible and effective control of behavior. It is not at present known whether these conscious phenomena involve potentialities of matter other than those we study in physics. They may well do so, and we must not beg this question in the absence of evidence. The

FEEDBACK

subtleties of the problem are explored at some length in the next chapter of this book.

Whatever the nature of the means or medium involved, the function of the central nervous system in the higher animals is clear. It is to provide a biologically more effective control of behavior under a combination of inner and environmental stimuli. An inner analogue or simulation of relevant aspects of the external world, which we are aware of as our idea of the environment, controls our responses, superseding mere instinct or reflex reaction. The world is still with us when we shut our eyes, and we use the "play of ideas" to predict the consequences of action. Thus our activity is adjusted more elaborately and advantageously to the circumstances in which we find ourselves.

This situation is strikingly similar in principle (though immensely more complex) to the introduction of a predictor in the control of a gun, for all predictors are essentially analogues of the external situation. The function of mind is to predict, and to adjust behavior accordingly. It operates like an analogue computer fed by sensory clues.

It is not surprising, therefore, that man sees the external world in terms of cause and effect. The distinction is largely subjective. "Cause" is what might conceivably be manipulated. "Effect" is what might conceivably be purposed.

Man is far from understanding himself, but it may turn out that his understanding of automatic control is one small further step toward that end.

IV. A CHESS PLAYING MACHINE

by Claude E. Shannon

ELECTRONIC COMPUTERS are designed primarily to carry out purely mathematical calculations, but their basic design is so general and flexible that they can be adapted to work symbolically with elements representing words, propositions or other conceptual entities.

One such possibility, which is already being investigated in several quarters, is that of translating from one language to another by means of a computer. The immediate goal is not a finished literary rendition, but only a word-by-word translation that would convey enough of the meaning to be understandable. Computing machines could also be employed for many other tasks of a semi-rote, semi-thinking character, such as designing electrical filters and relay circuits, helping to regulate airplane traffic at busy airports, and routing long-distance telephone calls most efficiently over a limited number of trunks.

Some of the possibilities in this direction can be illustrated by setting up a computer in such a way that it will play a fair game of chess. This problem, of course, is of no importance in itself, but it was undertaken with a serious purpose in mind. The investigation of the chess-playing problem is intended to develop techniques that can be used for more practical applications.

The chess machine is an ideal one to start with for several reasons. The problem is sharply defined, both in the allowed operations (the moves of chess) and in the ultimate goal (checkmate). It is neither so simple as to be trivial nor too difficult for satisfactory solution. And such a machine could be pitted against a human opponent, giving a clear measure of the machine's ability in this type of reasoning.

A preliminary attempt along these lines was made in 1914 by a Spanish inventor named L. Torres y Quevedo, who constructed a device that played an end game of king and rook against king. The machine, playing the side with king and rook, would force checkmate in a few moves however its human opponent played. Since an explicit set of rules can be given for making satisfactory moves in such an end game, the problem is relatively simple, but the idea was quite advanced for that period.

A CHESS PLAYING MACHINE

The problem of setting up an electronic computer in such a way that it will play a complete game of chess can be divided into three parts: first, a code must be chosen so that chess positions and the chess pieces can be represented as numbers; second, a strategy must be found for choosing the moves to be made; and third, this strategy must be translated into a sequence of elementary computer orders, or a program.

A suitable code for the chessboard and the chess pieces is shown in the diagram. Each square on the board has a number consisting of two digits, the first digit corresponding to the "rank" or horizontal row, the second to the "file" or vertical row. Each different chess piece also is designated by a number: a pawn is numbered 1, a knight 2, a bishop 3, a rook 4 and so on. White pieces are represented by positive numbers and black pieces by negative ones. The positions of all the pieces on the board can be shown by a sequence of 64 numbers, with zeros to indicate the empty squares. Thus any chess position can be recorded as a series of numbers and stored in the numerical memory of a computing machine.

A chess move is specified by giving the number of the square on which the piece stands and of the one to which it is moved. Ordinarily two numbers would be sufficient to describe a move, but to take care of the special case of the promotion of a pawn to a higher piece a third number is necessary. This number indicates the piece to which the pawn is converted. In all other moves the third number is zero. Thus a knight move from square 01 to 22 is encoded into 01, 22, 0. The move of a pawn from 62 to 72, and its promotion to a queen, is represented by 62, 72, 5.

The second main problem is that of deciding on a strategy of play. A straightforward process must be found for calculating a reasonably good move for any given chess position. This is the most difficult part of the problem. Even the high speeds available in electronic computers are hopelessly inadequate to play perfect chess by calculating all possible variations to the end of the game. In a typical chess position there will be about 32 possible moves with 32 possible replies—already this creates 1,024 possibilities. Most chess games last 40 moves or more for each side. So the total number of possible variations in an average game is about 10^{120}. A machine calculating one variation each millionth of a second would require over 10^{95} years to decide on its first move!

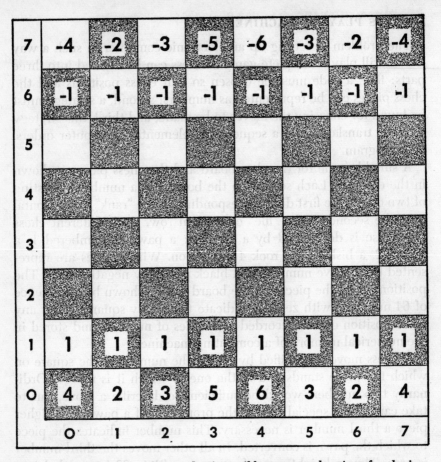

Code for chess-playing machine makes it possible to express location of each piece by number. Each square may be designated by two numbers, and the six different types of chess men each have a number. Three digits will thus locate a piece.

Other methods of attempting to play perfect chess seem equally impracticable; we resign ourselves, therefore, to having the machine play a reasonably skillful game, admitting occasional moves that may not be the best. This, of course, is precisely what human players do: no one plays a perfect game.

In setting up a strategy on the machine one must establish a method of numerical evaluation for any given chess position. A chess player looking at a position can form an estimate as to which side, White or Black, has the advantage. Furthermore, his evaluation is roughly quantitative. He may say, "White has a rook for a bishop,

540

an advantage of about two pawns"; or "Black has sufficient mobility to compensate for a sacrificed pawn." These judgments are based on long experience and are summarized in the principles of chess expounded in chess literature. For example, it has been found that a queen is worth nine pawns, a rook is worth five, and a bishop or a knight is worth about three. As a first rough approximation, a position can be evaluated by merely adding up the total forces for each side, measured in terms of the pawn unit. There are, however, numerous other features which must be taken into account: the mobility and placement of pieces, the weakness of king protection, the nature of the pawn formation, and so on. These too can be given numerical weights and combined in the evaluation, and it is here that the knowledge and experience of chess masters must be enlisted.

Assuming that a suitable method of position evaluation has been decided upon, how should a move be selected? The simplest process is to consider all the possible moves in the given position and choose the one that gives the best immediate evaluation. Since, however, chess players generally look more than one move ahead, one must take account of the opponent's various possible responses to each projected move. Assuming that the opponent's reply will be the one giving the best evaluation from his point of view, we would choose the move that would leave us as well off as possible after his best reply. Unfortunately, with the computer speeds at present available, the machine could not explore all the possibilities for more than two moves ahead for each side, so a strategy of this type would play a poor game by human standards. Good chess players frequently play combinations four or five moves deep, and occasionally world champions have seen as many as 20 moves ahead. This is possible only because the variations they consider are highly selected. They do not investigate all lines of play, but only the important ones.

Clearly it would be highly desirable to improve the strategy for the machine by including such a selection process in it. Of course one could go too far in this direction. Investigating one particular line of play for 40 moves would be as bad as investigating all lines for just two moves. A suitable compromise would be to examine only the important possible variations—that is, forcing moves, captures and main threats—and carry out the investigation of the possible moves far enough to make the consequences of each fairly clear. It is possible to set up some rough criteria for selecting important variations, not as efficiently as a chess master, but sufficiently well to reduce the number of variations appreciably and thereby permit a deeper investigation of the moves actually considered.

The final problem is that of reducing the strategy to a sequence of orders, translated into the machine's language. This is a relatively straightforward but tedious process, and we shall only indicate some of the general features. The complete program is made up of nine sub-programs and a master program that calls the sub-programs into operation as needed. Six of the sub-programs deal with the movements of the various kinds of pieces. In effect they tell the machine the allowed moves for these pieces. Another sub-program enables the machine to make a move "mentally" without actually carrying it out: that is, with a given position stored in its memory it can construct the position that would result if the move were made. The seventh sub-program enables the computer to make a list of all possible moves in a given position, and the last sub-program evaluates any given position. The master program correlates and supervises the application of the sub-programs. It starts the seventh sub-program making a list of possible moves, which in turn calls in previous sub-programs to determine where the various pieces could move. The master program then evaluates the resulting positions by means of the eighth sub-program and compares the results according to the process described above. After comparison of all the investigated variations, the one that gives the best evaluation according to the machine's calculations is selected. This move is translated into standard chess notation and typed out by the machine.

It is believed that an electronic computer programmed in this manner would play a fairly strong game at speeds comparable to human speeds. A machine has several obvious advantages over a human player: 1) it can make individual calculations with much greater speed; 2) its play is free of errors other than those due to deficiencies of the program, whereas human players often make very simple and obvious blunders; 3) it is free from laziness, or the temptation to make an instinctive move without proper analysis of the position; 4) it is free from "nerves," so it will make no blunders due to overconfidence or defeatism. Against these advantages, however, must be weighed the flexibility, imagination and learning capacity of the human mind.

A CHESS PLAYING MACHINE

Under some circumstances the machine might well defeat the program designer. In one sense, the designer can surely outplay his machine; knowing the strategy used by the machine, he can apply the same tactics at a deeper level. But he would require several weeks to calculate a move, while the machine uses only a few minutes. On an equal time basis, the speed, patience and deadly accuracy of the machine would be telling against human fallibility.

As described so far, the machine would always make the same move in the same position. If the opponent made the same moves, this would always lead to the same game. Once the opponent won a game, he could win every time thereafter by playing the same strategy, taking advantage of some particular position in which the machine chooses a weak move. One way to vary the machine's play would be to introduce a statistical element. Whenever it was confronted with two or more possible moves that were about equally good according to the machine's calculations, it would choose from them at random. Thus if it arrived at the same position a second time it might choose a different move.

Another place where statistical variation could be introduced is in the opening game. It would be desirable to have a number of standard openings, perhaps a few hundred, stored in the memory of the machine. For the first few moves, until the opponent deviated from the standard responses or the machine reached the end of the stored sequence of moves, the machine would play by memory. This could hardly be considered cheating, since that is the way chess masters play the opening.

We may note that within its limits a machine of this type will play a brilliant game. It will readily make spectacular sacrifices of important pieces in order to gain a later advantage or to give checkmate, provided the completion of the combination occurs within its computing limits. For example, in the position illustrated here the machine would quickly discover the sacrificial mate in three moves:

	White	*Black*
1.	R-K8 Ch	R X R
2.	Q-Kt4 Ch	Q X Q
3.	Kt-B6 Mate	

Winning combinations of this type are frequently overlooked in amateur play.

Problem that the Shannon machine could solve brilliantly might begin from this position. The machine, playing white, would sacrifice a rook and its queen, the most powerful piece on the board, and then checkmate in only one more move.

The chief weakness of the machine is that it will not learn by its mistakes. The only way to improve its play is by improving the program. Some thought has been given to designing a program that would develop its own improvements in strategy with increasing experience in play. Although it appears to be theoretically possible, the methods thought of so far do not seem to be very practical. One possibility is to devise a program that would change the terms and coefficients involved in the evaluation function on the basis of the results of games the machine had already played. Small variations might be introduced in these terms, and the values would be selected to give the greatest percentage of wins.

544

V. IMITATION OF LIFE

by W. Grey Walter

> "When we were little . . . we went to school in the
> sea. The master was an old Turtle—we used to call
> him Tortoise."
> "Why did you call him Tortoise if he wasn't one?"
> Alice asked.
> "We called him Tortoise because he taught us," said
> the Mock Turtle angrily. "Really you are very dull!"
> —Lewis Carroll,
> *Alice's Adventures in Wonderland*

IN THE DARK AGES before the invention of the electronic vacuum
tube there were many legends of living statues and magic
pictures. One has only to recall the importance of graven
images and holy pictures in many religions to realize how
readily living and even divine properties are projected into inanimate objects by hopeful but bewildered men and women.
Idolatry, witchcraft and other superstitions are so deeply rooted and
widespread that it is possible even the most detached scientific
activity may be psychologically equivalent to them; such activity
may help to satisfy the desire for power, to assuage the fear of the
unknown or to compensate for the flatness of everyday existence.

In any case there is an intense modern interest in machines that
imitate life. The great difference between magic and the scientific
imitation of life is that where the former is content to copy external
appearance, the latter is concerned more with performance and
behavior. Except in the comic strips the scientific robot does not
look in the least like a living creature, though it may reproduce in
great detail some of the complex functions which classical physiologists described as diagnostic of living processes. Some of the simpler
of these functions can be duplicated by mechanical contrivances.
But it was not until the electronic age that serious efforts were made
to imitate and even to surpass the complex performance of the
nervous system.

All the gradations of feeling and action of which we are capable
are provided by variations in the frequency of nerve impulses and
by the number of nerve cells stimulated. The brain cipher is even
simpler than Morse code: it uses only dots, the number of which

545

per second conveys all information. Communication engineers call this system "pulse-frequency modulation." It was "invented" by animals many millions of years ago, and it has advantages over other methods which are only just beginning to be applied. The engineers who have designed our great computing machines adopted this system without realizing that they were copying their own brains. (The popular term electronic brain is not so very fanciful.) In the language of these machines there are only two statements, "yes" and "no," and in their arithmetic only two numbers, 1 and 0. They surpass human capacity mainly in their great speed of action and in their ability to perform many interdependent computations at the same time, e.g., to solve simultaneous differential equations with hundreds of variables.

Magical though these machines may appear to the layman, their resemblance to living creatures is limited to certain details of their design. Above all they are in no sense free as most animals are free; rather they are parasites, depending upon their human hosts for nourishment and stimulation.

In a different category from computing machines are certain devices that have been made to imitate more closely the simpler types of living creatures, including their limitations (which in a computer would be serious faults) as well as their virtues. These less ambitious but perhaps more attractive mechanical creatures have evolved along two main lines. First there are stationary ones—sessile, the biologist would call them—which are rooted in a source of electric power and have very limited freedom. The prototype of these is the "homeostat" made by W. R. Ashby of Gloucester, England. It was created to study the mechanism whereby an animal adapts its total system to preserve its internal stability in spite of violent external changes.

The term "homeostasis" was coined by the Harvard University physiologist Walter B. Cannon to describe the many delicate biological mechanisms which detect slight changes of temperature or chemical state within the body and compensate for them by producing equal and opposite changes. Communication engineers, as Arnold Tustin has shown, rediscovered this important expedient in their grapplings with the problems of circuits and computers. They describe a system in which errors or variations from some desirable state are automatically neutralized as containing "negative or inverse feedback." In an animal most of what is called reflex activity has exactly this property.

In Ashby's homeostat there are a number of electronic circuits similar to the reflex arcs in the spinal cord of an animal. These are so combined with a number of radio tubes and relays that out of many thousands of possible connections the machine will automatically find one that leads to a condition of dynamic internal stability. That is, after several trials and errors the instrument establishes connections which tend to neutralize any change that the experimenter tries to impose from outside. It is a curious fact that although the machine is man-made, the experimenter finds it impossible to tell at any moment exactly what the machine's circuit is without "killing" it and dissecting out the "nervous system"; that is, switching off the current and tracing out the wires to the relays. Nevertheless the homeostat does not behave very like an active animal—it is more like a sleeping creature which when disturbed stirs and finds a comfortable position.

Another branch of electromechanical evolution is represented by the little machines we have made in Bristol. We have given them the mock-biological name *Machina speculatrix,* because they illustrate particularly the exploratory, speculative behavior that is so characteristic of most animals. The machine on which we have chiefly concentrated is a small creature with a smooth shell and a protruding neck carrying a single eye which scans the surroundings for light stimuli; because of its general appearance we call the genus "Testudo," or tortoise. The Adam and Eve of this line are nicknamed Elmer and Elsie, after the initials of the terms describing them—ELectro MEchanical Robots, Light-Sensitive, with Internal and External stability. Instead of the 10,000 million cells of our brains, Elmer and Elsie contain but two functional elements: two miniature radio tubes, two sense organs, one for light and the other for touch, and two effectors or motors, one for crawling and the other for steering. Their power is supplied by a miniature hearing-aid B battery and a miniature six-volt storage battery, which provides both A and C current for the tubes and the current for the motors.

The number of components in the device was deliberately restricted to two in order to discover what degree of complexity of behavior and independence could be achieved with the smallest

THE BRAIN AND THE MACHINE

number of elements connected in a system providing the greatest number of possible interconnections. From the theoretical standpoint two elements equivalent to circuits in the nervous system can exist in six modes; if one is called A and the other B, we can distinguish A, B, A + B, A → B, B → A and A ⇌ B as possible dynamic forms. To indicate the variety of behavior possible for even so simple a system as this, one need only mention that six elements would be more than enough to form a system which would provide a new pattern every tenth of a second for 280 years—four times the human lifetime of 70 years! It is unlikely that the number of perceptible functional elements in the human brain is anything like the total number of nerve cells; it is more likely to be of the order of 1,000. But even if it were only 10, this number of elements could provide enough variety for a lifetime of experience for all the men who ever lived or will be born if mankind survives a thousand million years.

So a two-element synthetic animal is enough to start with. The strange richness provided by this particular sort of permutation introduces right away one of the aspects of animal behavior—and human psychology—which *M. speculatrix* is designed to illustrate: the uncertainty, randomness, free will or independence so strikingly absent in most well-designed machines. The fact that only a few richly interconnected elements can provide practically infinite modes of existence suggests that there is no logical or experimental necessity to invoke more than *number* to account for our subjective conviction of freedom of will and our objective awareness of personality in our fellow men.

The behavior of Elmer and Elsie is in fact remarkably unpredictable. The photocell, or "eye," is linked with the steering mechanism. In the absence of an adequate light-stimulus Elmer (or Elsie) explores continuously, and at the same time the motor drives it forward in a crawling motion. The two motions combined give the creature a cycloidal gait, while the photocell "looks" in every direction in turn. This process of scanning and its synchronization with the steering device may be analogous to the mechanism whereby the electrical pulse of the brain known as the alpha rhythm sweeps over the visual brain areas and at the same time releases or blocks impulses destined for the muscles of the body. In both cases the function is primarily one of economy, just as in a television system the

548

scanning of the image permits transmission of hundreds of thousands of point-details on one channel instead of on as many channels.

The effect of this arrangement on Elmer is that in the dark it explores in a very thorough manner a considerable area, remaining alert to the possibility of light and avoiding obstacles that it cannot surmount or push aside. When the photocell sees a light, the resultant signal is amplified by both tubes in the amplifier. If the light is very weak, only a *change* of illumination is transmitted as an effective signal. A slightly stronger signal is amplified without loss of its absolute level. In either case the effect is to halt the steering mechanism so that the machine moves toward the light source or maneuvers so that it can approach the light with the least difficulty. This behavior is of course analogous to the reflex behavior known as "positive tropism," such as is exhibited by a moth flying into a candle. But Elmer does not blunder into the light, for when the brilliance exceeds a certain value—that of a flashlight about six inches away—the signal becomes strong enough to operate a relay in the first tube, which has the reverse effect from the second one. Now the steering mechanism is turned on again at double speed, so the creature abruptly sheers away and seeks a more gentle climate. If there is a single light source, the machine circles around it in a complex path of advance and withdrawal; if there is another light farther away, the machine will visit first one and then the other and will continually stroll back and forth between the two. In this way it neatly solves the dilemma of Buridan's ass, which the scholastic philosophers said would die of starvation between two bundles of hay if it did not possess a transcendental free will.

For Elmer hay is represented, of course, by the electricity it needs to recharge its batteries. Within the hutch where it normally lives is a battery charger and a 20-watt lamp. When the creature's batteries are well charged, it is attracted to this light from afar, but at the threshold the brilliance is great enough to act as a repellent, so the model wanders off for further exploration. When the batteries start to run down, the first effect is to enhance the sensitivity of the amplifier so that the attraction of the light is felt from even farther away. But soon the level of sensitivity falls and then, if the machine is fortunate and finds itself at the entrance to its kennel, it will be attracted right home, for the light no longer seems so dazzling. Once well in, it can make contact with the charger. The moment current flows in the circuit between the charger and the batteries the creature's own nervous system and motors are automatically disconnected; charging continues until the battery voltage has risen to its maximum. Then the internal circuits are automatically reconnected and the little creature, repelled now by the light which before the feast had been so irresistible, circles away for further adventures.

Inevitably in its peripatetic existence *M. speculatrix* encounters many obstacles. These it cannot "see," because it has no vestige of pattern vision, though it will avoid an obstacle that casts a shadow when it is approaching a light. The creature is equipped, however, with a device that enables it to get around obstacles. Its shell is suspended on a single rubber mounting and has sufficient flexibility to move and close a ring contact. This contact converts the two-stage amplifier into a multivibrator. The oscillations so generated rhythmically open and close the relays that control the full power to the motors for steering and crawling. At the same time the amplifier is prevented from transmitting the signals picked up by the photocell. Accordingly when the creature makes contact with an obstacle, whether in its speculative or tropistic mode, all stimuli are ignored and its gait is transformed into a succession of butts, withdrawals and sidesteps until the interference is either pushed aside or circumvented. The oscillations persist for about a second after the obstacle has been left behind; during this short memory of frustration Elmer darts off and gives the danger area a wide berth.

When the models were first made, a small light was connected in the steering-motor circuit to act as an indicator showing when the motor was turned off and on. It was soon found that this light endowed the machines with a new mode of behavior. When the photocell sees the indicator light in a mirror or reflected from a white surface, the model flickers and jigs at its reflection in a manner so specific that were it an animal a biologist would be justified in attributing to it a capacity for self-recognition. The reason for the flicker is that the vision of the light results in the indicator light being switched off, and darkness in turn switches it on again, so an oscillation of the light is set up.

Two creatures of this type meeting face to face are affected in a similar but again distinctive manner. Each, attracted by the light the other carries, extinguishes its own source of attraction, so the two systems become involved in a mutual oscillation, leading finally to a stately retreat. When the encounter is from the side or from behind, each regards the other merely as an obstacle; when both are attracted by the same light, their jostling as they approach the light eliminates the possibility of either reaching its goal. When one machine casually interferes with another while the latter is seriously seeking its charging light, a dog-in-the-manger situation develops which results in the more needy one expiring from exhaustion within sight of succor.

These machines are perhaps the simplest that can be said to resemble animals. Crude though they are, they give an eerie impression of purposefulness, independence and spontaneity. More complex models that we are now constructing have memory circuits in which associations are stored as electric oscillations, so the creatures can learn simple tricks, forget them slowly and relearn more quickly. This compact, plastic and easily accessible form of short-term memory may be very similar to the way in which the brain establishes the simpler and more evanescent conditioned reflexes.

One intriguing effect in these higher forms of synthetic life is that as soon as two receptors and a learning circuit are provided, the possibility of a conflict neurosis immediately appears. In difficult situations the creature sulks or becomes wildly agitated and can be cured only by rest or shock—the two favorite stratagems of the psychiatrist. It appears that it would even be technically feasible to build processes of self-repair and of reproduction into these machines.

Perhaps we flatter ourselves in thinking that man is the pinnacle of an estimable creation. Yet as our imitation of life becomes more faithful our veneration of its marvelous processes will not necessarily become less sincere.

PART 12

I. EYE AND CAMERA *by George Wald*

George Wald began to study the chemistry and physiology of vision 25 years ago in the laboratory of Selig Hecht, the great pioneer biophysicist. At that time he took on as his life work the investigation of the whole process of vision, from the retina of the eye to the optic centers of the brain. Today, he has analyzed large sections of this process to the point where he can duplicate their chemical and physical reactions in the test tube. Dr. Wald is professor of biology at Harvard and one of the great contemporary teachers at that university. Most of his work is with advanced students, but his introductory lectures on biology are an annual favorite with the freshman class. He is also one of the authors of the Harvard Report, "Education in a Free Society."

II. SMELL AND TASTE *by A. J. Haagen-Smit*

Since he teaches at California Institute of Technology at Pasadena and therefore lives near Los Angeles, it was almost inevitable that A. J. Haagen-Smit, an expert on smell and taste, should have taken up arms against the city's villainous smog. In addition to teaching biochemistry and conducting research on plant substances, he has done a great deal to identify the active irritants in the Los Angeles atmosphere and is today a member of the scientific board that advises the chamber of commerce on air pollution and other chemical problems. Haagen-Smit, a native of the Netherlands, came to this country in 1940.

III. EXPERIMENTS IN PERCEPTION
by W. H. Ittelson and F. P. Kilpatrick

William Ittelson and Franklin Kilpatrick are both assistant professors of psychology at Princeton University, where they teach and conduct research in social and perceptual psychology. They are former graduate students of Hadley Cantril at Princeton and have worked at the Ames Institute for Associated Research, where many of the experiments discussed in their essay were first developed.

IV. WHAT IS PAIN? *by W. K. Livingston*

When he was a young doctor on the staff of Massachusetts General Hospital in 1920, W. K. Livingston discovered that the customary—and painless—method of opening a hole in the intestine was to use a red hot soldering iron. Why did this seemingly Draconic treatment produce no sensation of pain in an organ that under other circumstances can be acutely sensitive? Dr. Livingston has been working on such questions ever since. Today he is professor of surgery at the University of Oregon and conducts a special clinic for the study of obscure, unusual or intractable pain.

SENSATION AND
PERCEPTION
Introduction

U P TO NOW this book has dealt with what science knows; the last
section raises the question of how science can know anything at all.
For all their tools and special training, scientists must in the end rely
like other men on the evidence of their senses. In doing so, they realize
uneasily that they have left one question unanswered: Since we can
experience the external world only through the agency of our senses, how
can we be sure that what exists "out there" corresponds to what our
senses report? For that matter, how can we be sure that objective reality
exists at all and is not just a figment of our perceptions?

It will be recognized that the problem is not peculiar to science; in
fact, it is the basic dilemma of epistemology and philosophers have
tussled with it at least since the time of Plato's idealism. Science and philosophy meet here on common ground, each trying to solve the same
problem by its own methods. Philosophy employs the tool of abstract
logic and builds broad and self-consistent structures to cope with the
problem. These structures—by Plato, Kant, Berkeley, James, Dewey and
others—are among the most impressive creations of the human mind, but
they derive authority only from their own presuppositions and can be
leveled by the same logic that erected them. Science attacks the sub-
jective-objective controversy with its customary instruments of analysis,
definition and experiment. It is trying to construct a theory that will be
subject to verification. The essays that follow discuss the groundwork
that has thus far been laid.

As a first step, science distinguishes between what the senses respond
to and what the brain receives. At this stage, the sense organs themselves
can be investigated by means not unlike those used to study the heart,
the kidney or any other physical organ of the body. The first two essays
are reports on the physiology of the eye and of the related senses of
smell and taste. These mechanisms are marvelously ingenious but they

553

are no longer baffling. The physical reactions that produce sight, smell, taste and the other sense responses have been broken down into their biochemical and biophysical elements; anyone sufficiently adept can obtain the same results by using the same methods. And, as George Gray showed in the previous chapter, the signals from the sensory receptors have been traced along their several nerve circuits to the proper terminals in the brain.

But what of the perceptions that the organism forms from the signals its central nervous system and particularly its brain receives? One can readily show that the identical sense stimulus is interpreted in a variety of ways, not only by different people, but by the same person at different times. The third article of this group describes a series of experiments in which subjects with normal senses were made to jump to wrong perceptions. What one perceives, it seems, is to a considerable extent a matter of what one has been trained to perceive and it is difficult to evade the conclusion that a person born with a different set of sense organs would perceive a different external reality.

Since, in fact, most people have the same sensations and interpret the external world in much the same way, science might content itself by saying that our reality is the only one possible to us and therefore the only one that need concern us. Unfortunately, the problem of correlating sensation with perception is more pressing than it may seem at first glance. Its urgency becomes clear when, in the final essay, W. K. Livingston raises the question, "What is Pain?" This study traces a well-nigh universal sensation from the original stimulus to the brain's appreciation of what has occurred and its decision as to what steps had best be taken. So far as we know, the sense reaction to a given wounding of the body is always the same, but the perception of pain varies constantly with the body's activity and the competing occupations of the brain. Dr. Livingston concludes that pain exists in the awareness of the sufferer; that without awareness there is no pain, however sharp the stimulus or strong the impulse carried by the nerves. To that extent at least reality is created by the individual who perceives it.

I. EYE AND CAMERA

by George Wald

O F ALL THE INSTRUMENTS made by man, none resembles a part of his body more than a camera does the eye. Yet this is not by design. A camera is no more a copy of an eye than the wing of a bird is a copy of that of an insect. Each is the product of an independent evolution; and if this has brought the camera and the eye together, it is not because one has mimicked the other, but because both have had to meet the same problems, and frequently have done so in much the same way. This is the type of phenomenon that biologists call convergent evolution, yet peculiar in that the one evolution is organic, the other technological.

Over the centuries much has been learned about vision from the camera, but little about photography from the eye. The camera made its first appearance not as an instrument for making pictures but as the *camera obscura* or dark chamber, a device that attempted no more than to project an inverted image upon a screen. Long after the optics of the camera obscura was well understood, the workings of the eye remained mysterious. Two notions were particularly troublesome. One was that radiation shines out of the eye; the other, that an inverted image on the retina is somehow incompatible with seeing right side up.

I am sure that many people are still not clear on either matter. I note, for example, that the X-ray vision of the comic-strip hero Superman, while regarded with skepticism by many adults, is not rejected on the ground that there are no X-rays about us with which to see. Clearly Superman's eyes supply the X-rays, and by directing them here and there he not only can see through opaque objects, but can on occasion shatter a brick wall or melt gold. As for the inverted image on the retina, most people who learn of it concede that it presents a problem, but comfort themselves with the thought that the brain somehow compensates for it. But of course there is no problem, and hence no compensation. We learn early in infancy to associate certain spatial relations in the outside world with certain patterns of nervous activity stimulated through the eyes. The spatial arrangements of the nervous activity itself are altogether irrelevant.

It was not until the 17th century that the gross optics of image formation in the eye was clearly expressed. This was accomplished by Johannes Kepler in 1611, and again by René Descartes in 1664. By the end of the century the first treatise on optics in English, written by William Molyneux of Dublin, contained several clear and simple diagrams comparing the projection of a real inverted image in a "pinhole" camera, in a camera obscura equipped with a lens and in an eye.

Today every schoolboy knows that the eye is like a camera. In both instruments a lens projects an inverted image of the surroundings upon a light-sensitive surface: the film in the camera and the retina in the eye. In both the opening of the lens is regulated by an iris. In both the inside of the chamber is lined with a coating of black material which absorbs stray light that would otherwise be reflected back and forth and obscure the image. Almost every schoolboy also knows a difference between the camera and the eye. A camera is focused by moving the lens toward or away from the film; in the eye the distance between the lens and the retina is fixed, and focusing is accomplished by changing the thickness of the lens.

The usual fate of such comparisons is that on closer examination they are exposed as trivial. In this case, however, just the opposite has occurred. The more we have come to know about the mechanism of vision, the more pointed and fruitful has become its comparison with photography. By now it is clear that the relationship between the eye and the camera goes far beyond simple optics, and has come to involve much of the essential physics and chemistry of both devices.

A photographer making an exposure in dim light opens the iris of his camera. The pupil of the eye also opens in dim light, to an extent governed by the activity of the retina. Both adjustments have the obvious effect of admitting more light through the lens. This is accomplished at some cost to the quality of the image, for the open lens usually defines the image less sharply, and has less depth of focus.

Sight and photography have many devices and operating principles in common. Both employ a lens to project an inverted image upon a light sensitive surface; both possess an iris to control the amount of light admitted to the chamber. Many more similarities are developed in the text. An important difference between the eye and the camera is indicated here. The camera is focused by moving the lens toward or away from the film; in the eye, a change in the thickness of the lens, controlled by the ciliary muscles, brings the image into focus on the retina.

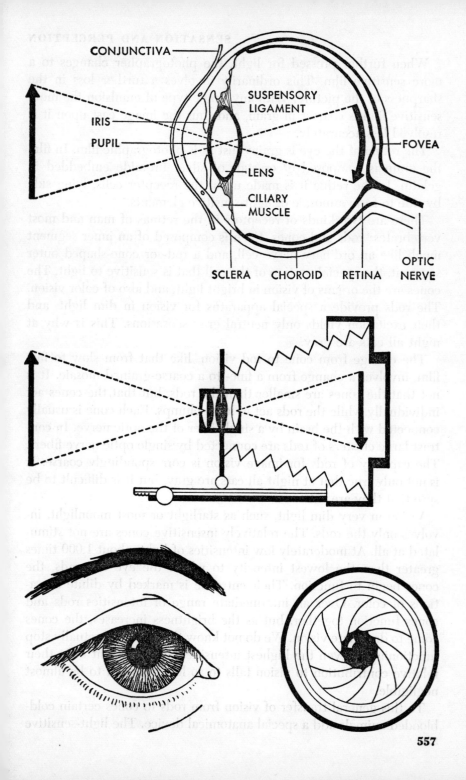

When further pressed for light, the photographer changes to a more sensitive film. This ordinarily involves a further loss in the sharpness of the picture. With any single type of emulsion the more sensitive film is coarser in grain, and thus the image cast upon it is resolved less accurately.

The retina of the eye is grainy just as is photographic film. In film the grain is composed of crystals of silver bromide embedded in gelatin. In the retina it is made up of the receptor cells, lying side by side to form a mosaic of light-sensitive elements.

There are two kinds of receptors in the retinas of man and most vertebrates: rods and cones. Each is composed of an inner segment much like an ordinary nerve cell, and a rod- or cone-shaped outer segment, the special portion of the cell that is sensitive to light. The cones are the organs of vision in bright light, and also of color vision. The rods provide a special apparatus for vision in dim light, and their excitation yields only neutral gray sensations. This is why at night all cats are gray.

The change from cone to rod vision, like that from slow to fast film, involves a change from a fine- to a coarse-grained mosaic. It is not that the cones are smaller than the rods, but that the cones act individually while the rods act in large clumps. Each cone is usually connected with the brain by a single fiber of the optic nerve. In contrast large clusters of rods are connected by single optic nerve fibers. The capacity of rods for image vision is correspondingly coarse. It is not only true that at night all cats are gray, but it is difficult to be sure that they are cats.

Vision in very dim light, such as starlight or most moonlight, involves only the rods. The relatively insensitive cones are not stimulated at all. At moderately low intensities of light, about 1,000 times greater than the lowest intensity to which the eye responds, the cones begin to function. Their entrance is marked by dilute sensations of color. Over an intermediate range of intensities rods and cones function together, but as the brightness increases, the cones come to dominate vision. We do not know that the rods actually stop functioning at even the highest intensities, but in bright light their relative contribution to vision falls to so low a level a to be almost negligible.

To this general transfer of vision from rods to cones certain cold-blooded animals add a special anatomical device. The light-sensitive outer segments of the rods and cones are carried at the ends of fine stalks called myoids, which can shorten and lengthen. In dim light the rod myoids contract while the cone myoids relax. The entire field of rods is thus pulled forward toward the light, while the cones are pushed into the background. In bright light the reverse occurs: the cones are pulled forward and the rods pushed back. One could scarcely imagine a closer approach to the change from fast to slow film in a camera.

The rods and cones share with the grains of the photographic plate another deeply significant property. It has long been known that in a film exposed to light each grain of silver bromide given enough developer blackens either completely or not at all, and that a grain is made susceptible to development by the absorption of one or at most a few quanta of light. It appears to be equally true that a cone or rod is excited by light to yield either its maximal response or none at all. This is certainly true of the nerve fibers to which the rods and cones are connected, and we now know that to produce this effect in a rod—and possibly also in a cone—only one quantum of light need be absorbed.

It is a basic tenet of photochemistry that one quantum of light is absorbed by, and in general can activate, only one molecule or atom. We must attempt to understand how such a small beginning can bring about such a large result as the development of a photographic grain or the discharge of a retinal receptor. In the photographic process the answer to this question seems to be that the absorption of a quantum of light causes the reduction of a silver ion to an atom of metallic silver, which then serves as a catalytic center for the development of the entire grain. It is possible that a similar mechanism operates in a rod or a cone. The absorption of a quantum of light by a light-sensitive molecule in either structure might convert it into a biological catalyst, or enzyme, which could then promote the further reactions that discharge the receptor cell. One wonders whether such a mechanism could possibly be rapid enough. A rod or a cone responds to light within a small fraction of a second; the mechanism would therefore have to complete its work within this small interval.

One of the strangest characteristics of the eye in dim light follows from some of these various phenomena. In focusing the eye is guided by its evaluation of the sharpness of the image on the retina. As the image deteriorates with the opening of the pupil in dim light, and as the retinal capacity to resolve the image falls with the shift from cones to rods, the ability to focus declines also. In very dim light the eye virtually ceases to adjust its focus at all. It has come to resemble a very cheap camera, a fixed-focus instrument.

In all that concerns its function, therefore, the eye is one device in bright light and another in dim. At low intensities all its resources are concentrated upon sensitivity, at whatever sacrifice of form; it is predominantly an instrument for seeing light, not pattern. In bright light all this changes. By narrowing the pupil, shifting from rods to cones, and other stratagems still to be described, the eye sacrifices light in order to achieve the utmost in pattern vision.

The use of a lens to project an image has created a special group of problems for both eye and camera. All simple lenses are subject to serious errors in image formation: the lens aberrations. Spherical aberration is found in all lenses bounded by spherical surfaces. The marginal portions of the lens bring rays of light to a shorter focus than the central region. The image of a point in space is therefore not a point, but a little "blur circle." The cost of a camera is largely determined by the extent to which this aberration is corrected by modifying the lens.

The human eye is astonishingly well corrected—often slightly over-corrected—for spherical aberration. This is accomplished in two ways. The cornea, which is the principal refracting surface of the eye, has a flatter curvature at its margin than at its center. This compensates in part for the tendency of a spherical surface to refract light more strongly at its margin. More important still, the lens is denser and hence refracts light more strongly at its core than in its outer layers.

A second major lens error, however, remains almost uncorrected in the human eye. This is chromatic aberration, or color error. All single lenses made of one material refract rays of short wave length more strongly than those of longer wave length, and so bring blue light to a shorter focus than red. The result is that the image of a point of white light is not a white point, but a blur circle fringed with color. Since this seriously disturbs the image, even the lenses of inexpensive cameras are corrected for chromatic aberration.

It has been known since the time of Isaac Newton, however, that the human eye has a large chromatic aberration. Its lens system

560

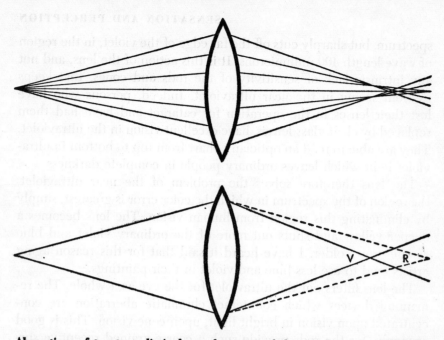

Aberrations of two types limit the performance of the vertebrate eye and make problems for the lens designer. Spherical aberration, at top, afflicts all lenses with spherical surfaces; the light which passes through outer portion of lens is brought to shorter focus than light which passes through center. The image of a point, as a result, is not a point, but a "blur circle." Chromatic aberration is characteristic of all lenses made of only one material, since any material refracts light of different wave lengths to different degrees and thus focuses them at different points. Result is a colored blur circle.

seems to be entirely uncorrected for this defect. Indeed, living organisms are probably unable to manufacture two transparent materials of such widely different refraction and dispersion as the crown and flint glasses from which color-corrected lenses are constructed.

The large color error of the human eye could make serious difficulties for image vision. Actually the error is moderate between the red end of the spectrum and the blue-green, but it increases rapidly at shorter wave lengths: the blue, violet and ultraviolet. These latter parts of the spectrum present the most serious problem. It is a problem for both the eye and the camera, but one for which the eye must find a special solution.

The first device that opposes the color error of the human eye is the yellow lens. The human lens is not only a lens but a color filter. It passes what we ordinarily consider to be the visible spectrum, but sharply cuts off the far edge of the violet, in the region of wave length 400 millimicrons. It is this action of the lens, and not any intrinsic lack of sensitivity of the rods and cones, that keeps us from seeing in the near ultraviolet. Indeed, persons who have lost their lenses in the operation for cataract and have had them replaced by clear glass lenses, have excellent vision in the ultraviolet. They are able to read an optician's chart from top to bottom in ultraviolet light which leaves ordinary people in complete darkness.

The lens therefore solves the problem of the near ultraviolet, the region of the spectrum in which the color error is greatest, simply by eliminating this region from human vision. The lens becomes a deeper yellow and shuts out more of the ordinary violet and blue as one grows older. I have heard it said that for this reason aging artists tend to use less blue and violet in their paintings.

The lens filters out the ultraviolet for the eye as a whole. The remaining devices which counteract chromatic aberration are concentrated upon vision in bright light, upon cone vision. This is good economy, for the rods provide such a coarse-grained receptive surface that they would be unable in any case to evaluate a sharp image on the retina.

As one goes from dim to bright light, from rod to cone vision, the sensitivity of the eye shifts toward the red end of the spectrum. This phenomenon was described in 1825 by the Czech physiologist Johannes Purkinje. He had noticed that with the first light of dawn blue objects tend to look relatively bright compared with red, but that they come to look relatively dim as the morning advances. The basis of this change is a large difference in spectral sensitivity between rods and cones. Rods have their maximal sensitivity in the blue-green at about 500 millimicrons; the entire spectral sensitivity of the cones is transposed toward the red, the maximum lying in the yellow-green at about 562 millimicrons. The point of this difference for our present argument is that as one goes from dim light, in which pattern vision is poor in any case, to bright light, in which it becomes acute, the sensitivity of the eye moves away from the region of the spectrum in which the chromatic aberration is large toward the part of the spectrum in which it is least.

The color correction of the eye is completed by a third dispensation. Toward the center of the human retina there is a small, shallow depression called the fovea, which contains only cones. While the retina as a whole sweeps through a visual angle of some 240 degrees, the fovea subtends an angle of only about 1.7 degrees. The fovea is considerably smaller than the head of a pin, yet with this tiny patch of retina the eye accomplishes all its most detailed vision.

The fovea also includes the fixation point of the eye. To look directly at something is to turn one's eye so that its image falls upon the fovea. Beyond the boundary of the fovea rods appear, and they become more and more numerous as the distance from the fovea increases. The apparatus for vision in bright light is thus concentrated toward the center of the retina, that for dim light toward its periphery. In very dim light, too dim to excite the cones, the fovea is blind. One can see objects then only by looking at them slightly askance to catch their images on areas rich in rods.

In man, apes and monkeys, alone of all known mammals, the fovea and the region of retina just around it is colored yellow. This area is called the yellow patch, or *macula lutea.* Its pigmentation lies as a yellow screen over the light receptors of the central retina, subtending a visual angle some five to 10 degrees in diameter. This pigment in the yellow patch takes up the absorption of light in the violet and blue regions of the spectrum just where absorption by the lens falls to very low values. In this way the yellow patch removes for the central retina the remaining regions of the spectrum for which the color error is high.

So the human eye, unable to correct its color error otherwise, throws away those portions of the spectrum that would make the most trouble. The yellow lens removes the rear ultraviolet for the eye as a whole, the macular pigment eliminates most of the violet and blue for the central retina, and the shift from rods to cones displaces vision in bright light bodily toward the red. By these three devices the apparatus of most acute vision avoids the entire range of the spectrum in which the chromatic aberration is large.

In 1876 Franz Boll of the University of Rome discovered in the rods of the frog retina a brilliant red pigment. This bleached in the light and was resynthesized in the dark, and so fulfilled the elementary requirements of a visual pigment. He called this substance visual red; later it was renamed visual purple or rhodopsin. This pigment marks the point of attack by light on the rods: the absorption of light by rhodopsin initiates the train of reactions that end in rod vision.

Boll had scarcely announced his discovery when Willy Kühne, professor of physiology at Heidelberg, took up the study of rhodopsin, and in one extraordinary year learned almost everything about it that was known until recently. In his first paper on retinal chemistry Kühne said: "Bound together with the pigment epithelium, the retina behaves not merely like a photographic plate, but like an entire photographic workshop, in which the workman continually renews the plate by laying on new light-sensitive material, while simultaneously erasing the old image."

Kühne saw at once that with this pigment which bleaches in the light it might be possible to take a picture with the living eye. He set about devising methods for carrying out such a process, and succeeded after many discouraging failures. He called the process optography and its products optograms.

One of Kühne's early optograms was made as follows. An albino rabbit was fastened with its head facing a barred window. From this position the rabbit could see only a gray and clouded sky. The animal's head was covered for several minutes with a cloth to adapt its eyes to the dark, that is to let rhodopsin accumulate in its rods. Then the animal was exposed for three minutes to the light. It was immediately decapitated, the eye removed and cut open along the equator, and the rear half of the eyeball containing the retina laid in a solution of alum for fixation. The next day Kühne saw, printed upon the retina in bleached and unaltered rhodopsin, a picture of the window with the clear pattern of its bars. Shortly thereafter, Kühne's interest in optograms led him into the following curious adventure. In the nearby town of Bruchsal on November 16, 1880, a young man was beheaded by guillotine. Kühne had made arrangements to receive the corpse. He had prepared a dimly lighted room screened with red and yellow glass to keep any rhodopsin left in the eyes from bleaching further. Ten minutes after the knife had fallen he obtained the whole retina from the left eye, and had the satisfaction of seeing and showing to several colleagues a sharply demarcated optogram printed upon its surface. To my knowledge it is the only human optogram on record. Kühne went to great pains to determine what this optogram represented. He says: "A search for the object which served as source for this optogram remained fruitless, in spite of a thorough inventory of all the surroundings and reports from many witnesses."

564

One of the triumphs of modern photography is its success in recording color. For this it is necessary not only to graft some system of color differentiation and rendition upon the photographic process; the finished product must then fulfill the very exacting requirement that it excite the same sensations of color in the human eye as did the original subject of the picture.

How the human eye resolves colors is not known. Normal human color vision seems to be compounded of three kinds of responses; we therefore speak of it as trichromatic or three-color vision. The three kinds of response call for at least three kinds of cone differing from one another in their sensitivity to the various regions of the spectrum. We can only guess at what regulates these differences. The simplest assumption is that the human cones contain three different light-sensitive pigments, but this is still a matter of surmise.

There exist retinas, however, in which one can approach the problem of color vision more directly. The eyes of certain turtles and of certain birds such as chickens and pigeons contain a great predominance of cones. Since cones are the organs of vision in bright light as well as of color vision, these animals necessarily function only at high light intensities. They are permanently night-blind, due to a poverty or complete absence of rods. It is for this reason that chickens must roost at sundown.

In the cones of these animals we find a system of brilliantly colored oil globules, one in each cone. The globule is situated at the joint between the inner and outer segments of the cone, so that light must pass through it just before entering the light-sensitive element. The globules therefore lie in the cones in the position of little individual color filters.

One has only to remove the retina from a chicken or a turtle and spread it on the stage of a microscope to see that the globules are of three colors: red, orange and greenish yellow. It was suggested many years ago that they provide the basis of color differentiation in these animals.

In a paper published in 1907 the German ophthalmologist Siegfried Garten remarked that he was led by such retinal color filters to invent a system of color photography based upon the same principle. This might have been the first instance in which an eye had directly inspired a development in photography. Unfortunately, however, in 1906 the French chemist Louis Lumière, apparently without benefit of chicken retinas, had brought out his autochrome process for color photography based upon exactly this principle.

To make his autochrome plates Lumière used suspensions of starch grains from rice, which he dyed red, green and blue. These were mixed in roughly equal proportions, and the mixture was strewn over the surface of an ordinary photographic plate. The granules were then squashed flat and the interstices were filled with particles of carbon. Each dyed granule served as a color filter for the patch of silver-bromide emulsion that lay just under it.

Just as the autochrome plate can accomplish color photography with a single light-sensitive substance, so the cones of the chicken retina should require no more than one light-sensitive pigment. We extracted such a pigment from the chicken retina in 1937. It is violet in color, and has therefore been named iodopsin from *ion*, the Greek word for violet. All three pigments of the colored oil globules have also been isolated and crystallized. Like the pigment of the human macula, they are all carotenoids: a greenish-yellow carotene; the golden mixture of xanthophylls found in chicken egg yolk; and red astaxanthin, the pigment of the boiled lobster.

Controversy thrives on ignorance, and we have had many years of disputation regarding the number of kinds of cone concerned in human color vision. Many investigators prefer three, some four, and at least one of my English colleagues seven. I myself incline toward three. It is a good number, and sufficient unto the day.

The appearance of three colors of oil globule in the cones of birds and turtles might be thought to provide strong support for trichromatic theories of color vision. The trouble is that these retinas do in fact contain a fourth class of globule which is colorless. Colorless globules have all the effect of a fourth color; there is no doubt that if we include them, bird and turtle retinas possess the basis for four-color vision.

Recent experiments have exposed a wholly unexpected parallel between vision and photography. Many years ago Kühne showed that rhodopsin can be extracted from the retinal rods into clear water solution. When such solutions are exposed to light, the rhodopsin bleaches just as it does in the retina.

It has been known for some time that the bleaching of rhodopsin in solution is not entirely accomplished by light. It is started by light, but then goes on in the dark for as long as an hour at room temperature. Bleaching is therefore a composite process. It is ushered in by a light reaction that converts rhodopsin to a highly unstable product; this then decomposes by ordinary chemical reactions—"dark" reactions in the sense that they do not require light.

Since great interest attaches to the initial unstable product of the light reaction, many attempts were made in our laboratory and at other laboratories to seize upon this substance and learn its properties. It has such a fleeting existence, however, that for some time nothing satisfactory was achieved.

In 1941, however, two English workers, E. E. Broda and C. F. Goodeve, succeeded in isolating the light reaction by irradiating rhodopsin solutions at about −73 degrees centigrade, roughly the temperature of dry ice. In such extreme cold, light reactions are unhindered, but ordinary dark processes cannot occur. Broda and Goodeve found that the exhaustive exposure of rhodopsin to light under these conditions produced only a very small change in its color, so small that though it could be measured one might not have been certain merely by looking at these solutions that any change had occurred at all. Yet the light reaction had been completed, and when such solutions were allowed to warm up to room temperature they bleached *in the dark*. We have recently repeated such experiments in our laboratory. With some differences which need not be discussed, the results were qualitatively as the English workers had described them.

These observations led us to re-examine certain early experiments of Kühne's. Kühne had found that if the retina of a frog or rabbit was thoroughly dried over sulfuric acid, it could be exposed even to brilliant sunlight for long periods without bleaching. Kühne concluded that dry rhodopsin is not affected by light, and this has been the common understanding of workers in the field of vision ever since.

It occurred to us, however, that dry rhodopsin, like extremely cold rhodopsin, might undergo the light reaction, though with such small change in color as to have escaped notice. To test this possibility we prepared films of rhodopsin in gelatin, which could be dried thoroughly and were of a quality that permitted making accurate measurements of their color transmission throughout the spectrum.

We found that when dry gelatin films of rhodopsin are exposed to light, the same change occurs as in very cold rhodopsin. The color is altered, but so slightly as easily to escape visual observation. In any case the change cannot be described as bleaching; if anything the color is a little intensified. Yet the light reaction is complete; if such exposed films are merely wetted with water, they bleach in the dark.

We have therefore two procedures—cooling to very low temperatures and removal of water—that clearly separate the light from the dark reactions in the bleaching of rhodopsin. Which of these reactions is responsible for stimulating rod vision? One cannot yet be certain, yet the response of the rods to light occurs so rapidly that only the light reaction seems fast enough to account for it.

What has been said, however, has a further consequence that brings it into direct relation with photography. Everyone knows that the photographic process also is divided into light and dark components. The result of exposing a film to light is usually invisible, a so-called "latent image." It is what later occurs in the darkroom, the dark reaction of development, that brings out the picture.

This now appears to be exactly what happens in vision. Here as in photography light produces an almost invisible result, a latent image, and this indeed is probably the process upon which retinal excitation depends. The visible loss of rhodopsin's color, its bleaching, is the result of subsequent dark reactions, of "development."

One can scarcely have notions like this without wanting to make a picture with a rhodopsin film; and we have been tempted into making one very crude rhodopsin photograph. Its subject is not exciting—only a row of black and white stripes. What is important is that it was made in typically photographic stages. The dry rhodopsin film was first exposed to light, producing a latent image. It was then developed in the dark by wetting. It then had to be fixed; and, though better ways are known, we fixed this photograph simply by redrying it. Since irradiated rhodopsin bleaches rather than blackens on development, the immediate result is a positive.

Photography with rhodopsin is only in its first crude stages, perhaps at the level that photography with silver bromide reached almost a century ago. I doubt that it has a future as a practical process. For us its primary interest is to pose certain problems in visual chemistry in a provocative form. It does, however, also add another chapter to the mingled histories of eye and camera.

568

II. SMELL AND TASTE
by A. J. Haagen-Smit

OUR SENSES of taste and smell constitute a most astonishing chemical laboratory. In a fraction of a second they can identify the chemical structure of compounds it would take a chemist days to analyze by the usual laboratory methods. A trained nose can recognize, for example, nearly every member of a series of homologous alcohols, aldehydes or acids. Moreover, from exceedingly small amounts of material it can analyze not only single compounds but complex mixtures of them in food.

Our chemical senses are of great importance for our well-being. They determine our reaction to foods and set the stage for digestion. The odor of broiled steak has an immediate effect on metabolism: it starts secretions of saliva and stomach juices even before eating begins. By setting up a favorable condition for digestion the flavor factors in food play a role in nutrition comparable to those of vitamins and hormones. On the other hand, our chemical senses readily recognize and enable us to reject substances produced by other organisms that spoil or rot the food. We have learned to associate these foul smells and bad tastes with adverse reactions of our digestive system. Actually the substances responsible for the bad flavors are harmless in the concentrations in which they generally occur, but they serve as warning signals against dangerous toxins the organisms produce at the same time.

Our perception of flavor depends on both taste and smell; indeed, it is often difficult to distinguish between the parts that odor and taste play in our food. Of the two, the sense of smell is by far the more sensitive, and it may be stimulated at a great distance. R. W. Moncrieff, in his classic book on the chemical senses, recounts how a female Great Peacock moth hatched in the laboratory attracted the same evening about 40 male specimens, which must have traveled several miles, because these insects were rare in the neighborhood. As everyone knows, the odor of a skunk is noticeable several hundred feet from where its glands are emptied. Because of the distances at which odors are perceived, some think the stimulation cannot be entirely chemical, on the ground that the odor substances must be too diluted when they reach the olfactory receptors; it has therefore been suggested that physical vibrations are responsible for odor.

569

But it can be shown that in a barely perceptible dilution of an odoriferous substance a single sniff still contains many millions of molecules. When the molecules arrive at the olfactory hairs and cells in the upper part of the nose, they must produce a reaction in the cells, giving rise to electrical impulses which are transmitted to the olfactory lobe of the brain.

To be smellable a substance apparently must fulfill two conditions: it must be volatile at ordinary temperature and must be soluble in fat solvents. All the known odor substances either are gases or have a high vapor pressure, boiling below about 300 degrees centigrade. Most inorganic substances, being salts of very low vapor pressure, have no discernible odor. Among the minority that do are the halogens (fluorine, chlorine, iodine, bromine), phosphorus, ozone and certain compounds such as hydrogen sulfide, sulfur dioxide, nitrogen oxides and ammonia. Their odor is usually rather unpleasant and often irritating.

In organic chemistry the situation is very different; the organic compounds are much more likely to be odoriferous and their odors have a vastly greater range. Of nearly half a million synthetic compounds listed in a well-known encyclopedia of organic substances, a large proportion have a high enough vapor pressure to make them odoriferous.

Carolus Linnaeus, the father of taxonomy, who in the 18th century began to establish laws of order in the living world by cataloguing plants, also attempted to classify substances according to their odors. But his classifications were necessarily subjective, and early odor taxonomists were severely handicapped by the lack of an assortment of pure organic chemicals to serve as standards of comparison. As is well known, odor receptors in the nose soon become used to a particular odor and cease to notice it—this is called fatigue. Consequently it should be possible to find out by a fatigue test whether one odor is essentially the same as another (*i.e.*, stimulates the same receptors). Such experiments have shown that the odors of camphor and of cloves, for example, produce fatigue for each other; hence they belong in the same subclass.

In recent years Ernest C. Crocker and Lloyd F. Henderson of the Arthur D. Little laboratories have reduced odors to four elementary classes, corresponding to four kinds of receptors. According to their system, all known odors are composites of these types: 1) fragrant or sweet, 2) acid or sour, 3) burnt or empyreumatic, 4) caprylic or goaty. Any odor is described by a formula which gives the strength of each component on a scale from one to eight: thus the odor of a rose is represented by 6423, meaning that it is strong in the fragrant component, has some acid odor and also has a little of the burnt and caprylic odors. The system describes ethyl alcohol as 5414 and vanillin as 7122. The authors of this system maintain that a trained observer can recognize to a certain extent the degree in which the four postulated basic odors are present. To try to describe the vast array of odors in terms of just four basic types may be an oversimplification, but the system has the virtue of emphasizing that every odor is a combination of impressions.

Once we are aware of this fact, we find that we are able to perform such analyses in our minds. It is then possible to overcome the difficulty that no substance smells exactly like another, and we can find the dominant odor for proper classification. In the numerous substances that have now been synthesized by the organic chemist we have material to test the different classification systems, and since the substances' chemical nature is well known, a search for correlations between chemical structure and odor is possible.

Let us concentrate on two distinctive and well-defined odors—those of camphor and mint. There are more than 200 known compounds with a camphorlike odor and nearly as many with the mint odor. In each group chemists have found a certain common characteristic of structure; when they synthesize a compound with this building principle in its structure, there is a fair certainty that the product will have the odor in question. For instance, one can produce a substance with a mint odor by building a molecule resembling menthol, the main constituent of the oil of peppermint. It is not necessary to follow the whole plan of the menthol molecule; apparently only a small part of its structure is responsible for the mint smell. The requirement seems to be a short carbon chain, with branches preferably not more than two or three carbon atoms long.

The camphor odor is closely related to that of mint; one can be converted to the other by slight changes in the molecule. Apparently the critical difference that transforms a minty compound into a camphorlike one is the substitution of a methyl group (CH_3) for one hydrogen atom in the mintlike substance. On the other hand, the odor reverts to mint when an ethyl group is substituted for one of

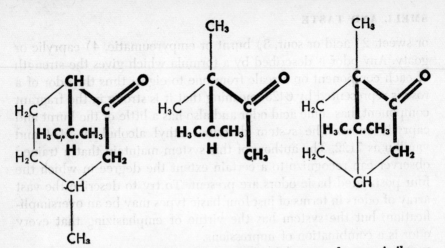

Molecular structure of substances that smell like mint or camphor are similar, as shown here by sections of diagrams drawn in heavier lines. At right is camphor; in center is a substance that smells like camphor. Substance at left smells like mint.

the methyl groups in the camphor structure. Similarities are shown the diagrams.

In general we observe that the camphor odor is characteristic of compounds which have a number of small groups crowding around a carbon atom. These do not necessarily have to be methyl groups; halogens and nitro groups are equally effective.

Closely related to the mint and camphor odors are those of cedar, wood, peach, musk and civet. As the number of carbon atoms is increased in the series of ring ketones, for example, the odor gradually changes from camphor to cedar to musk to civet. The odor evidently is governed by the size of the carbon ring. In a similar series of lactones, musk odor develops when the molecules reach a size corresponding to that of the cyclic ketones of musk odor. The typical musk smell is largely due to the structure of the carbon skeleton, and to a lesser degree to the oxygen atom.

These and similar considerations lead to the conclusion that for mint, camphor, turpentine, cedar, lemon, cineole, peach, musk and civet, the hydrocarbon part of the molecule is of dominant importance, and that there is reason to classify all these odors in one group, which corresponds to Crocker's class of "fragrant."

It is reasonable to assume that for each class of odorous materials there is a specific receptor mechanism in the smelling apparatus. For the class to which camphor and mint belong, the mechanism must enable us to detect small differences in the carbon skeleton of a large number of compounds. Just as a specific antibody meshes with an antigen in the body, so the active part of an odorous molecule may fit some part of a protein structure in the nasal receptors, thereby altering cell reactions and giving rise to electric impulses This theory would account for the fact that only a part of the molecule is of dominant importance for the smell impression. The grouping of odors in classes according to similarity of chemical structure, or, in other words, the fact that small changes in the active part of the molecule produce only slight alterations in the odor sensation, may be explained by the assumption that the various substances in a class fit the receptor molecules more or less closely.

The consulting chemist Jerome Alexander and George B. Kistiakowsky of Harvard University have suggested that odoriferous substances act by interfering with enzyme-catalyzed reactions in the odor receptors. Since enzymes are affected in their action by exceedingly small amounts of a variety of substances, this theory plausibly explains the high sensitivity of our sense of smell and the wide range of compounds that possess odors. The well-known reversibility of inhibitory effects on enzymes would account for the rapid recovery of the reception system to normal, thus enabling it to register new odor impressions.

As has been mentioned, odor strongly affects the flavor of food. We are suddenly reminded of this when a cold inactivates our olfactory system. The food tastes flat, since we are dependent on our sense of taste alone. We are left with a distinction between bitter, sweet, salt and sour.

The sensory apparatus of taste is located chiefly on the upper surface of the tongue, at the soft palate, on the epiglottis and at the beginning of the gullet. Here lie the so-called taste buds, estimated to number about 9,000. It is fairly certain that different tastes are located at different places. The bitter sensations are definitely located at the back of the tongue, whereas the sweet and salt receptors are at the tip and edges. To detect a substance by taste we need a far greater amount than we can detect by smell—about 3,000 times as much in some cases.

As in the odor field, attempts have been made to find the chemical relation among the members of each of the four taste groups. The sour taste is related to the acidity of the solution, though not in direct proportion; for example, a 1/200 normal solution of acetic acid tastes just as sour as a 1/800 normal solution of HCl, which is four to five times as acid.

Salty tastes are produced by inorganic salts. The anions of chlorine, bromine, SO_4 and NO_3 are especially effective in producing this saline taste when combined with the proper metal ions. For example, sodium chloride is salty, whereas cesium chloride has a dominating bitter taste.

The sweet taste is given by sugars, saccharin, dulcin and beryllium chloride. It is difficult to see what these substances have in common. As in the cases of odor substances, minor changes in these compounds do not remove the sweet taste. For example, compounds somewhat related to glucose, such as glycerol, are sweet.

The bitter taste, similarly, is exhibited by a wide variety of compounds—many alkaloids, certain glucosides, bile salts, magnesium and ammonium salts.

Bitter and sweet are closely related, and a slight modification of the molecule is sometimes enough to produce strongly bitter substances instead of the expected sweet ones. For example, when the oxygen in the sweet material dulcin is replaced by sulfur, we get the bitter compound p-ethoxy-phenylthiocarbamide.

This substance, and the material from which it is derived (phenyl thiourea), brings out an interesting point: just as people may have color blindness, some people have certain kinds of "taste blindness." To three or four out of ten people, this substance is not bitter but tasteless. Apparently it is a hereditary taste deficiency, which started among Caucasians and spread to other races. Since the discovery of this particular type of "taste blindness," many more substances have been found to exhibit similar taste properties. Sodium benzoate, tasteless to most people, tastes either sweet or bitter to one out of four. As in the odor field, we have to turn our attention to the receptor cells to find the mechanism of taste perception. Here, too, an enzyme theory seems to be the most promising approach. An investigator of the taste mechanism is in a more fortunate position than those studying the odor receptors, because the four basic tastes, though subjective, are a great deal more sharply defined than the basic odor types. In addition, experimental work is aided by the fact that the taste buds are more accessible and are located in different places for the four tastes. This allows one to carry out reactions on the tissues and to study the influence of a number of taste substances on the cell mechanism. By such histochemical methods it has been shown that the skin overlying the taste buds contains relatively high concentrations of certain enzymes such as alkali phosphatases and esterases, and that these enzymes are inhibited by substances having a well-defined taste and are not inhibited by others.

In the past, studies of odor and taste reception have concentrated on the stimulants and attempted to deduce the nature of the reception mechanism from them. At best such deductions will be vague and uncertain. For a real understanding of the basic processes, we must study the happenings in the receptor cells themselves. The enzyme theory and the other new working hypotheses about the tasting and smelling processes illustrate this shift in emphasis. They may stimulate the exceedingly difficult experimental work that is necessary before solutions of the odor and taste problems can be found.

III. EXPERIMENTS IN PERCEPTION

by W. H. Ittelson and F. P. Kilpatrick

WHAT IS PERCEPTION? Why do we see what we see, feel what we feel, hear what we hear? We act in terms of what we perceive; our acts lead to new perceptions; these lead to new acts, and so on in the incredibly complex process that constitutes life. Clearly, then, an understanding of the process by which man becomes aware of himself and his world is basic to any adequate understanding of human behavior. But the problem of explaining how and why we perceive in the way we do is one of the most controversial fields in psychology. We shall describe here some recent experimental work which sheds new light on the problem and points the way to a new theory of perception.

The fact that we see a chair and are then able to go to the place at which we localize it and rest our bodies on a substantial object does not seem particularly amazing or difficult to explain—until we try to explain it. If we accept the prevailing current view that we can never be aware of the world as such, but only of the nervous impulses arising from the impingement of physical forces on sensory receptors, we immediately face the necessity of explaining the correspondence between what we perceive and whatever it is that is there.

An extremely logical, unbeatable—and scientifically useless—answer is simply to say there is no real world, that everything exists in the mind alone. Another approach is to postulate the existence of an external world, to grant that there is some general correspondence between that world and what we perceive and to seek some understandable and useful explanation of why that should be. Most of the prominent theories about perception have grown out of the latter approach. These theories generally agree that even though much of the correspondence may be due to learning, at some basic level there exists an absolute correspondence between what is "out there" and what is in the "mind." But there is a great deal of disagreement concerning the level at which such innately determined correspondence occurs. At one extreme are theorists who believe that the correspondence occurs at the level of simple sensations, such as color, brightness, weight, hardness, and so on, and that out of these sensations are compounded more complex awarenesses, such as the recognition of a pencil or a book. At the other extreme are Gestalt psychologists who feel that complex perceptions such as the form of an object are the result of an inherent relationship between the properties of the thing perceived and the properties of the brain. All these schools seem to agree, however, that there is some perceptual level at which exists absolute objectivity; that is, a one-to-one correspondence between experience and reality.

This belief is basic to current thinking in many fields. It underlies most theorizing concerning the nature of science, including Percy W. Bridgman's attempt to reach final scientific objectivity in the "observable operation." In psychology one is hard put to find an approach to human behavior which departs from this basic premise. But it leads to dichotomies such as organism *v.* environment, subjective *v.* objective. Stimuli or stimulus patterns are treated as though they exist apart from the perceiving organism. Psychologists seek to find mechanical relationships or interactions between the organism and an "objectively defined" environment. They often rule out purposes and values as not belonging in a strictly scientific psychology.

The experiments to be described here arose from a widespread and growing feeling that such dichotomies are false, and that in practice it is impossible to leave values and purposes out of consideration in scientific observation. The experiments were designed to re-examine some of the basic ideas from which these problems stem.

During the past few years Adelbert Ames, Jr., of the Institute for Associated Research in Hanover, N. H., has designed some new ways of studying visual perception. They have resulted in a new conception of the nature of knowing and of observation. This theory neither denies the existence of objects nor proposes that they exist in a given form independently, that is, apart from the perceiving organism. Instead, it suggests that the world each of us knows is a world created in large measure from our experience in dealing with the environment.

Let us illustrate this in specific terms through some of the demonstrations. In one of them the subject sits in a dark room in which he can see only two star points of light. Both are equidistant from the observer, but one is brighter than the other. If the observer closes one eye and keeps his head still, the brighter point of light looks nearer than the dimmer one. Such apparent differences are related not only to brightness but also to direction from the observer. If two points of light of equal brightness are situated near the floor, one about a foot above the other, the upper one will generally be perceived as farther away than the lower one; if they are near the ceiling, the lower one will appear farther away.

A somewhat more complex experiment uses two partly inflated balloons illuminated from a concealed source. The balloons are in fixed positions about one foot apart. Their relative sizes can be varied by means of a lever control connected to a bellows; another lever controls their relative brightness. When the size and brightness of both balloons are the same, an observer looking at them with one eye from 10 feet or more sees them as two glowing spheres at equal distances from him. If the brightnesses are left the same and the relative sizes are changed, the larger balloon appears to nearly all observers somewhat nearer. If the size lever is moved continuously, causing continuous variation in the relative size of the balloons, they appear to move dramatically back and forth through space, even when the observer watches with both eyes open. The result is similar when the sizes are kept equal and the relative brightness is varied.

With the same apparatus the effects of size and brightness may be combined so that they supplement or conflict with each other. When they supplement each other, the variation in apparent distance is much greater than when either size or brightness alone is varied. When they oppose each other, the variation is much less. Most people give more weight to relative size than to relative brightness in judging distance.

These phenomena cannot be explained by referring to "reality," because "reality" and perception do not correspond. They cannot be explained by reference to the pattern in the retina of the eye, because for any given retinal pattern there are an infinite number of brightness-size-distance combinations to which that pattern might be related. When faced with such a situation, in which an unlimited number of possibilities can be related to a given retinal pattern, the organism apparently calls upon its previous experiences and assumes that what has been most probable in the past is most probable in the immediate occasion. When presented with two star-points of different brightness, a person unconsciously "bets" or "assumes" that the two points, being similar, are probably identical (*i.e.*, of equal brightness), and therefore that the one which seems brighter must be nearer. Similarly the observed facts in the case of two star-points placed vertically one above the other suggest that when we look down we assume, on the basis of past experience, that objects in the lower part of the visual field are nearer than objects in the upper part; when we look up, we assume the opposite to be true. An analogous explanation can be made of the role of relative size as an indication of relative distance.

Why do the differences in distance seem so much greater when the relative size of two objects is varied continuously than when the size difference is fixed? This phenomenon, too, apparently is based on experience. It is a fairly common experience, though not usual, to find that two similar objects of different sizes are actually the same distance away from us. But it is rare indeed to see two stationary objects at the same distance, one growing larger and the other smaller; almost always in everyday life when we see two identical or nearly identical objects change relative size they are in motion in relation to each other. Hence under the experimental conditions we are much more likely to assume distance differences in the objects of changing size than in those of fixed size.

Visual perception involves an impression not only of *where* an object is but of *what* it is. From the demonstrations already described we may guess that there is a very strong relationship between localization in space ("thereness") and the assignment of objective properties ("thatness"). This relationship can be demonstrated by a cube experiment.

Two solid white cubes are suspended on wires that are painted black so as to be invisible against a black background. One cube is about 3 feet from the observer and the other about 12 feet. The observer's head is in a headrest so positioned that the cubes are almost in line with each other but he can see both, the nearer cube being slightly to the right. A tiny metal shield is then placed a few inches in front of the left eye. It is just big enough to cut off the view of the far cube from the left eye. The result is that the near cube is seen with both eyes and the far cube with just the right eye. Under these conditions the observer can fix the position of the near cube very well, because he has available all the cues that come from the use of the two eyes. But in the case of the far cube seen with only one eye, localization is much more difficult and uncertain.

Now since the two cubes are almost in line visually, a slight movement of the head to the right will cause the inside vertical edges of the cubes to coincide. Such coincidence of edge is strongly related to an assumption of "togetherness." Hence when the subject moves his head in this way, the uncertainly located distant cube appears to have moved forward to a position even with the nearer cube. Under these conditions not only does the mislocated cube appear smaller, but it appears different in shape, that is, no longer cubical, even though the pattern cast by the cube on the retina of the eye has not changed at all.

The most reasonable explanation of these visual phenomena seems to be that an observer unconsciously relates to the stimulus pattern some sort of weighted average of the past consequences of acting with respect to that pattern. The particular perception "chosen" is the one that has the best predictive value, on the basis of previous experience, for action in carrying out the purposes of the organism. From this one may make two rather crucial deductions: 1) an unfamiliar external configuration which yields the same retinal pattern as one the observer is accustomed to deal with will be perceived as the familiar configuration; 2) when the observer acts on his interpretation of the unfamiliar configuration and finds that he is wrong, his perception will change even though the retinal pattern is unchanged.

Let us illustrate with some actual demonstrations. If an observer in a dark room looks with one eye at two lines of light which are at the same distance and elevation but of different lengths, the longer line will look nearer than the shorter one. Apparently he assumes that the lines are identical and translates the difference in length into a difference in position. If the observer takes a wand with a luminous tip and tries to touch first one line and then the other, he will be unable to do so at first. After repeated practice, however, he can learn to touch the two lines quickly and accurately. At this point he no longer sees the lines as at different distances; they now look, as they are, the same distance from him. He originally assumed that the two lines were the same length because that seemed the best bet under the circumstances. After he had tested this assumption by purposive action, he shifted to the assumption, less probable in terms of past experience but still possible, that the lines were at the same distance but of different lengths. As his assumption changed, perception did also.

There is another experiment that demonstrates these points even more convincingly. It uses a distorted room in which the floor slopes up to the right of the observer, the rear wall recedes from right to left and the windows are of different sizes and trapezoidal in shape. When an observer looks at this room with one eye from a certain point, the room appears completely normal, as if the floor were level, the rear wall at right angles to the line of sight and the windows rectangular and of the same size. Presumably the observer chooses this particular appearance instead of some other because of the assumptions he brings to the occasion. If he now takes a long stick and tries to touch the various parts of the room, he will be unsuccessful, even though he has gone into the situation knowing the true shape of the room. With practice, however, he becomes more and more successful in touching what he wants to touch with the stick. More important, he sees the room more and more in its true shape, even though the stimulus pattern on his retina has remained unchanged.

By means of a piece of apparatus called the "rotating trapezoidal window" it has been possible to extend the investigation to complex perceptual situations involving movement. This device consists of a trapezoidal surface with panes cut in it and shadows painted on it to give the appearance of a window. It is mounted on a rod connected to a motor so that it rotates at a slow constant speed in an upright position about its own axis. When an observer views the rotating surface with one eye from about 10 feet or more or with both eyes from about 25 feet or more, he sees not a rotating trapezoid but an oscillating rectangle. Its speed of movement and its shape appear to vary markedly as it turns. If a small cube is attached by a short rod to the upper part of the short side of the trapezoid, it seems to become detached, sail freely around the front of the trapezoid and attach itself again as the apparatus rotates.

All these experiments, and many more that have been made, suggest strongly that perception is never a sure thing, never an absolute revelation of "what is." Rather, what we see is a prediction—our own personal construction designed to give us the best possible bet for carrying out our purposes in action. We make these bets on the basis of our past experience. When we have a great deal of relevant and consistent experience to relate to stimulus patterns, the probability of success of our prediction (perception) as a guide to action is extremely high, and we tend to have a feeling of surety. When our experience is limited or inconsistent, the reverse holds true. According to the new theory of perception developed from the demonstrations we have described, perception is a functional affair based on action, experience and probability. The thing perceived is an inseparable part of the function of perceiving, which in turn includes all aspects of the total process of living. This view differs from the old rival theories: the thing perceived is neither just a figment of the mind nor an innately determined absolute revelation of a reality postulated to exist apart from the perceiving organism. Object and percept are part and parcel of the same thing.

This conclusion of course has far-reaching implications for many areas of study, for some assumption as to what perception is must underly any philosophy or comprehensive theory of psychology, of science or of knowledge in general. Although the particular investigations involved here are restricted to visual perception, this is only a vehicle which carries us into a basic inquiry of much wider significance.

IV. WHAT IS PAIN?

by W. K. Livingston

EVERY FEELING person knows from personal experience what pain is, yet scientists have found it extraordinarily difficult to agree on a satisfactory definition for it. The question is not a metaphysical one. It has profound bearing on the search for ways to relieve pain and on basic human fears. Probably no subject in medical science interests people more than this one. They like to hear about new anesthetic agents, analgesic drugs and nerve operations to control pain; about "pain clinics" for the study of rare pain phenomena; about laboratory investigations of the physiology and psychology of pain. They frequently question their physicians about the pains of cancer, heart disease and other feared maladies. They ask about their own immediate pains—how long they will last, how much worse they can get. In particular, they ask about the pain of dying.

There can be no definite answers to such queries until we learn the answers to certain more specific problems. The most penetrating questions about pain come from children rather than adults. A child's fear of the unknown is closely coupled with his experiences of pain. Whenever he has to face some new ordeal, his invariable query is, "Will it hurt?" If he sees some person with a disabling injury, he tries to imagine how it would feel to have the same injury. He wants to know how much it hurts to break a leg, to have a tooth pulled, to undergo a surgical operation, to be wounded in battle. His sympathy goes out spontaneously to injured people and injured animals. He wants to know how much it hurts a fish to be caught on a hook or to flop around in a boat, and he may even insist that the writhing of an angleworm impaled on the hook is evidence of great pain.

Naive as the child's questions may sound, they are fundamental. Transcribed into more formal terms, some of them might read: How far down the scale of animal life is there a conscious perception of pain? Are an animal's reactions to injury an accurate measure of pain? Is physical pain ever devoid of psychological factors? Can human pain be measured objectively? Why do certain emotional states make pain more tolerable, while others make it worse? Is pain compatible with unconsciousness? Does an anesthetic agent abolish pain or does it merely erase its memory? How often is death painful?

It is doubtful that these questions will ever be answered by anything better than speculation or personal opinion until there is some agreement as to what pain is. I have used the question, "What is pain?" for my title because I believe its answer is fundamental to understanding the phenomenon and is the only basis on which one can build his own philosophy about it.

First let us consider a couple of specific cases to clarify the nature of the problem. A young woman is giving birth to her first baby. Her labor pains have become so severe that her obstetrician orders an anesthetic. She is given just enough to keep her in a state of "analgesia," meaning that she feels no pain but remains conscious. The remarkable feature of her analgesic state is that her pain perception should be so profoundly depressed while other perceptions and the "thinking" part of her brain are still functioning. When the delivery is over, she reports that under the anesthetic she felt no pain. That settles the matter for her, but it does not answer the observer's question of whether or not pain was present. We can see the problem more clearly by imagining what would have happened to this woman if the anesthetic had been deepened.

With just a slight increase in dosage the woman might have entered a state of excitement in which she was practically unconscious but the body would make heroic efforts to escape the stimulus. She would scream and struggle each time her uterus contracted. She would no longer cooperate with the obstetrician and her talk would become incoherent. If the anesthetic were further deepened, she would stop talking and struggling, but each contraction of the uterus would be accompanied by a rise in her blood pressure, a quickening of the heartbeat and other physical responses. A light touch on the cornea of her opened eye would make the lids twitch. If the anesthetic were deepened still more, these body responses would disappear one by one in an orderly sequence. Finally, the vital centers controlling the heartbeat and respiration would become depressed and her heart and breathing would stop, though for a brief time her nerves might still be capable of transmitting sensory signals. She would not yet be "dead," for strenuous measures might revive her, but after some minutes all possibility of resuscitation would be gone.

At what particular stage in this progression from complete consciousness to death did her pain disappear? Was it after the first few whiffs of the anesthetic, after her complaints of pain ceased, after she stopped struggling, after the disappearance of her corneal reflex, after the cessation of her heartbeat or only after she was irretrievably dead?

The second case presents a similar question under different circumstances. A fisherman is sitting in one of a line of boats stretching from one sand spit to another at the mouth of a river. He suddenly feels a smashing strike, and as he lunges back to set the hook, a large salmon breaks out of the water, shaking the hook in its mouth. He realizes that his best chance for landing the salmon lies in getting ashore before his line runs out or becomes entangled with the lines of other fishermen in the neighboring boats. Fighting the salmon as he goes, he starts crossing from boat to boat to reach the spit. Once there, he runs far out on the beach and after a hard struggle lands his salmon. As he winds up his line, he looks down and sees that the wet sand under his right shoe is reddening. Then he notices a long rent in his trousers and is surprised to discover a deep cut in his left leg. He realizes that this injury must have been sustained while he was crossing the line of boats. Yet he cannot recall having felt the slightest pain at the time.

There is nothing particularly unusual about this incident: people often are injured in battle or automobile accidents without being aware of it until afterwards. I have selected this case because the man was not dazed or in shock. He says he had "no pain." I would agree with him. I am unwilling to call anything pain unless it is perceived as such. In my opinion the woman in childbirth had no pain of any consequence after the first few whiffs of anesthetic.

The two cases make plain the fact that to resolve the issue we need a clear-cut decision as to what we mean by the words "pain" and "perception." One reason pain is so difficult to define is that it has so many aspects. The interpretation varies with the point of view of the investigator or the sufferer. To the sociologist pain and the threat of pain are powerful instruments of learning and social preservation. To the biologist pain is a sensory signal which warns the individual when a harmful stimulus threatens injury. To a man with an incurable cancer, pain is a destructive force: his suffering began too late to serve as an effective warning and it did not stop after the warning had been given. To the physiologist pain is a sensation like sight or hearing, but he tends to ignore its conscious, perceptual aspects, because consciousness has, as yet, no physiological equivalents; one might say that he is studying the pain "signal." To the psychologist, on the other hand, the important thing about pain is the brain's translation of the signal into a sensory experience. He finds pain, like all perceptions, to be subjective, individual and modified by degrees of attention, emotional states and the conditioning influence of past experience.

To the layman the sensation of pain, which he has known all his life, seems a perfectly straightforward, noncontroversial matter. He knows that pain is caused by physical injury and believes that its intensity is proportional to the force of a blow, the heat of an iron or the depth of a wound. This concept of pain as a physical quantum, measurable in terms of stimulus intensity or the body's response to injury, is a reasonable everyday interpretation. But there are many situations where it does not apply. Bullet wounds are usually painless, partly because the impact of the missile can temporarily paralyze nerve conduction. Superficial wounds usually are more painful than deep ones, because the skin is much more richly supplied with sensory nerve endings than are the deeper tissues. The internal organs can be cut, crushed or burned without causing the slightest distress. Then also there are enormous individual variations in sensitivity to pain. At one extreme are patients with such conditions as causalgia, facial neuralgia or postherpetic pain—conditions in which the skin becomes so sensitive that the lightest touch or even a breath of air precipitates an acute exacerbation of pain. At the other extreme are those unfortunate children who are constantly injuring themselves because they were born without the normal susceptibility to pain. Such a child may lean casually against a hot stove without showing signs of distress.

In the majority of instances pain *is* proportional to the injury. Therefore we are surprised when it differs noticeably from what we would have expected. We attribute such exceptions to "psychic" causes and wish we had some reliable objective method for measuring pain.

Attempts have been made to develop an objective scale of pain in terms of stimulus intensity. The stimulus used for eliciting pain may be an electric current, heat or some kind of pressure. These experiments show that most normal people have about the same threshold for pain. For instance, the average normal person begins to feel pain when heat applied to the skin reaches around 220 millicalories per square centimeter per second. This amount of heat will redden the skin after repeated tests, and it is close to the level of heat at which cells are irreversibly damaged.

Although people are fairly uniform in their perception threshold, they vary greatly in their tolerance of pain—that is, the amount of heat above the threshold that they will bear before pulling away from the testing instrument. A stoical person may endure heat which actually burns the skin. Once this burning point has been passed, the pain lessens, even though the heat level is raised, because the burning process destroys the sensory fibers in the skin, and the deeper tissues have fewer such fibers. Thus one might call the burning level the pain "ceiling."

In one method for measuring pain the levels of stimulus intensity between the threshold and the ceiling have been divided into 10 equal steps, called "dols," from the Latin *dolor,* meaning pain. This "dol scale" in the hands of experts has proved of some value in testing the efficacy of analgesic drugs, since it provides a rough measure of the ability of a given drug to raise the threshold for pain perception. But the method is unreliable when attempts are made to measure human pains or to use test subjects who have not had intensive training in sensory discrimination. There are too many sources of error in the measurements and too many psychological factors involved to permit anyone to claim that this instrument truly measures pain.

The foregoing considerations raise the question: "Is awareness of pain essential to its being called by that name?" The answer that my colleagues and I have made to this question is based, in part, on our clinical investigations of pain. We conduct a pain clinic in which we study selected patients with pains that are peculiarly resistant to treatment. Whenever our treatment is successful, we search for objective evidence to confirm the patient's subjective sense of improvement. In the experimental laboratory we are trying to locate the site where analgesic and anesthetic drugs act on the nervous system and to trace pain signals from, for example, the tooth of an anesthetized cat to its brain. These investigations can be mentioned only in passing, and much that follows deals with the work of others in the general field of sensory perception. Here, however, are some of the considerations on which our interpretation of pain is based.

The sensory nerves carry a continuous stream of impulses to the central nervous system from all parts of the body. Most of the information we derive from the outside world comes by way of the "primary" sensations of sight, hearing, touch, taste and smell, each of which has its own system of nerves. The information is transmitted as a complex pattern of nerve impulses, which I shall refer to as a signal rather than a message. The impulses carried by any one nerve fiber have an amplitude characteristic of that fiber and can vary only in frequency.

Pain is frequently said to be a primary sensation. There was a time when heat and cold also were so considered. This idea developed after it was demonstrated that pressure on tiny spots in the skin could elicit discrete sensations of cold, heat, pain or touch. It was assumed that beneath each such spot would be found a specialized sensory ending for just one of these "four modalities of cutaneous sensibility." It was further assumed that all other sensations from skin stimulation must represent some combination of these primary modalities.

An examination of a bit of skin under a high-powered microscope indicates that the matter is not so simple. The deep layers of the skin contain a large number of sensory fibers of various sizes. Each fiber branches like a tree, and its branches interweave with the branches of many neighboring fibers. At the end of each branch is a sensory receptor characteristic of that particular fiber. These receptor endings range in complexity from highly organized structures of considerable size to "bare" undifferentiated fibrils with no more than a tiny knob at the tip. The intermingling of the fiber branches and the great number of different endings at any one skin spot suggests how difficult it would be to stimulate one ending or one fiber selectively. An ordinary stimulus, whether a pinprick, a light touch or pressure, invariably activates a large number of different sensory fibers. The evidence is inescapable that the sensations we describe as "touch" and "pain" must be derived from the concurrent activation of many different sensory fibers of various sizes and distribution.

588

The old idea that fine fibers with bare endings are exclusively responsible for pain sensation is no longer tenable. The pinna of the human ear is supplied solely by such receptors, yet from this area of skin it is possible to elicit sensations of heat, cold and touch. On the other hand, we know that pain may be carried by two different types of fibers. What is called "fast pain" is carried by relatively large, myelinated (sheathed) fibers of the type known as "A" fibers. The sensation they transmit is variously described as "bright," "sharp" and "pricking." Slow pain is carried by "C" fibers—very small fibers with little or no myelin covering. The sensation provoked by C fibers has been characterized as "lingering," "reverberating" and "burning." Anyone who has dropped a heavy object on his toes has felt both kinds of pain. First comes the sharp, well-localized fast pain, then an instant during which the pain seems to have passed over, then a throbbing slow pain which appears to spread beyond the toes into the whole foot.

The nerve impulses set up in different sensory fibers by a single skin stimulus do not remain together as they ascend the nerve. They travel at very different speeds, the fibers being of different diameters. The largest fibers of the A group carry impulses at more than 100 meters per second—about as fast as a DC-3 transport plane cruises. The smallest C-group fibers conduct impulses of very low amplitude at rates of little more than one meter per second—about as fast as a man walks. Thus the nerve signal set up by a single stimulus reaches the spinal cord as a complex pattern of many impulses, spread out in time and in spacing.

Within the cord the impulses are further altered when they transfer to the secondary neurons that carry them up to the brain. How much the pattern alters in character at this relay station depends not only on the fibers' connections with the secondary neurons but also on the activity going on within the spinal cord itself. The pattern can again be modified in another relay station in the thalamus. Finally, the perception eventually derived from the signal pattern depends in part on what is going on in the sensorium—the seat of sensation in the brain.

This complex mechanism explains why it is so important to distinguish sharply between a signal and the perception ultimately derived from it. Abnormal conditions of the skin (*e.g.,* a sunburn), inflammatory conditions within the spinal cord (as in infantile paralysis), anything that happens to be going on within the relay stations or the sensorium—all this can profoundly alter the interpretation of a given stimulus.

This does not mean that the neural mechanism is a hit-or-miss affair. Under normal conditions the impulses follow prescribed pathways, and their pattern is not distorted at the relay stations. If the arrangement were not orderly, we should never be able to tell what part of the body was being stimulated or to describe as accurately as we usually can the qualities of the stimulating agent.

On the other hand, when we experience a sensation that we have never encountered before, what we perceive of it represents the best interpretation we are able to make of the signal at that particular moment. This interpretation depends on the accuracy of the information conveyed by the signal, the state of the sensorium at that instant and the conditioning influences of experience, childhood training and our personal sense of values.

We predicate all of our actions on the assumption that the information we derive from our senses is accurate. But even though the information conveyed by each sensory system is as accurate as its momentary status can make it, it does not follow that our interpretation of the sensory impression is always correct. The conflicting descriptions that different people offer for the same object, whether it be an escaped criminal or a flying saucer, are commonplace illustrations of how visual perceptions can be misinterpreted.

Psychologists have produced convincing evidence to support the view that what we see is not an exact replica of "reality" but an assumption based on past experience. This is well illustrated by the experiments in visual illusions described by Ittelson and Kilpatrick earlier in this chapter.

I believe that the interpretation of all sensory information is modified by the same factors that apply to visual perception. I am sure that this is true of pain perception. The interpretation an individual

Nerve pathway, conducting sensation from skin to brain is shown in this schematic diagram. Thickness of spinal cord is exaggerated in sketch of complete circuit at right. First nerve conducts impulse from skin to spinal cord; second nerve, in main trunk of spinal cord, relays the impulse to the thalamus, and a third conducts it to the cerebral cortex. The detail drawing at left shows how first nerve carries impulse to gray matter of spinal cord, where it is picked up by ending of second nerve.

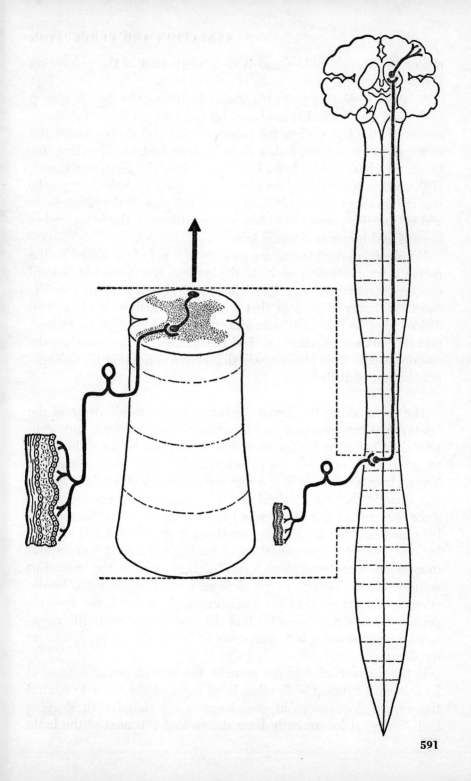

makes of a specific pain signal is an intrinsic part of the perception and a determining factor in its emotional tone.

A father is playing with his son at bedtime. The boy is almost undressed and the father holds out his pajamas. As the last garment comes off, the boy pushes the pajamas aside and dashes across the room. As he passes, the father slaps his bare bottom. The sting surprises the child and he looks back to see how the blow was meant. If the father is laughing, the chances are good that the boy laughs too, as if the slap were a pleasant part of the play. If the father looks and acts as if he meant the blow as a punishment, the boy clutches himself and howls as if badly hurt.

No physician doubts that the severity of a pain is modified by the patient's interpretation of it. If the patient has a morbid fear of cancer, every pain he develops is intensified because it suggests the onset of cancer. The pain that a child experiences is often conditioned by the fears, attitudes and afflictions of his parents. Indeed, parental influences may be decisive factors in determining the amount of pain their children will suffer from minor injuries throughout the rest of their lives.

The use of drugs to alleviate suffering makes us all aware of the relationship between pain and the state of consciousness. But "consciousness" is even harder to define than pain, because there are so many kinds and degrees of consciousness and unconsciousness. No one knows where or how consciousness occurs. For a long time it was thought that the cerebral cortex was the seat of consciousness, since one part of the cortex was known to be the receiving station for vision, another for hearing, another for speech and still another for sensory-motor representation of the body parts. It was natural to associate consciousness with this "highest level" in the brain. But we now know that any or all of these cortical centers can be destroyed without abolishing consciousness. At present, the best experimental evidence indicates that the area most essential to maintaining consciousness is a compartment below the cortex known as the diencephalon.

If the diencephalon is not actually the seat of consciousness, at least it must interact with other large areas of the brain to control the normal fluctuations in consciousness associated with sleeping and waking. It has recently been shown that this area of the brain contains an "arousal center." Destruction of this center in the brain of a cat or monkey will cause the animal to remain in an unconscious state that resembles normal sleep, though the operation does not interrupt the main sensory pathways for sight, hearing or pain. The animal can be awakened by a loud sound, by a bright light shined in its eyes or by the strong stimulation of a sensory nerve, but it soon lapses back into sleep. On the other hand, if the arousal center is left undisturbed, an animal sleeps and wakes in the normal sequence even when all the main sensory systems are cut so that the sensory signals, which we have always thought so important for the preservation of consciousness, cannot reach the sensorium by their usual direct route.

The seat of the arousal center is in a part of the brain known as the "reticular formation," a structure of relatively small nerve cells which lie outside the main pathways for motor and sensory conduction. It extends from the diencephalon downward throughout the length of the central nervous system. Until recently about all that was known about the function of the reticular formation was that its upper portion facilitates activities in many other parts of the brain and spinal cord, while a somewhat lower portion inhibits them. Now the arousal center has been found in the upper part of the formation. It acts upon the entire cerebral cortex at the same time. In contrast, any single sensory system influences only a limited area of the cortex. The arousal center's power to activate the cortex as a whole seems to be closely related to its capacity for waking a sleeping animal.

A sleeping individual is usually awakened quite readily by sensory stimulation, particularly if the stimulus is intense. Painful stimuli arouse the sleeper more effectively than either light or sound. It appears that the sensory signals do not themselves awaken the individual but activate the arousal center, by way of side branches from the main sensory pathways. It is possible that anesthetic drugs depress the arousal center of the reticular formation rather than the specific cortical centers. It is also possible that pain is related more closely to activity in the reticular formation than to a special "pain center" in the brain. What we need to explore these suggestions is a pain signal of sufficient intensity to be traceable into the brain.

For various reasons the best sensory fibers on which to experiment are the A fibers, which transmit pain signals of the fast type. The sensory fibers in the pulp of the teeth seem to meet the specifications, and their endings are all of the undifferentiated type. The tooth pulp is exquisitely sensitive to mechanical, electrical and thermal stimulation. Its nerve fibers can be activated by an electrical pulse measured in hundredths of a millisecond. Experiments in the stimulation of these fibers are now being carried out on cats under deep Nembutal anesthesia. We can assume that the signal we are studying would be perceived as pain by the animals if they were conscious, because when we apply these same short pulses to our own amalgam-filled teeth, we experience a sharp stab of pain.

To date the experiments have progressed only to the point where secondary discharges have been recorded from the thalamus and the sensory-motor cortex of the brain. We have not yet traced the discharges from these receiving centers into the reticular formation or into other parts of the brain. But it looks as if tooth-pulp stimulation will be a valuable tool for finding out how the secondary influences may spread to other parts of the brain.

This presentation has indicated some of the difficulties encountered in any investigation of pain. We are handicapped from the outset by the lack of a clear-cut definition of the entity we are to study. The few definitions for pain that have been proposed tend to emphasize its protective function and make no mention of its harmful potentialities. The best of these definitions was suggested years ago by the English physiologist Sir Charles Sherrington. He said that pain was "the psychical adjunct of an imperative protective reflex." This statement says a great deal in few words, but it hardly defines pain, since it tells nothing of its nature beyond the fact that it is "psychical." The statement also conveys the impression that the important protective factor is the reflex, to which pain is merely an "adjunct." As far as the lower animals are concerned, this is probably true. But for thinking human beings pain assumes a much greater significance than it does for lower animals.

What are some of man's defenses against injury and how intimately are they related to one another? The simplest and most familiar is the withdrawal reflex. A man standing beside a hot stove happens to touch it with one hand. His arm muscles jerk the hand away before he has time to feel any pain or to know what is happening. The muscular reflexes are the body's first line of defense

594

against injury. In many situations they effectively break contact w
the offending stimulus. The withdrawal reflex takes place over t
shortest possible route from the site of injury to the spinal cord an
back again to the local musculature. The speed of the reflex tend
to reduce tissue damage to a minimum.

A second line of defense is represented by visceral reflexes that
involve the vital organs and glands of internal secretion. We are all
familiar with the increase in the heartbeat and in respiration, the
dilation of the pupils and the sense of tension throughout the body
that accompanies pain. These are the more obvious manifestations
of a chain reaction that mobilizes all of our resources to meet what
may be an emergency situation.

The third line of defense is the voluntary response to a situation.
A noxious stimulus initiates a signal which is translated by the brain
into a pain perception. Having felt the pain, the individual can find
its source and decide on the basis of experience how to deal with
the situation.

In the sense that a reflex response is much faster than a voluntary
one, it affords a better protection against injury. But reflexes are
always stereotyped and often totally inappropriate to the situation.
They occur whether they can serve any useful purpose or not, and
they may waste the body's resources. Under conditions of sustained
or repeated injury the body may be so depleted that it no longer
can withstand infection and new stresses. As a matter of fact, actual
tissue injury need not be present to cause this exhaustion. Fear can
do exactly the same thing. Often the threat of pain does a person
more harm than the injuries that taught him to fear it.

The intimate relationship among these three lines of body defense, all activated by the same noxious stimulus, makes it easy
to confuse them with one another. In our desire to find a method
for measuring pain, we are tempted to identify pain with the
measurable associated mechanisms—the noxious stimulus, the body
response or the signal pattern on its way to the sensorium. To do
so, I believe, is an error. As Sherrington says, pain is a psychical
process. It represents some activity of the brain which cannot be
fully accounted for as yet by experimental observations. The signal
pattern is doubtless a part of the mechanisms from which perceptions are secondarily derived, but the process itself is beyond
the reach of our present recording techniques. It seems to me to be

just as misleading to identify pain with the signal pattern as it would be to identify an act directed by human intelligence with a reflex. Furthermore, clinical experience indicates that the psychic and physical factors that determine the intensity of a pain are inseparable components of a single sensory experience. It is so difficult to eliminate psychological factors from the simplest test for pain that I doubt we shall ever find a satisfactory method of objective measurement.

Returning to the original question—What is pain?—I believe that we can accept the "common sense" answer. Pain is a perception. To be "perceived" means to be "felt." Anything that depresses brain function impairs pain perception. It doesn't seem to make much difference whether the depression is due to drugs, excessive fatigue or any of the many factors that deprive the sensitive brain cells of their supply of oxygen. The brain cells involved in pain perception are selectively depressed by anesthetic and analgesic drugs, so that this sensation falls in intensity before other sensory perceptions are seriously impaired.

I am convinced that neither a dying man nor a person undergoing anesthesia feels any pain, though their groans and body movements, those physical manifestations which we so naturally associate with pain, may seem to support the contrary view.

With these convictions I can tell the man who fears death will be painful that dying is merely the closing event in a sequential loss of function which accompanies brain depression. Just as an exhausted mountain climber gratefully lies down on the rocks and goes to sleep under conditions that would be intolerable to him in his normal state, so a dying man may welcome death because it offers his exhausted body rest. Before all his senses fail, before he loses all power of speech and movement, before his heart stops beating, long before his nerves lose their capacity to transmit pain signals, the ability of his brain to translate these signals into pain perception has been lost. For pain is a product of consciousness in which the essential element is awareness.

INDEX

Dates given after titles of chapters refer to the issues of SCIENTIFIC AMERICAN in which the articles were originally published.

INDEX

Tswett, Michael, 235
Tube, electronic, 515–522, 533
Tuberculosis, hormone treatment, 385–386
Turpentine, odor, 572
Turtle, 282, 461, 565
    mechanical, 548–551
Tuscany, volcanic steam used in, 86
Tustin, Arnold, 546
    "Feedback" (Sept. 1952), 490, 528–537
    Mechanics of Economic Systems, 490
Tuve, Merle A., 104
Two-meson theory, 88, 110

Ulcers, 395
"Ultimate Particles, The" (June 1948), George W. Gray, 88, 92–105
Ultraviolet, 130
    effect on viruses, 340
    mutation and, 274, 276
    vision and, 562
Unified field theory, 526
Unit cell, 142
Universe:
    age, 8–9, 11, 31, 35n
    chemical elements, 10–12
    dust clouds in, 36–46
    maximum contraction period, 11–12
    structure of, 23
    theory of expanding, *see* Expanding universe
"Universe from Palomar, The" (Feb. 1952), George W. Gray, 2, 23–35
Unspecialized, law of the, 484, 489
Uranium, 11, 186
    atoms, 97, 99
    fission role, 157–166
Uranium 233, 161
Uranium 235, 11, 157–166
Uranium 236, 157
Uranium 238, 11, 157–166
Uranium 239, 161
Uranus, 41
Uranyl nitrate, 172
Urea, 247, 292
Urease, 247
Urey, Harold C., 71, 204
Uric acid, 206–209
Urine, 404
    alcapton in, 271–272

Vaccination, 344
Vacuum, 99–101
Vacuum tube circuits, 511
Vallois, Henri, 455, 461

Valve, electronic, *see* Tube, electronic
Vanderbilt University, 356
Vanillin, odor, 571
Van Vleck, John H., 129
Varicose veins, 475
Veksler, V., 94
Venning, Eleanor, 384
Venus, 44
Vertebrae, man's, 473
Vesalius, Andreas, 388
Vesuvius, 81, 84, 85
Vigneaud, Vincent de, 205
Violet light, 130
Virgo, 7
Virology, 335–346
"Virus, The" (May 1951), F. M. Burnet, 332, 335–346
Viruses, 232
    cell penetration, 337–341, 347, 349, 354
    genes, 278
    growth medium, 350–351
    hosts, 343, 345, 350, 358
    hybridization, 342
    immunization, 343–346
    mutation and selective survival, 339
    neuro-flu, 341–342
    "osmotic shock," 348
    "partial," 341
    reproduction, 339–340
    research in, *see* Virus research
    subclinical infections, 343
Virus research:
    bacterial virus multiplication, 347–354
    cell penetration, 337–341
    common cold, 366–372
    enzyme action, 337–338
    host animals, 338
    heredity, 340
    interference effect, 340
    laboratory requisites to, 335–336, 338
    life history of herpes, 355–359
    microscopy, 342
    polio, 361–365
    quantitative, 336
    recombination experiments, 342
    tissue-culture methods, 361, 364–365
Visual projection area, 503
Visceral reflexes, 595
Vision:
    perceptual experiments in, 577–581
    physiology of, 555–568
    see also Eye *and* Sight
Vitamins, 258
"Volcanoes" (Nov. 1951), Howel Williams, 52, 77–87
Volcanoes: